Nobel Laureates and Twentieth-Century Physics

In this richly illustrated book the author combines history with real science. Using an original approach he presents the major achievements of twentieth-century physics – for example, relativity, quantum mechanics, atomic and nuclear physics, the invention of the transistor and the laser, superconductivity, binary pulsars, and the Bose–Einstein condensate – each as they emerged as the product of the genius of those physicists whose labours, since 1901, have been crowned with a Nobel Prize.

Here, in the form of a year-by-year chronicle, biographies and revealing personal anecdotes help bring to life the main events of the past hundred years. The work of the most famous physicists of the twentieth century – great names, such as Bohr, the Curies, Einstein, Fermi, Feynman, Gell-Mann, Heisenberg, Rutherford and Schrödinger – is presented, often in the words and imagery of the prizewinners themselves.

The author uses plain language to avoid technical jargon as much as possible. He does not hesitate, however, to explain abstruse theories when necessary. With clear step-by-step explanations and lively down-to-earth examples, this engaging work will be of interest to working scientists, students, and the lay reader curious about the wonders of the universe of science.

MAURO DARDO is professor of experimental physics at the newly founded *Amedeo Avogadro* University of Eastern Piedmont at Alessandria, in Northern Italy. Here from 1992 to 1998 he also served as dean of the faculty of sciences. He studied physics at the University of Turin, and obtained his doctor's degree there in 1964. Soon afterwards he began his teaching and research career. This took him first to the USA for a one-year study period, and then, in 1980, to Cagliari in Sardinia as professor of physics. Three years later, in 1983, occasion took him back to his *alma mater*, in Turin, to a post in the department of physics. He has taken part in research programmes in a number of different fields of physics, including cosmic rays, elementary particle physics and high-energy astrophysics. His special interests have long lain in the area of history in general – and the history of science, and the wider diffusion of scientific culture, in particular. He lives in a small town, rich in local traditions, in rural Piedmont, close to the border of the ancient marquisate of *Monferrato*, and within sight of the Alps.

Nobel Laureates and Twentieth-Century Physics

Mauro Dardo
Università degli Studi del Piemonte Orientale
Amedeo Avogadro

PUBLISHED BY THE PRESS SYNDICATE OF THE UNIVERSITY OF CAMBRIDGE
The Pitt Building, Trumpington Street, Cambridge, United Kingdom

CAMBRIDGE UNIVERSITY PRESS
The Edinburgh Building, Cambridge, CB2 2RU, UK
40 West 20th Street, New York, NY 10011–4211, USA
477 Williamstown Road, Port Melbourne, VIC 3207, Australia
Ruiz de Alarcón 13, 28014 Madrid, Spain
Dock House, The Waterfront, Cape Town 8001, South Africa

http://www.cambridge.org

© M. Dardo 2004

This book is in copyright. Subject to statutory exception
and to the provisions of relevant collective licensing agreements,
no reproduction of any part may take place without
the written permission of Cambridge University Press.

First published 2004

Printed in the United Kingdom at the University Press, Cambridge

Typefaces Times NR 10/13 pt. and Universe   *System* LATEX 2_ε [TB]

A catalogue record for this book is available from the British Library

Library of Congress Cataloguing in Publication data
Dardo, M. (Mauro)
Nobel laureates and twentieth-century physics / Mauro Dardo.
    p.   cm.
Includes bibliographical references and index.
ISBN 0 521 83247 0 – ISBN 0 521 54008 9 (paperback)
1. Physics – History – 20th century.   2. Nobel Prizes.   3. Physicists – Biography.   I. Title.
QC7.D27   2004
530′.09′04 – dc22   2004049240

ISBN 0 521 83247 0 hardback
ISBN 0 521 54008 9 paperback

The publisher has used its best endeavours to ensure that the URLs for external websites referred to in this book are correct and active at the time of going to press. However, the publisher has no responsibility for the websites and can make no guarantee that a site will remain live or that the content is or will remain appropriate.

to my parents

Contents

Preface	page ix
Acknowledgments	x

Part I Introduction

1 Introduction	3
2 Founding fathers	7
3 Highlights of classical physics	17

Part II The triumphs of modern physics (1901–1950)

4 New foundations	33
5 The quantum atom	77
6 The golden years	125
7 The thirties	173
8 The nuclear age	213

Part III New frontiers (1951–2003)

9 Wave of inventions	237
10 New vistas on the cosmos	279
11 The small, the large – the complex	319
12 Big physics – small physics	369
13 New trends	409
Appendix Nobel Prizes for physics	469
Glossary of terms	475
Notes	495
Select bibliography	513
Further reading	515
Index	517

Preface

This book is about the Nobel Prizes for physics: how they were awarded each year, and for what particular merit; how the discoveries that they have honoured fit into the wider picture of the evolution of twentieth-century physics, enlarged our understanding of nature and, in terms of new technologies, changed and moulded our everyday lives. But above all it is about the prizewinners themselves, how they came to make the contributions to science for which they are renowned and, through personal details and anecdotes, it aims to tell us what sort of people they were, and indeed are.

The book is divided into three parts. The first part contains an introductory chapter which includes a short description of the Nobel Prize. Then follow two chapters which deal with classical physics, in so far as it constitutes the roots of modern physics. These chapters, through a rapid historical journey, will present the reader with some fundamental concepts in physics, together with information about the giants of classical science, so taking the reader up to the doorstep of twentieth-century physics.

The second and third parts form the core of the book. They contain ten chapters, which, year by year, describe the work for which the awards were given, with short biographical notes on each Nobel laureate. In parallel, in each year, are included concise descriptions of the principal achievements in physics during the year itself. Each chapter begins with an introduction, which summarises the major events during the period in question, and each ends with illustrations and descriptions of sites where the most famous events took place. Finally, the reader will find a glossary of terms which we believe will be of assistance, especially if he or she is a non-specialist. Simple sketches and diagrams will help in understanding certain important concepts.

The author has tried wherever possible to use plain language and to avoid technical jargon, whilst nevertheless maintaining scientific and historical rigour.

Nobel Laureates and Twentieth-Century Physics is addressed to scientists active in the worlds of research or teaching, to students, both undergraduates and graduates: and also, and by no means least, to the general reader who is eager to venture into the great scientific themes that have distinguished the last hundred years of the history of physics and science in general.

Acknowledgments

This book might never have seen the light of day – at least in this form – without the co-operation and helpfulness of Richard Izard. His particular care has been for the *Englishness* of the book. Its style owns much to him, as does its readability and its 'feeling'. I am deeply indebted to him for his constant and skilled assistance.

In the first place I wish to record my profound gratitude to all those Nobel prizewinners who have kindly read the pages concerning their Nobel Prizes, and offered me so many helpful criticisms, further information, and practical suggestions for improving the text: Zhores Alferov, Philip Anderson, Aage Bohr, Georges Charpak, Claude Cohen-Tannoudji, Leon Cooper, Pierre-Gilles de Gennes, Riccardo Giacconi, Antony Hewish, Brian Josephson, Wolfgang Ketterle, K. Alex Müller, William Phillips, Heinrich Rorher, Carlo Rubbia, Jack Steinberger, Gerardus 't Hooft, Charles Townes, and Martinus Veltmann.

I also wish to take this opportunity to thank all those numerous people – including the majority of all the above-cited Nobel prizewinners – who have given me permission to quote from their books and articles: Georg Bednorz, Hans Bethe, Nicolaas Bloembergen, Owen Chamberlain, Steven Chu, James Cronin, Paul Davies, Robert Marc Friedman, Murray Gell-Mann, Sheldon Glashow, George Johnson, Daniel Kleppner, Robert Laughlin, Leon Lederman, Simon van der Meer, Sir Brian Pippard, Norman Ramsey, Silvan S. Schweber, Daniel Tsui, Steven Weinberg, the late Victor Weisskopf, Kenneth Wilson, Chen Ning Yang.

At the same time my thanks are due to many institutions for permission to reproduce excerpts from their publications: I am particularly indebted to the Nobel Foundation, (holders of the copyright for the Nobel Lectures). I also wish to acknowledge help form the Cambridge University Press, the Hebrew University of Jerusalem, the Institute of Physics Publishing, Lucent Technologies/Bell Labs, the MIT Press, the Oxford University Press, the Princeton University Press, and Springer-Verlag. To these I must add those individuals and institutions which provided photographs and illustrations, together with their permission to publish (acknowledgements will be found in the figure captions or in the section 'Notes').

I am also deeply indebted to a large number of individuals for their help and encouragement. I wish to thank particularly Andrzej Stasiak, of the University of Lausanne, Switzerland, for his careful review of the whole manuscript, and for a notable contribution both of invaluable comments and of practical suggestions from which I have benefited greatly. Portions of the manuscript dealing with diverse topics have been read by colleagues and correspondents, and I think particularly of Paolo Allia, Polytechnic of Turin, Italy; Ugo Amaldi, Sergio Ferrara and Giorgio Stefanini, CERN, Geneva; Ferdinando Amman and Vito Svelto, University of Pavia, Italy; Joseph Avron, Technion, Haifa, Israel; Giorgio Parisi, University *La Sapienza* in Rome, Italy; Lucio Braicovich and Orazio Svelto, Polytechnic of Milan, Italy; Giulio Casati, University of Como, Italy; Russel J. Donnelly, University of Oregon, USA; Attilio Ferrari, University of Turin, Italy; Giorgio Frossati, University of Leiden, Holland; Leo Kadanoff, University of Chicago, USA; Daniel Kleppner, MIT, USA; Emilio Picasso, Scuola Normale Superiore, Pisa, Italy; Guido Pizzella, University of Rome at Tor Vergata, Italy; Sir Brian Pippard, University of Cambridge, England; Martin A. Pomerantz; University of Delaware, Newark, USA; Renzo Ricca, University College, London, England; USA; Michael Stone, University of Illinois at Urbana-Champaign, USA; Adrian Sutton, Oxford University, England; Valentine Telegdi, Caltech, USA, and CERN, Switzerland; Clifford M. Will, Washington University, USA; Dieter Vollhardt, University of Ausburg, Germany. These all deserve my thanks, and I am happy to be able to record my gratitude here.

Naturally any errors or misconceptions that still remain in the book are my responsibility, and I take this opportunity to apologise sincerely for them.

The technical aspects called into play during the preparation of this book have been attended to by too many people to thank individually, but I must acknowledge my debt to Michele Manzini and Aldo Masoero, each of whom have been of major assistance throughout all stages of the preparation of the work. A word of genuine appreciation is due to my friend Piero Bosso for his line drawings, and to Françoise Hayes for her helpfulness during the preparatory phase of this work. My university, the *Amedeo Avogadro* University of Eastern Piedmont (Italy), has supported the research that was undertaken in the preparation of this book, and I am glad to be able to record my appreciation accordingly.

Finally, it is with particular pleasure that I express my gratitude to Simon Capelin (Publishing Director – Physical Science and Engineering) and to the staff of the Cambridge University Press, whose professional competence, cordiality and patience have made the whole process of bringing this book to birth so smooth and effortless.

Part I
Introduction

Chapter 1
Introduction

The Nobel Prize awards entered upon their second century of life on 10 December of the year 2001. The event was celebrated in Stockholm, and in Oslo for the peace prize, with due pomp and circumstance when that year's prizewinners received their diplomas and medals in the presence of the Kings of Sweden and Norway, and of 200 past laureates who had been invited from all four corners of the earth to attend the ceremonies. The grandeur of the celebrations went to confirm what is universally agreed – that these prizes are considered everywhere to be the most prestigious honours of our times, far outstripping, indeed, all others.

The aim of the prizes is to reward outstanding contributions in the sciences of physics, chemistry, and physiology or medicine, as well as in literature and for peace. They were created by the Swedish inventor and industrialist Alfred Nobel, who decided to set up a special fund in his will for this purpose. The first awards were conferred on 10 December 1901, on the fifth anniversary of Nobel's death. Thereafter, the ritual has always taken place on the same date. Since its beginning in 1901, and up to the year 2003, there have been 758 individuals and organisations who have received a Nobel Prize: 171 of these were for physics, 143 for chemistry, 180 for physiology or medicine, 100 for literature, 111 for peace and 53 for economic sciences (this last prize was set up in 1968 by the Bank of Sweden, to further honour the memory of Alfred Nobel). The prizes in 1901 were each worth 150 000 Swedish crowns (a current value of close to 900 000 US dollars). In 2003 each prize was worth 10 million Swedish crowns (nearly 1.3 million US dollars); and so it can still be considered one of the world's most valuable awards.

The Nobel Foundation is a private institution which administers the prizes and manages the finances of the fund. As instructed by Alfred Nobel in his will, the prizes for physics and chemistry are awarded by the Royal Swedish Academy of Sciences; the prize for physiology or medicine by the Caroline Institute of Medicine in Stockholm; the prize for literature by the Swedish Academy, also in Stockholm; and the peace prize by a five-person committee elected by the Norwegian parliament. Finally, it is the Swedish Academy of Sciences that awards the prize for economic sciences.

Fig. 1.1. Alfred Nobel. (Portrait by E. Österman, © The Nobel Foundation, Stockholm.)

Alfred Bernhard Nobel (1833–96) was born in Stockholm, the son of an inventor and industrialist. In the early 1860s he built a small factory to manufacture nitro-glycerine, a powerful explosive then recently discovered; and a few years later he invented a method of mixing nitro-glycerine with an organic material, so reducing its volatility. The new explosive, named dynamite, was soon used in the building of roads, canals, railways and tunnels through mountains.

Nobel then built a network of factories throughout Europe. At his death on 10 December 1896, in San Remo on the Italian Riviera, his business empire was made up of more than ninety factories, and included more than three hundred patents. In his will Nobel provided that most of his estate, estimated at over 31 million Swedish crowns (with a current value of nearly 180 million US dollars), be set up as a fund to establish the five original annual prizes.

The prize for physics

The decision-making process for evaluating proposals, and for the final selection for the prize in physics, is in outline as follows. Invitations to nominate candidates are sent out by a Committee, whose five members are elected by the Swedish Academy of Sciences. Those who are invited to submit nominations are scientists who have already been awarded the Nobel Prize, members of academies and professors of physics from foreign universities. The Committee looks for suitable candidates and proposes one of them to the Academy. This proposal is then voted on, initially by its Physics Section. Lastly, the final decision is taken by the Academy in a plenary session.

And now for the prizewinners themselves, and a word on their places of origin. The very first Nobel Prize for physics went to the German physicist Wilhelm Röntgen for his discovery of X-rays. Since then the Royal Academy has awarded ninety-seven prizes (for six years they were not awarded) to 171 scientists (among them, only two women: Marie Curie in 1903, and Maria Goeppert Mayer in 1963). The work for which the prizes were awarded was mainly carried out at universities, but also at national and international research centres, and industrial laboratories: these institutions belonged to fourteen different countries in all.

The USA rose to prominence in the second half of the twentieth century; of the forty-four prizes bestowed on it, thirty-six are post-1950 (seventy-seven laureates, among whom eight before 1950). The United Kingdom follows with twenty-two laureates, Germany with nineteen, France with twelve; Russia has ten laureates; Holland eight; Switzerland six; Sweden and Japan each have four laureates; while Denmark and Italy have both arrived at three. Finally, Austria, Canada, and India each have one laureate.[1]

The Nobel Prize enhances the scientific prestige not only of nations, but also that of universities and research organisations. It is understood to be a measure of the performance of the institution responsible. In Europe, the University of Cambridge can boast fourteen out of the twenty-two prizewinners belonging to the UK; the Lebedev Institute of Physics in Moscow has six laureates; the Institute for Physical Problems in Moscow, the universities of Berlin, Leiden in Holland, Copenhagen in Denmark and the Sorbonne in Paris have three laureates each. In the USA the most honoured universities have been Columbia University with nine laureates and Harvard with seven. Next in order come Cornell, Stanford and the Massachusetts Institute of Technology (MIT) each with six; Berkeley and the University of Chicago with four laureates each; and finally, Princeton University, the University of Illinois at Urbana-Champaign and the California Institute of Technology (Caltech), each with three laureates.[2]

The largest national and international laboratories, all established after the Second World War, have competed in the field of particle physics: in the United States, the Brookhaven National Laboratory (BNL) was honoured with four Nobel Prizes (seven laureates); the Stanford Linear Accelerator Center (SLAC) has had three prizes (five laureates); and the Lawrence Berkeley National Laboratory (LBNL) three prizes (four laureates). In Europe, CERN, the international laboratory for particle physics near Geneva, Switzerland, has had two prizes assigned to it, and three laureates. Finally, the National Institute of Standards and Technology (NIST) in the USA has been twice rewarded during recent years (with two prizewinners) for discoveries in the field of atomic physics.

Industrial laboratories have contributed strongly to the progress of modern physics: in the lead we find the Bell Telephone Laboratories (commonly known as Bell Labs), which constitute the research branch of the American company

Lucent Technologies (formerly part of AT&T). Since 1937 it has been honoured with six Nobel Prizes for physics, bestowed on ten scientists. Next after it is the IBM Zurich Research Laboratory (Switzerland), with two prizes and four laureates.

And now, before retracing the paths taken by our Nobel laureates, we should spare a glance at where the roots of the science of physics are to be found. So let us go back a little in time and have a look at the masters of science of the past, those predecessors of our Nobelists. This will provide a necessary prologue to our chronicle of the events of the last hundred years, and our history of the Nobel awards and their place in the evolution of twentieth-century physics.

Chapter 2
Founding fathers

History tells us that science develops in a continuous and comparatively steady way. Nevertheless, certain periods are marked by dramatic discoveries or far-reaching new theoretical ideas. One of these crucial periods was the seventeenth century, when modern science was born.

The revolution began in the field of astronomy. In 1543 the Polish canon and astronomer Nicolaus Copernicus published his famous book entitled *On the Revolutions of the Celestial Spheres*. In it he replaced the old astronomical system of the Greek astronomer Claudius Ptolemy of Alexandria with a new one. Copernicus proposed that the sun, not the earth, is at the centre of the universe, and that all the planets (including the earth) revolve around the sun. Later, in the early 1600s, the German astronomer and mathematician Johannes Kepler perfected Copernicus' theory, and came up with three empirical laws regarding planetary motion. He based his findings on the work of the Danish astronomer Tycho Brahe, who had made careful observations on the positions of the planets in the sky. The theory of their circular motion was thus abandoned and replaced by Kepler's elliptical orbits.

The *coup de grâce* to the philosophy of nature that had been based on the teaching of the Greek philosopher Aristotle is associated with the name of a great Italian scientist – Galileo Galilei. His fame mostly rests on his having produced convincing evidence for the Copernican system of the Universe, and for having pioneered modern science through his studies on the motion of bodies.

The revolution initiated by Galileo was completed at the end of the seventeenth century in England by Isaac Newton, probably the greatest scientific intellect of all time. Using mathematical techniques that he himself had devised, Newton formulated his three famous laws of motion and the law of universal gravitation, and thereby managed to explain how objects move both on earth and in the heavens. This was the first synthesis in the history of science, and represented the culmination of the 'Scientific Revolution'. Such was Newton's physics that it became a model for the scientists of the following two centuries, and greatly influenced the whole course of science. In the eighteenth century physicists and mathematicians greatly developed Newton's work, and the first systematic studies of electrical and heat phenomena were undertaken.

Towards a new science

Galileo – he was the last great Italian to be called by this, his first name – was professor of mathematics at the University of the Venetian Republic, which was at Padua. Here, in the summer of 1609, he heard about a spyglass that a Dutch optician had invented. Working on the principles of this invention, he built a telescope, and pointed it towards the heavens. He saw satellites orbiting Jupiter, spots on the sun, and noted the phases of Venus. These observations confirmed the Copernican theory. He later presented convincing arguments for his discoveries in his book entitled *Dialogue Concerning the Two Chief World Systems, Ptolemaic and Copernican*, generally known more simply as *Dialogue*.

Although Galileo is widely remembered for his astronomical discoveries, it is in the field of mechanics that he made substantial contributions to our understanding of nature, for he laid the foundations for the science of motion, presented in his masterpiece, *Discourses and Mathematical Demonstrations Concerning Two New Sciences*, more frequently called *Two New Sciences*.

The science of motion

The *scientific method* is applied by all physicists in their attempts to understand how the natural world works, and it underlies the principles that guide all scientific research and experimentation. It can be summarised in the words of Richard Feynman (Nobelist in 1965): 'Observation, reason, and experiment make up what we call the scientific method.' Then physicists try 'to find the laws behind experiment', that is, the basic 'rules' that govern the phenomena of nature, and as Galileo argued, the language that these rules are written in is mathematics:

> [Natural] Philosophy is written in this grand book, the universe, which stands continually open to our gaze; [but this book] cannot be understood unless one first learns to comprehend the language and read the letters in which it is composed. It is written in the language of mathematics, and its characters are triangles, circles, and other geometric figures without which it is humanly impossible to understand a single word of it.[1]

A similar example of the scientific method was that adopted by Galileo himself in his studies on falling bodies. He was able to derive the laws of their motion, which he then proceeded to verify by experiments. These involved balls which he caused to roll down an inclined plane: so doing, he became the first to discover that all freely falling bodies, no matter what their mass, experience the same acceleration – provided that this happens at the same place near the earth's surface. Galileo is also given the credit for two basic principles of the science of motion. The first is the *principle of relativity*. In the *Dialogue* he wrote:

> Shut yourself up with some friend in the main cabin below decks on some large ship, and have with you there some flies, butterflies, and other small flying

Fig. 2.1. Galileo Galilei (1564–1642). (University of Rochester, courtesy AIP Emilio Segrè Visual Archives.)

animals ... With the ship standing still, observe carefully how the little animals fly with equal speed to all sides of the cabin ... When you have observed all these things carefully, ... have the ship proceed with any speed you like, so long as the motion is uniform and not fluctuating this way and that. You will discover not the least change in all the effects named, nor could you tell from any of them whether the ship was moving or standing still.[2]

So, assuming that the ship glides smoothly along, without any brusque movements, the passengers will be unable to notice the forward motion of the ship (provided, naturally, that they do not notice any apparent movement relative to their surroundings).

The second principle is the *principle of inertia*, which we may express as follows: *if a body has nothing acting on it, its velocity remains the same; thus a body which is at rest remains at rest, and one that is moving uniformly and rectilinearly continues to move uniformly and rectilinearly*. Galileo never stated this principle in its general form; it is implicit in his studies on motion, which he repeatedly referred to in his book, *Two New Sciences*. It was Isaac Newton who arrived at its definitive modern formulation.

Let Newton be!

> Nature, and Nature's Laws lay hid in Night.
> God said, Let Newton be! and All was Light.
> (Alexander Pope, English poet)

Isaac Newton, the 'great synthesizer' (as the biographer Gale Christianson called him), further elaborated the principles of the science of motion that had been outlined by Galileo. He also formulated the law of universal gravitation, and made great discoveries in the science of optics. In April 1665, Newton received his bachelor's degree from the University of Cambridge, England. That summer the university had to close for two years due to the outbreak of plague. So he returned to his home in Woolsthorpe, Lincolnshire, and spent this period ('the prime of my age for invention') devoting himself to 'mathematics and [natural] philosophy more than at any time since'. His extraordinary genius developed, and he succeeded in devising new and ingenious theories. He presented his new mechanics in his immortal *Philosophiae naturalis principia mathematica* (*Mathematical Principles of Natural Philosophy*), or *Principia* as it is universally known, which appeared in 1687. His optical researches were published in 1704 in his second masterpiece, entitled *Opticks*.

The *Principia*

> The [Principia] by Isaac Newton . . . is probably the most important single work ever published in the physical sciences.
>
> (Stephen W. Hawking)[3]

Let us briefly look into this monument of human intellect, whilst we review some of the basic notions of the science of mechanics. In Book I, following the preface, which provides a set of definitions (mass, momentum, inertia and force), Newton presents his famous 'scholium', in which he discusses the concepts of 'absolute time' and 'absolute space'.

Absolute time and space

> I. Absolute, true, and mathematical time, in and of itself and of its own nature, without reference to anything external, flows uniformly and by another name is called duration. Relative, apparent, and common time is any sensible and external measure . . . of duration by means of motion; such a measure – for example, an hour, a day, a month, a year – is commonly used instead of true time.
>
> II. Absolute space, of its own nature without reference to anything external, always remains homogeneous and immovable. Relative space is any movable measure or dimension of this absolute space . . . determined by our senses from the situation of the space with respect to bodies and is popularly used for immovable space . . .[4]

These concepts were not seriously challenged until the beginning of the twentieth century, when Albert Einstein's (Nobel 1921) revolutionary *special theory of relativity* (p. 51) rendered them obsolete, with the result that they were abandoned.

Following this discussion on the nature of time and space, Newton proceeds with the laws of motion, which form the starting point of every argument in classical mechanics.

> Law 1. Every body perseveres in its state of being at rest or of moving uniformly straight forward, except insofar as it is compelled to change its state by forces impressed.
>
> Law 2. A change in motion is proportional to the motive force impressed and takes place along the straight line in which that force is impressed.[5]

By 'motion', Newton means what we call *momentum* (*mass × velocity*). Hence, the second law states that the force acting on a body is equal to the *rate of change of momentum* of the body itself. And again, in a situation in which the mass of the body is constant in time, the force is equal to the product *mass × acceleration*. Today, in every textbook of physics, the second law of motion is expressed in the following form: *a force acting on a body causes an acceleration which is directly proportional to the force acting on it and inversely proportional to the mass of the body*.

It follows that, if the force is equal to zero, the acceleration of the body is zero. Therefore, we may conclude with Newton that, *in absence of applied forces a body will move with constant velocity in a straight line or be at rest* (Law 1). This is a restatement of Galileo's *principle of inertia*. It also defines what we mean by *inertial frame of reference*, a concept which played a special role in classical mechanics, and an even more dominant role in Einstein's special theory of relativity. (A *frame of reference* is a conceptual framework used in determining the position and the motion of a body in space.) So we can again say that: *inertial* (or *Galilean*) *frames of reference are those in which bodies not subject to forces move at constant speeds in a straight line*.

From his second law, Newton derived a corollary, which is a precise statement of the *principle of relativity* described in Galileo's example of the ship (p. 8). We can express it as follows: *all inertial frames of reference are equivalent for the formulation of the laws of mechanics*. It means that if the ship is moving at a constant speed in a straight line, in all physical experiments performed therein, *all laws of mechanics* will appear the same as they would if the ship were not moving. As a consequence, it is impossible to tell, by making mechanical experiments, whether the ship is moving or not. This principle had been used in physics for more than two hundred years, and it was only at the beginning of the twentieth century that it was questioned, because it seemed in conflict with the laws describing the phenomena of electricity, magnetism and light. In his special theory of relativity Einstein was to propose his own principle of relativity, with such profound repercussions on the concepts of space and time that Newton's laws of motion would have to suffer modification.

Newton's mechanics was also to be called into question by the second great revolution in physics of the twentieth century, known as *quantum mechanics* (p. 143). It was to be a new way of looking at nature in the microscopic world of molecules, atoms, nuclei and subatomic particles. Otherwise, Newton's mechanics adequately describes all physical phenomena in the macroscopic world. The whole physical theory continues to be regarded as forming a beautiful and harmonious mathematical framework, capable of resolving all problems of a classical type, such as the motions of stars, planets, moons, aeroplanes, artificial satellites, automobiles, golf balls, rockets and every other type of macroscopic body travelling at a speed considerably less than 300 000 kilometres per second (that is, at the speed of light in empty space).

Universal gravitation

At the beginning of Book III of the *Principia*, Newton wrote: 'In the preceding books I have presented principles of [natural] philosophy that are not, however, philosophical but strictly mathematical . . . It still remains for us to exhibit the system of the world from these same principles.'[6] Book III demonstrates the law of universal gravitation at work in the universe. Newton wrote (Proposition 7, Theorem 7): 'Gravity exists in all bodies universally and is proportional to the quantity of matter in each. We have already proved that all planets . . . [gravitate] toward one another and also that the gravity toward any one planet, taken by itself, is inversely as the square of the distance of places from the centre of the planet. And it follows . . . that the gravity toward all the planets is proportional to the matter in them.'[7] And again (Corollary 2): 'The gravitation toward each of the individual equal particles of a body is inversely as the square of the distance of places from those particles.'[8] Today, Newton's law of universal gravitation is thus expressed: *any two material bodies in the universe attract each other by a force that is directly proportional to the product of their masses and declines as the square of the distance between them.*

Gravitation governs the motion of falling bodies, and also the motions of the moon around the earth, and the planets around the sun; it also holds together the sun, and the other stars, of which there are some hundred billion, of our Milky Way galaxy. More than two hundred years later, Albert Einstein, after presenting a new description of space and time in his special theory of relativity, went on to develop his *general theory of relativity* (p. 106), which shows how the behaviour of gravity deviates from that predicted by Newton.

Opticks

Many of the most important contributions to the development of the science of optics were made by Newton. In one of his famous experiments Newton proved that white light is a mixture of a whole range of single (or homogeneous) colours, which are refracted by a glass prism at different angles.

But what is light? Is it a wave like sound, or is it a stream of particles? In the first years of his work, Newton seemed inclined to embrace both the wave and the corpuscular theories of light. Later on he became a strong opponent of the wave theory, and instead advocated the corpuscular hypothesis. His main reason for supporting the latter was the difficulty encountered by the wave theory in trying

Fig. 2.2. Sir Isaac Newton (1642–1727). (Portrait by Charles Jervas, 1703, by permission of The Royal Society, London.)

to explain the rectilinear propagation of light. In the last pages of *Opticks* he wrote:

> [Query 29]. Are not the Rays of Light very small Bodies emitted from shining Substances? For such Bodies will pass through uniform Mediums in right Lines without bending into the Shadow, which is the Nature of the Rays of Light. They will also be capable of several Properties, and be able to conserve their Properties unchanged in passing through several Mediums, which is another Condition of the Rays of Light.[9]

The wave theory

At about the same time that Newton was formulating his corpuscular theory of light, Christiaan Huygens (1629–95), on the continent of Europe, was working on the wave theory.

Huygens was born at The Hague, the Netherlands. He was educated at the University of Leiden and at Breda. In 1666 he went to Paris to the French Academy of Sciences; here he remained for fifteen years. He then returned to the Netherlands, and in 1690 he published his notable *Traité de la lumière* (*Treatise on Light*), where he laid the foundations of the wave theory of light. In Huygens' view light was transmitted through an ethereal medium, filling up all space and permeating all ordinary matter. A light pulse would be transmitted in the same way as a shock wave passing through a row of elastic balls.

Newton's great authority rested like a weighty stone over the wave theory during the eighteenth century. Despite this, prominent scientists such as Leonhard Euler and Benjamin Franklin still remained devotees of the wave theory.

The Age of Reason

During the eighteenth century Newton's natural philosophy spread throughout Europe. Theoretical mechanics developed in the wake of the *Principia*, whereas the experimental approach of the *Opticks* influenced the sciences of electricity, magnetism and light.

In Switzerland, Leonhard Euler used mathematics to resolve complex mechanical problems, while Johann Bernoulli and his son Daniel studied the mechanics of fluids and elastic media. In France, the mathematician and philosopher Jean Le Rond d'Alembert contributed to the development of Newton's theory with his *Treatise on Dynamics*, and the Franco-Italian mathematician Joseph-Louis Lagrange published his famous book, *Analytic Mechanics*, in which he transformed mechanics into a branch of mathematics. Starting from the end of the century, Pierre Simon de Laplace composed his masterly five volumes entitled *Treatise on Celestial Mechanics*, in which he summarised all the results obtained in mechanics and physical astronomy from Newton onward, and presented mathematical theories for the study of a wide range of physical phenomena.

A particularly significant experiment was performed in 1798 by the English scientist Henry Cavendish. In his private laboratory, he directly measured the gravitational attraction between two lead spheres, and from this he was able, using Newton's second law, to infer the mass and density of the earth from the way that bodies at its surface fall.

Even though magnetic and electrostatic effects had been known since antiquity, it was only in the eighteenth century that the sciences of electricity and magnetism began to receive serious attention. The American scholar Benjamin Franklin introduced the concepts of positive and negative electricity, and by the end of the century, Joseph Priestley in England had discovered that the force between electrical charges is governed by an inverse-square law, as is the law of universal gravitation. This was confirmed more directly in France by Charles Augustin de Coulomb, who in 1785 first measured electric attraction and repulsion. Coulomb's law is known in the following form: *the force between two particles carrying electric charges is directly proportional to the product of the charges and declines as the square of the distance between them.*

Fig. 2.3. Crane Court was acquired as a home for the Royal Society of London in 1710, while Isaac Newton was its president. (By permission of The Royal Society, London.) The Society was founded on 28 November 1660, and chartered in 1663 by King Charles II. Newton was its president from 1703 until his death. Among its presidents in the twentieth century we find Ernest Lord Rutherford (Nobel laureate in chemistry), and four Nobel laureates in physics: Lord Rayleigh, Sir Joseph John Thomson, Sir William Henry Bragg and Patrick Lord Maynard Blackett.

During the eighteenth century, heat phenomena were explained by postulating the existence of an imponderable fluid, called *caloric*. This concept was capable of describing some thermal processes, but it transpired that it was unable to stand up to the test of experiment. It was only in the first half of the nineteenth century that physicists began to realise the connection between heat and energy.

Chapter 3
Highlights of classical physics

In the nineteenth century the science of physics developed rapidly. It managed to account for a wide variety of hitherto unexplained phenomena in diverse fields such as heat, electricity, magnetism and light, as well as mechanics.

Early in the 1800s experiments proved that light propagates as waves. Thus, the wave theory replaced Newton's corpuscular hypothesis, so explaining all the optical phenomena then known.

In mid-century it became understood that heat is a form of energy. This greatly assisted the evolution of the new science of thermodynamics: in retrospect, its two laws were to become amongst the most far-reaching triumphs of nineteenth-century science.

Electricity and magnetism, until then seen as two distinct phenomena, were found to be manifestations of one and the same unified electromagnetic field: not only could electricity produce magnetism, but magnetism too could produce electricity. Their mutual interaction was beautifully synthesised in a set of mathematical equations by James Clerk Maxwell. He also showed that the electromagnetic field propagates through space in the form of waves, and that their speed coincides exactly with the speed of light. So light too is electromagnetic in nature. Maxwell's great synthesis of electricity and magnetism represents a milestone on the road towards the goal of the ultimate unification of all the forces of nature.

During the later decades of the century, the idea that all matter consists of atoms gained more and more ground. Physicists and chemists were able to identify chemical elements by the characteristic light that they emitted. And it was then discovered that the properties of all chemical elements varied systematically with their atomic weight.

But in spite of the many accomplishments in the field of *classical physics*, a number of vexing problems still remained unsolved: these included the question of the potential existence of a medium through which light propagates; the lack of a real understanding of cathode rays and atoms; and the failure to find a key for interpreting light emitted by hot bodies and incandescent gases. Then, toward the close of the century, a flood of discoveries swept in: these included X-rays, radioactivity, the electron and new radioactive elements. All this brought about a veritable revolution, which opened the door to twentieth-century physics.

Heat and energy

In the nineteenth century the science of thermodynamics (which concerns the transformation of one form of energy into another) evolved as a result of studies on heat phenomena.

In the very last years of the 1700s, the American scientist Benjamin Thompson (Count Rumford), while supervising the boring of cannon, had noticed that quantities of heat were being produced by friction, and so he became convinced that heat could be a form of motion. This and other experiments were disregarded until about mid-century, when the true nature of heat was finally recognised.

In the 1840s, the British amateur scientist James Prescott Joule performed ingenious experiments on heat. He determined a value for the amount of mechanical energy required to produce a unit quantity of heat, and he concluded that heat is a form of energy. This is expressed by the *first law of thermodynamics* – itself a part of a more general principle, the principle of *conservation of energy*. This principle was conceived by the German scientist Hermann von Helmholtz. He suggested that various forms of energy, associated with gravity, electricity, magnetism, heat and motion were the same: whenever one kind of energy disappears, an equivalent amount of other types appears.

Investigations made by the French military engineer Sadi Carnot regarding steam engines formed the roots for the *second law of thermodynamics*. In 1824, he published a paper in which he predicted the maximum possible efficiency for a heat engine. Carnot's views were later developed in Britain by the Victorian physicist and mathematician William Thomson (Lord Kelvin), who early in the 1850s enunciated the second law. This law was given more precise mathematical formulation by the German mathematical physicist Rudolf Clausius. He introduced a new concept, which he called *entropy* (this is an abstract quantity which reflects, for example, the portion of thermal energy that is not available for conversion into mechanical energy by a heat engine).

In the second half of the century the atomic hypothesis began to gain acceptance, particularly in the study of gases. By that time there were available several empirical laws describing their properties. So, physicists were able to develop what was called the *kinetic theory of gases*, where the particles (atoms or molecules) composing a gas were considered to be in continual motion, bouncing off one another and off the walls of their container. This theory showed that heat energy was equivalent to the average kinetic energy of the particles. By the 1880s, what came to be known as *statistical mechanics* was developed by the Austrian physicist Ludwig Boltzmann, and by the American theorist Josiah Willard Gibbs. Finally, Johannes van der Waals (Nobel 1910) developed, at the University of Amsterdam, his theory of liquids and gases.

Electricity and magnetism

In 1801 in Paris, before a select audience, presided over by Napoleon Bonaparte himself, Alessandro Volta, then professor of experimental physics at the University of Pavia in Italy, presented a device which he had invented one year earlier. It was a source of an electric current, called a *Voltaic pile*. Following that demonstration a flurry of experiments began to be carried out in laboratories, and new and unexpected phenomena were discovered.

The Danish physicist and chemist Hans Christian Oersted, then at the University of Copenhagen, observed in 1820 that an electric current flowing in a wire deflected a nearby compass needle. So he provided the first experimental evidence that electricity and magnetism are inseparably linked; soon afterwards, in France, André-Marie Ampère worked out the mathematical theory for this new phenomenon.

Meanwhile, Michael Faraday, the greatest experimentalist of the nineteenth century, was working at the Royal Institution in London. In 1831 he discovered that magnetism could generate an electric current in a circuit. (The same phenomenon had been discovered independently by Joseph Henry in the USA.) This remarkable discovery, which was to change our everyday life, is called *electromagnetic induction*: it is the foundation of the modern electrical industry. Faraday also tried to account for electromagnetic effects in terms of *lines of force*. These were fictitious lines radiating from magnets or electric charges, and filling the space around the sources. This led eventually to the concepts of *electric* and *magnetic fields*.

In 1865 the British scientist James Clerk Maxwell, in a famous memoir entitled *A Dynamical Theory of the Electromagnetic Field* (one of the greatest papers in theoretical physics of the nineteenth century) presented a set of four equations describing all phenomena involving electricity and magnetism. He wrote:

> The theory I propose may . . . be called a theory of the *Electromagnetic Field*, because it has to do with the space in the neighbourhood of the electric or magnetic bodies . . . The electromagnetic field is that part of space which contains and surrounds bodies in electric or magnetic conditions . . . In order to bring these results within the power of symbolical calculation, I then express them in the form of the General Equations of the Electromagnetic Field.[1]

These equations not only describe the evolution in space and time of electric and magnetic fields generated by charges, magnets and currents, but also show that the two cannot be separated. An electric field changing with time generates a magnetic field, which induces an electric field in adjacent regions of space, which in turn generates a magnetic field, and so on. There exists, in fact, only a single unified field: the *electromagnetic field*.

Changing electric and magnetic fields propagate outward in all directions, resulting in a wave disturbance travelling in empty space. Maxwell's equations thus predicted the existence of *electromagnetic waves*. He was able to calculate the speed at which these waves propagate, by taking into consideration the values of purely electric and magnetic measured quantities. The result turned out to be near enough exactly the same as the value of the speed of light in empty space, then known from optical measurements. The conclusion was inevitable:

> This velocity is so nearly that of light, that it seems we have strong reason to conclude that light itself (including radiant heat, and other radiations if any) is an electromagnetic disturbance in the form of waves propagated through the electromagnetic field according to electromagnetic laws.[2]

So 'light is an electromagnetic wave', and along with it there are other waves with wavelengths which can be either shorter or longer than those of visible light. Maxwell's complete formulation of electromagnetism was published in 1873 in his masterpiece entitled *A Treatise on Electricity and Magnetism*.

The unification of electricity, magnetism and light represented the crowning achievement of classical physics in the nineteenth century. Maxwell's equations give us the mathematical basis necessary for understanding electromagnetism in just the same way as Newton's laws of motion and of universal gravitation enable us to comprehend mechanics. The areas of applications covered by Maxwell's theory are remarkable. They include all electromagnetic and optical devices such as electric motors, electric generators, radio, television, radar, computers, microscopes, telescopes and telecommunication systems.

The first experimental evidence of electromagnetic waves came more than twenty years later. In 1887–8, the German physicist Heinrich Hertz, by generating oscillating currents in an electrical circuit, was able to produce and detect electromagnetic waves with wavelengths that were far longer than those of ordinary light (they came to be called *radio waves*). This great discovery led to the development of the wireless telegraph and the radio. Today we know that electromagnetic waves cover an extremely broad interval of wavelengths and frequencies. Radio and television waves, infrared rays, visible light, ultraviolet radiation, X-rays and gamma rays all form parts of the *electromagnetic spectrum*.

Towards the end of the century, physicists increasingly developed electromagnetic theories. The Dutch physicist Hendrik Lorentz (Nobelist in 1902), then at the University of Leiden, the Netherlands, refined Maxwell's equations, and developed a new theory, called *electron theory*. He studied microscopic particles, each carrying an electric charge, which he thought must be contained in all atoms of matter. He then assumed that these particles generated electric and magnetic fields in the *ether*, a medium which he regarded as immovable and permeating the interior of all material bodies. So Lorentz deduced the equations of the electromagnetic field and those of the motion of charged particles in the field. He also

Fig. 3.1. James Clerk Maxwell (1831–79). (Courtesy The Cavendish Laboratory, University of Cambridge, England.)

theorised that the oscillations of charged particles inside the atoms were the source of light.

Light

The controversy between Newton and Huygens on the nature of light (p. 13) was at last settled at the beginning of the nineteenth century, thanks to ingenious experiments performed by the English physician and physicist Thomas Young. The following extract presents Young's discovery of the *interference* of light. It is taken from his *Course of Lectures on Natural Philosophy and Mechanical Arts*, published in 1807:

> ... when a beam of homogeneous [monochromatic] light falls on a screen in which there are two very small holes or slits, ... the light is diffracted in every direction. In this case, when the two newly formed beams are received on a surface placed so as to intercept them, their light is divided by dark stripes into portions nearly equal ... The middle of the two portions is always light ...[3]

On the screen the two beams of light formed a series of bright and dark bands, called *interference fringes*, which cannot be accounted for by the corpuscular theory of light. However, they are easily explained by the wave theory. The bright band represented the reinforcement of two light waves coming from the slits.

In other words, the two waves were 'in phase', and their crests coincided. If, however, the waves were 'out of phase' – the crest of one eliminating the trough of the other – they cancelled each other out.

Ten years after Young's observations the French physicist Augustin-Jean Fresnel performed experiments that confirmed Young's results and led to the rejection of Newton's corpuscular hypothesis. He also developed the mathematical theory of light propagation, and was able to calculate the *diffraction patterns* arising from various obstacles and apertures. So the wave theory led to a complete elucidation of all the phenomena then known that were connected with the propagation of light, such as polarisation, refraction, interference, diffraction and double refraction in crystals.

Vexing problems

Thanks to its great achievements, many physicists of the late nineteenth century thought that classical physics (Newton's mechanics, thermodynamics, the kinetic theory of gases, Maxwell's electromagnetism, the wave theory of light) could explain all known physical phenomena in nature. Albert Michelson (Nobel 1907), who himself helped in dismantling one of the pillars of classical physics, exclaimed in 1894: '. . . it seems probable that most of the grand underlying principles [of physical science] have been firmly established and that further advances are to be sought chiefly in the rigorous application of these principles to all the phenomena which come under our notice . . .'[4] But on the horizon dark clouds of doubt had already begun to gather.

The ether wind

A first cloud was represented by the problem of the ether. As we have seen (p. 20), one of the consequences of Maxwell's equations is that light consists of waves, propagating in empty space at the speed of 300 000 kilometres per second. But if light is a wave, how can it travel through the voids of space, as it seems to do, in its long journey through the cosmos from the stars? Does there exist for it, perhaps, some medium of transmission? Sound waves, for instance, require a material medium. Trying to answer these questions, physicists postulated the existence of a *luminiferous* ('light-carrying') *ether*, which was thought to be like a tenuous substance pervading all space. This ether would have to possess rather strange properties: it would have to have a vanishingly-small density, and, at the same time, be extremely rigid and frictionless. If a motionless ether existed, the speed of light would naturally be interpreted as a speed relative to the ether itself: just as the speed of sound waves always refers to a medium, such as stationary air.

This hypothetical ether, although it seemed useful, did not survive the test of experiment. In the 1880s the American physicist Albert Michelson undertook the task of submitting it to experimental verification. In particular, Michelson set out to measure the speed with which the earth moved through the ether. To make the test he invented a device, called an *interferometer*, so sensitive as to be able to observe the effect of a possible *ether wind*. In 1887, with the help of his colleague the chemical physicist Edward Morley, Michelson, using his interferometer, carried out a very precise experiment, but the speed of the earth relative to the ether could not be detected. The result of the experiment was null. Other experiments, based on equally ingenious ideas, similarly failed to demonstrate the existence of any form of ether.

These results were very disturbing. The first idea for finding a way out of the puzzle came from the Irish physicist George Fitzgerald and, independently, from Hendrik Lorentz at Leiden. They suggested (1889–95) that material bodies contract in the direction of their motion when they are moving. So, if this contraction is applied to the interferometer apparatus, we would be able to understand why the Michelson–Morley experiment had given a totally null effect. Although the contraction hypothesis successfully accounted for the negative result of the experiment, it was open to the objection that its assumptions were not verifiable. So physicists at the end of the century had reached what seemed a dead end. In fact they had to wait for Albert Einstein, who in 1905 resolved the question brilliantly.

Cathode rays

Another controversial subject dominating the experimental physics of the late nineteenth century was one regarding the true nature of the rays which were produced when an electric current was driven through a vessel containing a gas at low pressure. In the late 1830s Faraday himself tried to investigate the electric discharge through an evacuated vessel but he was not able to provide a good enough vacuum in order to obtain significant results.

During the 1850s, the German glass-blower Johann Heinrich Geissler invented a powerful mercury air vacuum pump. He then contrived a glass tube enclosing two metal electrodes; and this tube contained a vacuum of unprecedented quality (a pressure of about one millimetre of mercury). During the next few years the German experimenters Julius Plücker and, after him, Johann Hittorf – while producing electric discharges in *Geissler tubes* – noticed the presence of rays spreading out from the negative electrode (Faraday had named this the *cathode*) and producing a green glow on the opposite wall of the glass tube. Moreover, Hittorf, in the late 1860s, clearly proved (thanks directly to their shadow-casting properties) the rectilinear propagation of the rays he had observed. And some years later another German physicist, Eugen Goldstein, suggested that the glow on the tube wall was caused by the impact of these rays – for which he then coined the name *cathode rays*.

Most of the German physicists were convinced that the cathode rays consisted of some sort of waves, somewhat after the fashion of light. This view was shared by the great Heinrich Hertz, persuaded by his own experimental results, which indicated that the rays were not deflected by electric forces. He undertook a new series of experiments (the last ones in his life) which also demonstrated that the rays passed through thin sheets of metal. Hertz' experimental programme on cathode rays was inherited by his young assistant, Philipp Lenard (Nobel 1905), who discovered many of their properties.

According to another view, which was shared mainly by British scientists, cathode rays consisted of material particles which were ejected from the cathode, and were electrically charged. The well-known physicist and chemist William Crookes, in his private laboratory in London, had designed improved versions of Geissler's tubes (then called *Crookes' tubes*). With these tubes he had been able to show, in the late 1870s, that the rays were deflected by a magnetic field in a manner consistent with their being streams of charged particles. In addition, Joseph John (J. J.) Thomson (Nobel 1906) started a series of experiments on cathode rays, after he had become third Cavendish Professor of physics at Cambridge University. Crookes, Thomson and a group of Victorian amateur scientists maintained that cathode rays were a stream of charged particles, so the conflicting interpretations as to the nature of cathode rays lasted until the end of the century, when results from experiments began to provide the clue to their real nature.

Light spectra

In 1666, Isaac Newton had discovered that white light could be split into what he called a *spectrum of colours* (p. 12). He had passed a beam of sunlight through a glass prism and had found that the beam spread out into a coloured band made up of red, orange, yellow, green, blue, indigo and violet light. (The phenomenon itself is familiar: for example, the beautiful colours of a rainbow are the result of sunlight passing through water droplets in the air, each one acting as a tiny prism.) In the 1810s the German master-optician Joseph von Fraunhofer repeated Newton's experiment of shining sunlight through a prism, but he passed the light beam through a narrow slit before it became refracted by the prism. He magnified the resulting spectrum and obtained a series of coloured images of the slit, melted together to form a rainbow spectrum. To his surprise, Fraunhofer discovered that some of the slit images were missing, so that the solar spectrum was crossed by dark lines, now called *spectral lines*. He counted over seven hundred such fine lines; today we know of as many as a million. (Incidentally, the first to observe dark lines in the solar spectrum was the Englishman William Wollaston in 1802.) Scientists then began to guess that these spectral lines were characteristic of the various chemical elements present in the sun, and that these same elements would absorb light at specific wavelengths.

Around 1860, the German chemist Robert Bunsen and his young colleague, the physicist Gustav Kirchhoff, invented a new device, called a *spectroscope*, consisting of a prism and lenses that magnified and focused the light spectra.

They sprinkled various substances on a gaseous flame, spread out their glow into spectra, and discovered that they consisted of patterns of bright-coloured lines against a dark background. They then found that each chemical element produced a specific pattern of spectral lines. The spectroscope was quickly applied to discovering new elements: it was thus that the science of *spectroscopy* (the science which studies light and other radiation spectra) was born. (Using this technique, Bunsen and Kirchhoff themselves were able to discover the elements cesium and rubidium. They also succeeded in explaining the origin of the solar spectral lines that had been observed more than forty years previously.)

Another device for obtaining the separation of light into its component colours, called a *diffraction grating*, had been invented by Fraunhofer himself. (A diffraction grating consists of a plate of glass or metal on which fine and equally spaced parallel scratches or grooves are formed. When a beam of light is projected on to the grating, it is diffracted in different directions, each direction giving a definite colour component. Given the spacing between the slits, a measurement of the angle by which a component is deflected determines its wavelength.) With this method, the Swedish astronomer Anders Ångström first identified in the early 1850s a group of spectral lines of hydrogen. Finally, in 1885, the Swiss schoolteacher Johann Balmer showed that hydrogen produced a whole series of visible spectral lines (known as *Balmer series*), and that the wavelengths of these lines could be easily calculated using a rather simple formula.

The spectra like those studied by Fraunhofer, Bunsen and Kirchhoff are called *line spectra*, because the light analysed is in the form of discrete wavelengths. (For example, an incandescent gas produces a set of bright lines against a dark background: this is called an *emission spectrum*; whereas, if white light is viewed through a cloud of transparent gas, dark lines appear among the colours forming the spectrum: this is called an *absorption spectrum*.)

In those years scientists were already thinking that matter consists of atoms: thus the line spectra could be interpreted as produced by atoms which absorb or emit light only at definite wavelengths (or frequencies). But why do atoms radiate or absorb light only at definite wavelengths? Classical physics could not answer this question. Scientists had to wait for more than fifty years before they could discover the key to understanding light spectra.

Black-body radiation

And now let us briefly describe one of the most vexing problems for classical physics. It concerns the radiation emitted by an object when it is heated. Everybody knows, for instance, how when a bar of iron is heated, it begins to glow deep red, then bright red-orange. At a still higher temperature, it shines with a yellowish-white light, and finally with a blue–white light. (When the radiation emitted by a glowing solid is analysed with a spectroscope, it shows a continuous distribution of wavelengths over a wide range; this is called a *continuous spectrum*.)

In 1879, the Austrian physicist Josef Stefan showed that a hot object emits radiant energy at a rate directly proportional to its absolute temperature raised to the power of four; that is, if one doubles the temperature of the object (measured in kelvins), then the radiant energy emitted increases by a factor of sixteen (2^4). A few years later Ludwig Boltzmann further proved that this law is best demonstrated by an ideal object called a *black body*; that is, an object which absorbs all the radiation falling upon it. He was able to deduce the same law from Maxwell's electromagnetic theory of light and from thermodynamics.

A hot object emits radiation with a continuous distribution of wavelengths, but there is a particular wavelength at which a predominant amount of radiation is emitted. This 'peak wavelength' gives the glowing object its characteristic colour. In 1893 the German physicist Wilhelm Wien (Nobel 1911) produced a formula which expressed a relationship between this peak wavelength and the absolute temperature of a black body; this is now known as *Wien's displacement law*. According to Wien's law, the peak wavelength of the emitted radiation decreases as the temperature of a black body increases; that is, the peak wavelength is 'displaced' towards smaller values. For example, a black body that glows with yellow light is hotter than one glowing with red light (and this is because yellow light has a shorter wavelength than red light).

The Stefan–Boltzmann and Wien laws are not enough to describe all the properties of the radiation emitted by a black body (called *black-body radiation*). A more complete description is given by the distribution of the radiant energy (called *black-body spectrum*); that is, the amount of energy radiated at each particular frequency, or wavelength, range, which includes both of the laws we have been discussing. An explanation for this energy distribution was the outstanding, unsolved problem in the last years of the nineteenth century. Quite a number of eminent physicists devised various theories based on classical electromagnetism and thermodynamics. But all these theories had only limited success. For instance, in 1896, Wien worked out a mathematical formula for the radiation spectrum which fitted the experimental data in a reasonably satisfactory way – but it was definitely not sufficiently exact.

Atoms

The concept of the *atom* (from the Greek word meaning 'indivisible') was introduced by Greek philosophers. Democritus of Abdera maintained that atoms were solid, hard, and indestructible, and that different substances were composed of different atoms or combinations of them.

The atomic hypothesis was reconsidered in the seventeenth century, during the Scientific Revolution of that time. The great Anglo-Irish chemist Robert Boyle believed in 'atomism' (what he called the 'corpuscular philosophy'), and Newton wrote in his *Opticks* that all matter is composed of 'solid and impenetrable' particles, expressing a view of the atom similar to that of Democritus and Boyle.

By the end of the eighteenth century chemists had begun to realise how chemicals combine. The French scientist Antoine-Laurent Lavoisier, one of the founders of modern chemistry, listed thirty-three *chemical elements*, substances that could not be broken down into still simpler ones. Then the British chemist and physicist John Dalton converted the atomic hypothesis into a quantitative theory. According to him, each chemical element is made up of identical atoms, and the difference between atoms of different elements is in their mass.

As experimental data were being collected regarding chemical elements, scientists started to look for some sort of pattern in their properties. In the late 1860s, two chemists, the Russian Dmitri Ivanovich Mendeleev and the German Julius Lothar Meyer, arranged the elements in the order of increasing atomic weight in a tabular form, called the *periodic system of elements* – so named because it showed that elements that were similar as regards their chemical properties recurred at regular, periodic, intervals.

Scientists working in the field of electricity also provided important clues about atoms. Faraday thought that atoms were bound together in molecules by electric forces, and that atoms or groups of atoms could bear a positive or negative electric charge (they are called *ions*). His investigations also anticipated the concept of a fundamental unit of electricity, which was named the *electron* by the Irish physicist George Johnstone Stoney. In spite, however, of the convincing but rather indirect evidence for the existence of atoms, scientists of the late nineteenth century had no real understanding of them.

Towards twentieth-century physics

The very last years of the nineteenth century saw a rapid succession of unexpected and revolutionary discoveries in experimental physics. These opened the way to a new world which no one had suspected. Let us follow the historical thread, starting from the first great discovery which marked the birth of twentieth-century physics.

In 1895 the German physicist Wilhelm Röntgen (Nobel 1901), while investigating the effects of cathode rays in a discharge tube, accidentally discovered a new form of penetrating radiation, which he called *X-rays*. The nature of this mysterious radiation was not clear. It was found to be more penetrating than cathode rays, and not prone to deflection by a magnetic field.

The association of X-rays with phosphorescence, a process in which certain substances absorb and then re-emit light, led to the search for other sources of X-rays. In 1896, the French physicist Henri Becquerel (Nobelist in 1903), while working in Paris, and investigating some phosphorescent samples of uranium salts, discovered that they emitted a penetrating radiation, different from X-rays. Successively, Becquerel himself, and J. J. Thomson at Cambridge, demonstrated that those mysterious rays could discharge electrified bodies, by ionising the air molecules (splitting them into negative electrons and positive ions), and so allowing the air to conduct an electric current.

Other substances emitting Becquerel's rays were identified. In 1898, the French physicists Pierre and Marie Curie (Nobelists in 1903) discovered *polonium* and *radium*: it was Madame Curie who coined the term *radioactivity* for the spontaneous emission of penetrating rays from chemical elements, and these were named *radioactive elements*.

In 1897 J. J. Thomson, in one of the greatest experiments of modern science, demonstrated conclusively that cathode rays were electrically charged particles, which are now called *electrons*. Thus he resolved the long-standing controversy regarding their nature, and revolutionised the knowledge of atomic structure.

During the years 1897–8 a brilliant young British physicist, named Ernest Rutherford (Nobel for chemistry 1908), made a series of experiments on radioactivity. He was working as a research student under J. J. Thomson at the Cavendish Laboratory in Cambridge. From his experiments Rutherford deduced that there were two separate components in the radiation emitted by radioactive substances: one component penetrated matter a hundred times more than the other did. He named *alpha rays* the component that was most easily absorbed, and *beta rays* the much more penetrating one. It was later found that beta rays were deflected by a magnetic field, so that it was concluded that they were negatively charged particles similar to cathode rays (electrons).

The last achievement in this history that should be mentioned is that of the Italian inventor Guglielmo Marconi (Nobelist in 1909), who in 1895 succeeded in making the first transmission of wireless electric signals over a distance of two kilometres. This was the beginning of wireless telegraphy.

Discoveries and inventions

A great turning point in technological innovation and economic development came in the second half of the eighteenth century, when the British engineer James Watt introduced the steam engine into industry. This constituted the first main wave of innovation in the modern era, commonly known as the Industrial Revolution. The steam engine then totally revolutionised transportation: this was the second wave of innovation, which came about in the 1840s and lasted for about fifty years. It also gave rise to the new science of thermodynamics, which expressed itself in many new technological developments, such as the internal combustion engine, refrigeration techniques and the chemical industry.

The discovery of the electric current at the dawn of the nineteenth century had vast repercussions on the social life of humanity. Here was a direct case of science leading the way to technology. Its foundations lay in the discoveries of the great physicists of the first half of the century: Volta, Oersted, Ampère, Faraday.

The first application of electricity to technology was the invention of the *telegraph* (in the late 1830s); this was, within a few decades, to link continents by submarine cables (the first success was achieved thanks to the perseverance and intelligence of Lord Kelvin). During the 1860s the first *electric generator* and the *electric motor* were invented, and soon these were used to power machinery. After the incandescent electric lamp had been developed by the American inventor Thomas Alva Edison in the 1880s, power plants were established for the wide distribution of electricity; electric light and power were to become indispensable parts of everyday life. The third wave of innovation was brought about by electric power and the motor-car, and it had its birth in the last decade of the century.

In the late 1870s, the telegraph was to lead directly to the *telephone* (this was invented by the British-born American audiologist Alexander Bell), and at its very end to the *wireless telegraph*: pioneers like Maxwell, Hertz and Marconi laid the foundation for wireless communication; and scientists such as Goldstein, Crookes, Lenard and J. J. Thomson pioneered electronics.

'Everything that can be invented has been invented', exclaimed the Commissioner of the US Office of Patents in 1899 – so spectacular had been the progress of science and technology. He did not know how wrong he was – as we shall soon see!

Part II
The triumphs of modern physics (1901–1950)

Fig. 4.1. The University of Göttingen, Germany, in the 1700s. (Courtesy Stadtarchiv Göttingen.)
The Georgia Augusta University of Göttingen was founded in 1737 by King George II of Great Britain. In the nineteenth century a great tradition of mathematics and physics was established in the university, principally connected with the names of Carl Friedrich Gauss and Wilhelm Weber.

Chapter 4
New foundations

As we have already noticed, in spite of its successes, classical physics failed to explain certain important phenomena occurring in the physical world. Then, unexpected and astonishing discoveries of the 1890s started a revolution which led to the amazing development of twentieth-century physics, the foundations of which were laid in the new century's first thirty years. Two new theories, which we shall be considering in the next chapters – *relativity* and *quantum mechanics* – are the major achievements of this period.

The first wave of the quantum revolution emerged, suitably enough, precisely in the year 1900, when Max Planck, directly inspired by experiment, announced his revolutionary idea of the *quantum of energy*. In 1905 Albert Einstein used the quantum concept more directly, proposing that light itself is a stream of energy quanta. In the same year Einstein went on to publish his *special theory of relativity*, which introduced radical changes into the classical concepts of space and time; not satisfied with this, in 1907 he started to look for a new theory of gravitation capable of including his special relativity.

During the first decade of the century, experimental studies developed greatly, especially in the recently discovered field of radioactivity. In only a few years after its discovery, the different types of radiations emitted from radioactive substances were identified. In 1902–3 Ernest Rutherford discovered how a radioactive element transforms itself into another element, and in 1908 he proved that alpha particles are helium nuclei. Meanwhile, techniques for producing very low temperatures were being developed; this led to the first successful liquefaction of the element helium (1908).

In the following pages we shall retrace the paths taken by the Nobel laureates in physics, year by year. During the first period that we are considering (1901–10), the Nobel Prizes highlight achievements of the late nineteenth century: of the ten prizes, two go to inventions, and seven to experimental discoveries – among them X-rays, radioactivity, the discovery of the true nature of cathode rays and new features regarding atomic spectra. Finally, two out of fourteen laureates receive the prize for their theoretical work on phenomena concerning the structure of matter, as well as on the atomic origin of light.

1901

Mysterious rays

Our chronicle starts suitably in Stockholm, on 10 December 1901. At the Old Royal Academy of Music, before the King of Sweden and in the presence of a throng of distinguished guests, C. T. Odhner, president of the Swedish Academy of Sciences, presents the first Nobel Prize for physics with these words:

> Now that the Royal Academy of Sciences has received from its Committees their expert opinion on the suggestions sent in, as well as their own suggestions, it has made its decision, and as current President I am here to make it known. The Academy has awarded the Nobel Prize in physics to Wilhelm Conrad Röntgen, Professor in the University of Munich, for the discovery with which his name is linked for all times: the discovery of the so-called Röntgen rays or, as he himself called them, X-rays.[1]

The Nobel Committee for physics had received nominations from eminent scientists all over the world. The majority (seventeen out of twenty-nine) had joined in proposing Röntgen. Among the other candidates there were: the chemical physicist Svante Arrhenius (he was then a member of the Physics Committee) from Sweden; the physicists Henri Becquerel and Gabriel Lippmann from France; J. J. Thomson from England; Johannes van der Waals and Pieter Zeeman from the Netherlands; Philipp Lenard from Germany; and the inventor Guglielmo Marconi from Italy. Arrhenius was to receive the Nobel Prize for chemistry in 1903, while all the others would be receiving the award for physics within a period of nine years.

Wilhelm Conrad Röntgen (1845–1923) was born at Lennep in the Rhineland, Germany, the only son of a textile manufacturer. He studied at the Swiss Federal Institute of Technology (best known by its German abbreviation, ETH) in Zurich. He was persuaded by his professors, Rudolf Clausius and August Kundt, to continue studying physics, and he obtained his doctorate in 1869 at the University of Zurich. He later became professor of physics at the University of Strasbourg (then in Germany), and in 1888 was invited to succeed Friedrich Kohlrausch as professor of physics and director of the new institute of physics at the University of Würzburg, where he made his great discovery, later rewarded with the Nobel Prize. In 1900, he accepted the chair of physics at the University of Munich, where he remained until 1920. Röntgen was an outstanding experimenter, with a talent for highly precise measurements, principally interested in specific heat of gases, and other physical properties of solids.

8 November 1895

Röntgen communicated his discovery to the Physical-Medical Society of Würzburg on 28 December 1895. He started with a description of the exceptional phenomenon:

Fig. 4.2. Wilhelm Röntgen. (Photo by Gen. Stab. Lit. Anst., courtesy AIP Emilio Segrè Visual Archives, Weber Collection.) Röntgen never patented his discovery nor gained personally from it. He gave his Nobel money to the University of Würzburg. A lifelong friend wrote of him: 'His outstanding characteristic was his integrity. Perhaps one can say that Röntgen was in every sense the embodiment of the ideals of the nineteenth century: strong, honest and powerful, devoted to his science and never doubting its value; . . . of a really rare faithfulness and sense of sacrifice for people, memories, and ideals . . .'.[2]

> If the discharge of a fairly large induction coil be made to pass through a Hittorf vacuum tube, or through a Lenard tube, a Crookes tube, or other similar apparatus, which has been sufficiently exhausted, the tube being covered with thin, black cardboard which fits it with tolerable closeness, and if the whole apparatus be placed in a completely darkened room, there is observed at each discharge a bright illumination of a paper screen covered with barium platinocyanide, placed in the vicinity of the induction coil . . .[3]

Shortly after his discovery, H. J. W. Dam, an American reporter, was granted an interview with Röntgen. Dam later described, in a paper called *McClure's Magazine*, what took place on the evening of 8 November 1895. Let us summarise what apparently happened.

Röntgen was in his laboratory, on the ground floor of the Institute of Physics. He had assembled an apparatus, composed of a discharge tube connected to a device for producing a high voltage between the electrodes of the tube. He had enclosed the tube in a black cardboard box. With the room completely darkened, Röntgen switched on the apparatus and, to his surprise, observed that a paper screen coated with a fluorescent chemical, placed on a nearby table, gave off light. He moved the screen back, a full two metres away from the discharge tube, far beyond the distance over which cathode rays travel in air, and the screen still glowed. Thus he came to the conclusion that the light effect on the fluorescent screen was caused by a new form of radiation, originating from the rays when they struck the glass wall of the tube. Röntgen in his report continued:

> If the cathode rays within the discharge-apparatus are deflected by means of a magnet, it is observed that the X-rays proceed from another spot – namely, from that which is the new terminus of the cathode rays. For this reason, therefore, the X-rays, which it is impossible to deflect, cannot be cathode rays simply transmitted or reflected without change by the glass wall . . . I therefore reach the conclusion that the X-rays are not identical with the cathode rays, but that they are produced by the cathode rays at the glass wall of the discharge apparatus.[4]

Röntgen spent a further two months of intense, isolated work, in an effort to determine the characteristics of these mysterious, invisible rays. He demonstrated that X-rays could pass through opaque material that was impenetrable to ordinary light, and could produce fluorescent light in several substances. He established that if the cathode rays were allowed to strike heavy materials, such as platinum, a far stronger radiation resulted than that arising from lighter ones. He also observed that X-rays were not deflected by a magnet and could blacken a photographic plate. When he chanced to take X-ray pictures of his wife Bertha's hand, he obtained an image showing the bones inside. This was the first radiography ever effected.

The new discovery, and particularly the dramatic X-ray photographs, provoked a great sensation. At the beginning of January 1896, Röntgen mailed his announcement and the photographs to his colleagues. 'And now the devil was let loose', he wrote. The news was immediately reported in the most important papers, like the *Wiener Press* (Vienna), followed by the *Frankfurter Zeitung* (Frankfurt) and the *New York Times*; and the British scientific journal *Nature* published Röntgen's December report at full length. Röntgen labelled the rays he had discovered with the symbol 'X', meaning 'unknown', because he did not yet understand their true nature. Physicists soon started to discuss what the mysterious X-rays might be. They performed ingenious experiments in the hope that they could reveal some enlightening characteristics, but the identification of X-rays with electromagnetic waves was only demonstrated more than fifteen years later.

Soon after Röntgen's discovery the popularity of X-rays grew. Commercial apparatuses quickly became available, and X-rays started to be used to detect foreign bodies in patients, and to help in setting fractured bones; they marked, in fact, the start of a revolution in medical diagnosis. Nowadays, X-rays are used in a wide range of applications, which extend from medicine, on to material science, biology, and so into industrial applications.

Quantum revolution

Now, let us turn our attention to what was happening in Berlin in December 1900. At the Friedrich-Wilhelm's University, Max Planck, while professor of theoretical physics there, made a great discovery, which solved the black-body problem (p. 25). All those who had so far tried to work it out (including the German Wilhelm Wien (p. 26) and the Englishman Lord Rayleigh) had failed. But what was astonishing was that, in finding the correct formula for the energy distribution of black-body radiation, Planck actually produced a revolution of ideas which marked the dawn of modern physics.

Planck argued that the atoms of an idealised heated solid (the *black body*) did not radiate energy continuously, but in 'discrete amounts'. On this basis he deduced a formula which was valid for all frequencies (or wavelengths) of the emitted light, and at each temperature of the black body; this was in perfect agreement with the experimental results obtained by physicists working at the famous Physikalisch-Technische Reichsanstalt (PTR) laboratory in Berlin–Charlottenburg. He announced his result at a meeting of the German Physical Society, held in Berlin on 14 December 1900, in a contribution entitled 'On the theory of the law of energy distribution in normal spectrum'. He then presented his radiation law in a paper which was published in the German physics journal *Annalen der Physik* in March 1901.[5]

In short, Planck conceived of a heated solid as being composed of oscillating atoms, which by their oscillations caused the emission of electromagnetic waves (like tiny elementary antennae). The oscillators also absorbed the radiation falling upon them – like the receiving antenna of a television set. But whereas the antenna absorbs the incoming waves in a continuous way (that is, at all frequencies), Planck supposed that the atoms absorbed the energy that is carried by the waves only in individual packets, or *quanta* (that is, at only definite frequencies), and that the energy of each *quantum* (Latin word for 'how much') had to be related to the frequency of the wave.

Planck's radiation law was rapidly accepted by the community of physicists, because it fitted the experimental data like a glove, but the quantum concept was subject to a great deal of scepticism. (As the Dutch physicist Peter Debye recalled, 'we did not know whether the quanta were something fundamentally new or not'.) Five years later young Albert Einstein arrived and took the next step: he said that, under certain circumstances, light itself behaves as if it consists of energy quanta (see p. 49). And after another eight years, Niels Bohr (Nobel 1922) developed the first quantum theory of atomic structure (see p. 89); so it was that quantum phenomena became steadily more and more important. Finally, the quantum concept was to become the basis of a new theory, named *quantum theory*, which explained all the phenomena of the atomic and subatomic world, and developed into one of the most revolutionary concepts in physics of the whole

Fig. 4.3. Max Planck. (Courtesy Archiv zur Geschichte der Max-Planck-Gesellschaft, Berlin-Dahlem.) Planck described his decision to introduce the quantum of action as 'an act of despair'. He wrote: '... I had already fought for 6 years [since 1894] with the problem of equilibrium between radiation and matter without arriving at any successful result... [An] interpretation "had" to be found at any price, however high it might be.'[6]

twentieth century. So the date of 14 December 1900 can well be considered the birth of the quantum era, and Max Planck recognised as its father.

The quantum of action

Planck assumed that the oscillating atoms of a black body emit or absorb energy only in quanta, each quantum containing an amount of energy (symbol: E) given by the formula

$$E = h \times \nu$$

where the Greek letter ν indicates the frequency of the radiation emitted or absorbed, and h corresponds to the *Planck constant* (or *elementary quantum of action*). This means that the energy of each quantum is proportional to the frequency of the radiation (if you double the frequency you double the energy too). The value of h, which is a fundamental constant of nature and permeates all quantum theory, is, in the customary units used in physics, as follows

$$h = 6.63 \times 10^{-34} \text{ joule} \times \text{second}$$

1902

Light and magnetism

From Germany we move to the Netherlands. The second annual Nobel Prize was awarded in 1902 jointly to Hendrik Lorentz of Leiden University and Pieter

Zeeman of Amsterdam University, for their pioneering research 'into the influence of magnetism upon radiation phenomena'. This research was both theoretical and experimental: Lorentz was awarded the prize for his theory of electromagnetic radiation, the so-called *electron theory* (p. 20); Zeeman was awarded the prize for his discovery of the effects of magnetic fields on atomic spectra.

The prizewinning researches had been carried out in Leiden during the last decade of the nineteenth century. The physical laboratory of that university had by this time become a renowned centre for experimental physics, like other famous establishments, such as the Cavendish Laboratory in Cambridge, England. Heike Kamerlingh Onnes (Nobel 1913), a colleague of Lorentz, was the director. It was here that Zeeman made his discovery in 1896, a discovery which greatly contributed towards an understanding of the light spectra of atoms. But now let us go back to Lorentz.

Hendrik Antoon Lorentz (1853–1928) was born at Arnhem, the Netherlands, the son of the owner of a plant nursery. During the years 1870–5 he studied physics at the University of Leiden, where he obtained his doctorate. In 1878, at the early age of twenty-five, he was appointed to the chair of mathematical physics at the same university. Here he spent most of his career, until his retirement in 1912. He then was appointed director of research at the Teyler Institute, Haarlem, retaining however an honorary position at Leiden, where he continued to lecture weekly.

Lorentz' work

The central aim of Lorentz' electron theory was to refine the electromagnetic theory of Maxwell, and so to explain the relationship between electromagnetism and light. He proposed that light waves were produced by oscillations of charged particles (later on known as *electrons*) inside the atoms of matter. A consequence of this theory was that a magnetic field would affect the electron oscillations, and thereby the frequencies of the light emitted. This was confirmed by Zeeman's experiments.

Lorentz was also well known for his work on the Lorentz–Fitzgerald contraction (p. 23) and for the so-called *Lorentz transformations*, which he introduced in 1904. These are a set of mathematical formulae, which explained the experiment of Michelson and Morley (p. 23), and described the shortening of lengths and the dilatation of time; they paved the way for Einstein's special theory of relativity.

Lorentz received a great number of honours and distinctions for his outstanding work, from all over the world. Here is an extract from his biography, written by the Nobel Foundation:

> ... Lorentz was regarded by all theoretical physicists as the world's leading spirit, who completed what was left unfinished by his predecessors and prepared the ground for the fruitful reception of the new ideas ... Lorentz was a man of immense personal charm. The very picture of unselfishness, full of genuine

interest in whoever had the privilege of crossing his path, he endeared himself both to the leaders of his age and to the ordinary citizen.[7]

The respect that the Dutch people had for Lorentz is seen in Owen Richardson's (Nobel 1928) description of his funeral:

> The funeral took place at Haarlem at noon on Friday, 10 February [1928].
> At the stroke of twelve the State telegraph and telephone services of Holland were suspended for three minutes as a revered tribute to the greatest man Holland has produced in our time. It was attended by many colleagues and distinguished physicists from foreign countries. The President, Sir Ernest Rutherford, represented the Royal Society; and made an appreciative oration by the graveside.[8]

The Zeeman effect

Investigations embracing the optical and magnetic properties of matter were one of the lines of research at Kamerlingh Onnes' laboratory. Here, Zeeman, an assistant to Lorentz, was busy carrying out spectroscopy experiments. In August 1896, he started to examine the yellow spectral lines of a sodium flame, which was held in the gap between the poles of a strong magnet. He used a diffraction grating recently acquired by the laboratory. Such a grating was far superior to a prism spectroscope in distinguishing closely spaced spectral lines. It was thus that Zeeman was able to find that the sodium spectrum was split into groups of lines, separated by a very small distance. (Since each position of the lines in a spectrum corresponds to a specific wavelength, one can say that, under the influence of a magnetic field, the sodium atoms were emitting light of different wavelengths.)

The effect, by then called the *Zeeman effect*, was explained by using Lorentz' electron theory of matter ('... a true explanation appears to me to be afforded by the theory of electric phenomena propounded by Professor Lorentz, ... [who] was good enough to show me how the motion of the [electrons] might be calculated ...', wrote Zeeman in his paper published in *Nature*, in the issue of 11 February 1897).

Lorentz assumed that oscillating charged particles (electrons) inside the atoms of the incandescent sodium gas emitted light. He then showed that a spectral line would split into two closely spaced lines (a doublet) or three lines (a triplet) depending on whether the emitted light was viewed in the direction parallel or perpendicular to that of the magnetic field. He then calculated the difference in the wavelengths of two adjacent lines, and obtained a satisfactory agreement with Zeeman's experimental results. This proved that the *electron* was the source of the radiation, and not the atom as others had assumed. From these experiments Lorentz also obtained the value of the ratio between the charge of the hypothetical electron and its mass (the so-called charge-to-mass ratio e/m). Thus the Zeeman effect may also be considered one of the experimental proofs of the existence of the electron. (Later on it was discovered that the spectral lines could also split into a greater number of closely spaced lines. This effect was explained only many years afterwards, following on the evolution of quantum mechanics.)

Pieter Zeeman (1865–1943) was born in Zonnemaire, Zeeland, the Netherlands, the son of a Lutheran minister. He studied at the University of Leiden under Lorentz and Kamerlingh Onnes. He became Lorentz' assistant in 1890, and obtained his doctorate in 1893. Zeeman was appointed professor of physics at the University of Amsterdam in 1900, and in 1908 succeeded Johannes van der Waals as director of the institute of physics there.

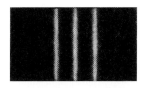

Fig. 4.4. The splitting of a spectral line in a magnetic field. (Top: in the presence of a magnetic field. Bottom: with no magnetic field.)

The Committe's experimentalist bias

In 1902 the Nobel Physics Committee received a total of six nominations in favour of Lorentz, but only one for Zeeman (this latter had received two nominations in 1901: '. . . it was Arrhenius who, at the last minute, handed in the one nomination of Zeeman that made it possible to consider the latter for the prize').[9] In spite of that, '. . . the Committee's recommendation of Lorentz [the theorist] depended largely on including Zeeman [the experimentalist] in the prize . . . Coupling Lorentz' name with that of Zeeman was an important condition for the awarding of the prize to the former, chiefly because of the Committee's experimentalist bias, but also because the Zeeman effect could be held up as a "discovery" that satisfied the conditions of the statutes.'[10] (In those early years the Committee was dominated by experimentalists from Uppsala University.)

1903

Radioactivity

The discovery of *natural radioactivity* in 1896 (p. 27) was not widely publicised. Physicists were very much occupied with the recent discovery of X-rays, which had become the focus of their attention.

Two years later interest in Becquerel's rays arose, when Marie Curie in France, and Gerhard Schmidt in Germany, independently discovered that *thorium*, besides uranium, emitted the mysterious radiation. In the same year, Marie and her husband Pierre began a systematic research into this subject, and soon announced the discovery of two more powerful radioactive elements, *polonium* and *radium*.

In the following years other new radioactive phenomena were discovered. In 1900, the French chemist Paul Villard discovered a third form of radioactive emission: this was an extremely penetrating form of radiation, which was named *gamma rays*; they were not deflected by a magnetic field and seemed analogous to X-rays.

The researcher who was to play a major role in the study of radioactivity was Ernest Rutherford, who was working at the Cavendish Laboratory (p. 28). After identifying alpha and beta rays, he left Cambridge in 1898 to become Professor of Physics at McGill University in Montreal, Canada. Here he began a research programme into radioactivity, soon achieving dominance in this field. Between 1902 and 1903 Rutherford and his young colleague, the English chemist Frederick Soddy, formulated the *spontaneous transformation* theory of radioactive decay; this explained how, when a radioactive atom breaks down ('decays'), it changes into a different atom with the emission of radiation.

Thus, by the end of 1903, we can say that radioactivity had already become the subject of scientific enquiry. And in fact the 1903 Nobel Prize for physics went jointly to Henri Becquerel and to Pierre and Marie Curie, precisely for their work on radioactivity.[11]

The Paris discovery

Soon after Röntgen's discovery of X-rays, Becquerel started to test a conjecture of his: that there must be some relationship between the mysterious rays and phosphorescence. He recalled in his Nobel lecture:

> At the beginning of 1896, on the very day that news reached Paris of the experiments of Röntgen and of the extraordinary properties of the rays emitted by the phosphorescent walls of Crookes' tubes, I thought of carrying out research to see whether all phosphorescent material emitted similar rays. The results of the experiment did not justify this idea, but in this research I encountered an unexpected phenomenon.[12]

Becquerel placed sheets of uranium salts on a photographic plate enveloped in black paper, and exposed it to sunlight (whose ultraviolet rays would stimulate phosphorescence). On developing the plate, he found that it bore images of the uranium compound. He also observed that the plate was affected even through foils of aluminium and copper. This could possibly have been the work of X-rays, for no radiation except X-rays was known to pass through metals. Becquerel decided to repeat the experiment in February, but as in those days the sun failed to make an appearance in Paris, he put his photographic plates away in a drawer, with the sheets of uranium salts lying on top of them. Sir William Crookes, who in those days was in Paris visiting Becquerel's laboratory, remembers:

> But the sun persistently kept behind clouds for several days, and, tired of waiting (or with the unconscious prevision of genius) Becquerel developed the plate. To his astonishment, instead of a blank, as expected, the plate had darkened as strongly as if the uranium had been previously exposed to sunlight . . . This was the foundation of the long series of experiments which led to the remarkable discoveries which have made 'Becquerel rays' a standard expression in Science.[13]

Some 'unknown radiation' other than X-rays was responsible for the darkening of the plates. Becquerel thought that this unknown radiation was produced by the

Fig. 4.5. Henri Becquerel. (Courtesy ACJC-Curie and Joliot-Curie fund, Paris.) Like his father Alexandre-Edmond (who had first photographed the solar spectrum in 1842) and his grandfather Antoine-César, Henri attended the *Ecole Polytechnique* in Paris. For several years Becquerel's research was concerned with optical phenomena. He further extended the work of his father by studying phosphorescent materials.

uranium in the salts. So these crucial experiments definitively marked Becquerel's discovery of natural radioactivity.

Antoine Henri Becquerel (1852–1908) was born in Paris, into a family of scientists extending right back through four generations. In 1878 he succeeded his father in the chair of applied physics at the *Conservatoire des Arts et Métiers*, and in 1892 he became a professor at the Department of Natural History of the Paris Museum. In 1889 Becquerel was elected to the French Academy of Sciences, and in 1895 was appointed to the chair of physics at the *Ecole Polytechnique*.

New radioactive elements

One of the scientists who was impressed by the mysterious 'uranic rays' discovered by Becquerel was a young Polish-born woman named Marie Curie. She resolved that this was to be the subject of her doctoral thesis, which was to be discussed at the Sorbonne University in Paris. Marie decided to measure the radiation emitted from uranium by means of an electrometer (an instrument for the measurement of very weak electric currents). Her husband, Pierre, gave up the research he was carrying out, and joined her in this promising new field.

The Curies began to study pitchblende (a natural uranium ore). To their surprise they found that some of the samples were more strongly radioactive than they had expected, and concluded that there must be other radioactive elements besides uranium in the pitchblende. By mid-1898 the Curies had been able to isolate a trace of a substance which was 400 times more radioactive than uranium itself.

In a paper published in July, they wrote: 'We thus believe that the substance that we have extracted from pitchblende contains a metal never known before . . . [We] suggest that it should be called *polonium* after the name of the country of origin of one of us.'[14]

After another few months of work, the Curies were able to produce samples even more radioactive than polonium. These contained still another element, which they called *radium*, because of its extremely intense radioactivity (it turned out to be three million times more radioactive than uranium!)

The Curies worked on for years to collect enough pure radium. By 1902 they had been able to isolate a tenth of a gram of nearly pure radium chloride, obtained from several tons of pitchblende residues. ('Sometimes I had to spend a whole day stirring a boiling mass with a heavy iron rod nearly as big as myself. I would be broken with fatigue at day's end',[15] Marie later wrote.) Marie presented the results of this work in her doctoral thesis on 25 June 1903. It was probably the greatest such dissertation in the history of physics!

Pierre Curie (1859–1906) was born in Paris, the son of a medical doctor. He studied at the Faculty of Sciences at the Sorbonne where, in 1895, he obtained his doctorate; here in 1900 he became a lecturer, and in 1904 a professor of physics. His early research activity was on crystallography, and together with his brother he discovered piezoelectricity in crystals. Later on he turned his attention to magnetism, and discovered that the magnetic properties of certain substances changed at a specific value of the temperature, now known as the *Curie point*. Pierre and Marie Curie had two daughters: Irène, who married Frédéric Joliot (they were jointly awarded the 1935 Nobel Prize for chemistry); and Eve, who married the American diplomat Henry R. Labouisse, winner of the Nobel Prize for peace in 1965. On 19 April 1906, Pierre Curie died in a street accident in Paris, when struck by the wheel of a horse-driven cart.

Madame Curie and the Nobel

> . . . [Gösta] Mittag-Leffler [a Swedish mathematician, and a member of the Swedish Academy of Sciences] played a role in assuring that Marie Curie was one of the three prizewinners . . . In 1903, only the names of Becquerel and Pierre Curie had been put forward in a proposal that clearly emanated from the French Academy of Sciences, since it was signed by . . . Poincaré together with several members of the academy who had not been invited to nominate. Marie Curie, of course, was not (and would never become) a member of the Academy . . .
> Mittag-Leffler [wrote] to Pierre Curie, . . . [who] replied: 'If it is true that one is seriously thinking about me, I very much wish to be considered together with Madame Curie with respect to our research on radioactive bodies.' . . . As it turned out, the committee recommended that the Curies together receive half of the physics prize.[16]

Fig. 4.6. The most famous scientific couple of all time, Pierre and Marie Curie. (Courtesy ACJC-Curie and Joliot-Curie fund, Paris.) 'In July 1895, they were married at the town hall at Sceaux, where Pierre's parents lived. They were given money as a wedding present which they used to buy a bicycle for each of them, and long, sometimes adventurous, cycle rides came to become their way of relaxing.'[17]

1904

The discovery of argon

In 1904 the Swedish Academy of Sciences honoured the Victorian 'old guard'. So we find them awarding the Nobel Prize for physics to the English physicist John William Strutt (Lord Rayleigh) for his investigation into gases, and for the successful isolation of argon, one of the rare gases of the atmosphere.

For centuries the nature and composition of atmospheric air had attracted the attention of scientists. The French chemist Antoine Laurent Lavoisier was the first to realise that air is a mixture of gases: in the 1770s he discovered in fact

Marie Curie (1867–1934)

Marie Skłodowska Curie was born in Warsaw in 1867 (the year Faraday died). She studied in local schools and received some training in physics and chemistry from her father, a secondary-school teacher. In 1891, she went to Paris to continue her studies at the *Sorbonne*:

> [She] ... rented a little attic in the Quartier Latin. There the cold was so intense that at night she had to pile on everything she had in the way of clothing so as to be able to sleep. But as compensation for all her privations she had total freedom to be able to devote herself wholly to her studies. 'It was like a new world opened to me, the world of science, which I was at last permitted to know in all liberty', [she wrote].[18]

In 1893 Marie obtained her degree in physics, and the following year in mathematics. In the same year she met Pierre Curie, and in 1895 they were married. In 1903 Madame Curie succeeded her husband as head of the physics laboratory at the Sorbonne, and following the tragic death of Pierre in 1906, she took his place as professor of general physics in the Faculty of Sciences.

The Curies' early researches were performed under miserable conditions, and their laboratory arrangements were appallingly inadequate. ('It was a cross between a stable and a potato shed, and if I had not seen the worktable and items of chemical apparatus, I would have thought that I was being played a practical joke', wrote Friedrich Wilhelm Ostwald, German Nobelist for chemistry, who came to Paris to see where radium had been discovered.) In 1911 Madame Curie received the Nobel Prize for chemistry, the first person, and the only woman, to receive two Nobel awards. In 1914 she was appointed director of the Laboratoire Curie at the recently founded Institut du Radium of the University of Paris.

During the First World War Madame Curie was intensively engaged in promoting the use of radium to alleviate suffering. She died of leukaemia, caused by working so long with radioactive substances, at Savoy, France, on 4 July 1934.

> Her strength, her purity of will, her austerity towards herself, her objectivity, her incorruptible judgement – all these were of a kind seldom found joined in a single individual.
>
> (Albert Einstein)[19]

that air is made up of oxygen and nitrogen. Some years later, the English scientist Henry Cavendish (well-known for being the first to measure the gravitational force, p. 14), while investigating atmospheric nitrogen, discovered that about one hundredth part of the air might be formed of some new inert constituent. His work was forgotten by the scientific community, so the nature of this small residue of air remained a secret for more than a hundred years.

In the early 1880s Lord Rayleigh compared the density of nitrogen, extracted from the air, with the density of nitrogen produced from one of its chemical compounds. He found that air nitrogen was about one half per cent denser. Stimulated by this anomaly and remembering earlier observations made by Cavendish, Rayleigh looked further into the matter. He thought that air nitrogen was heavier than chemical nitrogen because of the presence in the atmosphere of an unknown element. So he tried to isolate it. ('This was a task of considerable difficulty; and it was undertaken by [William] Ramsay [a Scottish chemist] and myself working at first independently but afterwards in concert', recalled Rayleigh in his Nobel lecture.) In 1895 they succeeded in isolating a small residue of gas, which constituted almost one per cent by weight of the atmosphere. They subjected this residue to an electric discharge and then examined it with the aid of a spectroscope; its spectrum in fact showed the presence of bright spectral lines that belonged to no known element. The newly discovered gas was named *argon* (from the Greek word meaning 'inactive'), because it seemed unable to form chemical compounds with any other elements.

Argon accounted for nearly all of the one per cent of unknown gas in the atmospheric air; but there were still traces of other gases present in there. A few years later Ramsay himself went on to discover four more inert gases: neon, krypton, xenon and helium. (The last had been discovered, in 1868, by the French astronomer Jules Janssen, in the spectrum of the sun's light. Ramsay, however, discovered helium on earth, in radioactive minerals.) For these discoveries he received, in the same year as Rayleigh, the Nobel Prize for chemistry.

The discovery of argon was considered Rayleigh's greatest contribution to science. But he also made other fundamental contributions in many different fields of physics. His early scientific activity dealt with such subjects as electromagnetism, acoustics and optics. Among his studies, the most significant one is his theory explaining how the blue colour of the sky is the result of the scattering of sunlight by air molecules in the atmosphere. The *Rayleigh scattering* law, derived from this theory, has since become a classic in the study of light propagation. In addition, he published in 1877–8 a two-volume masterpiece, entitled *Theory of Sound*, which has remained a notable landmark in the literature on acoustics.

John William Strutt, Third Baron Rayleigh (1842–1919). Not many members of the British aristocracy have achieved fame as scientists, but one who certainly did was Lord Rayleigh. He was born at Langford Grove, Maldon, Essex, England. In 1861 he began to study at Trinity College, Cambridge, where he graduated in 1865. He soon won fame as an outstanding physicist, and in the period 1879–84 he was second Cavendish Professor of Experimental Physics at Cambridge, in succession to James Clerk Maxwell. From 1887 to 1905 Lord Rayleigh was Professor of Natural Philosophy at the Royal Institution in London. He was elected a Fellow of the Royal Society in 1873, and was its president from 1905 to 1908,

Fig. 4.7. Lord Rayleigh (left) and Lord Kelvin, circa 1900. (From *John William Strutt, Third Baron Rayleigh*, by Robert John Strutt, Fourth Baron Rayleigh, New York, Longmans, Green & Co., 1924; Courtesy AIP Emilio Segrè Visual Archives.)

when he was elected Chancellor of Cambridge University, a position that he retained until his death.

1905

Cathode rays

Let us now turn to cathode rays, the subject for which the German physicist Philipp Lenard earned the 1905 Nobel Prize.

Lenard had started his research work on cathode rays in the early 1890s, when he was investigating how these rays could pass through thin metal plates (p. 24). As a result of his researches he had come closer to the view that cathode rays were identical to the electrons discovered by J. J. Thomson in 1897. In similar experiments, made in 1899, Lenard had proved that electrons were ejected when ultraviolet light struck metal surfaces. This phenomenon, known as the *photoelectric effect*, had first been discovered in 1887 by Heinrich Hertz. (Lenard's Nobel Prize mainly concerns this phenomenon.) In 1902 Lenard showed that no electrons can be emitted from an illuminated metal surface if the light frequency is less than a certain minimum value. He also showed that, for higher frequencies, electrons were in fact emitted, but with a maximum kinetic energy that did not depend on the intensity of the light (the amount of light falling on the surface); that is, an increase of light intensity boosted the number of electrons emitted, but not their energy (or speed). These observations indicated a kind of interaction between light and matter that could not be explained in terms of classical light waves. The explanation was later provided by Albert Einstein in the very year that Lenard earned his prize (see p. 50).

Philipp Lenard (1862–1947) was born at Pressburg in Hungary (now Bratislava, Slovakia). He studied physics successively at Budapest, Vienna, Berlin and Heidelberg. In 1886 he obtained his doctorate at Heidelberg, and in 1890 he accepted a position as assistant to Heinrich Hertz at the University of Bonn, where he began to experiment with cathode rays. In 1894 he became a professor of physics, in turn at the universities of Breslau, Aachen, Heidelberg and Kiel. In 1907 he returned to teach at Heidelberg, where he stayed until his retirement. Lenard's Nobel biography ends with these words:

> Lenard was an experimentalist of genius, but more doubtful as a theorist. Some of his discoveries were great ones and others were very important, but he claimed for them more than their true value. Although he was given many honours, . . . he believed that he was disregarded and this probably explains why he attacked other physicists in many countries. He became a convinced member of Hitler's National Socialist Party and maintained unreserved adherence to it. The party responded by making him the Chief of Aryan or German Physics.[20]

An exceptional year

The year 1905 was an exceptional year. While Pablo Picasso was painting in his so-called 'Rose Period', a German-born Swiss physicist named Albert Einstein was working at the Swiss patent office in Bern. He published, at the tender age of twenty-six, two revolutionary papers in the *Annalen der Physik*.

In May 1905, in a letter to his friend Conrad Habicht, Einstein wrote:

> I promise you four papers in return, the first of which I might send you soon, since I will soon get the complimentary reprints. The paper deals with radiation and the energy properties of light and is very revolutionary . . . The fourth paper is only a rough draft at this point, and is an electrodynamics of moving bodies which employs a modification of the theory of space and time; the purely kinematic part of this paper will surely interest you.[21]

Light quanta

In the first paper cited in this letter, and described as 'very revolutionary' (entitled 'On a Heuristic Point of View Concerning the Production and Transformation of Light'), Einstein advanced the hypothesis that light consists of *quanta of energy* (concentrated bundles of electromagnetic energy, later on called *photons*) which travel through space. In his paper, Einstein proposed: '. . . when a light

Fig. 4.8. Albert Einstein at the patent office. (Courtesy of the Albert Einstein Archives, The Jewish National & University Library, The Hebrew University of Jerusalem, Israel.) Einstein remained at the patent office for seven years (1902–9). He enjoyed his work there. Writing to his lifelong friend and colleague Michele Besso in 1919, he spoke nostalgically of it as '. . . that secular cloister where I hatched my most beautiful ideas and where we had such good times together.'[22]

ray is spreading from a point, the energy is not distributed continuously over ever-increasing spaces, but consists of a finite number of energy quanta that are localised in points in space, which move without dividing, and can be absorbed or generated only as a whole.'[23] According to him, the energy carried by each light quantum should be proportional to its frequency: $E = h\nu$. This famous formula had already been introduced five years earlier by Max Planck to explain the energy distribution from a black body (p. 38). Einstein, however, went further than Planck. He not merely assumed that the energy quanta were involved in the processes of absorption and emission of light by atoms, but also maintained that such quantum properties were inherent in the nature of light itself.

This revolutionary idea contradicted Maxwell's view that light is a continuous wave, and that the energy it carries is distributed throughout the space occupied by the electromagnetic field. Now, light was instead held to be made up of discrete particles. Einstein demonstrated the effectiveness of his light-quanta hypothesis in explaining the photoelectric effect, a phenomenon that classical physics had been unable to resolve. He wrote:

> The usual conception, that the energy of light is continuously distributed over the space through which it travels, meets with especially great difficulties when one attempts to explain the photoelectric phenomena; these difficulties are presented in a pioneering work by Mr. Lenard [p. 49]. According to the conception that the exciting light consists of quanta of energy [$h\nu$], the production of cathode rays by light can be conceived in the following way . . .[24]

> ### The photoelectric effect
>
> Einstein started by assuming that a light quantum penetrating a metal surface would collide with an atom, and would transfer all its energy to one of its electrons. This same electron would then be ejected from the metal with a kinetic energy expressed by the following equation:
>
> $$K = hv - P$$
>
> This equation states that a part (P) of the photon energy (hv) causes the electron to pass through the metal surface. The excess energy (K) is given to the electron in the form of kinetic energy.
>
> Lenard's experiments had shown that the maximum kinetic energy of the electrons ejected did not depend on the intensity of the light that was shone upon the metal surface. This conclusion is in complete agreement with Einstein's equation. Moreover, the existence of a minimum frequency, characteristic of the metal used, also follows immediately from the above equation.
>
> Einstein's equation for the photoelectric effect introduced certain additional predictions. It laid down firstly, that the electron kinetic energy is proportional to the light frequency; and secondly, that the slope of the plot representing that equation (a straight line), is given by h, the same universal constant introduced by Planck in 1900 (p. 38). None of this was known prior to 1905.

The September paper

The fourth paper cited in Einstein's letter to his friend Habicht was published on 26 September 1905 in the *Annalen der Physik*. It bears the title 'On the electrodynamics of moving bodies'. In this celebrated paper Einstein elaborated his *special theory of relativity*, which sparked off a revolution in physics, comparable to that in astronomy brought about by the ideas of Copernicus, Kepler, Galileo and Newton, a revolution which led to new and unexpected discoveries. He based his theory on a reinterpretation of Newton's principle of relativity (p. 11). Einstein's principle of relativity can be expressed in the following way:

> All the laws of physics have the same form for all observers moving in straight lines and at constant speeds relative to one another (inertial observers).

The second principle upon which special relativity is based concerns the constancy of the velocity of light. We may express it as follows:

> The velocity of light in empty space, when measured, is independent of the motion of the source and the motion of the observer.

These brief and simple-sounding propositions led to revolutionary changes to our most basic concepts of space and time, concepts that had gone unquestioned

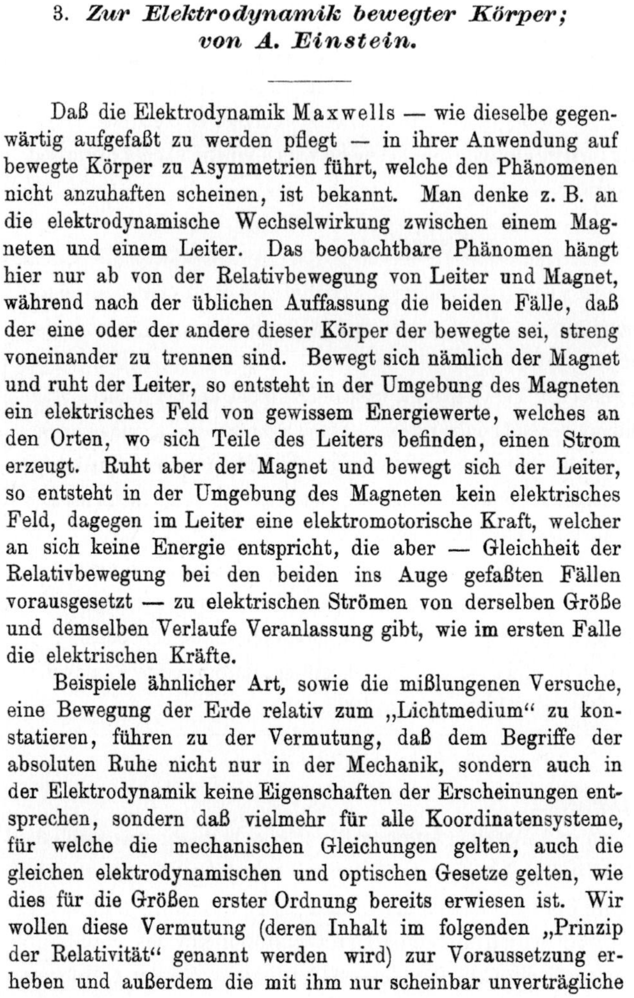

Fig. 4.9. Einstein's September paper on special relativity (opening page), a landmark in the development of modern physics.[25]

3. *Zur Elektrodynamik bewegter Körper;*
von A. Einstein.

Daß die Elektrodynamik Maxwells — wie dieselbe gegenwärtig aufgefaßt zu werden pflegt — in ihrer Anwendung auf bewegte Körper zu Asymmetrien führt, welche den Phänomenen nicht anzuhaften scheinen, ist bekannt. Man denke z. B. an die elektrodynamische Wechselwirkung zwischen einem Magneten und einem Leiter. Das beobachtbare Phänomen hängt hier nur ab von der Relativbewegung von Leiter und Magnet, während nach der üblichen Auffassung die beiden Fälle, daß der eine oder der andere dieser Körper der bewegte sei, streng voneinander zu trennen sind. Bewegt sich nämlich der Magnet und ruht der Leiter, so entsteht in der Umgebung des Magneten ein elektrisches Feld von gewissem Energiewerte, welches an den Orten, wo sich Teile des Leiters befinden, einen Strom erzeugt. Ruht aber der Magnet und bewegt sich der Leiter, so entsteht in der Umgebung des Magneten kein elektrisches Feld, dagegen im Leiter eine elektromotorische Kraft, welcher an sich keine Energie entspricht, die aber — Gleichheit der Relativbewegung bei den beiden ins Auge gefaßten Fällen vorausgesetzt — zu elektrischen Strömen von derselben Größe und demselben Verlaufe Veranlassung gibt, wie im ersten Falle die elektrischen Kräfte.

   Beispiele ähnlicher Art, sowie die mißlungenen Versuche, eine Bewegung der Erde relativ zum „Lichtmedium" zu konstatieren, führen zu der Vermutung, daß dem Begriffe der absoluten Ruhe nicht nur in der Mechanik, sondern auch in der Elektrodynamik keine Eigenschaften der Erscheinungen entsprechen, sondern daß vielmehr für alle Koordinatensysteme, für welche die mechanischen Gleichungen gelten, auch die gleichen elektrodynamischen und optischen Gesetze gelten, wie dies für die Größen erster Ordnung bereits erwiesen ist. Wir wollen diese Vermutung (deren Inhalt im folgenden „Prinzip der Relativität" genannt werden wird) zur Voraussetzung erheben und außerdem die mit ihm nur scheinbar unverträgliche

since Newton's time. They entail very radical and far-reaching implications. Here are four:

> *First.* Two events that are simultaneous to one observer may not be simultaneous to another.
>
> *Second.* When two observers move relative to each other with a constant velocity their clocks run differently; and if they measure the length of an object, they do not get the same results.
>
> *Third.* Moving objects gain mass. And again, no object can ever travel at a speed greater than that of light in empty space.
>
> *Fourth.* Mass and energy are two forms of the same thing, and one can be converted into the other.

Pursuing a light wave

> ... such a principle [concerning the constancy of the velocity of light] resulted from a paradox upon which I had already hit at the age of sixteen: if I pursue a beam of light with the velocity 'c' (velocity of light in a vacuum), I should observe such a beam of light as a spatially-oscillatory electromagnetic field at rest. However, there seems to be no such thing, whether on the basis of experience or according to Maxwell's equations. From the very beginning it appeared to me intuitively clear that, judged from the standpoint of such an observer, everything would have to happen according to the same laws as for an observer who, relative to the earth, was at rest.
>
> (Albert Einstein)[26]

On the electrodynamics of moving bodies – introduction

> It is well known that Maxwell's electrodynamics – as usually understood at present – when applied to moving bodies, leads to asymmetries that do not seem to attach to the phenomena. Let us recall, for example, the electrodynamic interaction between a magnet and a conductor . . .
>
> Examples of a similar kind, and the failure of attempts to detect a motion of the earth relative to the 'light medium', lead to the conjecture that not only in mechanics, but in electrodynamics as well, the phenomena do not have any properties corresponding to the concept of absolute rest, but that in all coordinate systems in which the mechanical equations are valid, also the same electrodynamic and optical laws are valid . . . We shall raise this conjecture (whose content will be called 'the principle of relativity' hereafter) to the status of a postulate and shall introduce, in addition, another postulate, only seemingly incompatible with the former one, that in empty space light is always propagating with a definite velocity [c] which is independent of the state of motion of the emitting body.[27]

The special theory of relativity

The special theory of relativity describes how events occurring at certain places and at definite times (such as the explosion of a star in the sky, or the emission of light by an atom) look to observers at rest in inertial frames of reference, which themselves are moving with a constant velocity relative to one another. In classical (or Newtonian) mechanics it was assumed that time and space were completely separate entities; and that spatial distances and time intervals between events were identical for all observers. Special relativity tell us that this is not so, but that time and space are intimately linked, and that distances and time intervals depend on the relative motion of the observers.

(*cont.*)

Newton's principle of relativity

In classical mechanics, in any inertial frame of reference, clocks, measuring the flow of time, run at the same rate. It means that the time of 'one hour' as measured by an observer in a certain frame (yourself on the earth, for example) corresponds precisely to 'one hour' as measured by all observers in inertial frames of reference (for example your astronaut friend, on board a uniformly moving spaceship). This is what Newton meant by *absolute time* in his *Principia* (see p. 10).

Newton's principle of relativity (p. 11) means that in all physical experiments performed in the spaceship, *all laws of mechanics* will have the same mathematical form as they would if the spaceship were not moving (exactly as in Galileo's ship, p. 8). This principle was called into question by Maxwell's electromagnetic theory of light, which postulated the existence of the *ether* as a medium through which light waves were transmitting (see p. 22). This esoteric substance was thought of as a *preferred* inertial frame of reference (like the *absolute frame* which is implied in Newton's concept of *absolute space*, p. 10), relative to which the speed of light was assumed to have a constant value in every direction.

Maxwell's equations describing electric, magnetic and optical phenomena, however, did not seem to obey Newton's principle of relativity. That is, if we use the rules of classical mechanics to transform such equations from one moving inertial frame to another, their form does not remain the same; in particular, the speed of light would have different values in the two frames. It follows that the speed of light should depend on the motion of an observer relative to the ether; by measuring this speed relative to an inertial frame, the spaceship, for example, your astronaut friend could determine the speed of the spaceship relative to the ether. Experiments based on this idea (the most famous of these was the Michelson–Morley experiment, p. 23) all failed to detect the slightest traces of any stationary ether.

Einstein's principle of relativity

At this point Einstein proposed his own principle of relativity: *all the laws of physics have the same form in all inertial frames of reference*. Einstein's intuition was to extend Newton's principle of relativity from mechanics to all physical phenomena. An immediate consequence is that we must abolish the hypothesis of any preferred frame of reference and, consequently, the existence of the ether.

The constancy of the velocity of light

Einstein's second principle states that *the velocity of light is independent of the motion of the source*. However, according to Einstein's principle of relativity, if a uniformly moving observer finds that light emitted from differently moving sources has the same speed, then all observers in inertial frames must obtain the same result. This leads to the conclusion that the speed of light is independent not only of the

(*cont.*)

motion of the source but also of the motion of the observer. In other words, *the speed of light in empty space is the same in all inertial frames*.

The relativity of times and distances

Einstein's two principles have astonishing consequences. First, two observers in different inertial frames will not agree on the answer to some questions, as for example whether two events which occur some distance apart occur at the same time or not. This means that *simultaneity* is not an absolute concept; it is in fact relative to the observer. After more than two hundred years, physicists had to change from Newton's *absolute time* to Einstein's *relativity of time*. In the same way, two observers in different inertial frames will not agree on the measured value of the distance between locations in space of two events: this means that spatial distances too are relative to the observer.

Lorentz transformations

The classical equations for converting measurements of spatial distances and time intervals between events from one observer to another, both of them in two uniformly moving frames of reference (the so-called *Galilean transformations*) are not in accordance with the fact that simultaneity is relative. They are replaced by a set of new equations which are derived in accordance with Einstein's two principles of special relativity, and which are called *Lorentz transformations*, because they were proposed by Hendrik Lorentz in 1904 (p. 39).

Time dilatation and length contraction

If an observer investigates the time between the ticks of a clock which is moving uniformly relative to him, he finds that the intervals are lengthened; in other words, the clock runs more slowly. The clock will run at the fastest rate when it is at rest, whereas the tick rate would be reduced to zero for a clock moving with a speed approaching that of light. This phenomenon is called *time dilatation*.

In the same way, the length of a rigid rod, moving uniformly relative to an observer, is foreshortened in the direction of its motion. This phenomenon is called *length contraction*. The rod will have the greatest length when it is at rest, whereas, measured in a frame in which it moves with a speed approaching that of light, its length will approach zero, and would become without physical meaning at speeds greater than that of light. This last result indicates that no material body can travel with a speed exceeding the speed of light. In this sense, *the speed of light in empty space is the ultimate speed*.

(cont.)

> ### Relativistic mechanics
>
> Whereas Maxwell's equations change their mathematical form when one uses Galilean transformations, they do not change if we use Lorentz transformations. On the contrary, Newton's laws of motion do change; they are not consistent with special relativity, and so they have to be rewritten in a different form. Consequently we have to use *relativistic mechanics*, with still more striking consequences. First, we find that the inertial mass of a body increases with its speed: it is least when the body is at rest, whereas it will become an ever-increasing mass as the body's speed approaches the velocity of light. Second, we discover that *mass and energy are equivalent*. The formula giving this is
>
> $$E = m \times c^2$$
>
> where E is the total energy of a body associated with its mass m. This is one of the most famous equations in modern physics, and it has been amply verified by experiments.
>
> The value of this energy content is enormous, it is about 100 000 billion joules per gram of matter. Where is this enormous energy supposed to reside, and how does it manifest itself? It resides, for example, in atomic nuclei, due to the action of the strong nuclear forces that hold them together, and it is released in nuclear reactions such as fusion and fission. It is liberated in collisions of high-energy particles, which take place in gigantic accelerators, and it is converted into the masses of exotic and ephemeral forms of matter which are therein produced.

Reactions to the September paper

Einstein's paper on relativity did not immediately startle the scientific community. The first to show some interest was Professor Planck, who wrote a note asking for more information on some points. During the winter of 1905, Planck himself introduced special relativity in a lecture at the University of Berlin. Other physicists came to Bern to speak to Einstein. Max von Laue (Nobel 1914) was the very first. He came in the summer of 1906, assuming that Einstein was at the University of Bern. Surprised to find him at the patent office, he later recalled: 'The young man who met me made such an unexpected impression on me that I could not believe he could be the father of the relativity theory.' Laue was soon converted to relativity, and in 1911 he published the first book on the subject. This was the beginning of a lifelong friendship with Einstein.

In 1907, Einstein was asked to write a review article on relativity, and this appeared in the *Journal of Radioactivity and Electronics*, edited by the well-known German physicist Johannes Stark (Nobel 1919). And one year later Hermann Minkowski, Einstein's old mathematics teacher at the ETH in Zurich, gave a lecture in which he presented his *space-time* interpretation of relativity (p. 71).

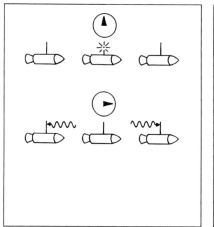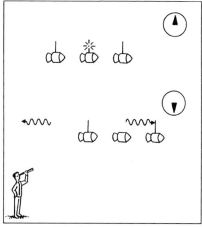

Fig. 4.10. Simultaneity is relative. The three spaceships are travelling at the same, constant velocity in single file. At a certain instant, the middle ship sends two flashes of light, and the astronaut riding it finds that the light signals arrive at the lead and rear ships simultaneously (left). But you, on the earth (right), estimate both the light signals travelling relative to you at the same speed: the *unchanging speed of light.* So you will find that the signal that has run after the lead ship takes longer to arrive than the signal that runs forward into the rear ship; they do not arrive simultaneously.

Einstein's theory, though still controversial, was nevertheless discussed at length within the community of German physicists. In 1912, Wilhelm Wien (Nobel 1911) recommended Einstein to the Physics Committee for the Nobel Prize. He wrote:

> I propose to award the Prize in equal shares to H. A. Lorentz in Leiden and A. Einstein in Prague . . . While Lorentz must be considered as the first to have found the mathematical content of the relativity principle, Einstein succeeded in reducing it to a simple principle. One should therefore assess the merits of both investigators as being comparable . . .[28]

Einstein was awarded the Nobel Prize for physics for the year 1921, but he never received a Nobel for his theories of relativity.

Testing special relativity

The Michelson–Morley experiment (p. 23) found no evidence for the existence of the ether. Thus, Einstein's principle of relativity had its experimental proof even before it was proposed.

Physicists used high-speed electrons (either cathode rays or beta rays – these last emitted from radioactive sources) to test the increase of mass with its speed. They measured electric and magnetic deflections of these electrons, and found that their mass increased exactly as foreseen by special relativity. These results were obtained during a period of approximately ten years (1909–19), and they verified Einstein's theory to a level of accuracy of one per cent. More accurate proofs were obtained by the study of the fine structure of the hydrogen spectral lines during the years 1915–20 (p. 10), and these led to a complete confirmation of the mass relativistic formula. These were the only experimental supports to special relativity prior to 1921, the year in which Einstein received the Nobel Prize.

The most important equation in modern physics, $E = mc^2$, expresses the equivalence of mass and energy. In a famous review article on relativity the Austrian physicist Wolfgang Pauli (Nobel 1945) wrote, in 1921:

> Already in his first published paper on the subject, Einstein pointed to the possibility of using radioactive processes to check the theory. But the looked-for mass defects in the atomic weights of the radioactive elements are too small to be measured experimentally . . . Perhaps the theorem of the equivalence of mass and energy can be checked at some future date by observations on the stability of nuclei.[29]

From the early 1920s a rapid succession of experimental discoveries contributed to confirming Einstein's relativity formulae (including the mass–energy formula): the *Compton effect* (1922–3) discovered by Arthur Compton (Nobelist in 1927); the *positron* (1932) by Carl Anderson (Nobelist in 1936); the first artificially produced nuclear disintegration created, in 1932, by John Cockcroft and Ernest Walton (Nobel 1951); the production of particles of matter from energy (1933) by Patrick Blackett (Nobel 1948); and the fission of atomic nuclei (1938) by Otto Hahn (Nobel for chemistry 1944).

One of the important consequences of the theory of special relativity is the effect of time dilatation. To prove it, physicists used unstable subatomic particles such as microscopic clocks. After the discovery of certain types of cosmic-ray particles called 'muons' (in the late 1930s, p. 195), the Italian physicist Bruno Rossi obtained the lifetime of these particles by using the time-dilatation formula. (Muons are unstable particles which live for a certain average time, called the 'lifetime'. They then break down, producing other particles. Due to the time-dilatation effect, muons moving at a relativistic speed have a lifetime longer than that of muons at rest, so they can be used as microscopic clocks.) Rossi's results were confirmed with even higher precision in the mid-1970s, in an experiment carried out at CERN, the international laboratory for particle physics near Geneva, Switzerland. Here physicists used muons produced in an accelerator, which could travel at the fantastic speed of approximately 298 000 kilometres per second and found that they possessed a lifetime more than 20 times longer than that of stationary muons.

Recent data concerning Einstein's principle on the constancy of the velocity of light were obtained from the Global Positioning System (in short GPS), the radio-navigation system consisting of a number of satellites, each carrying an atomic clock. Physicists analysed the radio signals from all the satellites and verified that their speed always possessed the same value (with an accuracy of about three parts in one billion).[30] This brilliantly confirmed the second principle of special relativity, which even today startles our non-relativistic common sense.

Einstein's conception of time

> Unexpectedly a friend of mine in Bern [Michele Besso] then helped me. That was a very beautiful day when I visited him and began to talk with him as follows: 'I have recently had a question [the constancy of the speed of light] which was difficult for me to understand. So I came here today to bring with me a battle on the question.' Trying a lot of discussions with him, I could suddenly comprehend the matter. Next day I visited him again and said to him without greeting: 'Thank you I've completely solved the problem.' My solution was really for the very concept of time, that is, that time is not absolutely defined but there is an inseparable connection between time and the signal velocity . . . Five weeks after . . . the present theory of special relativity was completed.
>
> (Albert Einstein)[31]

1906

The electron

For two consecutive years the Nobel Prize honoured discoveries in the realm of cathode rays. In 1905 the award was bestowed on Philipp Lenard (p. 48); and then in 1906 it went to J. J. Thomson, from the Cavendish Laboratory, Cambridge. (This prize was the second one to be conferred on a Cavendish man, the first having been awarded in 1904 to Lord Rayleigh, p. 45.)

As we know, Thomson had finally resolved, in 1897, a controversy about the nature of cathode rays which had lasted for half a century (p. 23). He had shown that these rays consisted of negatively charged particles, very much lighter than hydrogen (which itself is the lightest among atoms). This discovery suggested that the new particles, called *electrons*, might be constituent parts of the atom, so revolutionising contemporary knowledge of the structure of the atom itself.

Thomson's 1897 experiments

Experiments before 1897 had already indicated that cathode rays could be charged particles and not some sort of waves (as propounded by the physicists of the German school). In 1895, the French physicist Jean-Baptiste Perrin reported results demonstrating that negatively charged particles were emitted from the cathode of a discharge tube. In 1896, Hendrik Lorentz deduced (from the Zeeman effect) a value for the charge-to-mass ratio (e/m) for particles that were supposed to be bound within atoms and to be emitting light (p. 41). In the same year two German physicists, Walther Kaufmann in Berlin and Emil Wiechert

Fig. 4.11. A schematic diagram of Thomson's tube for the first experiment. (Source: J. J. Thomson, 'Cathode rays'. *Philosophical Magazine*, 44, 1897, p. 295, courtesy Taylor & Francis Ltd., http://www.tandf.co.uk/journals.)

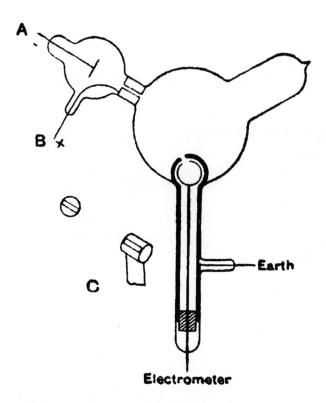

at Königsberg, had obtained estimates of the value of the same ratio for freely moving cathode rays.

Thomson carried out complete measurements, thus obtaining a precise value for the charge-to-mass ratio of cathode rays, and succeeded in indicating their true nature. Three experiments led Thomson to the conclusions that he reached.

> *First*. Thomson realised that Perrin's experiment was not conclusive. As he pointed out:

> The supporters of the theory that these rays are waves in the ether might say, and indeed have said, that while they did not deny that electrified particles might be shot off from the cathode, these particles were, in their opinion, merely accidental accompaniments of the rays, and were no more to do with the rays than the bullet has with the flash of a rifle.[32]

He succeeded in improving Perrin's apparatus, and so was able to prove that the cathode rays themselves were in fact charged particles.

> *Second*. In a second experiment, Thomson demonstrated that cathode rays could be deflected not only by a magnet but also by electric forces (the great Heinrich

Hertz, a very accomplished experimenter, had failed to obtain such a deflection, p. 24); he also discovered that they were negatively charged. Thomson had realised that Hertz' failure was due to a bad vacuum in the discharge tube:

> What to do? The obvious solution was easier said than done: reducing the neutralizing action of the ions by pumping on the gas for all the equipment was worth, exhausting the tube to an even higher state of rarefaction than hitherto believed possible . . . [By] running the discharge through the tube day after day without introducing fresh gas, . . . it was possible to get a much better vacuum. The deflection of the cathode rays by electric forces became quite marked, and its direction indicated that the particles forming the cathode rays were negatively electrified.[33]

> ***Third***. In his third experiment, Thomson sought to determine the charge-to-mass ratio for the cathode-ray particles, measuring their deflection in both an electric and a magnetic field. From his measurements, he was able to calculate precisely the ratio e/m of a single particle. The results were astonishing: this ratio turned out to be over a thousand times greater than that of a hydrogen ion (today's value is about 1840 times). He thus concluded: '. . . for the carriers of the electricity in the cathode rays m/e is very small compared with its value in electrolysis. The smallness of m/e may be due to the smallness of m or the largeness of e, or to a combination of these two.'[34] (The choice between these two possibilities was settled by Philipp Lenard. He showed that cathode rays must have a mass far smaller than the mass of any atom.)

Thomson made the first announcement of his discovery of these 'corpuscles', as he called them, at the Royal Institution in London on 30 April 1897. The full article was published later that year in the October issue of the *Philosophical Magazine*.[35] Between 1898 and 1900, Thomson found that electrons could also boil off from hot metals (what was called the *thermionic effect*), and that charged particles emitted in the photoelectric effect were identical to electrons (a result once again confirmed by Lenard). In his book entitled *Recollections and Reflections* (1936) he wrote that he had not been able to escape from the following conclusions:

> That atoms are not indivisible, for negatively electrified particles can be torn from them by the action of electric forces, impact of rapidly moving atoms, ultraviolet light or heat.
> That these particles are all of the same mass, and carry the same charge of negative electricity from whatever kind of atom they may be derived, and are constituents of all atoms.
> That the mass of these particles is less than one thousandth part of the mass of an atom of hydrogen. I at first called these particles corpuscles, but they are now called by the more appropriate name electrons.[36]

Fig. 4.12. J. J. Thomson giving a lecture demonstration of the cathode-ray tube (Courtesy The Cavendish Laboratory, University of Cambridge, England.)

J. J. Thomson (1856–1940)

Joseph John Thomson, called 'J. J.' by his friends and colleagues, was born in Cheetham Hill, Manchester, England. When he was only 14 years old he entered Owens College, Manchester. In 1876, he obtained a scholarship at Trinity College, Cambridge, where he remained a member of the college for the rest of his life, becoming Master in 1918.

In 1884, at only twenty-eight, Thomson was elected a Fellow of the Royal Society, and appointed third Cavendish Professor of Experimental Physics at Cambridge University, in succession to Lord Rayleigh. Thomson became professor of natural philosophy at the Royal Institution, London, in 1905; three years later he was knighted, and in 1909 was made president of the British Association for the Advancement of Science. In 1912, he received the Order of Merit, and in 1919, after retiring from his chair at the Cavendish, he became honorary professor at Cambridge University.

From 1916 to 1920, Thomson was president of the Royal Society, and during the First World War, he was engaged in defence activities. From 1884 to 1919, he directed the celebrated Cavendish Laboratory, gathering around him a group of brilliant physicists coming from all over the world. Seven Nobel Prizes were awarded to those who worked under him. The American physicist H. A. Bumstead, who worked at the Cavendish in 1904–5, thus described his experience:

> ... I have never seen a lab in which there seemed to be so much independence and so little restraint on the man with ideas ... The great admiration and reliance with which J. J. was regarded by everybody was unquestionable; yet in matters of detail there was no subservience. I saw a good many men, while I was there, following their own causes, and I remember well the frank tone prevailing at the meetings of the Cavendish Society [founded by J. J.], where the Prof's theories or experiments were not spared in the general criticisms.[37]

J. J., father of the electron

In those days, 'students from all over the world looked to work with him . . . Though the master's suggestions were, of course, most anxiously sought and respected, it is no exaggeration to add that we were all rather afraid he might touch some of our apparatus.' Thomson himself was well aware that his interaction with experimental equipment was not always felicitous: 'I believe all the glass in the place is bewitched.' [he said] (In his student days he once nearly lost his eyesight by causing a laboratory explosion.) . . . '[His] intuitive ability to comprehend the inner working of intricate apparatus without the trouble of handling it appeared . . . as something verging on the miraculous, the hallmark of a great genius.'[38]

Discoveries and inventions

In 1883, more than ten years before J. J. Thomson discovered the electron, Thomas Alva Edison, the American inventor of the incandescent lamp (p. 29), chanced to observe a curious effect. While testing carbon filaments for electric light bulbs, he noticed a tiny electric current flowing from the filament to a nearby electrode, situated in an evacuated tube. More than twenty years later, in 1904, the British engineer John Ambrose Fleming, an adviser to Marconi's Wireless Telegraph Company, applied the Edison effect to invent the *thermionic diode*, also known as the *vacuum tube*. This device restricted a current so that it flowed in one direction only, thus converting an alternating current into a direct one; it was widely used in radio receivers, to convert radio waves into the direct current signals needed to drive earphones. In 1906 the American inventor Lee De Forest developed a new vacuum tube, by adding a third electrode to the diode. It was called the *Audion*, but it was commonly known as the *triode valve*. It could amplify electric currents, and it was capable of more sensitive reception of even very feeble radio signals.

Fig. 4.13. Three of the original Fleming thermionic diodes. (Courtesy Marconi Corporation.)

(In 1910 De Forest was able to broadcast the voice of the famous Italian tenor Enrico Caruso from New York's Metropolitan Opera House.) These pioneer inventions opened the way to electronics, a field which developed in the following years around the new conception of electric current as a stream of electrons.

> I have found a method of rectifying electrical oscillations, that is, making the flow of electricity all in the same direction . . . I have been receiving signals on an aerial with nothing but a mirror galvanometer and my device . . . This opens up a wide field for work, as I can now measure exactly the effect of the transmitter. I have not mentioned this to anyone yet as it may become very useful.
> (From a letter from Fleming to Guglielmo Marconi, November 1904)[39]

1907

Michelson's interferometer

Light was the passion of his life! And thanks to this passion, Albert Michelson (who was actually described as 'the master of light') was awarded the 1907 Nobel Prize for physics. He was the first American scientist to receive such a prestigious accolade. It was for the invention of his remarkable optical instrument, the *interferometer* (p. 23), and its utilisation in the fields of metrology and astronomy. In addition, the Swedish Academy of Sciences emphasised the pioneering applications of the new instrument in the field of atomic spectra which, at that time, was considered to be at the very frontier of contemporary physics research. His famous experiment on the 'ether wind' was not mentioned either in the official annoucement made public by the Academy, nor in Michelson's Nobel lecture. (The Michelson–Morley experiment (p. 23) is now considered one of the most important experiments of the nineteenth century; but at that time the response of the physics community to its failure in detecting a stationary ether was not particularly enthusiastic.)

Albert Abraham Michelson (1852–1931) was born in Strelno, Prussia, Germany (now Poland). He went to the USA with his parents when he was two years old. At seventeen he entered the Naval Academy at Annapolis, where he graduated in 1873.

In the autumn of 1880 Michelson travelled to Europe to continue his studies and researches. He visited the universities of Berlin and Heidelberg, in Germany, and the Collège de France and the Ecole Polytechnique, in Paris. While in Berlin, he invented his interferometer (now called the *Michelson interferometer*), which he used not only for the experiment on the 'ether wind' (first in Germany and later with Edward Morley at Cleveland, Ohio, USA), but also for measuring wavelengths of light with an unprecedented accuracy.

While at the Case School of Applied Science in Cleveland (where he was a professor of physics in the period 1883–90), Michelson began an ambitious series of experiments which extended for over forty years, and the object of which was to obtain precise measurements of the speed of light, with an ever greater accuracy. In 1892 he moved to the University of Chicago, where he became the first head of the Department of Physics; here he remained until the year 1929, when he went to work at the Mount Wilson Observatory, Pasadena (California).

During their work in the field of atomic spectroscopy, Michelson and his co-worker Morley discovered that spectral lines emitted from certain elements are not simple, but on the contrary, have a complex structure. These are groups of extremely closely packed lines, which even the strongest spectroscope could not resolve. They discovered, for example, that the red line of the hydrogen atom (the so-called *Balmer-alpha line*) is in fact composed of two closely spaced lines (this is what is called the *fine structure* of the spectral lines). These findings were to become of the utmost importance for atomic physics; their significance was understood only many years later, when quantum mechanics was fully developed.

Michelson also used his interferometer to measure the standard metre in terms of the characteristic red light emitted by cadmium atoms. With this same kind of instrument he was able to determine the width of heavenly objects: in 1920, he succeeded for the first time in history in measuring the angular width of a star (Alpha Orionis). The president of the Swedish Academy of Sciences thus summarised Michelson's researches:

> Your interferometer has rendered it possible to obtain a non-material standard of length, possessed of a degree of accuracy never hitherto attained . . . Your contributions to [spectroscopy] embrace methods for the determination of the . . . [wavelengths] in a more exact manner than those hitherto known. Furthermore, you have discovered the important fact that the lines in the spectra, which had been regarded as perfectly distinct, are really in most cases groups of lines . . . Astronomy has also derived great advantage . . . from your method of measurements . . . The results you have attained are excellent in themselves and are calculated to pave the way for the future advancement of science.[40]

Towards general relativity

After developing his special theory of relativity, Albert Einstein turned his attention to gravity. In 1907 he published a paper reviewing relativity. In it he wrote:

> So far we have applied the principle of relativity, i.e., the assumption that the physical laws are independent of the state of motion of the reference system, only to *non-accelerated* [inertial] reference systems. Is it conceivable that the principle of relativity also applies to systems that are accelerated relative to each other?[41]

Einstein's answer was 'yes', and he at once set out his *principle of equivalence*:

Fig. 4.14. Albert Michelson. (Michelson Museum (originally from US Naval Ordnance Test Station, Chine Lake, CA), courtesy AIP Emilio Segrè Visual Archives.) In 1879 Michelson was attracted by a statement he read in a letter from Maxwell: '. . . in all terrestrial methods of determining the velocity of light, the light comes back along the same path again, so that the velocity of the earth with respect to the ether would alter the time of the double passage . . .'[42] This was the germ of the Michelson–Morley experiment.

> . . . we shall therefore assume the complete physical equivalence of a gravitational field and a corresponding acceleration of the reference system. This assumption extends the principle of relativity to the uniformly accelerated translational motion of the reference system . . . [It] permits the replacement of a homogeneous gravitational field by a uniformly accelerated reference system . . .[43]

Let us consider the example reproduced in Fig. 4.15. Newton would say that the apple falls to the floor of a spaceship standing on earth, because the force of gravity pulls it down (Fig. 4.15, left). Einstein's principle of equivalence, on the other hand, states that the apple would appear to behave in exactly the same way in intergalactic space (where gravitational forces are negligible) if the spaceship moved upward with a constant acceleration corresponding to that of the falling apple on earth (Fig. 4.15, right). In other words, in the spaceship, the downward pull of gravity is 'equivalent' to the upward acceleration of the spaceship itself. This principle allowed Einstein to focus entirely on motion, rather than force, in discussing gravity. It states that *acceleration is equivalent to gravity*, and it will

 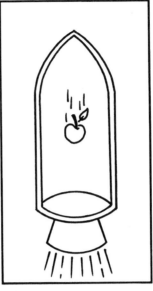

Fig. 4.15. The apple falls in the earth's gravity field with a downward constant acceleration (left). In a gravity-free space, the spaceship accelerates upward with the same acceleration (right): an observer who is inside the spaceships cannot distinguish one situation from the other.

be incorporated into the *general theory of relativity* that Einstein will develop in the following years.

The same principle is also related to the notion of equivalence between the inertial mass and the gravitational mass (this explains Galileo's celebrated experiments which demonstrated that different bodies fall with the same acceleration, see p. 8). (The inertial mass of a body determines its response to a force, while its gravitational mass determines the response to the gravitational field.) Newton, more than 200 years previously, had tested the equivalence between the two masses with pendulums made from different materials, obtaining an agreement within 1/100. In the early 1920s the Hungarian physicist Baron Roland von Eötvös established the equivalence with an accuracy of about one part in a hundred million. Forty years later the American physicist Robert Dicke obtained the same results with an accuracy improved to about three parts in a hundred billion and two Russian scientists, Vladimir Brazinsky and Vladimir Panov, bettered even this with a precision of one part in a thousand billion.[44]

Inertial and gravitational mass

Isaac Newton wrote in his *Principia*:

> Quantity of matter is a measure of matter that arises from its density and volume jointly . . . I mean this quantity whenever I use the term 'body' or 'mass' . . . It can always be known from a body's weight, for – by making very accurate experiments with pendulums – I have found it to be proportional to the weight . . .
> (Book 1, Definition I)[45]

Falling from the roof

> [For] '... an observer falling freely from the roof of a house there exists ... no gravitational field'. Indeed, if the observer drops some bodies then these remain relative to him in a state of rest or of uniform motion, independent of their particular chemical or physical nature ... The observer therefore has the right to interpret his state as 'at rest'. Because of this idea, the ... experimental law that in the gravitational field all bodies fall with the same acceleration attained at once a deep physical meaning. Namely, if there were to exist just one single object that falls in the gravitational field in a way different from all others, then with its help the observer could realise that he is in a gravitational field and is falling in it ... The experimentally known matter independence of the acceleration of fall is therefore a powerful argument for the fact that the relativity postulate has to be extended to coordinate systems which, relative to each other, are in non-uniform motion.
>
> (Albert Einstein)[46]

1908

Rutherford's metamorphosis

Thirteen years had passed since September 1895, when young Rutherford arrived at Cambridge as one of the first of Thomson's research students. Two years after the Nobel had been awarded to his professor, and when he was only thirty-seven, he too was granted the high distinction, 'for his investigations into the disintegration of the elements, and the chemistry of radioactive substances'. The prize was for chemistry, not for physics. A startling metamorphosis for Rutherford!

Rutherford had returned to Britain in 1907, to accept a chair at the University of Manchester. During his nine years at McGill University in Montreal he had made a succession of brilliant discoveries in radioactivity. The Nobel Prize was the just recognition for his extraordinary contribution to the understanding of the disintegration and transmutation of radioactive elements. In a collection of articles entitled *Nobel: The Man and His Prizes*, Arne Westgren, a chemist of the Swedish Academy of Sciences, wrote this about the Academy's proposal to award the Nobel Prize for chemistry to Rutherford:

> Rutherford had also been suggested by several nominators for the Physics Prize, but at a joint meeting the two Nobel Committees decided that it would be most suitable, considering the fundamental importance of his work for chemical research, to award him the Prize for Chemistry.[47]

And one of the members of the Physics Section of the Academy thus commented on the choice: 'The physicists had been tricked into handing over their best candidate, Rutherford, so to speak, against his own wishes, to the chemists.'[48] (At this time the chemists wanted to assert their right to make awards in the field of radioactivity.)

Fig. 4.16. Ernest Rutherford. (© The Nobel Foundation, Stockholm.) To his old student at McGill University, Otto Hahn, Rutherford wrote: '... I must confess it was very unexpected [the Nobel Prize] and I am very startled at my metamorphosis into a chemist.'[49] And again, in his Nobel banquet speech, on 11 December 1908, he noted that he had '... dealt with many different transformations with various periods of time, but that the quickest he had met was his own transformation in one moment from a physicist to a chemist.'[50]

At Manchester, Rutherford continued his researches on alpha particles. In the same year (1908), with his student Thomas Royds, he finally proved that alpha particles are doubly charged helium ions. They sealed a quantity of radon (a radioactive gas) in a glass tube: this had sufficiently thin walls for the alpha particles that were emitted by radon to be able to penetrate them. After a time, this vessel was found to contain a gas showing helium spectral lines (each alpha particle had picked up two electrons and become a neutral helium atom). In the same period, Rutherford and the German physicist Hans Geiger succeeded in measuring the charge of alpha particles; they then obtained a value for the electron charge which agreed very well with the one obtained in 1900 by Max Planck from his radiation law. (Later on Rutherford declared that this agreement had caused him to become an early adherent to Planck's theory, see p. 114.)

The physics prize

Instead of Rutherford, the Physics Committee, on the basis of a report made by one of its more influential members, Svante Arrhenius, came to the decision to recommend Max Planck for the 1908 Nobel Prize. The Physics Section of the Academy having subsequently accepted Planck, it was reported in the international press that he was the winner of the year's physics prize. But the Academy in its plenary section – strongly influenced by the mathematician Gösta Mittag-Leffler – passed Planck over in favour of the French physicist Gabriel Lippmann. (Mittag-Leffler's argument in his attack on Planck concerned the hypothesis of the quantum of energy. He said during the Academy meeting: 'Planck's derivation [of his radiation law] is based on a totally new hypothesis which can hardly be considered plausible, namely that of the elementary quantum of energy . . . an evaluation of its worth at present raises great difficulties . . . the deferment of a definitive judgement is therefore to be preferred.')[51] It was thus that the Academy refused the Committee's proposal – as well as a proposal for dividing the prize between Planck and Wilhelm Wien – and chose Lippmann instead. ('Mittag-Leffler . . . had [thus also] the satisfaction . . . of seeing Arrhenius [his eternal rival] go down to defeat.')[52] Lippmann was awarded the prize, 'for his method of reproducing colours photographically based on the phenomenon of interference.'

Gabriel Lippmann (1845–1921) was born in Hollerich, Luxembourg. He studied at the *Ecole Normale Supérieure* in Paris. In 1883, he was appointed professor of mathematical physics at the Sorbonne, and three years later he became head of the Laboratories of Physical Research there.

Lippmann's invention which was rewarded with the Nobel Prize dated back to the early 1890s, when he developed a new process of colour photography which used the natural colours of light instead of colour pigments. He had spread a fine grain-emulsion with colloidal silver bromide on a flat sheet of glass. To this he had then applied a layer of mercury which served as a mirror. The mercury reflected the light waves back through the emulsion, and these interfered with the waves striking the plate, forming standing waves. Thus, a latent image was formed in the emulsion, which varied in depth according to the colour (or wavelength) of each wave. Once the plate was developed a brilliant coloured image was obtained.

Lippmann's ingenious method only aroused interest in the 1960s, when the new technique of *holography* was invented. In his Nobel lecture in 1971, Dennis Gabor wrote: 'In 1962, just before the "holography explosion", the Soviet physicist Yu. N. Denisyuk published an important paper in which he combined holography with the ingenious method of photography in natural colours, for which Gabriel Lippmann received the Nobel Prize in 1908.'[53]

Space-time

Now, let us turn back to relativity. In a lecture given in September 1908 at Cologne, Germany, Hermann Minkowski (p. 56), at that time professor of mathematics at the University of Göttingen, suggested a geometric interpretation of special relativity, by introducing a new mathematical picture. He proposed that the universe was *four-dimensional*, three dimensions being the ordinary *spatial dimensions* (length, breadth and height), and the fourth being *time*. This fusion of four dimensions is named, in the technical language of physicists, *Minkowski space-time*. Points in the space-time are *events*, and the histories of particles appear as *world-lines*. In addition, physical laws can be regarded as relations between world-lines. Thus we can speak of a geometric representation of physics, and space-time becomes the alternative to Newton's separate concepts of absolute space and absolute time.

1909

Wireless telegraphy

The year 1909 saw the Swedish Academy of Sciences honouring a technical invention for the second time (the first had been Lippmann's invention, which had been acknowledged in 1908, p. 70). This award was shared by Guglielmo Marconi and the German physicist Karl Braun, for their contributions to the development of wireless telegraphy. The recognition arrived eight years to the day after Marconi's success in transmitting radio waves across the Atlantic. The event had created a world-wide sensation, for it had demonstrated that the curvature of the earth did not limit communication using radio waves, even when these had to travel over great distances.

Marconi and Braun

Heinrich Hertz, who made a famous experiment on radio waves (p. 20), did not believe that these waves could propagate over more than short distances. In 1890 the French physicist Édouard Branly had invented a device known as a 'coherer'. The English physicist Sir Oliver Lodge then perfected this instrument to such an extent that he was able to detect Morse Code signals and transcribe them on paper. However, it was Marconi who succeeded in transmitting radio waves over long distances.

Guglielmo Marconi (1874–1937) was born in Bologna, Italy. His father was an Italian country gentleman and his mother an Irish noblewoman. He was educated first in Bologna and later in Florence. He studied physics and acquired extensive skills in electromagnetic wave techniques, building on the work of Maxwell, Hertz, Lodge and the Italian Augusto Righi.

In 1894, Marconi began to carry out experiments at Pontecchio, near Bologna. After experimenting over short distances, he improved his apparatus by connecting one side of the generator and receiver (a Branly coherer) to the ground, and the other to a wire – later on called an *antenna* – directed towards the sky. Thus, he succeeded in 1895 in sending wireless signals over a distance of about 2.4 kilometres. In 1900 he patented his invention (the famous patent No. 7777); and on 12 December 1901 he was able to transmit the first wireless telegraphic signal across the Atlantic. After the First World War Marconi continued his investigations, and in 1931 he began research into the propagation of waves of very short wavelengths. This resulted in the opening in 1932 of the world's first microwave radiotelephone link between the Vatican City and the Pope's palace at Castel Gandolfo, near Rome. This was the predecessor of modern radio communication.

Karl Ferdinand Braun (1850–1918) was born at Fulda, Hesse–Kassel (Germany). He received his doctorate from the University of Berlin in 1872. He spent the next years teaching in various universities at Marburg, Strasbourg, Karlsruhe and Tübingen. In 1895 he returned to Strasbourg, where he became director of the institute of physics. Braun improved Marconi's transmitting system, greatly increasing its broadcasting range. His method was later applied to radio, radar and television. He was also known for his discovery in the 1870s of crystalline materials; these led to the development of crystal detectors, which were used in early radio receivers.

Braun also developed the cathode-ray oscilloscope. In 1897 he succeeded in producing, in an evacuated tube, a narrow beam of electrons, which, guided by

Fig. 4.17. Guglielmo Marconi, circa 1896. (Courtesy Marconi Corporation.)

means of a changing electric voltage, produced luminous patterns on a fluorescent screen. This invention, the precursor of the television tube, also became one of the most-used instruments in research laboratories.

> ... shortly before midday [of 12 December 1901, in St John's, Newfoundland, Canada] I placed the single earphone to my ear and started listening. The receiver on the table before me was very crude – a few coils and condensers and a coherer – no [vacuum tubes], no amplifiers, not even a crystal . . . The answer came at 12:30 when I heard, faintly but distinctly, 'pip-pip-pip'. I handed the phone to Kemp: 'Can you hear anything?' I asked. 'Yes,' he said, 'the letter S' – he could hear it. I knew then that all my anticipations had been justified. The electric waves sent out into space from Poldhu [Cornwall, England] had traversed the Atlantic – the distance, enormous as it seemed then, of 1700 miles – unimpeded by the curvature of the earth.
>
> (Guglielmo Marconi)[54]

1910

Van der Waals' laws

Henri Poincaré was a renowned French mathematician who made significant contributions to relativity. He had been nominated for the Nobel Prize on several occasions, and especially in 1910.

As Abraham Pais remarked: 'The exceptionally high number (thirty-four) of signatories to letters nominating Poincaré in 1910 was the result of a campaign mounted by Mittag-Leffler . . .'[55] who was claiming 'that the time has come to give the physics prize to a pure theorist.' But, continues Pais: 'In its report, the Committee noted that neither Poincaré's brilliant mathematical contributions nor his mathematical-philosophical essays . . . could be designated discoveries or inventions within physics . . .'.[56] The favourite of the Committee was Knut Ångström (Anders Ångström's son, p. 25), who, however, died just before the end of the decision-taking process. And so the Academy decided to award the prize to Johannes van der Waals, from the University of Amsterdam, 'for his studies of the physical state of liquids and gases'.

Johannes Diderik van der Waals (1837–1923) was born in Leiden, the Netherlands. In 1873 he obtained his doctorate at the local university there, with a thesis in which he put forward mathematical formulae explaining the behaviour of both gases and liquids. He also demonstrated that these two states of aggregation are in fact of the same nature. To describe their behaviour he introduced into his equations two new physical quantities representing the volumes of molecules and the attractive forces between them; so doing he obtained a new law, known as the *van der Waals equation of state*. This law describes gases at high densities and pressures more accurately than the so-called 'ideal-gas law' does.

Van der Waals made his second discovery in 1880, when he was at the University of Amsterdam. Using special physical quantities, he worked out a more general equation, known as the *law of corresponding states*, which provided a complete description of a great variety of real gases. It was this law which was used in 1898 by the Scottish chemist James Dewar in order to achieve the liquefaction of hydrogen. In 1908 van der Waals' pupil Heike Kamerlingh Onnes (Nobel 1913) used the same law as a guide in his experiments which led to the liquefaction of helium – the last of the chemical elements to be condensed. It was thus that van der Waals' pioneering work, which the Nobel Physics Committee had in earlier times always considered 'too old', took on new importance and earned him the 1910 Nobel award.

Fig. 4.18. A photograph of the frontage of the old Cavendish Laboratory, taken from Free School Lane. (Courtesy The Cavendish Laboratory, University of Cambridge, England.)

Fig. 4.19. The main building of the Friedrich-Wilhelms University on Unter den Linden, Berlin, circa 1900. (Courtesy Archiv zur Geschichte der Max-Planck-Gesellschaft, Berlin-Dahlem.) This was the most famous German university during the eighteenth century and the first decades of the twentieth century.

Fig. 4.20. The Sorbonne University in Paris, circa 1910. (Courtesy ACJC-Curie and Joliot Curie fund, Paris.) Here Marie Skłodowska Curie undertook her studies in the early 1890s, and taught from 1906 onwards.

The atmosphere at the Cavendish

> The rooms on the ground floor all opened into one another, and their occupants wandered to and fro as they felt inclined. J. J. occupied the room at the end, and for a time his assistant, Everett, tried to establish the convention that it was private, indeed I think there was a notice to this effect on the door. But in practice little attention was paid to it, and when Rutherford . . . and others were established to work there as well as Everett, the game was up. Rutherford and J. J. talked on and off at all times, discussing the papers in the latest number of . . . [the] *Annalen*, or the *Philosophical Magazine*. I myself began work upstairs, but when I had acquired some seniority among the research workers, I petitioned to be removed downstairs, where the heating was less miserably inadequate . . . The picture that remains of the laboratory in those days is of a score of individuals scattered about in various rooms, two or three in a larger room – but each working at his own particular problem . . . The research workers had their own habits for commencing work and finishing it at times suited to their own individual temperaments . . .
>
> (Lord Rayleigh)[57]

Chapter 5
The quantum atom

The very first person that we meet as we swing into the second decade of the century is an old acquaintance – Ernest Rutherford. He claims our attention with his revolutionary model of the *nuclear atom* (1911). A year later a crucial experiment was performed at the University of Munich, as a result of which the wave nature of X-rays was definitely established. This discovery was immediately acknowledged by the community of physicists, and was soon exploited for the various uses to which X-rays could be put, notably the utilisation of X-ray diffraction techniques to determine the unknown structure of crystals, and investigations into the inner structure of atoms. In fact, X-rays had now become so important that in the following years of 1914, 1915 and 1917 the Nobel Prizes for physics is all concerned with them.

New developments in quantum physics took off in 1913 when the Danish scientist Niels Bohr published his pioneering papers on the constitution of atoms and molecules. Using Rutherford's nuclear atom and Planck's quantum of action, Bohr worked out a revolutionary model for the hydrogen atom that was successful in explaining its spectral lines. His theory soon received strong support from experiments; his model was extended in the following years as physicists attempted to explain the spectra of complex atoms. Meanwhile, the number of scientists working in the field of quantum physics increased rapidly.

In 1914 the Great War broke out, and research in many laboratories ground to a halt. Amidst the wartime disruptions, however, Albert Einstein in Berlin, and Ernest Rutherford in Manchester, undertook what would one day be recognised as their crowning scientific achievements. In 1915–16 Einstein came out with his *general theory of relativity* – a completely new view of gravitation. (This theory passed its first test three years later when starlight was observed to be bent by the action of the sun.) And then in 1919 it was Rutherford again, this time as he succeeded in observing the first artificial transmutation of atomic nuclei.

Besides the Nobel Prizes awarded for X-rays we must mention two other awards of supreme importance: that of 1913, given for the liquefaction of helium and the discovery of superconductivity; and that of 1918, when the prize was awarded to Max Planck, and so quantum physics received recognition for the first time.

1911

Wien's radiation laws

The matter of black-body radiation was at last recognised as being worthy of reaching the Nobel scene! The Nobel Physics Committee decided in fact to award the 1911 Nobel Prize to Wilhelm Wien precisely for, in their own words, 'his discoveries concerning the laws of heat radiation'.

As we have seen (p. 26), Wien had devised, in the 1890s, a theory that gave birth to a mathematical formula – the *displacement law* – which explained how the colour composition of the radiation from a black body changes with the temperature. He then worked out another formula that accurately reproduced the energy distribution of a heated black body at the violet end of the spectrum (short wavelengths) – but this formula could not be extended to the red end. Since Wien's formula failed for longer wavelengths, Max Planck began to investigate further into the matter; finally, he succeeded in deriving a distribution which agreed very well with the experimental data (p. 37).

In 1908 Wien had already been proposed for a Nobel Prize in physics, to be shared with Planck; and we know how Planck's *affaire* ended (p. 70). What happened in 1911? Here is a passage from a member of the Nobel Foundation, quoted by Abraham Pais:

> In 1911 the original suggestion from the Committee was that the prize should be given to Professor A. Gullstrand [a prominent ophthalmologist from Uppsala University], 'for his work in geometrical optics.' Gullstrand had become a member of the [Physics] Committee the same year . . . However, it turned out that the Committee for Physiology and Medicine had had the same good idea, giving Gullstrand their prize 'for his work on the dioptrics of the eye.' So Gullstrand declined the prize in physics, and the Committee wrote an extra report (now including Gullstrand among the signers) suggesting Wien for the prize.[1]

The case of Planck's award had had the effect of stiffening opposition within the Committee: 'After the defeat in 1908, the Committee had gotten "cold feet" as far as Planck was concerned'.[2] The Committee's decision to occupy itself with the controversial field of quantum theory was probably considered premature. In his Nobel lecture, Wien threw light on the status of what was a new theory at the time:

> It is the merit of Planck to have introduced new hypotheses which enable us to avoid Rayleigh's radiation law. For long waves, this law is undoubtedly correct, and the right radiation formula must have a form such that, for very long waves, it passes into Rayleigh's law, and for short waves into the law formulated by me. Planck . . . subjects [the] distribution of energy to a restriction by introducing the famous hypothesis of elements [quanta] of energy . . .[3]

Fig. 5.1. Wilhelm Wien. (Courtesy AIP Emilio Segrè Visual Archives, Weber and E. Scott Barr Collections.)

But, Wien said, there is 'great difficulty for the understanding of these energy elements [quanta]', and concluded:

> It will be seen from the few observations I am able to offer in this context how great are the difficulties that remain in radiation theory. But the reference to these difficulties, which it is the duty of the scientific approach to emphasise, must not prevent us from paying tribute to the great positive achievements which the Planck theory has already accomplished.[4]

Wilhelm Wien (1864–1928) was born at Gaffken, Prussia (now in Poland), the son of a landowner. In 1882, he went to the University of Göttingen to study mathematics and science. He then moved to the University of Berlin, where he worked in Hermann von Helmholtz' laboratory, obtaining his doctorate in 1886. In the period 1890–6 he worked as Helmholtz' assistant at the PTR in Berlin-Charlottenburg, where he was engaged on problems connected with radiation – the work for which he later received the Nobel Prize. He was successively Professor of Physics at Aix-la-Chapelle, Giessen and Würzburg. In 1920 he was appointed Professor of Physics at the University of Munich, where he remained throughout the rest of his career. Wien was one of those rare twentieth-century physicists who worked with success in both experimental and theoretical physics. He also carried out research into thermodynamics and hydrodynamics, and made pioneering studies on cathode rays, X-rays and canal rays.

Fig. 5.2. The nuclear atom possesses a tiny central core in which more than 99 per cent of the atom's mass is concentrated. Around this positively charged core (the nucleus) a swarm of negatively charged electrons are moving, under the influence of electric forces. The nucleus is some 10 000 times smaller than the atom itself.

The nuclear atom

After more than fifteen years of outstanding research activity, Ernest Rutherford continued to ride a wave of brilliant successes in physics. In 1911 he made one of the greatest discoveries of his career, and announced his revolutionary idea regarding the structure of the atom. This idea laid the foundation for modern atomic and nuclear physics.

This event took place at the start of 1909. Rutherford suggested to Hans Geiger and Ernest Marsden (one of Rutherford's research students, who also came from New Zealand) that it would have been of interest to see whether fast-moving alpha particles were deflected through very large angles when traversing a sheet of metal: 'One day', Marsden recalled later, 'Rutherford came into the room where we were counting the alpha particles [and said]: "see if you can get some effect of alpha particles directly reflected from a metal surface." '[5]

The two men soon began to make measurements on different materials. To their astonishment, they found that, on average, one in every 8000 of the alpha particles, when striking a thin sheet of gold metal, actually bounced back from the target. Describing how incredulous he was at the unexpected result, Rutherford is said to have exclaimed: 'It was quite the most incredible event that ever happened to me in my life. It was almost as incredible as if you had fired a 15-inch shell at a piece of tissue paper and it came back and hit you.'[6]

Thanks not only to genius, but to persistence too, Rutherford was able to explain the strange results obtained by Geiger and Marsden, by coming out with a new theoretical model for the atom. Charles Galton Darwin, a member of Rutherford's group in Manchester, thus remembers a memorable evening, just before Christmas 1910: 'One of the greate experiences of my life was that on one Sunday evening the Rutherfords had invited some of us to supper, and after supper the nuclear theory came out . . . [Rutherford] assumed a central charge in the atom – it was indeed a year or two before it was renamed the nucleus – repelling the alpha particle according to the ordinary laws of electricity . . .'[7] Not long after this Sunday supper Rutherford visited Geiger in his laboratory. Geiger recalled the event in these words: 'One day Rutherford, obviously in the best of spirits, came into my room and told me that he now knew what the atom looked like and how to explain the large deflections of the alpha particles.'[8] Finally, in May 1911, Rutherford's paper, which put the nuclear atom before the world, was published in the *Philosophical Magazine*. It was entitled 'The scattering of alpha and beta particles by matter and the structure of the atom'.[9] Here are the salient points as summarised by Marsden:

> Rutherford had shown . . . that the scattering of alpha . . . particles by matter could be explained by assuming that the atom consisted of 'a strong positive . . . central charge concentrated within a sphere of less than [three-million-millionths] centimetre radius, and surrounded by electricity of opposite sign distributed throughout the remainder of the volume of the atom of about [one hundred-millionth] centimetre radius.'[10]

At that time most physicists believed in a model that had been suggested by J. J. Thomson as far back as 1904. In this hypothesis the positive charge of the atom was thought to be spread out throughout the whole atom. The electrons were imagined as vibrating about fixed centres inside the atom itself. Rutherford showed that this model of the atom, called the 'plum pudding model', was not consistent with the alpha-scattering experiments, and proposed instead his *nuclear atom*, which is the one that we now accept.

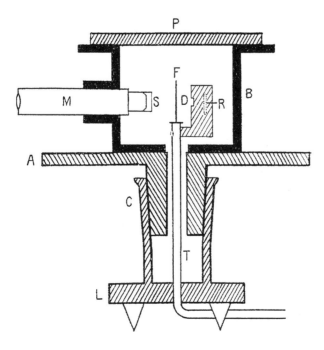

Fig. 5.3. An apparatus used by Geiger and Marsden for studying the scattering of alpha particles. (Source: *Philosophical Magazine*, 25, 1913, p. 607,[11] courtesy Taylor & Francis Ltd., http://www.tandf.co.uk/journals.) The particles, coming from the source R, are directed at a metal foil F. The scintillations that are produced by them are viewed on the fluorescent screen S, through the microscope M, which can rotate to detect particles at different angles.

Ernest Rutherford (1871–1937)

Ernest Rutherford was born at Bridgewater, a small town close to Nelson, New Zealand, where his parents had emigrated from Great Britain in the early 1840s. In 1887 Rutherford won a scholarship to Nelson College, where he was a popular student and a keen footballer. The following year he entered Canterbury College, Christchurch, where his interest for science developed. He graduated with first-class degrees in mathematics and physics in 1893. In 1895 he won a scholarship to Cambridge University. So he left New Zealand for England, where he carried out postgraduate research under J. J. Thomson at the Cavendish Laboratory in Cambridge.

From 1898 to 1907 he was professor of physics at McGill University in Montreal, Canada, and then at the University of Manchester, England. At Manchester he remained until 1919, when he became fourth Cavendish Professor and director of the Cavendish Laboratory. In 1914 he was knighted, and in 1931 became Baron Rutherford of Nelson and Cambridge, in recognition of his scientific work. He was awarded the Order of Merit in 1921, and from 1925 to 1930 was president of the Royal Society of London.

Lord Rutherford was a pioneer researcher in nuclear physics, and one of the greatest experimentalists of all time. At the universities of McGill, Manchester and Cambridge he inspired and led two generations of physicists. Numerous future Nobel laureates worked under him: Edward Appleton, Patrick Blackett, Niels Bohr, James Chadwick, John Cockcroft, Peter Kapitza, Cecil Powell, George Paget Thomson, Ernest Walton (all Nobelists in physics); and in addition Francis Aston, Otto Hahn, George de Hevesy, Frederick Soddy (all Nobelists in chemistry). He died on 19 October 1937, and was buried in Westminster Abbey close to Isaac Newton.

> He had . . . a volcanic energy and an intense enthusiasm – his most obvious characteristic – and an immense capacity for work. A 'clever' man with these advantages can produce notable work, but he would not be a Rutherford. Rutherford had no cleverness – just greatness. He had the most astonishing insight into physical processes, and in a few remarks he would illuminate a whole subject. There is a stock phrase – 'to throw light on a subject'. This is exactly what Rutherford did. To work with him was a continual joy and wonder. He seemed to know the answer before the experiment was made, and was ready to push on with irresistible urge to the next. He was indeed a pioneer – a word he often used – at his best in exploring an unknown country, pointing out the really important features and leaving the rest for others to survey at leisure. He was, in my opinion, the greatest experimental physicist since Faraday.
>
> (James Chadwick)[12]

The first Solvay conference

In the spring of 1910 a German professor of physical chemistry at the University of Berlin, Walther Nernst, met the Belgian industrialist Ernest Solvay, who had invented a special process for producing soda ash (sodium carbonate). The latter had founded the firm Solvay & Company, and established companies in several foreign countries. Solvay had become a rich man, and he used a large part of his wealth for educational and social purposes.

Nernst knew of Solvay's enthusiasm for science, and so suggested that he should organise an international conference on the new-born physics. Solvay responded favourably and offered his financial support. At the end of October 1911 twenty-three leading physicists arrived in Brussels (Belgium) to take part in the first Solvay Conference, which was devoted to problems regarding 'Radiation theory and the quanta'. Reports were submitted by Nernst himself, Albert Einstein, Max Planck and Arnold Sommerfeld (this last was a brilliant theoretical physicist and teacher at the University of Munich). All the participants realised that some fundamental principles at the very basis of classical physics were liable to be challenged. One subject treated in detail (principally by Nernst and Einstein) was that of the *specific heat* of solids, an old and vexing problem that classical physics had been unable to resolve.

In the mid-1800s physicists had found that in order to raise the temperature of a unit mass of a solid by one degree, one has to supply the same quantity of heat (in other words, the specific heat of solids was considered a quantity with a constant value). This rule (called the *Dulong–Petit rule*) had a secure foundation in the kinetic theory of matter. However, in the 1870s experimenters found that some substances had a smaller specific heat at very low temperatures. This clearly violated the assumptions of the classical theory.

The problem was attacked by Einstein in 1906–7 on the basis of quantum theory. Using Planck's hypothesis relative to the energy quanta, he obtained a formula for the specific heat which provided an explanation for the observed anomalies. (He had extended Planck's idea to the atoms in a crystalline solid, when they undergo oscillations caused by thermal motion.) Three years later Nernst carried out experiments and studied the specific heat of substances at very low temperatures, obtaining results which gave support to Einstein's theory.

At the Solvay conference Nernst presented new, more accurate data which showed deviations from Einstein's formula when used to describe the specific heat at temperatures near absolute zero. Einstein's theory, though valid, still required improvements. These were made independently a year later by two physicists, the Dutch Peter Debye, then working at the University of Zurich, and the German Max Born, a young professor at the University of Göttingen.

Fig. 5.4. The participants at the first Solvay Conference, held in Brussels at the Hotel Métropole in October 1911. (Reproduced by kind permission of the International Solvay Institutes for Physics and Chemistry, Brussels, Belgium.) From left to right, standing: O. Goldschmidt, M. Planck, H. Rubens, A. Sommerfeld, M. de Broglie, J. Jeans (eleventh), E. Rutherford, H. Kamerlingh Onnes, A. Einstein, P. Langevin; seated: W. Nernst, E. Solvay (third), H. Lorentz, J. Perrin (sixth), W. Wien, M. Curie, H. Poincaré.

Ernest Solvay's invitation

Dear Sir, To all appearances, we are at the moment in the midst of new developments regarding the principles on which the classical molecular and kinetic theory of matter has been based . . . Although not involved in special questions of this sort, but nevertheless moved by genuine enthusiasm for all problems the study of which expands and improves our knowledge of nature, the undersigned has thought that, even if not leading to a definite decision, a written and oral exchange of views between researchers occupied more or less directly with these questions might, through a preparatory critique, at least clear the way for the solution of these problems . . . To that end, the undersigned proposes to you to participate in a 'scientific Congress', which will be held in Brussels . . .

> To make participation possible for all those invited, I offer to pay 1000 francs for travel expenses to each of them . . . I hope that I can count on your collaboration, and I beg to assure you, dear Sir, of my highest esteem.[13]

1912

The year's Nobel Prize

The Swedish Academy of Sciences considered practical inventions, as opposed to discoveries, in awarding the 1912 prize for physics. The winner was an outsider, unknown to the community of physicists. He was the Swedish engineer **Nils Gustaf Dalén** (1869–1937), managing director of the Gas Accumulator Company, Stockholm. His invention was an automatic sun valve, called 'solventil', which depended on the action of sunlight to regulate a gaslight source. It extinguished the light at sunrise and lit it automatically when night had fallen, or at other periods of darkness. This invention soon achieved a certain fame, being used for buoys and unmanned lighthouses. Dalén won the Nobel on the basis of a single nomination by a member of the Academy's Technology Section (he had received no nominations previously), even though he was competing with prominent scientists like Max Planck, Albert Einstein and Heike Kamerlingh Onnes. Briefly, here is the story concerning Dalén's award.

The Nobel Prizes had then been in existence for eleven years. At this point the members of the Academy representing the field of technology had grown thoroughly dissatisfied, because they felt that the awards had become more and more to reflect a research orientation that had become too academic, so they manoeuvred in the Academy to assert their influence. It was thus that, while the Physics Committee and the Physics Section recommended the award to the experimentalist Kamerlingh Onnes for his achievement in the liquefaction of helium, the Academy, inspired by the engineers, rebelled against this recommendation and decided otherwise. They consequently awarded the prize to Dalén, the inventor. (Kamerlingh Onnes' turn for the prize would come a year later.)

The cloud chamber

In 1895 the Scottish physicist Charles T. R. Wilson began studying clouds at the Cavendish. He tried to produce artificial clouds in the laboratory, similar to those he was accustomed to seeing on the Scottish hill-tops. Wilson recalls how he had had the idea:

> In September 1894 I spent a few weeks in the Observatory which then existed on the summit of Ben Nevis, the highest of the Scottish hills. The wonderful optical phenomena shown when the sun shone on the clouds surrounding the hill-top, and especially the coloured rings surrounding the sun (coronas) or surrounding the shadow cast by the hill-top or observed on mist or cloud (glories), greatly excited my interest and made me wish to imitate them in the laboratory.[14]

Fig. 5.5. One of the original Wilson cloud chambers, 1911. (Courtesy The Cavendish Laboratory, University of Cambridge, England.)

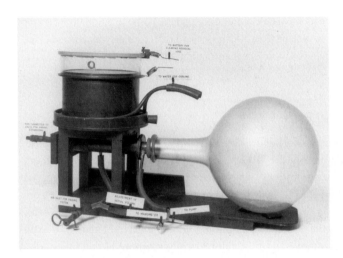

Wilson devised a way of causing air, saturated with water vapour, to expand in a glass chamber, so that it would cool and the vapour would condense into droplets on dust particles, so producing a cloud inside the chamber. He then discovered that under special conditions the same kind of clouds could be produced even in the absence of dust. He thought that these clouds were formed by condensation on ions (charged atoms or molecules) in the air.

When news of Röntgen's discovery of X-rays reached Cambridge, Wilson thought that ion formation, as a result of such radiation, might bring about a more intense cloud in his chamber. He experimented and found that the radiation did actually leave a trail of minuscule water droplets, which then made visible its path through the chamber. Wilson continues thus:

> At the beginning of 1896 J. J. Thomson was investigating the conductivity of air exposed to the new rays – and I had the opportunity of using an X-ray tube of the primitive form then used which had been made by Prof. Thomson's assistant Mr Everett in the Cavendish Laboratory. I can well recall my delight when I found at the first trial that . . . a fog . . . was produced . . .[15]

The *cloud chamber*, as it was called, was perfected by Wilson between the years 1911 and 1912, and was used for observing fast-moving charged particles, like the alpha and beta particles emitted from radioactive sources, which produced ions in the gas contained in the chamber.

Rain from the heavens

At six o'clock on the morning of 7 August 1912, a balloon ascended from a field near the town of Aussig in Austria. In the gondola of the balloon were three men: a navigator, a meteorologist, and a physicist. During the next two and half hours the balloon rose to an altitude of 13 000 feet while drifting rapidly northward. For another hour it floated between 13 000 and 16 000 feet. At noon the balloon

Fig. 5.6. Victor Hess after one of his balloon flights. (Courtesy Professor Martin A. Pomerantz.)

touched down near the German town of Pieskow, 30 miles east of Berlin and some 125 miles from Aussig.

The physicist and leader of the flight was Victor F. Hess. He had taken with him three electroscopes of the kind then being used to detect and measure the radiation emitted by radium and other radioactive substances. While his companions took care of the navigation and measured altitude and temperature, Hess watched his instruments and recorded their readings . . . In the November 1912 issue of the German journal *Physikalische Zeitschrift*, Hess summarised his work with the statement: 'The results of my observations are best explained by the assumption that a radiation of very great penetrating power enters our atmosphere from above.

(Bruno Rossi)[16]

1913

Near absolute zero

The Nobel Prize was finally awarded to Heike Kamerlingh Onnes in 1913 for two remarkable achievements: the *liquefaction of helium* and the discovery of *superconductivity*. Kamerlingh Onnes worked at the University of Leiden, as had Hendrik Lorentz and Pieter Zeeman, who had been awarded the Nobel Prize in 1902. Three distinguished exponents of the Leiden school had thus joined the restricted circle of Stockholm laureates.

Heike Kamerlingh Onnes (1853–1926) was born in Groeningen, the Netherlands, the son of a prominent entrepreneur, the owner of a tile factory. After graduating from the University of Groningen, he studied under Robert Bunsen for two years (1871–3) in Heidelberg, Germany, and was one of the students allowed to work in Gustav Kirchhoff's private laboratory. He then returned to Groningen, where in 1879 he obtained his doctorate in physics. In 1882 he was appointed Professor of Experimental Physics at the University of Leiden (where he stayed until 1923). Here, at the Department of Physics, he started a new line of research in the field of low-temperature physics. Keeping to his working dictum 'through measurement to knowledge', he organised his establishment so well that it became one of the best international laboratories of his time (it is still well known as the 'Kamerlingh Onnes Laboratory'), and Leiden became the world centre for low-temperature research, and so the 'coldest spot on earth'.

Liquefying helium
As we have mentioned (p. 75), in 1898 James Dewar succeeded in liquefying hydrogen. However, the element *helium*, which is the lightest of the noble gases, had held out steadfastly against liquefaction. But on 10 July 1908, Kamerlingh Onnes finally succeeded in liquefying it. He had investigated van der Waals' equation (p. 75), and studied the thermodynamic properties of liquids and gases over a wide range of pressures and temperatures. To liquefy helium, Kamerlingh Onnes followed Dewar's method, but carried it a step further. He used liquid hydrogen to cool helium gas under pressure; when he let it expand through a nozzle the gas cooled further, until it was liquefied. He then let the liquefied helium evaporate, so that he managed to reach the temperature at which helium remains in its liquid state under ordinary atmospheric pressure; that is, the temperature of –268.8 °C, or 4.2 degrees over absolute zero (4.2 kelvin).

Superconductivity
After his brilliant success in liquefying helium, Kamerlingh Onnes and his team went on to initiate a new research programme. They thought of using the new liquid as a refrigerant, and then they proceeded to investigate the electrical resistance of various metals at temperatures near absolute zero. It is typical of a metal that its resistance to the flow of electric current decreases with temperature. It was predicted, even by classical physics, that if absolute zero could be achieved the resistance in a pure metal, whose inter-atomic structure was perfectly regular, should disappear.

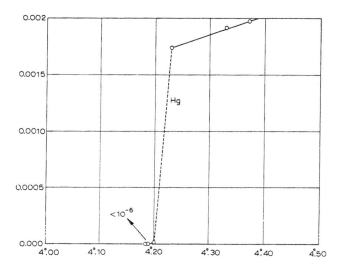

Fig. 5.7. The abrupt falling to zero of the electrical resistance of a sample made of mercury at the temperature of 4.2 kelvin. (Source: Heike Kamerlingh Onnes' Nobel lecture, 1913. © The Nobel Foundation, Stockholm.)[17]

One day, three of Kamerlingh Onnes' co-workers (two assistants, Gilles Holst and Cornelius Dorsman, and a technician, Gerrit Flim) were testing the electrical resistance of a sample of mercury, when to their surprise they found that it suddenly fell to zero at −268.8 °C; this was a very low temperature, but still above absolute zero. It was Kamerlingh Onnes himself who reported this initial result to the Royal Academy of Amsterdam on 21 April 1911; soon after this discovery, he and his co-workers observed the same phenomenon in other metals: in lead below −265.8 °C, and in tin below −269.3 °C.

He gave the extraordinary phenomenon that had been discovered in his laboratory the name of *superconductivity*. Electric currents, once they have been created in closed circuits made of *superconducting materials* (also called *superconductors*), persist for a long time without diminishing, even when there is no energy source in the circuit. It was immediately apparent that if superconductivity could be developed in materials at less drastically low temperatures, the technological potential would be almost unbounded.

The Bohr atomic model

Meanwhile, across the English Channel, important news came in September 1913 from a meeting of the British Association for the Advancement of Science in Birmingham. As reported in the November issue of *Nature*, the British physicist James Jeans referred to 'the recent work by Dr. Bohr, who has arrived at a convincing and brilliant explanation of the laws of spectral lines'.

Niels Bohr, the physicist who had been able to find a solution to the vexing old problem of atomic spectral lines (p. 25), had come to England from the University of Copenhagen. In 1911 he had been a research student under J. J. Thomson, at

Fig. 5.8. H. Kamerlingh Onnes (left) and J. van der Waals. (Courtesy Archives of the Kamerlingh Onnes Laboratory – Professor Rudolf de Bruyn-Ouboter, Leiden, the Netherlands.) 'It was a wonderful sight when the liquid, which looked almost unreal, was seen for the first time. It was not noticed when it flowed in. Its presence could not be confirmed until it had already filled up the vessel. Its surface stood sharply against the vessel like the edge of a knife. How happy I was to be able to show condensed helium to my distinguished friend van der Waals, whose theory had guided me to the end of my work on the liquefaction of gases.' (Heike Kamerlingh Onnes, 1913).[18]

the Cavendish Laboratory. And in the spring of 1912 he had joined Rutherford's group in Manchester. As soon as he had arrived at the Manchester laboratory, he had begun an experimental course in techniques for investigating radioactivity; but, as he recalled, 'just a few weeks later on I said to Rutherford that . . . I would better like to concentrate on the theoretical things'.

George de Hevesy, a young Hungarian nobleman, who was to become a close friend of Bohr, drew the newcomer into the social life of Rutherford's laboratory. Hevesy, a man of the world, made a great impression on Bohr, not only for his uncommon knowledge about science (he was to receive the 1943 Nobel Prize for chemistry), but also for his gifts as a charmer with the ladies. On the scientific side, Bohr soon saw that the *nuclear atom*, as proposed by Rutherford a year earlier (p. 80), offered a basis for a comprehensive theory on the structure of atoms. In a letter to his brother Harald, he wrote:

> Perhaps I have found out a little about the structure of atoms. Don't talk about it to anybody, for otherwise I couldn't write to you about it so soon. If it should

be right it wouldn't be a suggestion of the nature of a possibility (i.e., an impossibility, as J. J. Thomson's theory) but perhaps a little bit of reality . . . You understand that I may yet be wrong; for, it hasn't been worked out fully yet (but I don't think so); also, I do not believe that Rutherford thinks that it is completely wild . . . Believe me, I am eager to finish it in a hurry, and to do that I have taken off a couple of days from the laboratory (this is also a secret).[19]

In the course of the summer of 1912, Bohr started from Rutherford's nuclear model and Planck's quantum hypothesis (p. 37), to work out his ideas about the structure of atoms and molecules. He did a tremendous amount of work. ('I am working day and night', he wrote to his fiancée at the beginning of June.)

At the end of July he returned to Copenhagen, gave lectures at the university, and during the first months of 1913 was able to work out the basis of his theory. The first part of his work appeared in the July issue of the *Philosophical Magazine* under the title 'On the constitution of atoms and molecules'.[20]

Bohr first attacked the problem of the dynamics of the atomic structure represented by Rutherford's nuclear atom. Then he proposed a model for hydrogen (the simplest among the atoms), which perfectly explained the experimental data that had been obtained, and accounted for its observed spectrum. In the second part, he investigated the theory of atoms that are heavier than hydrogen. Bohr's atomic model was the first step in the development of a quantum theory of the atomic structure. It was the beginning of two of the most adventurous decades in the history of physics.

Recollections

Dear Dr. Bohr, I received this morning the amended manuscript of your paper, which I have read again. I think the additions are excellent and appear quite reasonable; the difficulty is, however, that your paper is already rather full and long for a single paper. I really think it desirable that you should abbreviate some of the discussions to bring it within more reasonable compass. As you know, it is the custom in England to put things very shortly and tersely in contrast to the Germanic method, where it appears to be a virtue to be as long-winded as possible . . . As I mentioned in my last letter, it is very desirable not to publish too long papers, as it frightens off practically all the readers.
(From a letter from Ernest Rutherford to Niels Bohr, 25 March 1913)[21]

First reactions

Let us start with Rutherford, Bohr's mentor, and the supervisor of his 1913 papers: he declared that now he could at last believe that his nuclear atom model was correct! At the Birmingham meeting in September, J. J. Thomson, however, behaved rather coolly, as he did not believe in the new quantum theory of atoms ('Thomson went out of physics', Bohr remarked later). When the veteran Lord Rayleigh was asked his opinion about Bohr's model he replied: '. . . a man who has passed his

> ### Bohr's hydrogen atom
>
> In Bohr's model of the hydrogen atom, a negatively charged electron moves around a heavier positive nucleus, since it is attracted by the electric force – just as the earth moves around the sun, because it is attracted by the gravitational force.
>
> Bohr started by assuming that the electron describes stable circular orbits of fixed radius, which he called *stationary orbits*; here the electron is held by the *centripetal force* provided by the attracting electric force. Then, from the classical laws of motion, he obtained the orbit radius, the orbital frequency (the number of times the electron circles the nucleus per second) and the energy of the electron on that orbit.
>
> At this point Bohr was faced with a serious difficulty: centripetal force means *centripetal acceleration*, and the classical theory of electromagnetism predicts that accelerated charges should radiate energy. Therefore, the electron should lose its energy, while spiralling into ever smaller orbits, and it should finally fall into the nucleus. As a consequence, the atom should be unstable and radiate electromagnetic energy of a continuously increasing frequency. Thus, classical physics again failed to explain phenomena of the microscopic world.
>
> Bohr circumvented this difficulty by an *ad hoc* hypothesis. He had no doubt; the solution lay in Planck's hypothesis of the quantum of action. He wrote in his paper:
>
>> [There] seems to be a general acknowledgement of the inadequacy of the classical electrodynamics in describing the behaviour of systems of atomic size. Whatever the alteration in the laws of motion of the electrons may be, it seems necessary to introduce in the laws in question a quantity foreign to the classical electrodynamics, i.e. Planck's constant, or as it often is called, the elementary quantum of action.[22]
>
> Bohr assumed that, like Planck's atomic oscillators, the atom can exist only in certain *stationary quantum states* which correspond to specific energy values and orbits, and that in these states the atom should not radiate. ('[We] are led to assume that these configurations will correspond to states of the system in which there is no radiation of energy; states which consequently will be stationary as long as the system is not disturbed from outside', he wrote.)[23] He then assumed that the electron energy increases by finite steps: this immediately led to the formulae which give the radii and the energies corresponding to the different quantum states of the atom.
>
> At this point Bohr applied Planck's quantum concept to determine the frequency of the radiation emitted when the atom makes a transition between two stationary states. Here are his two famous assumptions:
>
>> – ... the dynamical equilibrium of the systems in the stationary states can be discussed by help of the ordinary mechanics, while the passing of the systems between stationary states cannot be treated on that basis.

> — ... the latter process is followed by the emission of a *homogeneous* [monochromatic] radiation, for which the relation between the frequency and the amount of energy emitted is the one given by Planck's theory.[24]

Therefore, according to Bohr, a monochromatic light wave can be emitted or absorbed only when the atom makes a transition (a *quantum jump*) between two stationary states possessing both different orbit radii and different energy values. For example, if E_2 is the energy of an excited state (atom with a greater orbit radius), and E_1 the energy of a less excited state (smaller radius), the frequency of light emitted in a quantum jump from state 2 to state 1 is:

$$\nu = (E_2 - E_1)/h$$

where h is the Planck constant. Since only certain values of the energy are allowed, the radiation emitted can have only certain allowed frequencies (and wavelengths). Therefore, the emission and absorption spectra will not be continuous spectra, but *line spectra*, as was in fact known from experiments. Bohr's model also enabled scientists to calculate the size of the lowest orbit of the hydrogen atom (about 5/100 of a nanometre in diameter), and the value of the so-called Rydberg constant (a fundamental number which appears in the formula that gives the values of the energies of the stationary states of atoms).

Quantum expressions

Quantum states. Atoms can exist only in states of *discrete energy*. The state of lowest energy is called *ground state*; the other allowed states of higher energies are called *excited states*. The set of allowed energy values are called *energy levels*.

Quantum jump. An atom can absorb light of the right frequency and make a *quantum jump* from the ground state to an excited state. When it drops again to the ground state, it emits light of the same frequency.

sixtieth year ought not to express himself about modern ideas.'[25] In Göttingen, the German Mecca for mathematics, Bohr's ideas were received with much more scepticism. His brother Harald strove to assure his friends there that Niels' work was important. He wrote to Niels: 'People here are still exceedingly interested in your papers, but I have the impression that most of them ... do not dare to believe that they can be objectively right; they find the assumptions too "bold" and "fantastic."'[26] Finally, from Vienna his friend de Hevesy wrote: 'I spoke to Einstein this afternoon ... I asked him about his view of your theory. He told me, it is a very interesting one, important one if it is right and so on and he had very similar ideas many years ago but had no pluck to develop it ... [Hardly] anything else could make me such a pleasure than this spontaneous judgement of Einstein.'[27]

Fig. 5.9. Niels Bohr (right) and his brother Harald, circa 1906. (Courtesy Niels Bohr Archive, Copenhagen.)

[Although] Niels was two years older, Harald matriculated the year after him when only seventeen. At the early age of 22 years . . . he [obtained his doctorate], and this was the prelude to a brilliant contribution to mathematics which gained him great international respect. At that time Harald Bohr was by far the more famous of the two brothers, not so much because of his scientific gifts as for the fact that he was among Denmark's best footballers. For a number of years he played half-back in the first team of the AB football club, and he won a silver medal for Denmark at the Olympic Games in London in 1908 . . . Niels too was a keen footballer, but he did not get any further than becoming reserve goalkeeper for the AB team . . . 'Yes, Niels was quite good; but he was too slow in "coming out", teasingly explained his younger brother in later years.'[28]

1914

X-ray diffraction

At last, seventeen years after their discovery, the true nature of X-rays was revealed. An experiment was carried out in 1912 under the direction of the German physicist Max von Laue, who was working at the University of Munich. Using crystals as a kind of three-dimensional diffraction grating, Laue's young co-workers discovered that X-rays do not behave like particles, as some physicists then thought, but that they have wave-like characteristics. In fact, they are electromagnetic waves like light. Von Laue's brilliant achievement was recognised by the award of the 1914 Nobel Prize for physics.

Von Laue's idea

In 1912 Max von Laue was a lecturer at the University of Munich. He was working in the Institute of Theoretical Physics, which was under the direction of Arnold Sommerfeld. Wilhelm Röntgen (Nobel 1901) was also in Munich at that time; he was director of the physics laboratory. So it was quite natural for von Laue to become interested in X-rays. He thus recalled:

> On my arrival in Munich in 1909 my attention was drawn constantly – first owing to the influence of Röntgen's work at this University and subsequently by Sommerfeld's active interest in X-rays and [gamma]-rays – . . . back to the question of their actual nature. To this was added a further important circumstance . . . [The] basic crystallographic law of rational indices had been explained simply and visually by the mineralogists through space-lattice arrangement of the atoms.[29]

Von Laue had heard of the theory of crystals from Paul von Groth – a world authority on crystallography – who was a professor at the University of Munich, and curator of the minerals section of the Bavarian National Museum. Von Laue remembered clearly what had happened one evening in February 1912:

> . . . P. P. Ewald [one of Sommerfeld's assistants] came to visit me. On Sommerfeld's instigation he was working on a mathematical investigation into the behaviour of long electromagnetic waves in a space-lattice . . . [During] the conversation I was suddenly struck by the obvious question of the behaviour of waves which are short by comparison with the lattice-constants of the space-lattice. And it was at that point that my intuition for optics suddenly gave me the answer: lattice spectra would have to ensue . . . I immediately told Ewald that I anticipated the occurrence of interference phenomena with X-rays.[30]

Previous research had indicated in fact that, if X-rays had the same characteristics as light waves, their wavelength would be of about a tenth of a nanometre (for convenience physicists call a nanometre the distance of one-billionth of a metre): that is, less than one thousandth of the shortest wavelength of light. In order to obtain X-ray interference similar to that made by light when it passes through a diffraction grating, the distance between the slits would have to be of the order of the radiation wavelength; this is more or less the distance between atoms in a crystalline solid.

X-rays as waves

Von Laue discussed his idea with Sommerfeld and others during a skiing expedition in the Alps, but all of them raised objections to the idea. In spite of this, Walter Friedrich, one of Sommerfeld's assistants, and a research student called Paul Knipping tested it out experimentally. They sent an X-ray beam through a crystal so that the scattered X-rays could be observed on a photographic film. The result showed that the scattered X-rays did actually form a diffraction pattern of bright spots. Von Laue recalled:

> Immediately from the outset the photographic plate located behind the crystal betrayed the presence of a considerable number of deflected rays [see Fig. 5.10] . . . These were the lattice spectra which had been anticipated . . . and, on June 8, 1912, Sommerfeld was able to submit to the Munich Academy the joint work of Friedrich, Knipping, and myself on X-ray interferences, which work, apart from the theory itself, also contained a series of very characteristic exposures.[31]

The conclusions of this experiment were that X-rays do have wave-like characteristics similar to those of light, and that the atomic structure of crystals is a regularly repeated arrangement. Immediately after Laue's discovery, a certain

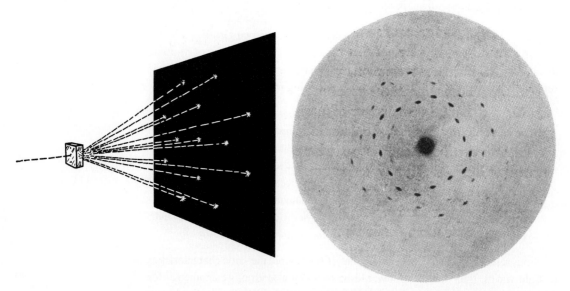

Fig. 5.10. Left: X-ray diffraction scheme. Right: one of the first photographs showing X-ray diffraction. (Source: Max von Laue's Nobel lecture, 1915; © The Nobel Foundation, Stockholm.)[32]

Fig. 5.11. Max von Laue was enthusiastic about sports cars and motorcycles. Here he is with his high speed car! (Courtesy Archiv zur Geschichte der Max-Planck-Gesellschaft, Berlin-Dahlem.)

William Lawrence Bragg (p. 100), a student at Cambridge, showed how diffraction patterns could be determined by the structure of the crystals. These discoveries later formed the basis of *X-ray spectroscopy*, and marked the origin of what now is called *crystal-structure analysis* – an important field in the study of matter in bulk, and also in the development of modern electronics.

Max von Laue (1879–1960) was born at Pfaffendorf, near Koblenz, Germany, the son of a military official. In 1903 he obtained his doctorate from the University of Berlin, where two years later he was appointed as Max Planck's assistant, and the following year as a lecturer. After his discovery, von Laue became professor of physics at the University of Zurich, and then at Frankfurt. In 1917, when the Kaiser Wilhelm Institute of Physics was established at Berlin–Dahlem, with Einstein as its director, von Laue was appointed deputy director; in 1919 he became Professor of Physics at the University of Berlin. During his life von Laue exerted great influence on the direction of German physics. He had a strong sense of justice and fair play. During the 1930s he defended scientific views, such as the theory of relativity, which were not approved by the German political world of the time, or by physicists such as Philipp Lenard and Johannes Stark. (He was the only member who vigorously protested when, in 1933, Einstein was forced to resign from the Prussian Academy of Sciences.)

The nuclear charge

Returning to the series of events in their chronological order, we must mention Henry Moseley, a young English physicist who was working with Rutherford at Manchester. Moseley decided to undertake experimental work on the new physics resulting from von Laue's discovery.

In 1908 the British physicist Charles Barkla (Nobel 1917) and his co-worker Charles Sadler had discovered that when X-rays of continuous frequencies (or energies) passed through a material they caused the emission of secondary radiation of fixed energy, which was characteristic of the material that they had passed through. Barkla called this homogeneous radiation *characteristic X-rays*.

Some years later the Dutch Antonius van der Broek organised the elements according to their weights and chemical properties. He attributed to each element a number which he asserted to be identical with the number of the electrons in an atom (now called the *atomic number Z*), starting from one for hydrogen. This number seemed to be connected with the characteristic X-rays discovered by Barkla.

In the summer of 1913 Moseley began by systematically investigating the characteristic X-rays produced by various elements. He used an experimental technique devised some months earlier by the Braggs, father and son (see Nobel 1915), in order to make accurate measurements of the frequencies of the emitted X-rays.

From his measurements, Moseley showed that the frequency of the characteristic X-rays of the elements he had investigated increased in a regular manner as one passed from one element to the next in the periodic system; that is, the frequency depended on the atomic number Z (rather than on the atomic weight) of the elements. Moseley went on to report his results in two papers published, between autumn 1913 and spring 1914, in the *Philosophical Magazine*. In these papers he wrote: '. . . there is in the atom a fundamental quantity which increases by regular steps as we pass from one element to the next. This quantity can only be the charge on the central positive nucleus . . . We are therefore led by experiment to the view that . . . [Z – the nuclear charge] is the same as the number of the place occupied by the element in the periodic system.'[33] Moreover, the values that he had obtained for the frequencies of the characteristic X-rays agreed remarkably well with those calculated from Bohr's theory of the atom.

Moseley had thus given the first experimental support to Rutherford's nuclear atom, and confirmed Bohr's atomic theory; he had also opened the door to a new research field, that of *X-ray spectroscopy*. (After the outbreak of the First World War, Moseley had to enlist for military service, as a second lieutenant in the Royal Engineers. He was killed in action on 10 August 1915, during the British Gallipoli campaign at the Dardanelles, Turkey.)

The Franck–Hertz experiment

Let us turn our attention to events of the year 1914 as they occurred at the University of Berlin. There, two gifted experimenters, James Franck and Gustav Hertz, had secured important results which further confirmed Bohr's hypotheses of the quantum excited states of the atom. Here is a brief description of their experiment.

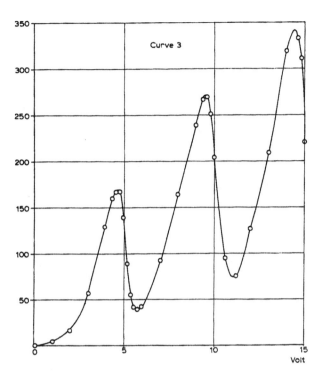

Fig. 5.12. Franck and Hertz' results. The peaks correspond to the excited states of the atoms. (Source: James Franck's Nobel lecture, 1925, © The Nobel Foundation, Stockholm.)[34]

Franck and Hertz observed that an electric current passing through mercury vapours contained in a vessel rose, on increasing the voltage between two electrodes, until a value of 4.9 volts was reached. Then the current dropped suddenly. When the voltage was increased beyond 4.9 volts, the current again rose up, but dropped once more at about 9.8 volts. The sudden 'drop-offs' of the current occurred at all voltages with a value of an integer multiple of 4.9. They also observed that, when the voltage exceeded 4.9 volts, mercury began to emit ultraviolet light of a definite frequency, which was exactly equal to the frequency defined by Bohr's formula $v = (E_2 - E_1)/h$ (p. 93).

The physical interpretation of the phenomenon can be given as follows: the mercury atoms were supplied with energy by bombarding them with the electrons of the current. The atoms could only take up such a portion of energy as exactly corresponded to a higher energy level. Thus, if the atoms were bombarded with electrons whose kinetic energy was less than the first excited level, they remained in the ground state. When the energy of the electrons became greater than the first excited level, but remained less than the second level, the atoms were brought into the first excited state, and so on. Thus the Franck–Hertz experiment provided one of the first experimental foundations to Bohr's atomic model.

1915

X-rays and crystals

Our chronicle now shifts to England. Here, in 1912, William Lawrence Bragg, who was twenty-two years old, was studying physics at Trinity College, Cambridge. During that summer Lawrence discussed with his father, William Henry, a professor of physics at the University of Leeds, the recent discovery made by von Laue concerning the diffraction of X-rays by crystals (p. 95). Upon his return to Cambridge in the autumn, he started studying the new phenomenon, and soon had a fresh idea. He remembered this in his Nobel lecture:

> I tried to attack the problem from a slightly different point of view, and to see what would happen if a series of irregular [X-ray] pulses fell on diffracting points arranged on a regular space lattice. This led naturally to the consideration of the diffraction effects as a reflexion of the pulses by the planes of the crystal structure.[35]

Before the year was out, he had found the solution and immediately published it in the *Proceedings of the Cambridge Philosophical Society*.[36]

The Bragg reflection method

This is young Bragg's explanation of how he treated X-ray diffraction as a reflection by successive parallel layers of atoms in a crystal. The X-rays as they arrive are reflected from each of the layers. The reflected waves add up constructively at certain special angles, because they are in phase with one another, and produce a strong reflected beam. At other angles, the waves cancel each other out, and the reflected beam is consequently very weak. The Braggs found the position and intensity (amount of X-rays) of the reflected beams by using an ionisation chamber (a gas-filled detector which produced a current pulse when X-rays ionised the gas contained in it), which was connected to an electrometer. But at this point we may let young Bragg himself speak:

> Although it seemed certain that the von Laue effect was due to the diffraction of very short waves, there remained the possibility that they might not be the X-rays. My father, in order to test this, examined whether the beam reflected by a crystal ionised a gas; this he found to be the case. He examined the strength of the reflexion at various angles, and the instrument which was first used for the purpose was developed later into the X-ray spectrometer with which we have done the greater part of our work.[37]

Father and son used this instrument, the so-called *X-ray spectrometer*, to analyse the spectral distribution of the radiation emitted from X-ray sources. By using a known X-ray wavelength they were also able to determine the distance between atoms, and hence the structure of crystals. (They were the first to determine the

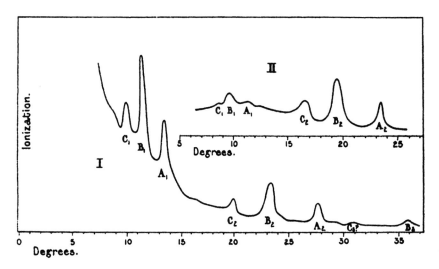

Fig. 5.13. Example of spectral distributions of X-rays obtained by the Braggs with their X-ray spectrometer. (Source: W. H. Bragg and W. L. Bragg, *Proceedings of the Royal Society of London*, A88, 1913, p. 431. Reprinted by permission of the Royal Society, London.)[38] Peaks in the intensity of X-rays occur at certain specific angles exactly as predicted by Lawrence Bragg's formula.

lattice constant of sodium chloride – that is, of rock salt.) The results of these investigations, made during the period 1912–14, were reported in a book entitled *X-rays and Crystal Structure*, which was published in 1915 and which earned them both the 1915 Nobel Prize for physics. Following their work, the X-ray diffraction technique became a powerful tool for the study of the structure of matter.

Bragg and Bragg

William Henry Bragg (1862–1942) was born at Westward, Cumberland, England. He studied mathematics at Trinity College, Cambridge, and physics at the Cavendish Laboratory. At the end of 1885, although only twenty-three years old, he was appointed Professor of Mathematics and Physics at the University of Adelaide, Australia. In 1909 he was nominated Cavendish professor of physics at the University of Leeds, England. Later he became Professor of Physics at University College, London (1915–25), and Fullerian professor of chemistry at the Royal Institution. He was knighted in 1920, he received the Order of Merit in 1931, and he was elected president of the Royal Society in 1935. Sir Henry received many honorary degrees from British and foreign universities, and was awarded the Rumford Medal and the Copley Medal of the Royal Society. He was a lover of tradition, gentle and free of pretence. He was also a popular scientific lecturer and writer: his Christmas Lectures for children at the Royal Institution were famous.

William Lawrence Bragg (1890–1971), Sir Henry's son, was born in Adelaide. He attended university there, and took his degree in mathematics in 1908. After coming to England in 1909, he studied for his degree in physics at Cambridge.

Fig. 5.14. William Lawrence Bragg at the time when he was awarded the Nobel Prize. (Courtesy AIP Emilio Segrè Visual Archives, Weber Collection.)

Lawrence received the Nobel Prize at the very early age of twenty-five, so becoming the youngest-ever Nobel laureate. After the First World War he was appointed Professor of Physics at the University of Manchester (1919–37), succeeding Sir Ernest Rutherford. From 1937 to 1938 he was director of the National Physical Laboratory at Cambridge, and from 1938 to 1953 fifth Cavendish Professor of Experimental Physics, his predecessors being James Clerk Maxwell, Lord Rayleigh, Sir J. J. Thomson and Lord Rutherford. Like his father he served as director of the Royal Institution in London between 1954 and 1965. Knighted in 1941, Sir Lawrence's main interests later lay in the application of X-ray analysis to proteins, opening the way to the new science of molecular biology. He participated in the work of Max Perutz and John Kendrew (Nobelists for chemistry in 1962) on the structure of proteins, and supported the work of Francis Crick and James Watson (Nobelists for medicine in 1962) on the structure of DNA (p. 244).

Einstein was right

Across the Atlantic, at the University of Chicago, Professor Robert Millikan carried out research work, in the period 1912–15, on a series of measurements relating to the photoelectric effect. In so doing he irrefutably verified Einstein's equation of 1905 (see p. 51).

Millikan used plates made of alkali metals, like sodium or potassium, placed in a vacuum tube and exposed to light of different frequencies. When the metal plates were illuminated with light that possessed a frequency greater than the threshold, electrons (called 'photoelectrons') were emitted and collected by an electrode placed in front of the emitting plate, so producing an electric current. Millikan was able to measure the kinetic energy of the photoelectrons for different frequencies of light, and then compare the results with Einstein's equation.

As you can easily see in Fig. 5.15, the experimental points, represented in the diagram by the dots, exactly fit with the straight line that represents graphically Einstein's equation – so demonstrating a perfect agreement between theory and experiment. (In spite of Einstein's success in explaining the photoelectric effect, most physicists, including Millikan himself, still had remained sceptical. Many years later Millikan remarked candidly: 'I spent ten years of my life testing that 1905 equation of Einstein's and contrary to all my expectations, I was compelled in 1915 to assert its unambiguous verification in spite of its unreasonableness, since it seemed to violate everything we knew about the interference of light.')[39] With the same experimental data Millikan was able also to obtain the first direct value of the Planck constant h, with an uncertainty of about two-thousandths.

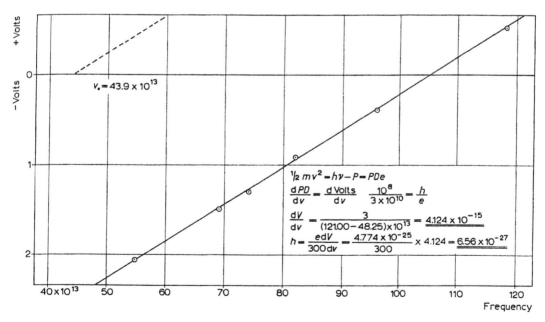

Fig. 5.15. Results of Millikan's experiment. (Source: Robert Millikan's Nobel lecture, 1924, © The Nobel Foundation, Stockholm.)[40]

1916

Curved space-time and gravitation

Due to the Great War (which broke out on 1 August 1914), the Swedish Academy of Sciences decided not to award the 1916 Nobel Prize. Meanwhile, however, some scientific work continued – in particular Einstein's researches (he was then living in Berlin).

Let us return to Einstein's career. After 1905, which had seen the birth of the special theory of relativity (p. 51), Einstein gradually began to be recognised as one of the most important scientists of his time. During the year 1909 he received an honorary degree from the University of Geneva, Switzerland, he was invited to Salzburg, Austria, to report on quantum theory and was offered a professorship by the University of Zurich. And the year after he was proposed (by Friedrich Wilhelm Ostwald), for the first time, for the Nobel Prize.

In 1911 he was appointed Professor of Theoretical Physics at the German university in Prague, the Karl-Ferdinand University, where he remained for eighteen months. Later on he returned to the Swiss Federal Institute of Technology (ETH) in Zurich, where he had earlier studied and graduated in the summer of 1900. In July 1913 he was made a member of the prestigious Prussian Academy of Sciences. The following year he left Zurich to become a professor at the University of Berlin. He was then thirty-five years old. It was in these two cities that Einstein developed his new theory of gravitation.

After the discovery of the equivalence principle in 1907 (p. 65), Einstein's scientific interests became focused principally on quantum theory. He returned to considering the gravitation problem when he was in Prague. It had become clear to him that a new geometry was needed for its description: the mathematical tool to be used could no longer be Euclidean geometry, but rather a new geometry developed in the nineteenth century by two German scientists; the 'prince of mathematicians', Karl Friedrich Gauss, and Bernhard Riemann. In a conference held in Kyoto in 1922, Einstein said:

> If all [accelerated] systems are equivalent, then Euclidean geometry cannot hold in all of them. To throw out geometry and keep [physical] laws is equivalent to describing thoughts without words. We must search for words before we can express thoughts. What must we search for at this point? This problem remained insoluble to me until 1912, when I suddenly realised that Gauss' theory of surfaces holds the key for unlocking this mystery . . . I realised that the foundations of geometry have physical significance.[41]

In 1923 he added: 'I had the decisive idea . . . only in 1912 . . . after my return to Zurich, without being aware at that time of the work of Riemann, Ricci, and Levi-Cività. This [work] was first brought to my attention by my friend Großmann . . .'[42]

The development of his theory took him deep into new fields of geometry and mathematics. He was greatly aided by his old friend and fellow student, the mathematician Marcel Grossmann. This collaboration resulted in a joint paper, published in 1914. Its mathematical roots are found in Gauss, Riemann, the Swiss mathematician Erwin Bruno Christoffel, and the Italians Gregorio Ricci-Curbastro and Tullio Levi-Civita. Towards the end of 1915, after two years of hard work, he finally saw the solution. In November he wrote to Arnold Sommerfeld:

> This last month I have lived through the most exciting and the most exacting period of my life ... When all my confidence in the old theory [developed in 1912–15] vanished, I saw clearly that a satisfactory solution could ... be reached ... The wonderful thing that happened then was that not only did Newton's theory result from it ... but also the perihelion motion of Mercury ... For the deviation of light by the sun I obtained twice the former amount.[43]

Einstein presented the findings of his theory at three consecutive sessions of the Prussian Academy in November 1915. And in March 1916 he finally sent, to the journal *Annalen der Physik*, his paper entitled 'The foundation of the general theory of relativity',[44] which contained a complete exposition of his ideas on gravitation.

In 1933, in concluding a lecture on the origins of general relativity at the University of Glasgow, he recalled those years:

> The years of searching in the dark for a truth that one feels but cannot express, the intense desire and the alternations of confidence and misgiving until one breaks through to clarity and understanding are known only to him who has himself experienced them.[45]

Einstein's ideas on gravitation soon caused a great sensation. After the initial excitement, however, interest rapidly decreased. This was because of a lack of experimental techniques sensitive enough to measure the tiny effects predicted by the theory.

Starting in the 1960s there was a rebirth of general relativity, due to the impressive development of new astronomical and astrophysical devices such as radio telescopes, new radiation detectors and extraterrestrial satellites. This led to dramatic discoveries such as quasars, pulsars, neutron stars, gravitational lenses and the cosmic microwave background, and the search for possible black holes and gravitational waves (see for example the 1993 Nobel Prize). All these exotic cosmic objects and phenomena opened a new era for general relativity, which has thus become an active area of research, both experimental and theoretical, whilst playing an important role both in astronomy and in physics.

The general theory of relativity

Einstein's goal was to produce a new theory of gravitation that might incorporate both his 1905 special theory of relativity, and his 1907 principle of equivalence.

Curved space-time

In Newton's theory of gravitation (p. 12) the attraction between two massive bodies is taking place instantaneously (it is an 'action at a distance', as physicists say); it means that the gravitational effects should travel with infinity velocity. But, according to special relativity, these effects cannot spread faster than the speed of light in empty space (p. 55). Hence Einstein tried to revise the space-time concept of special relativity (p. 71) in order to extend it to gravitation.

He assumed that there exists a *gravitational field*, in the space around massive bodies (something analogous to the electromagnetic field around electrical and magnetic bodies). This field is responsible for the gravitational attraction, and is represented in terms of a *curved space-time*.

This is a completely new concept. In fact, the space-time of special relativity is 'flat'. This means that, if we imagine that we are living in a two-dimensional flat space, we will assert that the shortest distance between two points in such a space is part of a straight line, or, that the angles of a triangle add up to 180 degrees. If, on the other hand, we imagine that we are living in a two-dimensional 'curved space' (for example, on the surface of a sphere), we shall find that Euclidean geometry – the one that we all learned in school – no longer holds good. In such a non-Euclidean space the shortest distance between two points becomes an arc of a circle, and the angles of a triangle add up to more than 180 degrees.

Gravitation as a distortion of space-time

Einstein thought that the sources of space-time curvature are *matter* and *energy*. For example, the gravitational field generated by a massive body causes space-time to curve; that is, it causes space to become curved and time to slow down. As a consequence the larger, the denser and the closer is the mass of a body, the greater is the resulting space-time curvature.

The general theory of relativity provides a set of equations (which are called *Einstein's field equations*) which enable one to calculate how much curvature is generated in a point of space-time around a massive body. Once this curvature is known, one can determine how matter responds to it, because the space-time curvature manifests itself as a gravitational attraction. As a result, general relativity can be considered to be an explanation of gravitation as a *distortion of space-time*.

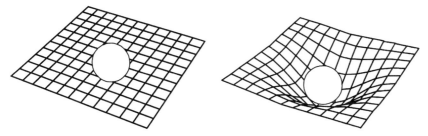

Fig. 5.16. Before Einstein's general relativity, space in the vicinity of a massive body was held to be flat. After 1916, space became curved: 'space acts on matter, telling it how to move. In turn, matter reacts back on space, telling it how to curve. Thus matter here influences matter there.'

Order and beauty

> What is there in places almost empty of Matter, and whence is it that the Sun and Planets gravitate towards one another, without dense Matter between them? Whence is it that Nature doth [does] nothing in vain; and whence arises all that Order and Beauty which we see in the World?
>
> (Isaac Newton, 1704)[46]

Three historic tests

The perihelion advance of Mercury. In Newton's theory of gravitation the elliptical orbits of a body are fixed in space, the major axis of the ellipse pointing always in the same direction. This result does not hold for the planets of the solar system, because their mutual interactions disturb their orbits. Their elliptical paths rotate in a plane. This rotation is called *perihelion advance* (or *perihelion precession*). (The perihelion is that point in its orbit where the planet finds itself nearest to the sun.)

In the case of the planet Mercury, its perihelion advance, calculated using Newton's law of gravitation, left, unaccounted for, an excess of forty-three seconds of arc per century. Now, in general relativity, the distortion of space-time near the sun can account for this extra advance. Thus, Einstein's calculations agreed almost perfectly with the observed value. Since the mid-1960s radar observations of Mercury have been carried out with great precision; again, the results of the planet's perihelion advance have agreed with general relativity. (Modern measurements agree with general relativity with an accuracy of one part in a thousand.)

Bending of starlight by the sun. General relativity predicts that light rays deviate from rectilinear motion near massive objects; they are bent towards the object itself. Einstein used his theory to calculate how much the light rays, passing as close as possible to the sun, would be deflected. He obtained a deflection just twice what would have been predicted by Newton's theory; it was precisely 1.75 seconds of arc. This was the most famous test of general relativity; it was first

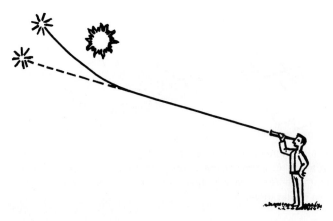

Fig. 5.17. Light from a distant star (solid line). The broken line shows the direction from which the light appears to be coming.

performed by the British astronomer Arthur Eddington in 1919, during a total eclipse of the sun (see p. 119).

In the 1960s and 1970s radio astronomers went on to measure with great accuracy the deflection of radio waves emitted by quasars when they pass near the sun, so further confirming general relativity. (A recent analysis of data accumulated over the last decade, and which used hundreds of radio sources and quasars located all over the sky, confirmed general relativity to four parts in ten thousand.)

In addition, the American physicist Irwin Shaprio discovered not only that light rays are bent by the curved space near the sun, but that they are delayed too. This time delay was actually measured for light making a round trip from earth to Mars and back, and passing close by the sun on its way; the time taken agreed with general relativity – to within an accuracy of one in a thousand. And even more, in the summer of 2002 Italian astrophysicists sent radio waves from the earth to the *Cassini* spacecraft (which is flying to Saturn) and back again. They measured the 'Shapiro delay' with an extraordinary precision of 2.3 parts in 100 000 (Einstein would surely be happy to claim that he is now exactly 99.998 per cent right!).[47]

Gravitational red shift. According to general relativity, the rate of a clock is slowed down when it is in the vicinity of a large mass. This can also be understood with the 'equivalence principle': a clock at ground level ticks more slowly than it would at the top of a tower. Atomic transitions function in effect like microscopic clocks. Thus, the wavelengths of the radiation when emitted during atomic transitions will change according to altitude: waves emitted on the ground will have a longer wavelength when compared with the same waves emitted, for example, at the top of a tower (this phenomenon is called *gravitational red shift*).

The first experiments designed to observe a gravitational red shift on earth were carried out in the first half of the 1960s. Gamma-ray photons from a certain isotope of iron were emitted at ground level and were detected at the top of a tower at Harvard University: a red shift was measured with an uncertainty of about one per cent (p. 282). A more precise experiment was carried out in 1976. A hydrogen maser (a device first developed by Norman Ramsey, see Nobel 1989), functioning as an atomic clock, was flown on a rocket, at an altitude of 10 000 kilometres; its frequency was compared to a similar maser on the ground. The result confirmed Einstein's prediction with an uncertainty of less than one in ten thousand.

Finally, physicists have calculated that the atomic clocks aboard the satellites of the Global Positioning System advance faster, by about 38 millionths of a second per day, than if the same clocks were on the surface of the earth; more than eighty per cent of this is due to the gravitational red shift (the remaining part is due to the time-dilatation effect of special relativity). So, to ensure the correct functioning of the Global Positioning System, this effect *must* be taken into account. (We might well say, in the words of Clifford M. Will, that Einstein was not only 'right', he was 'practical', too!!)[48]

1917

Characteristic X-rays

The 1917 Nobel Prize was bestowed upon Charles Barkla for his research on *characteristic X-rays*. Moseley's discovery regarding X-ray spectroscopy (p. 98) was worthy of a Nobel Prize; however, he unfortunately died before he could be awarded it. The 1924 Nobel Prize, as we will see, was awarded for X-ray researches: during the Nobel ceremony presentation speech it was thus stated that, 'Moseley [died] at the Dardanelles before he could be awarded the prize, but his researches had directed attention to the merits of Barkla, who consequently in 1918 was proposed for the [1917] Nobel Prize, which was awarded to him without delay'.[49]

Barkla began his research on X-rays at J. J. Thomson's suggestion while at the Cavendish Laboratory, and he continued them after his return to the University of Liverpool. In 1906 he was able to obtain polarised X-rays (selected X-ray waves that always vibrate in the same direction), thus demonstrating that X-rays were transverse waves, as is ordinary light. As we have seen (p. 98), in 1908 Barkla had proved the emission of characteristic X-rays from chemical elements. And in 1911 he had found two types of these rays, which he had named *K*- and *L*-radiation.

Soon afterwards, it was found that the same kind of radiation was also emitted under the action of cathode rays, when the element in question was used as a target in an X-ray tube. A few years later, in 1913–14, using this same method, the Braggs, father and son, measured the spectra of X-rays emitted by different materials. Their results indicated that these rays were made up of two types, a continuous spectrum (a continuum of frequencies), and a line spectrum of single frequencies, the latter being characteristic of the target material in the tube.

Barkla's results were thus confirmed. In turn, Moseley studied the spectra of the characteristic X-rays for most chemical elements, and found a relationship between their frequency and the atomic number Z of the elements themselves (p. 98).

Charles Glover Barkla (1877–1944) was born at Widnes, Lancashire, England. In 1894 he began to study physics at University College, Liverpool. Five years later he went to Cambridge, and began to work at the Cavendish Laboratory. He then moved to King's College, and in 1902 returned to Liverpool. He then moved back to London University in 1909, and finally, in 1913, he became Professor of Natural Philosophy at the University of Edinburgh, Scotland.

New quantum numbers

In the period 1914–17 Arnold Sommerfeld carried out important research on the structure of atoms. He was a central figure at the University of Munich, where he had held the chair of theoretical physics since 1906. Thanks to his studies and his charisma as a teacher, he became head of one of the most important schools of atomic physics of the first decades of the twentieth century. The list of his students includes four Nobel laureates: Werner Heisenberg (Nobel 1932), Wolfgang Pauli (Nobel 1945), Hans Bethe (Nobel 1967) and Peter Debye (Nobel for chemistry 1936). (Sommerfeld could be named an 'uncrowned laureate': although considered a peer of Nobel prizewinners in every respect, he was never actually summoned to Stockholm!)

In the mid-1910s Sommerfeld worked intensively on Bohr's model of the hydrogen atom. He went further than Bohr, who had examined only circular orbits of the electron, by dealing with the more realistic elliptical orbits. He thought of the quantum states of the atom as being characterised by two quantum numbers: the so-called *principal quantum number n* (related to the size of the orbit), and the *orbital quantum number l* (related to its shape).

He then went further. He introduced a third quantum number, m, which he called the *magnetic quantum number*, related to the allowed orientations of the orbital plane relative to some fixed direction. This is called *spatial quantization*, and it means that the plane containing the orbit of the electron can assume only specific orientations. The same specific orientations are assumed by a physical quantity which represents the orbital motion of an electron around the nucleus; it is called *angular momentum*, and it is represented by an arrow with its direction pointing perpendicularly to the orbital plane, see Fig. 5.18.

Finally, Sommerfeld added to his theory a treatment of the electron, which took into account the relativistic variation of its mass. With this more refined model he was able to explain the *fine structure* of the hydrogen spectral lines, a long-known phenomenon that had first been observed by Albert Michelson in the late 1880s (p. 65). Thanks to his theory, Sommerfeld, and independently Debye, were also

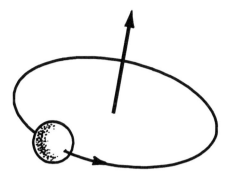

Fig. 5.18. The arrow emerging from the plane of the orbit represents the orbital angular momentum of the particle (for instance, an electron) orbiting anticlockwise.

able to explain the Zeeman effect (p. 40), while Paul Epstein (a former Russian student of Sommerfeld's) and the German astrophysicist Karl Schwarzschild went on to explain the *Stark effect* (the splitting of atomic spectral lines in an electric field, see p. 116). Sommerfeld's theory also contains a number, called the *fine structure constant*, which defines the ratio between the speed of the electron in the smallest atomic orbit and the speed of light. This number contains three fundamental constants of nature (the elementary charge e; the Planck constant h; and the speed of light c) and has a numerical value equal to about 1/137. (Today this number is a kind of symbol of modern electrodynamics.)

A splendid flash

After general relativity, Einstein returned to quantum theory. In August 1916 he wrote to his friend Michele Besso: 'A splendid flash came to me concerning the absorption and emission of radiation . . . A surprisingly simple derivation of Planck's formula, I would say *the* derivation. Everything completely quantum.'[50] He published the result of his 'splendid flash' in three papers (1916–17). The novelty of this work was the introduction of 'probability' into quantum physics.

Einstein started by considering a gas of light quanta in equilibrium with a gas of atoms. Then he used Bohr's assumption that atoms exist in stationary states with discrete energies. At this point he calculated the probabilities involved in the emission and absorption of light quanta by the atoms. He considered three processes. The first he called *spontaneous emission*: when atoms are excited to a higher energy level, they can drop back spontaneously to a lower level by emitting energy in the form of randomly directed photons. In the second case – *absorption* – photons can be absorbed by the atoms, so arousing them to a higher, excited state. Finally, he recognised the existence of a third new process, called *stimulated emission*. Here, when atoms are excited to a higher energy level, it is possible to force them to return to a lower level by shining electromagnetic radiation on them. This radiation must be of the right frequency – that is, the same frequency as the one that has to be emitted. A remarkable feature of this last process is that the emitted radiation is in phase with, and propagates in the same direction as, the radiation impinging on the atoms. As we will see, this process is a key to the operation of the *laser*, a device that produces an intense beam of light of a very pure single colour (see p. 292).

1918

Planck's radiation law

After the passing over of Planck for the 1908 Nobel Prize, the Nobel Physics Committee and the Academy seemed to harden their resistance to his candidacy further. Abraham Pais reports that, '... the importance but also the contradictions of quantum theory came more into focus from around 1910 on, so the award to Planck was postponed in the hope that the difficulties of the quantum theory could be sorted out.'[51]

Ten years had to pass before Stockholm could again give support to Planck's candidacy. Meanwhile, Planck received nominations almost every year. Finally, in 1919 he was awarded the prize for the year 1918. The Academy had found a solution, as can be seen by reading some passages from the presentation speech made by its president A. G. Ekstrand at the Nobel ceremony:

> Experience had to provide powerful confirmation . . . before Planck's radiation theory could be accepted. In the meantime this theory has had unheard-of success. The specific heat of substances, Stokes' Law for phosphorescence and fluorescence phenomena and the photoelectric effect provide, as Einstein has first suggested, most powerful support for Planck's radiation theory as against the older, usual concept. A still greater triumph was enjoyed by Planck's theory in the field of spectral analysis, where Bohr's basic work . . . provided an explanation for the enigmatic laws ruling within this part of science.[52]

Planck was presented with the prize on 1 June 1920. He began the traditional Nobel lecture with these words:

> When I look back to the time . . . when the concept . . . of the physical quantum of action began, for the first time, to unfold from the mass of experimental facts, . . . the whole development seems to me to provide a fresh illustration of the long-since proved saying of Goethe's that man errs as long as he strives. And the whole strenuous intellectual work of an industrious research worker would appear, after all, in vain and hopeless, if he were not occasionally through some striking facts to find that he had, at the end of all his criss-cross journey, at last accomplished at least one step which was conclusively nearer the truth.[53]

Through his long scientific career, Planck's main interest lay in thermodynamics. The second law of thermodynamics was the subject of his doctoral thesis, and it lay at the core of the researches that led him in 1900 to the revolutionary discovery of the quantum of action (see p. 37). Here is a brief story of that discovery.

Various attempts had been made to determine the energy spectrum for a given temperature of a black body. Planck was attracted to this subject by hearing of the investigations of a group of gifted experimentalists – Otto Lummer and Ernst Pringsheim, Heinrich Rubens and Ferdinand Kurlbaum – who were working at the PTR in Berlin–Charlottenburg during the late 1890s.

Some years before, Wilhelm Wien had published his energy distribution law (p. 78); Planck made a series of attempts to derive Wien's law by connecting the second law of thermodynamics to Maxwell's classical electromagnetism. However, in a meeting of the German Physical Society, in February 1900, Pringsheim showed that, in accord with findings made by himself and Lummer, Wien's law did not fit experimental data for long wavelengths. Moreover, results obtained by Rubens and Kurlbaum in the infrared region (long wavelengths) showed that here the intensity of radiation increased in proportion to the temperature – a result which Lord Rayleigh had arrived at on theoretical grounds at about the same time.

Planck learned of the results of Rubens and Kurlbaum on Sunday 7 October 1900, when Rubens himself and his wife were visiting Planck's house for tea. On receiving this information, he immediately started studying the theoretical implications, and in a few hours obtained a new formula. As a student of his recalled later:

> The same evening still . . . [Planck] reported this formula to Rubens on a postcard, which the latter received the following morning . . . One or two days later Rubens again went to Planck, and was able to bring him the news that the new formula agreed perfectly with his observations.[54]

So, in that very same month, Planck had a new radiation law in his pocket, one that correctly reproduced the entire spectrum of black-body radiation. He now wanted to find the physical explanation of his law, and so he adopted Boltzmann's statistical interpretation of the second law of thermodynamics. As Planck wrote:

> . . . even if the radiation formula should prove itself to be absolutely accurate, it would still only have . . . a strictly limited value. For this reason, I busied myself . . . with the task of elucidating a true physical character for the formula, and this problem led me automatically to . . . Boltzmann's trends of ideas; until after some weeks of the most strenuous work of my life, light came into the darkness, and a new undreamed-of perspective opened up before me.[55]

Thus Planck was able to explain his famous formula for the black-body spectrum at the meeting of the German Physical Society on 14 December 1900 (p. 37).

Max Planck, father of the quantum theory

> It is difficult to realise today, when the quantum theory is successfully applied in so many fields of science, how strange and almost fantastic this new conception of radiation appeared to many scientific men thirty years ago. It was difficult at first

Fig. 5.19. From left to right: W. Nernst, A. Einstein, M. Planck, R. Millikan, and M. von Laue. (Courtesy Archiv zur Geschichte der Max-Planck-Gesellschaft, Berlin–Dahlem.)

to obtain any convincing proof of the correctness of the theory and the deductions that followed from it. In this connection I may refer to experiments made by Professor Geiger and myself in 1908. On my side, the agreement with Planck's deduction of [the electron charge] . . . made me an early adherent to the general idea of a quantum of action. I was in consequence able to view with equanimity and even to encourage Professor Bohr's bold application of the quantum theory propounded by Planck.

(Ernest Rutherford)[56]

Scarcely any other discovery in the history of science has produced such extraordinary results within the short span of our generation as those which have directly arisen from Max Planck's discovery of the elementary quantum of action. This discovery has been prolific, to a constantly increasing degree of progression, in furnishing means for the interpretation and harmonising of results obtained from the study of atomic phenomena, which is a study that has made marvellous progress within the past thirty years. But the quantum theory has done something more. It has brought about a radical revolution in the scientific interpretation of natural phenomena . . . For having placed in our hands the means of bringing about these results the discoverer of the quantum theory deserves the unqualified gratitude of his colleagues!

(Niels Bohr)[57]

Max Planck (1858–1947)

Max Karl Planck was born in Kiel, Germany. He came from an academic family, his father having been a distinguished Professor of Constitutional Law at the University of Kiel, and both his grandfather and great-grandfather having been professors of theology at the University of Göttingen.

In 1874 Planck entered the University of Munich. He then studied in Berlin, under Hermann von Helmholtz and Gustav Kirchhoff. Later on he returned to Munich, and received his doctorate in 1879. He then began his academic career at the same university, and six years later he was appointed associate professor at the University of Kiel. In 1889, after the death of Kirchhoff, Planck succeeded him in the chair of theoretical physics at the world-famous University of Berlin, where he remained until his retirement.

In 1914 Planck succeeded in bringing Albert Einstein to Berlin, and later Max von Laue (Nobel 1914), his favourite student. When Planck retired in 1927, Erwin Schrödinger (Nobelist in 1933), the founder of wave mechanics, was chosen as his successor. So, for more than thirty years, Berlin was one of the most important centres of theoretical physics in the world – this remained so in fact until 1933 when the Nazis ascended to power and every sector of life, not excluding science, felt their deplorable influence.

Among German physicists Planck was in a position of great authority. From 1912 to 1938 he was permanent secretary of the mathematics and physics sections of the Prussian Academy of Sciences. 'When he was seventy-five years old, Planck saw Hitler coming to power. For a German patriot such as Planck, not blinded by the dictator's propaganda and parades, this was a severe blow.'[58] Furthermore, during the period 1930–7 (and subsequently in 1945–6) he also served as president of the Kaiser Wilhelm Society, now called the Max Planck Society, an institution that supports many sectors of German science ('[the] burden was heavy and, under the circumstances, most unpleasant, but Planck thought it was his duty to try to save what he could).'[59]

Planck was a man of deep philosophical and religious conviction. He withstood with fortitude many tragedies during a long life: his elder son, Carl, was killed in action during the First World War; his twin daughters both died in childbirth, and his younger son Erwin was accused of complicity in the plot to kill Hitler in July 1944 and was executed. In the same year Planck's house was completely destroyed in an air-raid on Berlin, and he lost all his possessions. He then moved to Göttingen, where he spent the remainder of his life: he died there on 4 October 1947, in his eighty-ninth year.

1919

New effects in atoms

The 1919 Nobel Prize for physics was awarded to Johannes Stark, a German physicist. He received the prize for two discoveries. The first was the observation of the Doppler effect for light in atoms (1905). His second discovery, in 1913, was his observation of the splitting of atomic spectral lines in an electric field.

Johannes Stark (1874–1957) was born in Schickenhof, Bavaria, Germany, the son of a landowner. He graduated in 1897 from the University of Munich. Between 1906 and 1922 he taught successively at the universities of Aachen, Hanover, Greifswald and Würzburg. In 1930 Stark joined the Nazi party, and in 1933 he became president of the PTR in Berlin–Charlottenburg; here he remained until the year 1939, when he was forced to resign. After the Second World War, he was sentenced to four years in a labour camp by a German de-Nazification court. (Stark and Philipp Lenard were among the most virulent opponents to Einstein's theories of relativity, which they considered 'Jewish physics'.)

Doppler and Stark effects

The pitch (frequency) of the whistle of a train is higher when the train is approaching us than when it has gone by. This well-known effect also applies to electromagnetic waves. When a light source is approaching an observer, the observer finds a higher frequency than that measured when the source is at rest, and sees the light colour shifted towards the violet end of the visible spectrum. On the other hand, if the light source is receding from the observer, the observer finds a lower frequency, and sees how the light colour has shifted towards the red. This phenomenon was first discovered in the 1840s by physicist, Christian Doppler, and it is called the *Doppler effect*.

There are differences, however, in the Doppler effect formulae for light and for sound. With light, the frequency shift is determined by the relative motion of source and observer, because light does not require a material medium for its transmission, and the speed of light relative to the source or the observer always has the same value (remember, the famous second principle of special relativity!) (With sound, however, the frequency shift depends on how the source or the observer is moving relative to the medium, for it is this which determines the speed of sound.) The Doppler effect for light is important when one considers astronomical or atomic sources, which, having high velocities, show pronounced frequency shifts.

Stark was the first to observe the Doppler effect for light in atoms. He used atomic sources called *canal rays*. These had been discovered in the 1880s by Eugen Goldstein, when he had used a discharge tube with a perforated cathode. He had seen a new radiation passing through the holes (canals) of the cathode in the direction opposite to the motion of the cathode rays themselves. This new radiation he therefore decided to call 'canal rays'. It was successively proved by Wilhelm Wien and J. J. Thomson that these rays were composed of positively charged ions of the gas in the discharge tube. In their collisions with the molecules contained inside the tube, they can be excited and emit light. Consequently, if the rays are approaching the observer, their spectral lines are displaced to the violet end of the spectrum. In 1905 Stark succeeded in observing this effect using a tube containing hydrogen. The same effect was later established for the canal rays of several other chemical elements.

The second series of experiments began in 1913, when Stark thought of investigating the influence of an electric field on the frequencies of the atomic spectral lines. He observed that when different kinds of canal rays (hydrogen or helium) are made to pass through a strong electric field, their spectral lines split into several components (different wavelengths). This effect is analogous to the Zeeman splitting of spectral lines when atoms are placed in a magnetic field (p. 40). The result of Stark's investigation was the discovery that an electric field causes changes in the motions of the atomic electrons and hence in the energy levels, but these changes differ from the analogous alteration under the influence of a magnetic field: as a consequence the spectral lines are split in a different way. As we have seen, the *Stark effect* was explained using Sommerfeld's theory (p. 111), and constituted one of the experimental evidences in support of Bohr's atomic model.

Modern alchemy

The artificial transmutation of atomic nuclei, the 'modern alchemy' as he named it, was Rutherford's last discovery at Manchester in 1919. He carried out these experiments alone, helped by his laboratory steward, as practically all the younger members of his team were engaged elsewhere in the war effort. The work was a continuation of the alpha-particle scattering experiments of Hans Geiger and Ernest Marsden (p. 80). Now, however, Rutherford wanted to investigate what happened when alpha particles collided with light atomic nuclei.

He mounted a radioactive source of alpha particles inside a vessel filled with a gas. A zinc-sulphide screen was used to observe scintillations from particles striking it. This screen was located far enough away to be beyond the distance over which alpha particles could travel in the gas contained inside the vessel, so that only scintillations from long-range particles, similar to hydrogen nuclei, would be observed.

Rutherford at first used oxygen as a gas, and he observed that the number of scintillations due to long-range particles decreased with the increase of the gas concentration. However, after admitting air into the vessel, he observed that the number of scintillations went up. He thought that this increase could only be due to the nitrogen in the air; it must mean that alpha particles had penetrated nitrogen nuclei, and had transformed them into oxygen nuclei, meanwhile knocking out

Fig. 5.20. Tracks of alpha particles in a cloud chamber. You can see (arrow) the disintegration of a nitrogen nucleus by an alpha particle: a proton of long range is emitted. (Courtesy The Cavendish Laboratory, University of Cambridge, England.)

hydrogen nuclei. Rutherford published his results in four papers, which appeared in June 1919 in the *Philosophical Magazine*, under the common title 'Collision of alpha particles with light atoms'.[60]

Rutherford's experiments provided the first example of an *artificially induced transmutation* of a chemical element. Moreover, they constituted the experimental proof that among the constituents of atomic nuclei there were hydrogen nuclei, or *protons* as they were subsequently named. The same results were obtained in the early 1920s in a Wilson cloud chamber. Here, collisions were occasionally seen, where the track of the striking alpha particle suddenly disappeared, and the tracks of new particles appeared.

Rutherford at the Cavendish

In 1919 J. J. Thomson retired after having directed the Cavendish Laboratory for thirty–five years. Rutherford succeeded him, thus becoming the fourth Cavendish Professor, following on James Clerk Maxwell, Lord Rayleigh and J. J. Thomson himself. Here are excerpts from two letters, one from Joseph Larmor and the other from J. J. Thomson. (Thomson replies to a letter in which Rutherford mentioned a possibility of accepting Larmor's invitation to take the Cavendish chair.)

Sir Joseph Larmor wrote to Rutherford . . . 'Our Vice-Chancellor, who is a man of rapid action, had ordered us to elect a new Cavendish Professor on April 2nd. It appears to be tacitly assumed that there would be no prospect of attracting you back to the scenes of your earlier youth. I wish there were.' . . . And he concluded, 'I confess to a strong personal longing that you might be available . . . to help make this the Imperial University that it is expected to be in the new scheme of things, and if I get any encouragement at all, I will not let it rest.' . . . Meanwhile Larmor was lobbying hard . . . [Days] later he wrote that he had been 'to see all the chief people concerned' and found that 'if you could come you would have a unanimously enthusiastic welcome'.

(From Joseph Larmor to Rutherford)[61]

J. J., however, was proving remarkably amenable. He replied promptly to Rutherford's letter: 'I am very glad you are still entertaining the possibility of coming to Cambridge . . . If you do you will find that I shall leave you an absolutely free hand in the management of the laboratory' . . . He repeated his promise to leave Rutherford entirely independent and emphasised that he would ' . . . never dream . . . of expressing any opinion about matters of policy'. And he ended with a postscript: 'There is a very keen hope that you may see your way to come to Cambridge. Nothing would give me so much pleasure as to have for my successor my most distinguished pupil.'

(From J. J. Thomson to Rutherford)[62]

The 1919 solar eclipse

In 1916, when Einstein published his theory of general relativity, Germany and Britain were at war. Nevertheless, the Dutch astronomer Willem de Sitter succeeded in smuggling a copy of Einstein's paper out to his colleague Arthur Eddington, who was working at the Royal Observatory at Greenwich. Eddington proposed that the new theory should be tested. Sir Frank Dyson, then Astronomer Royal of Great Britain, suggested the date of 29 May 1919, when a solar eclipse was to occur. On that day the sun was to be in a region of the sky (the Hyades) populated by bright stars that would be distinctly observable against the solar corona, during the total eclipse of the sun. The light of these stars passed close to the sun on its way to the earth, so its bending could be detected. Eddington organised two expeditions, one to Príncipe Island in the Gulf of Guinea, and the other to Sobral in Brazil (the path of the total eclipse ran across Africa and South America).

Both expeditions obtained photographs showing the apparent positions of the stars as modified by the deflection of their light near the sun. These were then compared with the positions of those same stars as observed when the sun was in a different place in the sky. The results of the analysis of the photographs gave strong support to Einstein's general relativity. Against his predicted deflections of 1.75 seconds of arc the observations yielded 1.98 (at Sobral) and 1.61 (at Príncipe Island) with an uncertainty of about 30 per cent. The news of the success reached Einstein in a telegram from his friend Hendrik Lorentz. In a postcard to his mother, dated 27 September 1919, Einstein wrote: 'Dear Mother, good news today. H. A. Lorentz has wired me that the British expeditions have actually proved the light deflection near the sun . . .'[63]

Newtonian ideas overthrown

> Yesterday afternoon in the rooms of the Royal Society, at a joint session of the Royal and Astronomical Societies, the results obtained by British observers of the total solar eclipse of May 29th were discussed. The greatest possible interest had been aroused in scientific circles by the hope that rival theories of a fundamental physical problem would be put to the test, and there was a very large attendance of astronomers and physicists. It was generally accepted that the observations were decisive in the verifying of the prediction of the famous physicist, Einstein, stated by the president of the Royal Society [J. J. Thomson] as being the most remarkable scientific event since the discovery of the predicted existence of the planet Neptune. But there was difference of opinion as to whether science had to face merely a new unexplained fact, or to reckon with a theory that would completely revolutionise the accepted fundamentals of physics . . . Sir Frank Dyson, the Astronomer Royal, described the work of the expeditions sent respectively to Sobral in North Brazil and the island of Principe, off the West Coast of Africa.
>
> (From the *Times*, London, 7 November 1919)

Einstein and the Physics Committee

Starting from 1910, the number of recommendations to the Nobel Physics Committee in favour of Einstein grew ever more – mostly regarding his *special relativity*. In 1917 nominations began to arrive also for his *general relativity*. Then, after the 1919 solar eclipse experiment, gravitation became one of the main motivations in the nominations in his favour (among others, Hendrik Lorentz, Pieter Zeeman and Heike Kamerlingh Onnes emphasised the successes both of the Mercury perihelion and of the bending of light near the sun). In 1920 Svante Arrhenius, one of the members of the Committee, was given the responsibility for preparing a report concerning the results that the English astronomers had obtained from the 1919 eclipse of the sun. In his report Arrhenius raised strong objections with regard to the accuracy of those observations. Thus the Committee concluded that for the time being *general relativity* was not deserving of an award. (Today, we could well imagine Svante Arrhenius agreeing that now Einstein is 99.998 per cent right (see p. 108), and he would have been quite certain to have proposed him for a Nobel Prize for relativity!)

1920

Metrology

Besides Einstein's candidacy, other theoretical physicists were nominated for the 1920 Nobel Prize (Niels Bohr and Arnold Sommerfeld for example, who had already received nominations in 1917 and 1918). But it was the French physicist Charles Guillaume who won the race. He had received only one nomination, from Bernhard Hasselberg, an experimentalist from Uppsala University, and a member of the Physics Committee. ('[In 1920] Hasselberg . . . lay gravely ill. Having championed precision measurement, especially in spectroscopy, Hasselberg now favoured his colleague on the International Bureau of Weights and Measures, Guillaume, whose work in metallurgy was of importance for metrology. To express appreciation for Hasselberg's work on the Committee since 1901, the Academy . . . [was able] to pay homage to him in this manner.')[64]

Charles Edouard Guillaume (1861–1938) was born at Fleurier, in the Swiss Jura, and obtained his doctorate in physics at the Swiss Federal Institute of Technology (ETH) in Zurich. He then joined the International Bureau of Weights and Measures at Sèvres near Paris as an assistant; later becoming associate director in 1902, and then director in 1915.

In 1890 he started a search to find a substitute for the platinum–iridium alloy for the standard metre prototypes. His investigations on nickel–iron alloys led him to the discovery of *invar*, a nickel alloy with a coefficient of expansion far lower than any previously known metal. Apart from metrology, the new material was widely used in the construction of precision instruments.

Atomic nuclei

Isotopes
Rutherford and Frederick Soddy discovered in 1902–3 that when radioactive atoms emit particles (alpha particles or electrons), they are changed into other varieties of atoms (p. 42). This inspired scientists to search for more such transformations. They discovered that there exist atoms of the same chemical element having different atomic weights. These atoms, which have very nearly the same chemical properties, share the same place in the Mendeleev periodic system: they were called *isotopes* (from the Greek words meaning 'equal').

In 1913 Soddy, working at Manchester under Rutherford, gave clear expression to the concept of the isotope. He showed that when a radioactive atom emits an alpha particle (involving a loss of two units of positive electric charge), it changes into an element two places lower in the periodic system (the atomic number Z decreases by two); and when an atom emits a beta particle (loss of one unit of negative charge), it changes into an element one place higher (Z increases by one). This was called the *law of radioactive displacement*.

Not long after the existence of radioactive isotopes had been demonstrated, there came indications that these might exist among ordinary stable elements as well. Proof of their existence came from J. J. Thomson; these types of investigations resulted in the development of the *mass spectrograph*, a device constructed by an assistant of Thomson's, named Francis Aston. (This instrument separated isotopes according to the differences in deflection of their ions by a magnetic field.) Both Soddy and Aston received the Nobel Prize for chemistry, Soddy in 1921 and Aston in 1922.

The constituents of the nucleus

To continue our brief review of the concept of the atomic nucleus, let us read some passages of the *Bakerian Lecture* given by Rutherford at the Royal Society in 1920. After describing the concept of the nuclear atom, he examined his experiments of 1919 on the transmutation of nuclei (p. 117), and considered the problem of their constitution. First, he proposed the *proton–electron* model; that is, he thought of an atomic nucleus as built up of hydrogen nuclei (protons) and electrons. He then hypothesised the existence of a new constituent of the nucleus, a neutral particle with about the same mass as the proton, which was later named a *neutron*: '. . . it seems very likely that one electron can . . . bind one H nucleus [proton] . . . [It] involves the idea of the possible existence of an atom of mass [one] which has zero nucleus charge.'[65]

In the early 1920s Rutherford, and his pupil James Chadwick, found that alpha particles, when approaching very close to hydrogen nuclei, were in no way scattered as predicted by the formula he had derived more than ten years before to explain Geiger and Marsden's results (p. 80). Did there perhaps exist another unknown force, acting in the domain of the atomic nucleus? Chadwick, in a paper written in 1921 with his colleague Etienne Bieler, thus remarked: 'The present experiments do not seem to throw any light on the nature of the law of variation of the forces at the seat of an electric charge, but merely show that the forces are of very great intensity.' (As noted by Abraham Pais, this statement can be considered as marking the birth of the discussion about the strong nuclear force.)[66] Attempts to understand this new force during the 1920s suffered from confusion over what particles actually constitute the nucleus. The answer came in the 1930s, after Chadwick himself had discovered the neutron (see Nobel 1935).

Einstein in Berlin

> In the spring of 1913 Planck and Nernst had come to Zurich for the purpose of sounding out Einstein about his possible interest in moving to Berlin. A combination of positions was held out to him: membership in the Prussian Academy with a special salary . . . a professorship at the University of Berlin . . . and the directorship of a physics institute to be established . . . Encouraged by Einstein's response, Planck, Nernst, Rubens, and Warburg joined in signing a formal laudatio . . . which was presented to the academy on June 12 1913. On July 3, the physics–mathematics section voted on the proposal. The result was twenty-one for, one against.[67]

The following excerpt from the *Times*, London (1920), reflects the attacks that Einstein suffered – due to his relativity theories and his political position – soon after the 1919 eclipse results.

Fig. 5.21. The Prussian Academy of Sciences on Unter den Linden, Berlin (1914–49). (Courtesy Berlin–Brandenburgische Akademie der Wissenschaften.) Einstein was a member of the Academy in the period 1913–33.

Professor Einstein is so much disgusted by attacks made upon him by his scientific colleagues that he proposes, says the Tageblatt, to leave Berlin altogether. The newspaper makes a strong protest against the annoyance to which Professor Einstein has been subjected, which it describes as disgraceful. It is the duty of the Berlin University to do all in its power to keep Professor Einstein. Everyone who desires to maintain the honour of German science in the future must now stand by this man.

Chapter 6
The golden years

Just as the 1920s are sometimes described as the 'roaring twenties', so these are truly 'golden years' for science, their glory reaching a peak in the second towering achievement of twentieth-century physics (the first one being *relativity*) – namely, the theory of *quantum mechanics*, which emerges in the same period.

At the beginning of the decade it became increasingly clear that the 'old quantum theory' of Niels Bohr and Arnold Sommerfeld was in a critically serious state, and that it had to be replaced. In spite of its success in describing the hydrogen spectrum, it was inadequate for explaining many other phenomena, including the light emitted and absorbed by complex atoms and by molecules, and the nature of light itself. 'We are speaking a language that does not adequately match the simplicity and beauty of the world of quanta', noted Wolfgang Pauli.

Then the second wave of the quantum revolution arrived, exploding in a flurry of discoveries: in 1922–3 Arthur Compton got experimental evidence of the existence of *light quanta* (Einstein's old hypothesis, which had been for long doubted by most physicists), and in 1924 Louis de Broglie put forwards his bold idea of *matter waves*. Within the space of a few years, 1925–7, an entirely new mechanics, called 'quantum mechanics', was formulated by Werner Heisenberg, Erwin Schrödinger, Paul Dirac, Max Born and others. The principal locations for the study of quantum mechanics were then Göttingen and Copenhagen, and Niels Bohr was its leading interpreter.

Quantum mechanics is truly revolutionary; it provides a new language for describing nature. It is highly successful in accounting for most phenomena concerning atoms and molecules, but is equally successful in dealing with atomic nuclei and bulk matter. It also provides a basis for a new theory of radiation. Additionally, the combination of quantum mechanics and relativity yields new and unexpected results.

The 1921 and 1922 Nobel Prizes, awarded respectively to Einstein and Bohr, show that quantum physics has now successfully come of age. Other awards are bestowed upon eight experimentalists, rewarding discoveries or inventions, the majority of them being made in the preceding decade; they are principally in the fields of atomic physics and spectroscopy. The discovery of the 'Compton effect' was promptly recognised in 1927, and in 1929, with the award to de Broglie, the new quantum theory began to play an increasingly important role on the Nobel stage.

1921

Einstein honoured

The Nobel Prize for physics was at long last awarded to Einstein for the year 1921. But the announcement was actually made in November 1922. The Nobel Physics Committee and the Academy had finally responded to the ever-growing number of recommendations that were in his favour. The prize was quoted as being given 'for his services to theoretical physics, and especially for his discovery of the law of the photoelectric effect'. Relativity was not mentioned in the official citation. It was still regarded as too controversial. But let us go and see what had been happening in the Committee.

After Arrhenius' 1920 report on general relativity (p. 120), Allvar Gullstrand (the ophthalmologist who had won the 1911 Nobel Prize for physiology, p. 78) wrote in 1921 a fifty-page report for the Committee concluding that '... neither the general nor the special theory of relativity warranted a Nobel Prize.'[1] Thus, '... claiming that they could not find any grounds for awarding a prize to Einstein, the Committee, followed by the Academy, voted to reserve the 1921 prize.'[2] In 1922 Gullstrand '... brought his report up-to-date and came to the same conclusion. Acceptance of these theories remained simply "a matter of faith (*trossak*)". He resolved that Einstein must never receive a prize.'[3] The Committee then asked the young theoretical physicist Carl Oseen from Uppsala University (he was an old friend of both Bohr brothers) for a report on Einstein's 1905 paper concerning the concept of light quanta and the photoelectric effect. Oseen presented a very favourable report on this matter: so the Committee proposed Einstein for the deferred 1921 prize, and subsequently the Academy voted in the same way. It was thus that the Einstein *affaire*, too, was happily concluded.

Let us now have a look at Arrhenius' presentation speech at the 1922 Nobel ceremony, to see if we can learn a bit more about Einstein's award. He started with these words:

> There is probably no physicist living today whose name has become so widely known as that of Albert Einstein. Most discussion centres on his theory of relativity. This pertains essentially to epistemology and has therefore been the subject of lively debate in philosophical circles. It will be no secret that the famous philosopher Bergson in Paris has challenged this theory, while other philosophers have acclaimed it wholeheartedly. The theory in question also has astrophysical implications which are being rigorously examined at the present time.[4]

And he concluded thus:

> Einstein's law of the photoelectric effect has been extremely rigorously tested by the American Millikan and his pupils and passed the test brilliantly. Owing to these studies by Einstein the quantum theory has been perfected to a high degree

Fig. 6.1. Albert Einstein, 1920. (Courtesy Niels Bohr Archive, Copenhagen.)

and an extensive literature grew up in this field whereby the extraordinary value of this theory was proved.[5]

Thus relativity was rigorously excluded. Let us read the letter that Christopher Aurivillius, secretary to the Swedish Academy of Sciences, wrote to Einstein on 10 November 1922:

> ... the Royal Academy of Sciences decided to award you last year's [1921] Nobel Prize for physics, in consideration of your work on theoretical physics and in particular for your discovery of the law of the photoelectric effect, but without taking into account the value which will be accorded your relativity and gravitation theories after these are confirmed in the future.[6]

Einstein was not in Berlin at that time. He and his wife were on their way to Japan: he would not be back until March 1923. Shortly afterwards, the Swedish ambassador to Germany personally took the diploma and medal to Einstein in Berlin. When, in July of the same year, Einstein gave his Nobel lecture to the Nordic Assembly of Naturalists at Göteborg, Sweden, in the presence of the King of Sweden, he spoke on his theories of relativity. ('On a very hot day in July, Einstein, dressed in black redingote, addressed an audience of about two thousand in the Jubilee Hall in Göteborg on "basic ideas and problems of the theory of relativity". King Gustav V, who was present, had a pleasant chat with Einstein afterwards.')[7]

Albert Einstein (1879–1955)

Albert Einstein, one of the giants of science, was born on 14 March 1879 at Ulm, Württemberg, Germany, the son of Jewish parents. In 1896 he entered the Swiss Federal Institute of Technology (ETH) in Zurich to study physics and mathematics.

After his graduation in 1900, he was appointed to the post of technical expert at the patent office in Bern (1902–9). In 1905 he received his doctorate from the University of Zurich, becoming in 1909 Extraordinary Professor there. In 1911 he was appointed full professor at the University of Prague, and in the following year he came back to Zurich to fill a similar position at the ETH. In 1914 Einstein left Zurich to move to Berlin, where he was appointed professor of theoretical physics at the university, member of the Prussian Academy of Sciences and director of the Kaiser Wilhelm Institute of Physics. When the Nazis gained power in 1933, he left Germany and never returned. He emigrated to the USA, where he accepted a position of professor of theoretical physics at the Institute for Advanced Study in Princeton, New Jersey; here he remained until his retirement in 1945.

In the first two decades of the twentieth century Einstein advanced two remarkable theories – the *special theory of relativity* (1905) and the *general theory of relativity* (1915–16) – in which he revolutionised our way of thinking about space, time, motion and gravitation. These theories played a central role in the development of twentieth-century physics. He also made important contributions to quantum theory, in particular his *light-quanta hypothesis* (1905), the concept of *stimulated emission* in atoms (1916–17), and the development of the *Bose–Einstein statistics* (1924–5).

In the 1920s Einstein became the target of strong attacks by anti-Semitic and right-wing circles, angered mainly by his full support to the pacifist and Zionist movements; even his scientific work was publicly called into question, especially his theories of relativity. Following on those years his interests centred on the search for a unification of electromagnetism and gravitation, in what came to be called a *unified field theory*. This remained for the rest of his life an unsuccessful quest. Einstein died on 18 April 1955 at Princeton.

> Einstein was a unique personality. He was not attracted by fame or fortune nor swayed by the opinions of the majority. He knew his talent and guarded it jealously against outside interference. Although fearless in support of any cause he considered worthy, he gave only so much of himself and no more. Physics was his life, and he lived it according to his own lights, with complete objectivity and integrity.
>
> (Isidor Rabi)[8]

The Stern–Gerlach experiment

Electrons orbiting in an atom are equivalent to tiny current loops, each producing a magnetic field. Hence a resultant magnetic field is generated, and the atom can be imagined as a *magnetic dipole*, whose strength is measured by what is called its *magnetic moment*. Arnold Sommerfeld had shown (p. 110) that the angular momentum of an atom (the sum of all the angular momenta of its orbiting electrons) can have only certain allowed orientations in space (the so-called 'spatial quantization'). As a consequence the associated magnetic dipole too – whose magnetic moment points in the opposite direction – can have only specific orientations.

If you put a magnetic dipole between the poles of a magnet, the magnetic field there creates forces on it. A uniform magnetic field (when there is the same value at every point) does not move the dipole. However, if the field is not uniform, the dipole itself is slightly deflected. Otto Stern (Nobel 1943) and Walther Gerlach, two German physicists, began a historic experiment in 1921 at the University of Frankfurt. They directed a beam of silver atoms through a slit, then through a non-uniform magnetic field produced by a wedge-shaped magnet. Just beyond the magnet the beam of atoms could strike a glass plate, and then be observed on a photographic film. In the absence of the magnetic field, they saw a thin line, in the shape of the slit, on the film. When they turned the non-uniform magnetic field on, the line split into two distinct traces (see Fig. 6.2). This meant that the atoms had been deflected according to the orientation of their magnetic moments with respect to the field (the two traces correspond to just two orientations in space).

Thus the Stern–Gerlach experiment demonstrated in a direct way the idea of spatial quantization. At the same time it showed up an anomaly in the

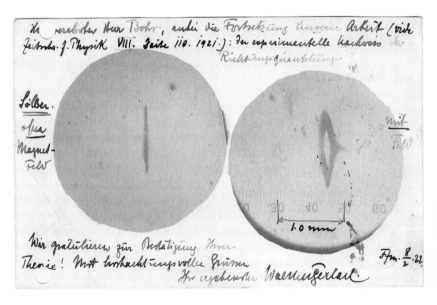

Fig. 6.2. The first splitting of a beam of atoms in the Stern–Gerlach experiment.[9] (Courtesy Niels Bohr Archive, Copenhagen.)

Fig. 6.3. A. Sommerfeld (left) and N. Bohr, 1919. (Courtesy Niels Bohr Archive, Copenhagen.)

Bohr–Sommerfeld theory, because the beam was split into two traces, instead of the three traces that had been predicted.

1922

Bohr honoured

The theory of atomic structure that had been evolving ever since 1913, known as the 'old quantum theory', was recognised in 1922 when the Nobel Prize was awarded to Niels Bohr.

A national hero

> From the 1920s on Bohr was not only a man of renown in scientific circles but also a national hero. 'There is a story that a young man arrived in Copenhagen and took a taxi to Bohr's institute; and the taxi man wouldn't take any money because it was to Bohr's that he had driven.'[10]

The Nobel Prize. At the Nobel ceremony, the veteran Arrhenius thus concluded his second speech (the first was for Einstein, p. 126):

> Your great success has shown that you have found the right roads to fundamental truths, and in so doing you have laid down principles which have led to the most splendid advances, and promise abundant fruit for the work of the future. May it be vouchsafed to you to cultivate for yet a long time to come, to the advantage of research, the wide field of work that you have opened up to Science.[11]

Bohr's trilogy of 1913 marked the birth of the new branch of quantum theory that concerned atomic structure. Experimental evidence strongly supported this theory as early as 1913 and 1914 (Moseley's work on X-ray spectra, the Franck–Hertz experiment, and the Stark effect, see pp. 98, 116). On the theoretical side, Arnold Sommerfeld made decisive progress during the period 1914–17 (p. 110), when he extended and perfected Bohr's model of the atom. These developments played an important role in the interpretation of many details in the structure of the atomic spectra.

While the so-called 'old quantum theory' had given satisfactory results in explaining several atomic phenomena, it seemed to encounter serious difficulties in the treatment of others. For instance, the theory failed to give a satisfactory description of the spectrum of the helium atom. Serious difficulties were also caused by the so-called *anomalous Zeeman effect* (a more complex splitting of spectral lines in a magnetic field, different from the *normal Zeeman effect* discovered in 1896, p. 40); and, as we have seen (p. 129), there was the two-trace anomaly in the Stern–Gerlach experiment. Moreover, studies on the spectra of complex atoms showed that it was necessary to change some of the rules that were being used. These difficulties, in fact, could only be overcome by a radical change in the theory. Bohr himself was aware of the fact that his theory contained simplifications. He concluded his Nobel lecture with these words:

> I have attempted to show how the development of atomic theory has contributed to the classification of extensive fields of observation, and by its predictions has pointed out the way to the completion of this classification. It is scarcely necessary, however, to emphasise that the theory is yet in a very preliminary stage, and many fundamental questions still await solution.[12]

The correspondence principle

In searching among the laws of atomic theory, Bohr was guided by the *correspondence principle* (so named by him in a lecture given at the University of Berlin in 1920). His idea was that the difference between the formulae of quantum theory and those of classical mechanics becomes less relevant in proportion as the physical system in question grows larger in size. It becomes negligible for systems of ordinary size. In other words, classical mechanics is just as good as quantum mechanics when applied to the motion of macroscopic objects.

An example can clarify the concept. The angular momentum of a spinning golf ball is about a million billion billion billion times greater than the angular momentum of an atom (this last is of the order of h, the Planck quantum of action). This means that the energy levels of the golf ball's quantum states lie so close together that they almost form a 'continuum', and that therefore the effect of quantization is meaningless. As a consequence, any value of the angular momentum is permissible for a golf ball, as required by classical mechanics.

Bohr used his correspondence principle, quantum numbers and the results derived from spectroscopic experiments to devise a building-up principle (the so-called *Aufbauprinzip*) for the periodic system of the elements. He presented his findings at the *Bohr Festival* at the University of Göttingen in June 1922; among other things, he was able to foresee the properties of a new element, which would have the atomic number $Z = 72$. This element was in fact discovered at the end of 1922 at Bohr's institute in Copenhagen, in an experiment carried out by his old friend George de Hevesy and by Dirk Coster (it was called *hafnium*, from *hafnia*, which means 'Copenhagen'). Bohr himself in Stockholm announced this discovery. ('While Coster telephoned these results through to Bohr, Hevesy took the train to Stockholm to be in time for Bohr's announcement at [his] Nobel lecture.')[13]

The Niels Bohr Institute

In September 1916 Bohr became professor of theoretical physics at the University of Copenhagen. In 1920 a new research institute was established for him. The official inauguration of the University Institute for Theoretical Physics (in 1965 it was renamed the Niels Bohr Institute) took place on 3 March 1921. Bohr concluded his address with these words, which were to become the institute's main task; that is, '. . . to introduce a constantly renewed number of young people into the results and methods of science . . .'.[14] About forty years later he thus recalled the first days of the institute:

> An auspicious start was given . . . by a visit . . . of James Franck [Nobelist in 1925], who . . . most kindly instructed the Danish collaborators in the refined techniques of excitation of atomic spectra by electron bombardment . . . The first among the many distinguished theoretical physicists who stayed with us for a longer period was [Hendrik] Kramers who as a quite young man came to Copenhagen during the war and proved to be such an invaluable asset to our group . . .[15]

And now for the names of a few brilliant young physicists who came to Copenhagen during the 1920s and 1930s: Pascual Jordan, Wolfgang Pauli and Werner Heisenberg from Germany; Paul Dirac and Nevill Mott from Cambridge, England; Victor Weisskopf from Austria; George Gamow and Lev Landau from Russia; J. Robert Oppenheimer, John Slater, Harold Urey, Isidor Rabi and Linus Pauling from the USA; George Uhlenbeck from the Netherlands; Oskar Klein from Sweden; and Yoshio Nishina from Japan. In the first twenty years following its foundation, sixty-three physicists from seventeen countries visited Bohr's institute. And no less than thirteen of them were Nobel prizewinners.

Niels Bohr (1885–1962)

Niels Henrik Bohr was born in Copenhagen, Denmark. His father was a famous physiologist, and a professor at the university; his mother came from a wealthy Danish-Jewish family. Bohr's scientific interests, his faculties and skills, grew up in an intellectually open and stimulating environment. Bohr studied mathematics, philosophy and physics at the University of Copenhagen; he took his master's degree there in 1909 and his doctorate in 1911.

In the autumn of the same year he went to Cambridge to study under J. J. Thomson, and in the spring of 1912 he moved to Manchester to join Rutherford's group. At Manchester, Bohr started to investigate the theoretical implications of the nuclear atom proposed by Rutherford, and he worked out a new and original description of atomic structure, known as the *Bohr atomic model*. In 1916 he was appointed to a chair of theoretical physics, which was created expressly for him at the University of Copenhagen, and from 1921 on he served as director of the Institute for Theoretical Physics, which remained for a long time the most important meeting point for the world's young theoretical physicists.

During the middle years of the 1920s a new quantum mechanics evolved. Bohr participated actively in discussions about the foundations of this new description of nature, which took place with the numerous visitors to his institute. During the 1930s he continued to study the methods and concepts pertaining to the quantum theory, and he also devoted himself to the new field of nuclear physics (he developed a new model of the atomic nucleus, and actively promoted a fuller understanding of the mechanism of nuclear fission.)

During the Second World War Bohr and his family escaped from German-occupied Denmark to Sweden, afterwards to England, and then later still to the USA. There he and his son Aage (Nobelist in 1975), code-named 'Mr Nicholas Baker and Jim Baker', became associated with the Manhattan Project. On his return to Copenhagen in 1945 Bohr helped to contribute to a wider diffusion of the peaceful applications of atomic and nuclear physics. He was a founder of CERN, the international laboratory for particle physics near Geneva, Switzerland, and later on he became interested in molecular biology. The number of honours that Bohr received was enormous: he was a member of more than twenty academies, and received *honoris causa* degrees from more than thirty universities.

> By the rigour of his rational thinking, the universality of his outlook and his deep humanity, Bohr will for ever rank, in the judgement of history, among the few fortunate men to whom it was given to help the human mind to a decisive step towards a fuller harmony with nature.
>
> (L. Rosenfeld)[16]

1923

American science

The scene of operations, where physics was concerned, began to shift from Europe to the USA during the second and third decades of the twentieth century. Professor Robert Millikan, director of the Norman Bridge Laboratory of Physics, at the California Institute of Technology (Caltech) in Pasadena, was awarded the Nobel Prize for physics in 1923, so becoming the second American Nobelist, after Albert Michelson (Nobel 1907). The work for which the Nobel was conferred upon him concerned his experimentation on the photoelectric effect (made in the early 1910s), and also his accurate determination of the electron charge (research dating from 1909 to 1917, when Millikan was at the University of Chicago).

Robert Andrews Millikan (1868–1953) was born in Morrison, Illinois, USA, the son of a Congregational minister. In 1895 he obtained his Ph.D. from Columbia University, New York City. He then went to Europe to study physics at Göttingen and in Berlin. In 1896 he returned to the USA, to the University of Chicago, where in 1910 he became a full professor of physics.

In 1909 Millikan undertook an experiment aimed at determining the electric charge carried by an electron. He began to measure the rate at which charged water droplets fell in an electric field. But the experiment was fraught with too many uncertainties to be convincing; however, he did obtain more precise results in a series of experiments that he made later on between 1910 and 1917.

From 1912 to 1915 Millikan also carried out, with his customary prowess, the experiments that verified Einstein's 1905 equation for the photoelectric effect, and, from an accurate analysis of the data, he obtained a very precise value for the Planck constant h (see p. 38). These experiments paved the way to Einstein and Bohr's Nobel Prizes, as one can see from the conclusion of the presentation speech at Millikan's Nobel ceremony: '. . . if these researches of Millikan had given a different result, the law of Einstein would have been without value, and the theory of Bohr without support. After Millikan's results both were awarded a Nobel Prize for physics . . .'.[17]

In 1921 Millikan left the University of Chicago to go to Caltech, where he undertook a major study of the cosmic radiation that Victor Hess had discovered in 1912 (see p. 86); he proved that this radiation is indeed of extraterrestrial origin. During Millikan's chairmanship of the executive council (1921–45), Caltech became one of the leading scientific institutions in the USA.

The oil-drop experiment

Electric charge is one of the fundamental properties of the particles of which matter is made. (The interactions responsible for the structure and properties of atoms and molecules, and indeed of all ordinary matter, are primarily electric interactions between the electrons of the atoms.) The negative charge of each electron is found by experiment to have the same magnitude, and this is also equal to that of the positive charge of each proton in the atomic nucleus. Electric charge exists in natural units equal to the charge of an electron or a proton, called the *elementary charge*, and denoted by the letter e. Like h (Planck's quantum of action) and c (the speed of light in empty space), the elementary charge is a universal physical constant, and each amount of any observed electric charge is always an integer multiple of this basic unit. In about 1910 Millikan devised a new technique for determining the electron charge, and soon undertook an experiment, which is still known as the *oil-drop experiment* by all students of physics. Let us describe it succinctly.

Millikan measured the electric charge that was present on tiny oil drops (their diameter being of the order of one thousandth of a millimetre). He sprayed these droplets into an electric field between two charged plates of a capacitor (see Fig. 6.4): the air in the chamber was ionised by exposing it to X-rays, or to radiation from a radioactive source, so that the droplets would pick up free electrons from the air and become electrically charged. (There are two vertical forces acting on a charged droplet placed in an electric field: the electric force and the weight of the droplet itself. These forces act in opposite directions, so that the droplet will be in equilibrium when the two forces are balanced.)

Millikan monitored the vertical motion of a droplet with a microscope, and, by adjusting the voltage of the capacitor, he was able to balance the two forces, and so hold the droplet stationary. He was then able to determine the minute charge on each droplet by measuring the balance voltage, and he proved that the charge on the many droplets he had examined consisted always of integers – multiples of an elementary quantum of electricity e, with values equal to e, $2e$, $3e$, and so on. This proved without ambiguity to be one of the most important facts in physics, namely that the 'electric charge is quantised'. (By 1913 Millikan had measured the elementary charge e with an uncertainty of one in a hundred, and by 1917 with an uncertainty of one in a thousand. Today's precision is better than one part in a million.) Millikan thus wrote:

> I have observed . . . the capture of many thousands of ions in this way, and in no case have I ever found one, the charge of which did not have either exactly the value of the smallest charge ever captured, or else a very small multiple of that value. Here, then, is direct unimpeachable proof that . . . the electrical charges found on ions all have either exactly the same value, or else small exact multiples of that value.[18]

The Compton effect

Experiments carried out by the American physicist Arthur Compton between 1922 and 1923 gave evidence for the particle nature of light. Compton projected an X-ray beam through a thin layer of matter, and discovered that some of the scattered X-rays emerged with a longer wavelength (that is, a lower frequency).

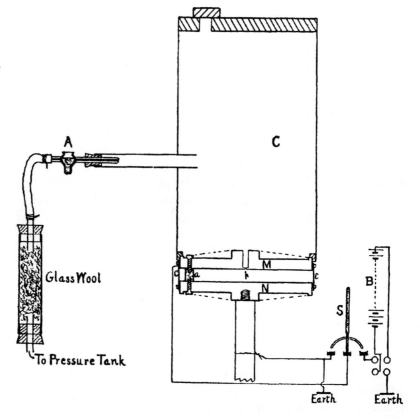

Fig. 6.4. A schematic diagram of Millikan's apparatus. (Source: R. A. Millikan, *Physical Review*, 32, 1911, p. 352. ©1911 by the American Physical Society.)[19]

Fig. 6.5. Mrs Greta Millikan and Professor Millikan. (Courtesy AIP – Emilio Segrè Visual Archives, Francis Simon Collection.)

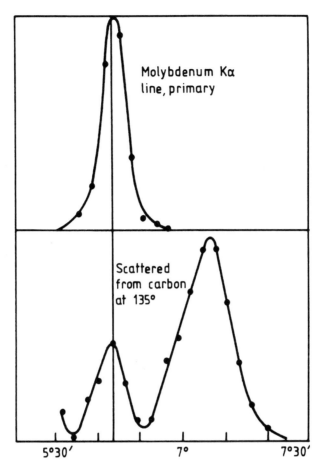

Fig. 6.6. Compton's results. The upper curve represents the intensity of X-rays of different wavelengths, striking the metal foil. The lower curve represents X-rays scattered back at 135 degrees. (Source: Arthur Compton's Nobel lecture, 1927; © The Nobel Foundation, Stockholm.)[20]

From the upper curve in Fig. 6.6, one can see the distribution of wavelengths of X-rays which are not scattered. In the lower curve, the right-hand peak represents X-rays scattered at large angles; these rays emerge from the metal foil with longer wavelengths. How can this fact be explained?

Classical electromagnetism gives no explanation of it. Compton himself (as well as Peter Debye in Zurich) interpreted the effect using Einstein's 1905 light-quanta hypothesis and special relativity. Compton thought of it as the effect of a collision between an X-ray quantum (a photon) bouncing off an electron of an atom contained in the target, like the collision between two billiard balls. In the collision the photon gives some of its energy and momentum to the electron, which is pushed forward. The scattered photon veers off, loses energy, and emerges with a lower frequency – that is, a longer wavelength – than the striking photon. (Einstein's formula, $E = h\nu$, says in fact that if, for example, you halve the energy, you obtain a halved frequency.)

Fig. 6.7. Karl Manne Siegbahn. (© The Nobel Foundation, Stockholm.) In addition to the Nobel Prize, Siegbahn received the Hughes and Rumford medals from the Royal Society of London (of which he was a member). He was also a member of the French Academy of Sciences, and from 1923 to 1950 he served as a member of the Nobel Physics Committee.

1924

X-ray spectroscopy

Manne Siegbahn, from Uppsala University, Sweden, won the 1924 Nobel Prize for physics. He was the second Swede to receive the prestigious accolade, after Nils Dalén in 1912. His work was done in the field of X-ray spectroscopy, and carried on the line of research that had begun with Max von Laue's notable discovery in 1912 (p. 95); it was then successively developed by the Braggs (Nobel 1915) and Henry Moseley (p. 98); and by Maurice de Broglie (Louis' brother) in Paris.

Karl Manne Siegbahn (1886–1978) was born in Örebro, Sweden. He studied at Lund University, where he obtained his doctorate in 1911. He became a research assistant to Johannes Rydberg there, and later, in 1920, succeeded him in the chair of physics. He was professor of physics at Uppsala University from 1923 until 1937, when he moved to the University of Stockholm. It was at that time that the Swedish Academy of Sciences created the Nobel Institute of Physics in Stockholm, and they appointed Siegbahn as its first director.

After Moseley's experiments in 1913–14, it became clear that the characteristic X-rays discovered by Charles Barkla in earlier years (p. 98) arose from the inner parts of the atom. ('Chemistry and the majority of physical effects, in particular optical spectra, happen in the outer courts of this structure [the atom]. Characteristic Röntgen rays, on the contrary, come from the innermost chamber, the "holy of holies"': with these words was the atom metaphorically described by Arnold Sommerfeld.)[21] X-ray spectroscopy, the study of X-ray messages precisely from this 'holy of holies', provided information on the atomic electron structure close to the atomic nucleus, and played a significant role in its understanding. Siegbahn and his co-workers developed innovative instrumentation that significantly improved the precision of measurement of X-ray wavelengths. They discovered a large number of new wavelengths in Barkla's series K and L of characteristic X-rays; and also found a new series, which was dubbed with the letter M.

A French aristocrat

The first signs heralding a second wave of the quantum revolution came from the work of a French aristocrat, Prince Louis de Broglie. In the autumn of 1924 he discussed his doctoral thesis in physics at the Sorbonne in Paris. This thesis contained a comprehensive account of his research, which he had presented in the period 1923–4 to the French Academy of Sciences. De Broglie suggested that just as light waves could act like particles in certain circumstances (Einstein's light quanta), so too could particles manifest a wavelike behaviour. In particular he proposed that electrons, which up to then had been regarded as minute particles, could in fact undergo diffraction and interference phenomena, just as light waves do.

Let us try to explain de Broglie's thinking. He first considered light as having a dual nature; it behaves as an electromagnetic wave, characterised by its frequency (ν); and it behaves as a particle, a photon, characterised by its energy $E = h\nu$ (h always corresponds to the Planck constant). Hence, by symmetry, de Broglie thought that a material particle, which is characterised by its momentum ($p =$ mass × velocity), behaves also as a wave. This *matter wave*, as he called it, possesses a wavelength (λ), called the *de Broglie wavelength*, related to the momentum (p) of the particle, exactly as the wavelength of a light wave is related to the momentum of the corresponding photon. This is written as

$$\lambda = h/p$$

He thus concluded that a matter wave guides the motion of a particle as an electromagnetic wave guides the propagation of light quanta: 'The rays of the ... waves coincide with the trajectories that are dynamically possible', he wrote. This brilliant and revolutionary idea provided the first well-grounded basis for Bohr's quantum atom. Indeed, in de Broglie's view, the permitted Bohr stationary circular orbits are just the very ones that exactly contain a whole number (1, 2, 3, and so on) of 'de Broglie wavelengths'.

1925

Atoms on stage

Let us jump back, for a moment, to some ten years earlier, to the time when the Franck–Hertz experiment was being carried out at the University of Berlin (see p. 98). This represented one of the first experimental proofs of Bohr's atomic model. Firstly, it showed that energy could in practice be either absorbed or emitted by atoms only in discrete amounts. Secondly, it confirmed Bohr's hypotheses concerning the quantum states of the atom, and the connection between these states and the radiation emitted or absorbed, that is, the connection between the energy levels and the frequency of this radiation.

As time went by, more and more experiments supported the concept of the quantum atom, and Franck's and Hertz' achievements were to represent a landmark on the forward path of quantum physics. It was thus that the Swedish Academy of Sciences decided to award the 1925 Nobel Prize for physics to James Franck and Gustav Hertz, the two authors of that seminal experiment. (As a matter of fact, in 1925 the Nobel Physics Committee had agreed to reserve the year's prize until 1926. It was indeed a memorable meeting of the Academy on 11 November 1926, when Franck and Hertz were at last awarded the 1925 Nobel Prize.)

James Franck (1882–1964) was born in Hamburg, Germany, the son of a Jewish banker. He studied at the universities of Heidelberg and Berlin, receiving his doctorate from the latter in 1906. In 1920 he was appointed Professor of Experimental Physics at the University of Göttingen, where he gathered around him a group of young physicists, all of whom were in later years to become renowned, each one in his own field. In 1933 Franck, who was Jewish, resigned his post for political reasons, and went to Denmark, to work with Bohr. He emigrated to the USA in 1935, and was appointed professor at Johns Hopkins University, Baltimore. In 1938 he became professor of physical chemistry at the University of Chicago. The most important of Franck's researches were in the fields of photochemistry and atomic physics. During the Second World War he worked on the Manhattan Project, which was responsible for developing the atom bomb.

Gustav Ludwig Hertz (1887–1975) was also born in Hamburg, a nephew of the great Heinrich Hertz (p. 20). He studied at the universities of Göttingen and Munich, and later in Berlin, where he was appointed to an assistantship in 1913, and began to work with Franck. From 1920 to 1925 Hertz worked in the physics laboratory of Philips at Eindhoven, and in 1928 he was appointed director of the physics institute of the Berlin–Charlottenburg Technological University. He resigned from this post in 1935 for political reasons, and returned to industry as director of a research laboratory for the Siemens Company. After the Second World War he was engaged for about ten years in research in the Soviet Union. He subsequently returned to Germany, and was appointed professor of physics at the University of Leipzig.

Waiting for a new mechanics

Quantum mechanics, the second revolutionary theory of twentieth-century physics, was conceived in the summer of 1925, further developed during the year

Fig. 6.8. James and Ingrid Franck (left) together with Gustav and Ellen Hertz in Stockholm, December 1926. (Courtesy Niels Bohr Archive, Copenhagen.)

1926 and brought to completion in 1927. Before looking at these new events, a glance should be spared for the last attainments of the 'old quantum theory'.

The electron spin

The old Bohr–Sommerfeld theory identified the quantum states of the hydrogen atom by three whole numbers (n, l, m), acting something like the fingerprints of its orbiting electron. In 1924 the Austrian physicist Wolfgang Pauli, then working at the University of Hamburg in Germany, proposed that a fourth quantum number, taking on only half-integer values, was necessary to identify the quantum states of the alkali atoms. Pauli thought of this new number as related to the external electron of these atoms. (Alkali atoms, such as sodium, have one electron outside a 'core', this last consisting of the nucleus and the inner electrons: it is just that electron which is responsible for their spectrum.)

During the summer of 1925 a young Dutch physicist, Georg Uhlenbeck, was working on his doctoral thesis at the University of Leiden. Samuel Goudsmit, a fellow graduate student, told him about the new quantum number introduced by Pauli. Soon Uhlenbeck had an idea: '. . . it occurred to me that, since (as I had learned) each quantum number corresponds to a degree of freedom of the electron, the fourth quantum number must mean that the electron had an additional degree of freedom – in other words the electron must be rotating!'[22]

Uhlenbeck and Goudsmit soon published their idea in the scientific journal *Naturwissenschaften*. In short they claimed that the electron has not only

Fig. 6.9. The earth has an intrinsic angular momentum as it spins continuously on its axis every day. In the 'old quantum theory', the electron was supposed to be a tiny spinning sphere and, like the earth, it spun. Electrons can spin clockwise (spin *down*) or anticlockwise (spin *up*).

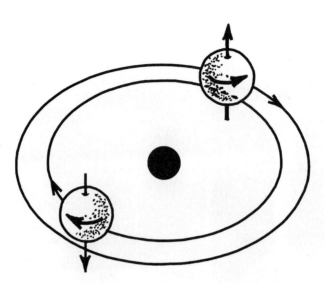

an orbital angular momentum round the nucleus, but also an *intrinsic angular momentum* or *spin*. Since a spinning charged particle acts as a tiny magnet, the electron too acts as a magnetic dipole, characterised by its own magnetic moment. Two quantum restrictions were imposed by Uhlenbeck and Goudsmit on the electron spin: it should have the unique value $(h/4\pi)$; and it could have only two orientations. In this way many observed effects till then unexplained could all be understood in terms of both the orbital angular momentum and the spin of the electrons circling around the nucleus. These effects included, for instance, the anomalous Zeeman effect, the alkali spectra, and the even number of traces in the Stern–Gerlach experiment.

The exclusion principle

After the electron spin hypothesis, the quantum states of atoms could be completely identified by a set of four quantum numbers (n, l, m, s). The first three are whole numbers, whereas the new spin quantum number (symbol: s) can be either $+\frac{1}{2}$ or $-\frac{1}{2}$ (in units of $h/2\pi$).

To understand the structure of all other atoms with more than one electron, an additional principle was introduced in the same year (1925) by Pauli himself, and it was named the *exclusion principle*. Pauli stated that, 'There cannot exist two or more equivalent electrons in the atom for which ... [the values of all four] quantum numbers coincide. If the atom contains an electron for which these quantum numbers have certain values then this state is "occupied".'[23] In other words, this principle says that no two quantum states of an atom can have the same fingerprint.

Pauli's exclusion principle was used to derive the structure and the chemical properties of atoms with many electrons so leading to a full explanation of the periodic system of the elements. When an atom has several electrons, these are all prevented from huddling down into the lowest quantum state – the reason being that they are forbidden by the exclusion principle. Instead, some electrons are forced to rise into the other states possessing higher energies. Each value of the principal quantum number n corresponds to a region of space around the nucleus in the form of a *shell*; thus physicists speak of a K-shell ($n = 1$), an L-shell ($n = 2$), an M-shell ($n = 3$) and an N-shell ($n = 4$), with a maximum number of electrons respectively equal to 2, 8, 18 and 32. (Apart from atomic physics, the Pauli exclusion principle has many important applications in areas as diverse as quantum chemistry, nuclear and particle physics, condensed-matter physics and astrophysics.)

Quantum mechanics

In the words of Victor Weisskopf *quantum mechanics* marked 'a turning point in man's understanding of nature comparable to Newton's discovery of the universal nature of gravity, Maxwell's electromagnetic theory of light, and Einstein's relativity theory'. Let us speak now about the three principal creators of the new mechanics.

A young man of genius
In 1925 Werner Heisenberg, a twenty-four-year-old assistant to Professor Max Born at the University of Göttingen, wrote a paper containing conclusions which completely reformulated the existing quantum theory. After showing the manuscript to Pauli, he handed it over to his professor. Born sent Heisenberg's work to the German journal *Zeitschrift für Physik*, where it was published in its September issue.[24] The paper, entitled 'Quantum theoretical re-interpretation of kinematic and mechanical relations', is one of the foundation stones of modern physics. The interpretation of Heisenberg's new ideas was undertaken by Born himself two months later, using abstract mathematical entities which students of physics and mathematics know as *matrices*. That same autumn Heisenberg joined Born and Pascual Jordan (another of Born's assistants), and they collaborated in writing a paper (it became known as the famous 'three-man paper')[25] in which they set forth the details of the matrix-based quantum mechanics. Meanwhile Heisenberg's theory had acquired the title of *matrix mechanics* or *Göttingen theory*.

Heisenberg to Pauli
On 9 July 1925 Heisenberg wrote to Pauli, his friend and critic: '... I dare to send you this brief preliminary manuscript of my work because I believe that it ... contains actual physics ... I must beg you to return it to me in 2–3 days, since I must either make its existence known in the next few days or burn it.'[26] Pauli

Fig. 6.10. Werner Heisenberg. (Max-Planck-Institut für Physik, courtesy Emilio Segrè Visual Archives.)

answered favourably: 'It was the first light of the dawn in quantum theory', he said.

Heisenberg's paper

In the introduction to his paper Heisenberg first described its content:

> In the present paper [we] seek to obtain the foundations of a quantum-theoretical mechanics based exclusively upon relationships between quantities which are observable in principle.

He then concluded by explaining his new idea:

> [It] seems more appropriate to discard all hope of observing the erstwhile unobservable quantities (like position, period of revolution of the electron) and simultaneously to concede that the partial agreement of the quantum rules in question is more or less fortuitous; one should try to establish a quantum-theoretical mechanics, analogous to classical mechanics, in which only the relations between observable quantities occur.[27]

1926

Atoms again

In the autumn of 1926 the Nobel Physics Committee proposed, once again, to reserve the year's prize until 1927. But some months later, in the previously mentioned meeting of the Academy, that of 11 November (see p. 140), the Committee's proposal was bitterly attacked. Carl Benedicks, a professor of metallurgy and a

Matrix mechanics

Heisenberg started by questioning the old planetary model of the atom, because he judged it to be inadequate, and moreover without sufficiently firm foundations. For instance, the Bohr–Sommerfeld theory of the atom spoke of orbits for the electrons circling around the nucleus. Heisenberg noted that these were entities that could not be observed directly: you cannot in any way determine the position of an electron in the atom, he claimed, without first breaking it up. (In fact, if you want to know where the electron is in space, you must shine an appropriate kind of radiation on it in order to see it. If you want to define its position with any accuracy within the atom, the dimension of which is about a tenth of a nanometre, you must use light of an extremely short wavelength; that is, you must irradiate the electron, for example with X-rays, with the result that the electron can be chipped away from the atom itself.)

Thus Heisenberg thought that the old Bohr–Sommerfeld theory had failed, because the idea on which it was based, the orbit picture, could never be put to the test of experiment. He moreover decided that any attempt to represent the atomic world by means of intuitive models based on quantities which cannot be determined experimentally was doomed to failure. Instead, he constructed a theory which describes the atomic structure in terms of *physically observable quantities*, such as frequencies and intensities of the light emitted or absorbed by atoms.

To manipulate these quantities he used, as Born had suggested, mathematical devices called *matrices*. (Matrices had been known since the 1850s, but they had not yet found any application in the field of physics.)

Heisenberg's idea was that a matrix is associated with every physical quantity. You can form a frequency matrix, a light intensity matrix, and so on. He then invented certain mathematical equations relative to these matrices, and used them to obtain highly accurate predictions of the characteristics of an atom.

A magic square

$$\begin{matrix} v_{11} & v_{12} & v_{13} \\ v_{21} & v_{22} & v_{23} \\ v_{31} & v_{32} & v_{33} \end{matrix}$$

This table of numbers is a *matrix*. The number placed, say, in the third (horizontal) row and second (vertical) column is associated with the frequency of the light emitted (or absorbed) by an atom when it jumps from an energy level with a principal quantum number $n = 3$ (2) to a level with $n = 2$ (3).

---

member of the Physics Section, led the challenge. He expressed the discontent of many members of the Academy, and proposed that rather than reserving the 1926 prize, it should be given to the French physicist Jean-Baptiste Perrin, for his work of more than fifteen years earlier; this had been related, when he was studying the Brownian motion, to proving the existence of atoms and molecules. This *affaire* has been thus described:

> Although Perrin had consistently received firm support from French and Dutch nominators, the Committee had long since closed the book on the Frenchman. During the past few years, the Committee had merely responded to repeated nominations for Perrin by claiming that nothing new could change the group's earlier judgement. His research was not of sufficient importance. Moreover, Perrin's last significant work, *Les Atomes*, was published in 1913; hence, according to the Committee, his achievements were too far back in time to be considered.[28]

But, at the Academy's November meeting, Perrin's case came up again. Then, '[discussion] ensued; disagreement prompted more than one round of voting. It appeared at one point that Perrin would not receive a majority, but in the end Benedicks won against the Committee's objections. A previously buried and forgotten candidate had now been resurrected.'[29]

Jean Baptiste Perrin (1870–1942) was born in Lille, France. He studied at the *Ecole Normale Supérieure* in Paris, where he then became a teacher of physics from 1894 to 1897. Later on he was a lecturer at the *Sorbonne* and in 1910 he was appointed Professor of Physical Chemistry there. He kept this post until 1940, when he emigrated to the USA. In 1895 Perrin proved that negatively charged particles were shot off from the negative electrode of a cathode-ray tube, so opening the way to the discovery of the electron by J. J. Thomson (p. 59). He is best known for his work, done from about 1908 to 1911, on the study of the random movements and the distribution of microscopic particles suspended in a liquid. Using an ultra-microscope that he himself had developed during the earlier years of the century, Perrin carefully observed how these particles were distributed in space under the action of gravity. The experiments that he then carried out earned him the Nobel Prize.

Brownian motion
Brownian motion was discovered by the British botanist Robert Brown, who had noted, in the late 1820s, that pollen grains of plants, suspended in water, oscillated with irregular motions. At first these motions were thought of as being due to the vitality of the pollen grains, but equally tiny particles of inorganic materials did the same. It was then suggested that the movement was due to the suspended particles being bombarded by the molecules that made up the liquid: a particle, hit on all sides by many molecules, would be compelled to jiggle around. The erratic movement of the tiny particles should have been a striking proof of the graininess of the liquid, so confirming the atomic structure of matter.

In a remarkable paper published in 1905, the very same year as those on special relativity and the light-quanta hypothesis, Einstein worked out a theoretical analysis of Brownian motion; he entitled it 'On the motion of particles suspended in liquids at rest required by the kinetic theory of heat'. He showed that the gas laws, obtained from kinetic theory, should govern the behaviour of the suspended particles. ('I considered that, according to atomistic [kinetic] theory, there would have to be a movement of suspended particles . . .', he wrote.) He also showed how the extent of these random movements could determine several important physical quantities, for example the masses of atoms and molecules, and the magnitude of the Avogadro number. This number takes its name from Amedeo Avogadro, a scientist of the first half of the nineteenth century, who lived in Turin, Piedmont (now a part of Italy). It expresses the number of atoms or molecules in one unit of the amount of a substance (called a *mole*).

Perrin was able to verify Einstein's theory of Brownian motion experimentally. He also obtained the first estimate of the size of molecules, and the value of the Avogadro number. In his Nobel lecture, after examining his results and those of experiments performed by other physicists, Perrin thus concluded:

> Such a collection of agreements between the various pieces of evidence, according to which the molecular structure is translated to the scale of our observations, creates a certitude at least equal to that which we attribute to the principles of thermodynamics. The *objective reality of molecules and atoms* which was doubted twenty years ago, can today be accepted as a *principle* . . . Nevertheless, however sure this new principle may be, it would still be a great step forward in our knowledge of matter . . . if we could perceive directly these molecules . . . Without having arrived there, I have at least been able to observe a phenomenon where the discontinuous structure of matter can be seen directly.[30]

New highlights

In the months following the Göttingen papers (p. 143), quantum mechanics led to an explosion of new ideas. Let us follow up some of these ideas, from two brilliant theoreticians, one coming from Vienna, Austria (Erwin Schrödinger), and then serving as Professor of Theoretical Physics at the University of Zurich; the other a Cambridge University man (Paul Dirac).

A professor from Zurich

Schrödinger's manner of attacking the problem of the structure of atoms differed from the line of reasoning adopted by Heisenberg. He started to develop his theory, once known as *wave mechanics*, at the end of 1925, and he published it in four famous papers in the *Annalen der Physik*[31] during the first half of 1926. (After reading Schrödinger's papers, Einstein wrote to him: 'the idea of your work springs from true genius!')

Fig. 6.11. Erwin Schrödinger. (Courtesy Archiv zur Geschichte der Max-Planck-Gesellschaft, Berlin–Dahlem.)

The first paper began with these words, which summarise his idea:

> In this communication, I wish first to show in the simplest case of the hydrogen atom . . . that . . . [Bohr's] rules for quantization can be replaced by another requirement, in which mention of 'whole numbers' no longer occurs. Instead the integers occur in the same natural way as the integers specifying the number of nodes in a vibrating string. The new conception can be generalised, and I believe it touches the deepest meaning of the quantum rules.[32]

Soon after its appearance Schrödinger's theory was acknowledged as marking significant progress from the Bohr–Sommerfeld theory, and as coming quite close to the profoundest nature of the atomic world. This was because wave mechanics does not postulate the existence of stationary orbits, and hence of discrete energy levels; the latter emerge in the same natural way as that by which it can be shown that a vibrating violin string can produce only discrete notes.

With his equation Schrödinger was able to calculate the hydrogen spectrum (this had also been obtained by Pauli in the autumn of 1925, using Heisenberg's matrix mechanics). In addition, Schrödinger calculated the Zeeman and Stark effects, and others too, consistently obtaining results in agreement with those obtained by Heisenberg's matrix method. The two methods represented two different mathematical forms of the same theory, and led to identical results. (Schrödinger himself soon showed that the two methods were equivalent.) So both Heisenberg and Schrödinger, from different starting points, had created a

Wave mechanics

Just like sunlight filtering through trees, light rays appear to travel in straight lines. The concept of rectilinear light rays is very useful in geometrical optics to represent the propagation paths of light.

Schrödinger started with the observation that classical (or Newtonian) laws governing the motion of a material particle can be put into a form analogous to the laws of geometrical optics. For instance, a particle without any force upon it moves at a constant speed along a straight line, just like light in a transparent homogeneous medium. Now, every student of physics knows know that the laws of geometrical optics fail in phenomena like interference or diffraction. In these cases one must employ wave optics, instead of geometrical optics, the latter representing only an approximation. Similarly, according to Schrödinger, the classical laws of mechanics represent only an approximation of more general laws.

Schrödinger introduced a theory describing the behaviour of particles by an equation, known as the *Schrödinger equation*. This was very similar to the classical equations, which had for long been known as effective in describing many wave phenomena such as sound waves, the vibrations of a string or electromagnetic waves. The protagonist of Schrödinger's equation is an abstract entity, called the *wave function*, and it is symbolized by the Greek letter ψ (psi).

Schrödinger's symbolic seal: ψ

Schrödinger obtained the wave functions for the hydrogen atom, and also found the same energy levels as those that Bohr and Sommerfeld had found with their orbits theory.

But what does the wave function ψ mean? What does it represent? Its interpretation was suggested, once again, by Max Born. In his view, the wave function for the hydrogen atom, for example, represents each of its possible physical states, and can be used in order to calculate the probability of finding the electron at a certain point in space. If, for example, the wave function is nearly zero at a certain point, it means that the probability of finding the electron there is extremely small, whereas the same electron is most likely to be found where the wave function is large. Thus, wave mechanics cannot serve to determine the motion of a particle, that is, its position and velocity at one moment. Instead its equations tell us how the wave function evolves in space and time; and the value of the wave function at a certain point in space tells us the probability that the particle is near that point. (Fig. 6.12 shows sketches of the probability distribution corresponding to two wave functions for the hydrogen atom.)

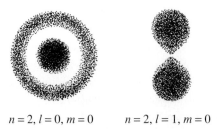

$n = 2, l = 0, m = 0$ $n = 2, l = 1, m = 0$

Fig. 6.12. A representation of the probabilities, obtained using the wave function ψ, that an electron will be found at a given point near a hydrogen nucleus, corresponding to two different quantum states. The densest parts of the clouds represent the places where the electron is most likely to be located.

comprehensive theory of quantum mechanics that reproduced all the positive results of the old Bohr–Sommerfeld theory.

Recollections

The roots of Schrödinger's theory are to be found neither in the work of Bohr nor in the recent discovery of Heisenberg. He was in fact more inspired by de Broglie's ideas. In April 1926 he wrote to Einstein:

> The whole business would not have been produced now, and perhaps never (I mean, not by me), if [one of your papers] had not impressed me with the importance of de Broglie's idea[33]

And again, he had this to say:

> My theory was stimulated by de Broglie and brief but infinitely far-seeing remarks by Einstein. I am not aware of a generic connection with Heisenberg. I, of course, knew of his theory but was scared away, if not repulsed, by its transcendental algebraic methods which seemed very difficult to me.[34]

A student from Cambridge

The third key individual of the quantum revolution was Paul Dirac. A quiet, shy student, he went to St John's College at Cambridge University to study mathematics, under the supervision of the mathematical physicist Ralph Fowler. He soon came to the fore as a highly original physicist, and in 1925, at the age of 23, he wrote his first revolutionary paper. Here, briefly, is the story.

At the end of July 1925, Heisenberg visited the Cavendish Laboratory. He gave a seminar on quantum theory, and told Fowler privately about his new work. After his return to Göttingen, Heisenberg sent Fowler a copy of his paper, and Fowler passed it on to Dirac. When Dirac looked at it, he soon realised that the essential point of the paper was the strange mathematical entities used by Heisenberg (the famous matrices). They had odd multiplication rules, depending on the order in which the multiplication was carried out. (We all learned at primary school that if we have two numbers, for example 2 and 3, then $2 \times 3 = 3 \times 2$; the contradictory, however, is the case with two matrices; in fact, *matrix a* × *matrix b* is in general different from *matrix b* × *matrix a* – that is to say that 2×3 in some cases does not equal $3 \times 2(!)$) Dirac pondered this puzzle, and discovered the key for unlocking the mystery:

> I went back to Cambridge at the beginning of October 1925, and resumed my previous style of life, intense thinking about these problems during the week and relaxing on Sunday, going for a long walk in the country alone . . . It was during one of the Sunday walks . . . [that] I remembered something that I had read up previously in advanced books of dynamics about these strange quantities . . . The idea first came in a flash, I suppose, and provided of course some excitement, and then of course came the reaction 'No, this is probably wrong.' . . . I just had to hurry home and see what I could then find about [them]. I looked through my notes, the notes that I had taken at various lectures, and there was no reference there . . . There was just nothing I could do . . . I just had to wait impatiently through that night without knowing whether this idea was really any good or not . . . The next morning I hurried along to one of the libraries as soon as it was open, and then I looked up [those quantities] in Whitaker's *Analytical Dynamics*, and I found that they were just what I needed.[35]

Thus Dirac, after a sleepless night, was able to reformulate Heisenberg's matrix mechanics in a new form. This led him to his first paper, which was published late in 1925 in the *Proceedings of the Royal Society*.[36] In the spring of the year 1926 Dirac wrote his doctoral thesis, in which he developed his original ideas. In September he went to Copenhagen to Bohr's institute, and then on to Göttingen. During those months he prepared his major paper presenting 'his' quantum mechanics. This paper also showed that Heisenberg's matrix mechanics and Schrödinger's wave mechanics were special cases of a more general mathematical formalism. Within a year, Dirac had become a leading figure in the world of international physics.

Early reactions

Soon after matrix mechanics was published, many physicists felt themselves somewhat disinclined to accept the new theory. They found it too complicated, and they wondered if it could actually explain empirical facts. Schrödinger's wave mechanics was instead welcomed with enthusiasm. It was favoured mainly by the older generation of physicists – Planck, Einstein, von Laue, Wien – both because its mathematics was simpler, and because it accorded with a semiclassical interpretation of the atomic phenomena. Planck wrote to Schrödinger: 'You can imagine with what interest and enthusiasm I plunged into the study of

Fig. 6.13. Paul Dirac at the University of Göttingen, 1928. (Courtesy Niels Bohr Archive, Copenhagen.) In Göttingen Dirac met Robert Oppenheimer, who lived in the same pension . . . and spent much time reading Dante in the original. It is said that Dirac once asked him: 'How can you do both physics and poetry? In physics we try to explain in simple terms something that nobody knew before. In poetry it is the exact opposite.'[37]

this epoch-making work . . .'[38] And Einstein added: 'I am convinced that you have made a decisive advance, . . . just as I am convinced that the Heisenberg–Born method is misleading.'[39] In July 1926 Schrödinger gave a lecture in Munich. Heisenberg was present, and asked him how his theory could explain quantum phenomena such as the photoelectric effect and black-body radiation. Wilhelm Wien then interrupted, exclaiming angrily: 'Young man, Professor Schrödinger will certainly take care of all these questions in due time. You must understand that we are now finished with all that nonsense about quantum jumps.'[40]

In September Schrödinger went to Copenhagen, to argue with Bohr about quantum mechanics. As Heisenberg recalled, 'Bohr . . . was able . . . to insist fanatically and with almost terrifying relentlessness . . . in all arguments. He would not give up, even after hours of struggling . . . It was perhaps from over-exertion that after a few days Schrödinger became ill and had to lie abed as a guest in Bohr's home.'[41]

1927

Light as particles

Arthur Compton and Charles T. R. Wilson shared the 1927 Nobel Prize for physics. Compton received the prize (the third to be awarded to the USA) for the discovery of the so-called *Compton effect* (see p. 135); and Wilson for his invention of the *cloud chamber* (see p. 86). Let us speak briefly about the importance of these scientific achievements for the development of modern physics.

Compton

You will remember how the effect discovered by Compton between 1922 and 1923 concerns the change in wavelength (called the *Compton wavelength shift*) of X-rays when these are scattered off by atomic electrons. We have also seen how Compton explained the observations that were puzzling him by speaking of a collision between a photon and an almost free atomic electron (the term *photon* was coined by the American physical chemist Gilbert Lewis in 1926). This process also results in a recoil electron, which was actually seen in a cloud chamber experiment performed by Compton himself, and by Alfred Simon.[42] The two particles, the scattered photon and the recoil electron, were also detected 'simultaneously' in a counter experiment by Walther Bothe and Hans Geiger at the PTR in Berlin-Charlottenburg (see p. 247). These experiments gave a valid support to Compton's vision of his effect, and to Einstein's old idea of the corpuscular nature of light – the 'light-quanta'. (This idea had never been accepted by most representatives of the physics community – including Planck and Bohr.)

The idea of light-quanta would come to be accepted in the following years, though not without some reluctance. The Nobel Physics Committee itself was cautious about this matter, in awarding the Nobel Prize to Compton, and in his presentation speech at the award ceremony, Manne Siegbahn carefully avoided taking a clear stand about the particle nature of radiation. (As recently noted by Gösta Ekspong, a member of the Swedish Academy of Sciences, '... at the time of Compton's Nobel Prize ... there were theories based on the wave picture, treating both the electrons and the X-radiation as waves, also giving the correct wavelength shift'.)[43] In his Nobel lecture (December 1927) Compton was more explicit; he said:

> We are ... confronted with the dilemma of having before us a convincing evidence that radiation consists of waves, and at the same time that it consists of corpuscles. It would seem that this dilemma is being solved by the new wave mechanics. De Broglie has assumed that associated with every particle of matter in motion there is a wave ... As applied to the motion of electrons, Schrödinger has shown the great power of this conception in studying atomic structure. It now seems, through the efforts of Heisenberg, Bohr, and others, that this conception of

the relation between corpuscles and waves is capable of giving us a unified view of the diffraction and interference of light, and at the same time of its diffuse scattering and the photoelectric effect.[44]

But how quickly things change! In the same year that Compton received his Nobel Prize, experiments in the USA and in Britain (pp. 156–7) demonstrated that matter, too, has a wave nature (as had been predicted by Louis de Broglie in 1923–4, p. 139), and Niels Bohr himself proceeded to explain the meaning of the dual nature both of radiation and of matter (p. 159). So in 1929, if you look at the presentation speech awarding the Nobel to de Broglie, you can read, among other things: 'It thus seems that light is at once a wave motion and a stream of corpuscles. Some of its properties are explained by the former supposition, others by the second. Both must be true.'[45]

Newton vindicated?

Let us go back some three hundred years, to read *Opticks* again (Query 29, p. 13): 'Are not the Rays of Light very small bodies emitted from shining Substances?'

> Einstein's intuition and insights guided him to the view that, despite the fact that light is waves, light must also be particles: indeed Nature is preposterous! It is not inconceivable to me that a few of the many worries that so haunted Einstein in the early twentieth century might already have been vaguely discernible to Newton over two hundred years earlier!
>
> (Roger Penrose)[46]

Arthur Holly Compton (1892–1962) was born in Wooster, Ohio, USA, the son of a professor of philosophy in the local college. In 1916 he received his Ph.D. in physics from Princeton University, and in 1920 he became Professor of Physics at Washington University, St Louis. In 1923 Compton moved to the University of Chicago, where he remained until 1945. During the 1930s he performed a world-ranging series of measurements on cosmic rays, showing that the intensity of these rays is correlated with latitude. In 1941 he was appointed chairman of a committee of the US National Academy of Sciences for the valuation of the use of nuclear energy in war, and with Ernest Lawrence (Nobel 1939) he initiated the Manhattan Project, which created the first atomic bomb. From 1942 to 1945 he was director of the Metallurgical Laboratory at the University of Chicago, where in 1942 Enrico Fermi (Nobel 1938) constructed the first nuclear reactor (p. 214). After the war Compton returned to St Louis as Chancellor of the University, and from 1953 on he was also Professor of Natural Philosophy there.

Wilson's invention

C. T. R. Wilson won the Nobel Prize for having invented the *cloud chamber*, an invention that dated back to the period 1911–12 (p. 86). Making use of Wilson's

Fig. 6.14. Arthur Compton. (Courtesy AIP Emilio Segrè Visual Archives.)

invention, physicists were able to see the tracks and the interactions of subatomic particles in the gas contained inside the chamber. In the 1920s Patrick Blackett (Nobel 1948), then a research student at the Cavendish, was able to photograph and study the interactions of alpha particles, and succeeded in confirming Rutherford's dramatic discovery of the year 1919, concerning the disintegration of the atomic nucleus. During the 1930s and 1940s the cloud chamber became, worldwide, the work-horse detector used in the newly developing fields of nuclear and cosmic-ray physics.

Charles Thomson Rees Wilson (1869–1959) was born in Glencorse, near Edinburgh, Scotland, the son of a farmer. He studied at Owen's College in Manchester (as did J. J. Thomson) and at Cambridge University, where he graduated in 1892. He then taught at Bradford Grammar School in Bradford, England, but he returned to the Cavendish in 1896 as a researcher. From 1916 on Wilson became

Fig. 6.15. A schematic diagram of the Davisson and Germer experiment. (Source: Clinton Davisson's Nobel lecture, 1937, © The Nobel Foundation, Stockholm.)[47]

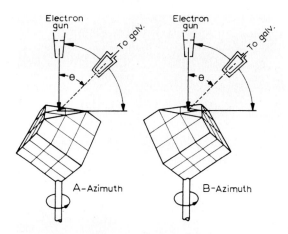

involved in the study of atmospheric electricity, and in 1925 he was appointed Jacksonian Professor of Natural Philosophy at the University of Cambridge. A Fellow of the Royal Society, Wilson received many honours for his research work. In 1956, at the age of 87, he published his last paper on a theory of thunderstorm electricity.

Electrons as waves

In 1926 Walter Elsasser, a graduate student at the University of Göttingen, suggested that, like X-rays, electrons could be diffracted by a crystal. So the wave nature of matter, as predicted by de Broglie, was capable of being tested in the same way that the wave nature of X-rays had first been tested by Max von Laue in 1912 (p. 95).

De Broglie's waves, which correspond to high-speed electrons, have, in fact, short wavelengths, comparable with the spacing between atomic layers in crystals. So the crystal lattice serves as a three-dimensional diffraction grating, and sharp peaks in the intensity of the diffracted beam should occur at specific angles. This idea was tested during the spring of 1927 by Clinton Davisson and Lester Germer, working at the Bell Labs in New York city, and also some months later by George Paget Thomson (the son of J. J. Thomson) at the University of Aberdeen, Scotland.

The Davisson–Germer experiment

Electrons from a heated filament reach high speeds in an electron gun (Fig. 6.15), and are directed at a nickel crystal. A detector, set at a particular angle, then reads the number of scattered electrons reaching it within a certain time. Davisson and Germer found that strong reflected beams occurred at specific angles, and noticed the similarity of this behaviour to X-ray diffraction. Simply following Bragg's reflection method (p. 100), they deduced the wavelengths for the electron waves, which, compared with those calculated from de Broglie's formula, appeared to be very similar. This similarity constituted a direct confirmation of the hypothesis regarding the wave nature of matter.

Electron diffraction patterns

In the summer of 1926 George Paget Thomson heard of Davisson's preliminary results at a meeting in London. Helped by his research student, Alexander Reid, Thomson then began a series of experiments, involving the scattering of electrons as they passed through thin foils of celluloid, aluminium and gold. (Peter Debye had used a similar technique several years earlier to study X-ray diffraction from polycrystalline specimens.) Thomson was thus able to obtain diffraction patterns produced by the electrons; these consisted of intensity peaks forming concentric rings around the direction of the incident beam. These patterns constituted *ad oculos* confirmation of the existence of de Broglie's waves (see an example in Fig. 6.19).

Quantum uncertainty

After introducing matrix mechanics (in 1925, see p. 143) Werner Heisenberg went on to study in depth the concepts of *wave* and *particle*. According to classical mechanics, a particle could be thought of as existing at a definite point in space, and as possessing a perfectly definite velocity. But, asked Heisenberg, can we actually determine both the exact position and the exact velocity of a particle at any given instant of time? And furthermore, how can such measurements be made?

For example, the particle in question could be stopped, and so we could determine its position. But in that case, we could not know its velocity. On the other hand, we might measure its velocity, but it would not be possible to determine where exactly the particle is located. Heisenberg showed that there is no way of determining the position of a particle with as great precision as we desire, unless the value of its velocity becomes completely uncertain. To measure both exactly is impossible, because it runs contrary to the laws of quantum mechanics. From our experiments, said Heisenberg, we can only infer that the particle is somewhere within a certain volume at a point of space, and is moving with a velocity which lies within a certain interval. The reason is that, in the microscopic world of atoms and particles, our measurements interfere with, and so have a decisive effect upon the objects which we observe. He elucidated his ideas in a paper published in 1927 in the *Zeitschrift für Physik*.[48] In this paper Heisenberg established what is now universally known as the *uncertainty principle* – the key to all quantum physics, and one of the fundamental principles in our whole comprehension of nature.

Walking under the stars

> Our evening discussions quite often lasted till after midnight . . . Still deeply disquieted after one of these discussions I went for a walk in the Fælledpark, which lies behind the Institute, to breathe the fresh air and calm down before going to bed. On this walk under the stars, the obvious idea occurred to me

that ... one could not simultaneously know the position and velocity of a particle ... [In February 1927, Bohr] left for a skiing holiday in Norway ... Left alone in Copenhagen I too was able to give my thoughts freer play, and I decided to make the above uncertainty the central point in the interpretation [of quantum theory] ... When Bohr returned from Norway, I was already able to present him with the first version of a paper ...

(Werner Heisenberg)[49]

Contraria sunt complementa

During the spring of 1927, following on from the publication of Heisenberg's paper on the uncertainty principle, discussions took place at Bohr's institute which concentrated on the physical interpretation of quantum mechanics.

Über den anschaulichen Inhalt der quantentheoretischen Kinematik und Mechanik.

Von W. Heisenberg in Kopenhagen.

Mit 2 Abbildungen. (Eingegangen am 23. März 1927.)

In der vorliegenden Arbeit werden zunächst exakte Definitionen der Worte: Ort, Geschwindigkeit, Energie usw. (z. B. des Elektrons) aufgestellt, die auch in der Quantenmechanik Gültigkeit behalten, und es wird gezeigt, daß kanonisch konjugierte Größen simultan nur mit einer charakteristischen Ungenauigkeit bestimmt werden können (§ 1). Diese Ungenauigkeit ist der eigentliche Grund für das Auftreten statistischer Zusammenhänge in der Quantenmechanik. Ihre mathematische Formulierung gelingt mittels der Dirac-Jordanschen Theorie (§ 2). Von den so gewonnenen Grundsätzen ausgehend wird gezeigt, wie die makroskopischen Vorgänge aus der Quantenmechanik heraus verstanden werden können (§ 3). Zur Erläuterung der Theorie werden einige besondere Gedankenexperimente diskutiert (§ 4).

Eine physikalische Theorie glauben wir dann anschaulich zu verstehen, wenn wir uns in allen einfachen Fällen die experimentellen Konsequenzen dieser Theorie qualitativ denken können, und wenn wir gleichzeitig erkannt haben, daß die Anwendung der Theorie niemals innere Widersprüche enthält. Zum Beispiel glauben wir die Einsteinsche Vorstellung vom geschlossenen dreidimensionalen Raum anschaulich zu verstehen, weil für uns die experimentellen Konsequenzen dieser Vorstellung widerspruchsfrei denkbar sind. Freilich widersprechen diese Konsequenzen unseren gewohnten anschaulichen Raum–Zeitbegriffen. Wir können uns aber davon überzeugen, daß die Möglichkeit der Anwendung dieser gewohnten Raum—Zeitbegriffe auf sehr große Räume weder aus unseren Denkgesetzen noch aus der Erfahrung gefolgert werden kann. Die anschauliche Deutung der Quantenmechanik ist bisher noch voll innerer Widersprüche, die sich im Kampf der Meinungen um Diskontinuums- und Kontinuumstheorie, Korpuskeln und Wellen auswirken. Schon daraus möchte man schließen, daß eine Deutung der Quantenmechanik mit den gewohnten kinematischen und mechanischen Begriffen jedenfalls nicht möglich ist. Die Quantenmechanik war ja gerade aus dem Versuch entstanden, mit jenen gewohnten kinematischen Begriffen zu brechen und an ihre Stelle Beziehungen zwischen konkreten experimentell gegebenen Zahlen zu setzen. Da dies gelungen scheint, wird andererseits das mathematische Schema der Quantenmechanik auch keiner Revision bedürfen. Ebensowenig wird eine Revision der Raum—Zeitgeometrie für kleine Räume und Zeiten notwendig sein, da wir durch Wahl hinreichend schwerer Massen die quantenmechanischen Gesetze den

Fig. 6.16. The opening page of Heisenberg's May 1927 paper entitled 'On the intuitive content of quantum-theoretical kinematics and mechanics'. This is where the uncertainty principle appeared for the first time. (© Springer-Verlag.)

> **Heisenberg's uncertainty principle**
>
> Let us imagine that we are using a powerful microscope to determine, as accurately as possible, the position of an electron. We must illuminate the electron with photons of the shortest wavelength (for example X-rays), and observe the scattered radiation. This implies, however, that the electron experiences a Compton recoil when the photon bounces from it, so that its position is completely altered. Hence, the particle position can only be determined subject to an uncertainty Δx, and its momentum is undetermined to the extent Δp. This was expressed mathematically by Heisenberg in 1927, by requiring the product of the two uncertainties to be of the order of the Planck constant. Heisenberg's uncertainty principle is
>
> $$\Delta x \times \Delta p \approx h$$

In Bohr's view, Heisenberg's principle followed from the concept of *wave–particle complementarity*, according to which, in quantum mechanics, every atomic phenomenon can be described using either the particle picture or the wave picture; these two are 'complementary' to each other. However, a certain thing (take, for example, an electron) cannot be both a wave and a particle at the same time; it must be either one or the other. This means that it is impossible to devise any kind of experiment capable of revealing simultaneously both the wave and the particle aspect of the electron. This limitation is implied in the laws of quantum mechanics, and it is expressed by Heisenberg's uncertainty principle. (In the same way as the concepts of position and velocity are complementary in Bohr's sense, so too are those of orbit and quantum state.) Bohr's complementarity, Heisenberg's uncertainty and Born's statistical interpretation of Schrödinger's wave function ψ are all related. Together, they form the physical interpretation of

Fig. 6.17. Bohr's coat-of-arms as a member of the Danish Order of the Elephant (normally only given to members of royal families and the presidents of foreign states). The Chinese symbol for Yin and Yang represents the idea of complementarity: *contraria sunt complementa* (opposites are complementary). (Courtesy Niels Bohr Archive, Copenhagen.)

quantum mechanics, as it was advocated by Bohr, Born, Heisenberg and Pauli; it is known as the *Copenhagen interpretation* of quantum theory.

Bohr presented his idea of 'complementarity' at the Volta Centenary Congress at Como, Italy, and, a month later, in October 1927, to the fifth Solvay conference in Brussels, the theme of which was 'Electrons and photons'. The new ideas were generally accepted by most of the physics community. Einstein, Schrödinger and de Broglie, however, were among the most notable dissenters. And in fact until the end of their lives these three never accepted the 'Copenhagen interpretation'.

Wave–particle dualism

Particles as particles

Fig. 6.18. Alpha-particle tracks.[50]

Particles as waves

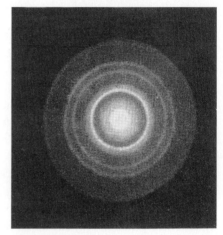

Fig. 6.19. Electron diffraction.[51]

Waves as waves

Fig. 6.20. X-ray diffraction.[52]

Waves as particles

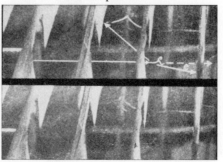

Fig. 6.21. X-ray scattering (e = recoil electron).[53]

Fig. 6.22. The participants at the fifth Solvay conference, Brussels 1927. (Reproduced by kind permission of the International Solvay Institutes for Physics and Chemistry, Brussels, Belgium.) From left to right, third row (standing): E. Schrödinger (sixth), W. Pauli (eighth), W. Heisenberg, R. Fowler; second row: P. Debye, W. L. Bragg (third), H. Kramers, P. Dirac, A. Compton, L. de Broglie, M. Born, N. Bohr; first row: M. Planck (second), M. Curie, H. Lorentz, A. Einstein, two others C. T. R. Wilson, O. Richardson.

The Copenhagen interpretation

The Solvay conference in Brussels in the autumn of 1927 closed this marvellous period in the history of atomic theory. Planck, Einstein, Lorentz, Bohr, de Broglie, Born and Schrödinger, and from the younger generation Kramers, Pauli, and Dirac, were gathered here and the discussions were soon focused on a duel between Einstein and Bohr on the question as to what extent atomic theory in its present form could be considered to be the final solution of the difficulties which had been discussed for several decades . . . These discussions continued even at the next Solvay meeting in 1930, and it was probably on this occasion that Einstein . . . proposed the famous experiment . . . in which the colour [frequency or energy] of a light quantum is to be determined by weighing the source before and after the emission of the quantum . . . It was a particular triumph for Bohr that

> he was able ... by using just Einstein's own formulae from general relativity, that even in this experiment the uncertainty relations are valid, and that Einstein's objections were unfounded. With this the Copenhagen interpretation of quantum theory seemed from now on to stand on solid ground.
>
> (Werner Heisenberg)[54]

1928

The thermionic effect

Owen Richardson, then director of research at King's College, University of London, was awarded the 1928 Nobel Prize in physics for his studies regarding the emission of electrons from heated metals. This phenomenon, called the *thermionic effect*, became the basis for vacuum tube research and technology. Let us see what it is about.

As you certainly know, most metals are good conductors of electricity. This is because they contain one or more outer electrons in each atom which can move almost freely through the material (the attracting electric forces are not strong enough to bind them to particular atoms). However, the same forces prevent these 'free' electrons from leaving the surface of the metal; the minimum amount of energy which has to be supplied to remove an electron from its surface is called the 'extraction energy'. When a metal is heated, the energies of its free electrons increase, and some of them acquire sufficient energy to overcome the extraction energy and escape from the surface. This is the process called 'thermionic emission'.

Richardson began to investigate this phenomenon in the early 1900s at the Cavendish Laboratory. He carried out experiments to see whether electric discharges in vacuum tubes were due to electrons coming up from the heated metal filament, and were not caused by any other phenomena. As he recalled in his Nobel lecture:

> This was not so easy at the beginning of this [twentieth] century as it would be at the present time [1929]. Largely owing to the technical importance of the phenomena under consideration the art of evacuating gases has advanced enormously since then. In those days the gas [in the interior of the tube] had all to be got away by hand pumps.[55]

In 1901 Richardson was able to show that a platinum filament in a vacuum tube actually emitted electrons. He then calculated the number of such electrons, and discovered a formula which explains how this number increases with the temperature of the filament.

He went on working on this subject at Cambridge until 1906, and then continued in the USA. In 1913 he carried out an important experiment, which showed that thermionic emission was not a secondary effect of some chemical reaction between the hot filament and the surrounding gas – as was commonly held by other physicists – but was in fact a real physical phenomenon in the hot metal. ('[I was able] . . . to show that the mass of electrons emitted exceeded the mass of the chemicals which could possibly be consumed. This experiment, I think, ended that controversy so far as it could be regarded seriously', Richardson commented later.)[56]

Owen Willans Richardson (1879–1959) was born in Dewsbury, Yorkshire, England. He studied at the University of Cambridge, where he graduated in 1900, and in 1902 became a Fellow of Trinity College. He then began research at the Cavendish Laboratory on the emission of electricity from glowing metals. In 1906 he transferred to the USA, and was appointed Professor of Physics at Princeton University, New Jersey; here he remained until the end of 1913. At Princeton, Richardson worked on thermionic emission, and on photoelectric and magnetic effects. In 1914 he returned to England as Professor of Physics at the University of London. He was appointed director of King's College, London, in 1924, and was knighted in 1939.

Discoveries and inventions

John Ambrose Fleming's invention of the first *vacuum-tube diode*, and then Lee De Forest's invention of the *vacuum-tube triode*, marked the transition to the age of electronics (see p. 63). In the years following 1910 the triode evolved into a *vacuum-tube amplifier*, used in both radio and telephone systems. The first vacuum-tube amplifier was invented in 1913 by a former student of Millikan's, by the name of Harold Arnold; he was working at AT&T's Western Electric laboratories, which were to become Bell Telephone Laboratories (Bell Labs) in 1925. (In 1915, using vacuum-tube amplifiers, AT&T's engineers were able to transmit speech for the first time by wireless from Arlington, Virginia, to both Paris and Honolulu.)

At that time Marconi's wireless communication had become a well-established technique. Every major nation began equipping its ships with wireless facilities (especially following the tragedy of the *Titanic* in 1912), and wireless communication with aircraft was also established. The 1920s saw an explosion in radio broadcasting, with commercial stations being opened throughout the world. In the same years the first experimental television systems started to be developed.

Vacuum tubes also made possible long-distance telephone lines. Since the end of the nineteenth century, it had been the American Telephone and Telegraph Company (AT&T) which had had the responsibility for running the American telephone network. AT&T bought De Forest's patent rights to use the triode, and developed a vacuum-tube amplifier, which made possible transcontinental telephone lines. In January 1915, the first line from New York to San Francisco was inaugurated in time for the Panama–Pacific Exposition.

1929

Prince Louis de Broglie

From 1927 the Nobel Physics Committee began to be aware of the importance of the recently born quantum mechanics. But at that time they were reluctant to consider it worthy of a prize. It was still too new, and as claimed by Carl Oseen (the only Committee member somewhat familiar with theoretical physics), the new theory had not yet led to any experimental discovery of fundamental importance.

But in 1929 things began to change. Oseen himself declared that 'abundant experimental proof of wave–particle dualism had been obtained', and so the Committee was able to recommend to the Academy that 'if the discovery of the wave nature of electrons is to be honoured this year, the prize should be given undivided to L. de Broglie'. And thus the Frenchman earned the 1929 Nobel Prize for physics.

Prince Louis Victor de Broglie (1892–1987) was born in Dieppe, France. He was the second member of a noble family, originally resident in Piedmont (then French, but now a part of Italy). He studied at the *Sorbonne* and, after taking his degree in history in 1910, chose instead to undertake research in physics. He received a degree in science in 1913. During the First World War de Broglie served in a radio station on the Eiffel Tower in Paris, where he was brought into contact with the technical aspects of physics. At the end of the war he resumed his studies with his elder brother, Maurice, who worked on experimental physics in a well-equipped spectroscopic laboratory in the family house in Paris. But it was the purely theoretical aspect of physics that attracted Louis. He described himself in these words:

> Having much more the state of mind of a pure theoretician than that of an experimenter or engineer, loving especially the general and philosophical view, I was drawn towards the problems of atomic physics . . . It was the conceptual difficulties which these problems raised; it was the mystery which surrounded that famous Planck's constant, *h*, which measures the quantum of action; it was the disturbing and badly-defined character of the dualism of waves and corpuscles, which appeared to assert itself more and more in the realms of physics . . .[57]

In 1923 de Broglie had the brilliant idea of bringing together the concepts of the particle and the wave: 'After long reflection in solitude and meditation, I suddenly had the idea, during the year 1923, that the discovery made by Einstein in 1905 should be generalised by extending it to all material particles and notably to electrons.'[58] In 1924, at the *Sorbonne*, he delivered his doctoral thesis entitled 'Recherches sur la théorie des quanta' ('Research on the quantum theory'). It contained the idea of matter waves, which were later used as a basis for wave mechanics as created by Schrödinger (p. 147).

In a speech given in 1945 de Broglie thus recalled the origins of his discovery:

Fig. 6.23. Prince Louis Victor de Broglie. (Courtesy AIP Emilio Segrè Visual Archives, Weber Collection.)

... the intervention of quanta and of Planck's constant h, as much in the theory of photons as in that of the quantization of the electronic movements, seemed to me to show clearly that the link between the two terms of the wave–corpuscle dualism took place through the intermediary of the quantum of action, and must in consequence be expressed mathematically for formulae in which the constant h would appear. This was already the case for the relations which, in the theory of the photon, expressed the energy and momentum of the corpuscle of light as a function of the frequency and of the wavelength of the luminous wave, and the form of these relations gave an indication of the interconnection that had to be established in the general case of any corpuscle whatever ... [see p. 139][59]

In the 1920s de Broglie taught at the *Sorbonne*. He was then appointed professor of Theoretical Physics at the *Institut Henri Poincaré*, which had just been built in Paris. In 1932 he was appointed to the chair of theoretical physics at the *Sorbonne*. Between 1930 and 1950 de Broglie's work was chiefly devoted to the study of the various extensions of wave mechanics, as well as to Dirac's electron theory, the new theory of light and applications of wave mechanics to nuclear physics. In 1933 he was elected a member of the French Academy of Sciences.

Expanding universe

Since the nineteenth century, astronomers have been analysing the light emitted by stars by studying their spectra – the patterns of lines which occur at wavelengths characteristic of atoms emitting light or absorbing it. In particular, they have been able to calculate the velocity of stars moving towards or away from the earth, by measuring the displacement of their spectral lines. This method is based on the Doppler effect for light (p. 116): when a source (a star) is moving away from an observer (an astronomer on earth) all the spectral lines in its spectrum are shifted towards the longer-wavelength (red) end of the visible spectrum; when the source is approaching the observer, the light shifts towards the violet end of the spectrum.

The Doppler effect was not restricted only to stars of our galaxy; it was used also to study objects beyond the Milky Way. In 1917 the American astronomer Vesto Slipher published a paper in which he described observations of twenty-five galaxies. He found that the spectral lines of most of those galaxies were red shifted, indicating that they were all moving away from our earth, and that the faintest – and presumably the most distant – galaxies presented a larger red shift, which demonstrated that they were receding at an even greater speed.

During the 1920s the American astronomer Edwin Hubble photographed the spectra of many galaxies with a 2.5-metre telescope at the Mount Wilson Observatory near Los Angeles. He estimated the distances of two dozen galaxies for which velocities had been determined by their red shifts, and found that there was a regular increase in the velocity with which galaxies rushed away, and that this was in proportion to their distances. (For example, if galaxy 2 is twice as far from earth as galaxy 1, then galaxy 2 is moving away from us at twice the speed of galaxy 1.) The straight-line relationship between the recession velocity of a galaxy and its distance was propounded by Hubble in 1929, and it is now known as the *Hubble law*.

How can we explain this? Cosmologists say that galaxies behave as though the universe is expanding in an identical way throughout all space. To an observer standing in any place in the universe, all the galaxies seem to be moving away, and the farther away from the observer a galaxy is, the faster it moves. (It does not matter where the observer is, the effect is always the same.) The first person to point out this hypothesis was the Russian mathematician Alexandre Friedmann. In 1922, making use of Einstein's general relativity, he worked out a model of an expanding universe. In 1927 the Belgian priest and mathematician Georges Lemaître came out with a similar model in which the universe and its expansion must have started off at a time in the distant past with a gigantic explosion.

1930

An Indian Nobelist

The 1930 Nobel Prize for physics was awarded for the first time to a scientist from Asia: an Indian 'native', Sir Chandrasekhara Raman was the Nobel laureate for this year. The research work which earned him the sought-after award had been carried out some years earlier, and it concerned the scattering of light by molecules. (In the year 1930 quantum mechanics continued to receive strong support for a Nobel Prize, as was demonstrated by the increasing number of nominations in favour of two of its creators – Heisenberg and Schrödinger. But in the Physics Committee, Carl Oseen continued to be reluctant to acknowledge the new theory, so that the Committee recommended the Academy to reward an experimentalist – in this case Raman, who had been nominated by Nobelists such as de Broglie, Rutherford and C. T. R. Wilson.) Raman had discovered that when an intense beam of monochromatic light traverses a transparent medium, a small fraction of this light is scattered; that is, it emerges in directions other than that of the incoming beam. Moreover, part of this scattered light is shifted into different wavelengths.

Like the Compton effect (p. 135), this phenomenon – which is called the *Raman effect* – is easily understood if light is considered as consisting of photons. When these photons strike the molecules of the sample, most of the collisions are elastic, so that the photons are scattered in such a way that their energy and wavelength remain unchanged (this process is known as *Rayleigh scattering*). On some occasions, however, the photons gain energy from or lose energy to the molecules in the medium, while these same molecules undergo vibrational or rotational motions. As a consequence, the photons are scattered with an increased or diminished energy – hence with a correspondingly shorter or longer wavelength; these wavelength shifts are a measure of the amount of energy involved in the transition between different energy levels of the molecules themselves.

The scattered light can then be analysed by use of special light-detecting techniques: the pattern of the spectral lines so obtained (the so-called *Raman spectrum*) gives useful information concerning the nature of the medium that has scattered the light. The Raman spectrum is used as an experimental aid to the solution of a wide range of problems in physics and chemistry. It assumes practical importance in spectrographic chemical analysis, and in the determination of the structure of molecules.

Chandrasekhara Venkata Raman (1888–1970) was born in Trichinopoly (now Tiruchehirappalli), India. In 1907 he earned a master's degree in physics at the University of Madras, and ten years later became Professor of Physics at the University of Calcutta. Knighted in 1929, he moved in 1933 to the Indian Institute of Science at Bangalore, as head of the Department of Physics, and in 1947 was appointed director of the Raman Research Institute there. A Fellow of the Royal Society of London, Sir Chandrasekhara contributed in his time to the building up of nearly every Indian research institution. He founded the *Indian Journal of Physics* and the Indian Academy of Sciences, and trained hundreds of students who later occupied important positions in universities all over India.

The blue Mediterranean sea

In the history of science, we often find that the study of some natural phenomenon has been the starting-point in the development of a new branch of knowledge. We

Fig. 6.24. Chandrasekhara Raman. (© The Nobel Foundation, Stockholm.) Raman spent ten years (1907–17) of his life as an officer in a Finance Department in Calcutta. However, he found time to carry out experimental research in the laboratory of the Indian Association for the Cultivation of Science there. His success attracted attention, and in 1917 he was invited to become Professor of Physics at Calcutta University.

have an instance of this in the colour of sky light, which has inspired numerous optical investigations . . . Even more striking, though not so familiar at all, is the colour exhibited by oceanic waters. A voyage to Europe in the summer of 1921 gave me the first opportunity of observing the wonderful blue opalescence of the Mediterranean Sea. It seemed not unlikely that the phenomenon owed its origin to the scattering of sunlight by the molecules of the water. To test this explanation, it appeared desirable to ascertain the laws governing the diffusion of light in liquids, and experiments with this object were started immediately on my return to Calcutta in September 1921.

(Chandrasekhara Raman)[60]

Distressing puzzles

In the 1920s, following on Rutherford's discovery of the *proton* (p. 118), physicists imagined the atomic nucleus to be made up of protons and electrons (the so-called *proton–electron model*). However, there were serious difficulties confronting this idea. Two sorts of problems arose. Firstly, it had been found in experiments that certain nuclei had an integer spin, and accordingly they should obey a set of rules called *Bose–Einstein statistics* (see p. 175). On the other hand, following the electron–proton model of the nucleus, those same nuclei should have had an odd number of particles (protons plus electrons), each with a spin equal to $\frac{1}{2}$ (in units of $h/2\pi$). Now, it will be obvious that by combining an odd number of $\frac{1}{2}$ spins, it is possible to obtain only a half-integer overall spin, such as $\frac{1}{2}$ or $\frac{3}{2}$, but never an integer spin – as experiments required. As a consequence, those same nuclei ought to obey not Bose–Einstein statistics, but another different set of rules called *Fermi–Dirac statistics* (see p. 176).

Secondly, for an electron in a region of nuclear dimension (about some million-billionths of a metre), Heisenberg's uncertainty principle (p. 159) shows that its momentum (and consequently its energy) is too large for the electron to be confined in such a small space. These were the doubts that theorists had about the proton–electron model of the nucleus. But as long as those were the only subatomic particles known, they were unable to find another model.

But there was still another puzzle that troubled the physicists. It was the emission of beta rays (fast electrons) by radioactive nuclei (the so-called *beta decay*). The main problem here was: where did the energy of the emitted electron come from? Einstein had taught that energy can be created by the conversion of a tiny amount of mass. And physicists had long known that these electrons emerge from the nucleus with a wide spectrum (or range) of energies. (The discovery was first made by James Chadwick, one of Rutherford's assistants, in 1914, when working at the PTR in Berlin–Charlottenburg on a research studentship.) This means that many electrons are not fast enough to account for the fraction of the mass of the nucleus which is converted into kinetic energy. So what has happened to this missing energy?

Niels Bohr, to resolve the enigma, was ready to abandon the 'sacred' principle of the conservation of energy – at least as far as it applied to systems as small as nuclei (he thought that it was valid only statistically). However, Wolfgang Pauli, in order to save that principle, put forward a bold solution. This is what he said: there probably ought to exist another particle in the beta decay, which carries the missing energy, and which comes out of the nucleus along with the electron. He first stated his idea in a letter that he sent to the participants at a meeting on radioactivity which took place in Tübingen, Germany, on 4 December 1930. He wrote:

> Dear Radioactive Ladies and Gentlemen, . . . I have considered, in connection with the 'wrong' statistics of the [nitrogen and lithium-6 nuclei] as well as with the continuous [spectrum of beta rays], a way out for saving the 'law of change' of statistics and the conservation of energy: i.e. the possibility that inside the nuclei there are particles electrically neutral . . . which have spin $\frac{1}{2}$ and follow the exclusion principle . . . I do not consider it advisable, for the moment, to publish something about these ideas and first I apply to you with confidence, dear Radioactives . . . Unfortunately I cannot come personally to Tübingen, because I am necessary here for a ball that will take place in Zurich the night from 6 to 7 December.[61]

Did Pauli's *neutrino* (the name has been attributed to Enrico Fermi)[62] solve the problems of beta decay? Did electrons really exist in atomic nuclei? The answers would become clear in a few years' time.

Fig. 6.25. An annual conference at Bohr's famous institute (1936). First row (from left to right): W. Pauli, P. Jordan, W. Heisenberg, M. Born, L. Meitner, O. Stern, J. Franck, G. de Hevesy. Third row: O. Richardson (second), P. A. M. Dirac (sixth). Fifth row: V. Weisskopf (second), M. Delbrück. Standing from left: N. Bohr, L. Rosenfeld, E. Amaldi, G. Wick. (Courtesy Niels Bohr Archive, Copenhagen.)

The Copenhagen Spirit

It was his great strength to assemble around him the most active, the most gifted, the most perceptive physicists of the world. At that time, we find with Bohr at his famous Institute for Theoretical Physics, in Copenhagen, people such as [Oskar] Klein, Kramers, Pauli, Heisenberg, Ehrenfest, Gamow, Bloch, Casimir, Landau, and many others. It was at that time, and with those people, that the foundations of the quantum concept were created . . .

In lively discussions, in groups of two or more, the deepest problems of the structure of matter were brought to light. One can imagine what atmosphere, what life, what intellectual activity reigned in Copenhagen at that time.

Here was Bohr's influence at its best. Here it was that he created his style, the 'Kopenhagener Geist' [Copenhagen Spirit], the style of a very special character that he imposed onto physics. We see him, the greatest among his colleagues, acting, talking, living as an equal in a group of young, optimistic, jocular, enthusiastic people, approaching the deepest riddles of nature with a spirit of attack, a spirit of freedom from conventional bonds, and a spirit of joy that can hardly be described . . . In this great period of physics, Bohr and his men touched the nerve of the universe . . .

(Victor Weisskopf)[63]

The Göttingen tradition

The quantum revolution of the 1920s had its power-houses in two centres – Göttingen and Copenhagen. The Copenhagen school was practically created from nothing by one man – Niels Bohr. But the Göttingen school emerged from an ancient tradition of mathematics and physics, established in the nineteenth century, and associated with the names of Carl Friedrich Gauss and Wilhelm Weber. This great tradition continued until the first decades of the twentieth century, borne forward by such famous mathematicians as Bernhard Riemann, David Hilbert, Felix Klein, Hermann Minkowski and Hermann Weyl.

Then its glory began to be equalled by that of physics. The first famous Göttingen physicist of the twentieth century was Peter Debye, who became professor there in 1914. Early in the 1920s, after Debye had moved to the ETH in Zurich, two new professors were appointed: they were James Franck and Max Born. Franck became the director of the Second Institute of Physics, where he was engaged on experimental atomic physics, while Born became director of the Institute of Theoretical Physics. It was thus, from this tandem, that the Born–Franck era of Göttingen physics had its birth and origin. It was here that in 1925 the quantum revolution broke out – with the development of Werner Heisenberg's matrix mechanics. This revolution soon grew apace, particularly thanks to the contributions of Erwin Schrödinger in Zurich, and Paul Dirac in Cambridge. Göttingen and Copenhagen continued to be famous for many years to come by attracting brilliant young physicists from all over the world. The Göttingen school, for example, was honoured with such celebrities as: Wolfgang Pauli and Victor Weisskopf (Austria); Patrick Blackett (England); Werner Heisenberg, Pascual Jordan. Fritz London and Maria Mayer (Germany); George Uhlenbeck (the Netherlands); Leo Szilard and Eugene Wigner (Hungary); George Gamow (Russia); Enrico Fermi (Italy); Arthur Compton and J. Robert Oppenheimer (USA).

Chapter 7
The thirties

The Nobel Prizes to Werner Heisenberg, Erwin Schrödinger and Paul Dirac at the beginning of the 1930s show how fully quantum mechanics had become recognised as the right way to a perfect understanding of the phenomena occurring in the atomic and subatomic world.

Young, talented physicists created research groups in Europe and America (at Leipzig, in Germany; at Bristol, in England; at Rome, in Italy; at MIT, at Columbia University, and at Harvard and Caltech, in the USA; to mention just a few). These were then the centres where the new theory was fruitfully applied to the fields of atomic and molecular physics, solid-state physics, nuclear physics, astrophysics and biochemistry.

These same years witnessed the construction of the first machines capable of accelerating subatomic particles to very high energies. Then 1932 – *annus mirabilis*, as it was named – saw four amazing advances: the first nuclear disintegration produced by accelerated particles; then the discoveries of the hydrogen isotope *deuterium*, of the first antiparticle – the *positron* – and finally, of the *neutron*. Two years later, *artificial radioactivity* was discovered, and the neutron became a new experimental tool for penetrating the secrets of nuclear matter. Physicists were now able to understand the structure of the nucleus and its composition of protons and neutrons. They also devised new theories in order to explain nuclear phenomena, such as beta decay, and the strong forces that act inside the atomic nuclei.

Major events in physics before the outbreak of the Second World War include: the discovery of *superfluidity* in helium; the discovery in cosmic rays of the first of many surprising new subatomic particles; the recognition of nuclear reactions as the source of the energy produced in stars; and the development of refined methods for studying atomic nuclei, atoms and molecules through their magnetic properties. But the most spectacular of all was the discovery, at the end of the year 1938, that uranium nuclei can undergo *fission* reactions and so produce huge amounts of energy.

As we reach the year 1940, we see that, over and above the two Nobel awards to quantum mechanics, four other Nobel Prizes were awarded for major discoveries. These appertained to the neutron, the positron, artificial radioactivity and neutron physics and last, but by no means least, to a revolutionary instrument, the cyclotron.

1931

Atom smashers

In 1931 the Nobel Physics Committee, after a long debate, still on the subject of quantum mechanics, decided to postpone the physics award until the following year. But even by 1932 no conclusive decisions had been reached, so the Committee, once again under the influence of Carl Oseen, proposed to the Academy that no prize should be given for 1931, and that the 1932 prize should be reserved for another twelve months. So, while waiting for new events, let us have a look at the year 1931 in so far as nuclear physics is concerned.

New tools

As we have learned, during the preceding years, all research into the disintegration of atomic nuclei was performed with alpha particles emitted by radioactive sources. This method was first used in 1919 by Rutherford in his experiments on the transmutation of nitrogen nuclei (p. 117). It soon became clear that certain important questions concerning nuclear phenomena could only be answered by experimenting with intense beams of high-energy (or very fast) particles, produced by artificial means. As the particles to be accelerated were electrically charged, it was obvious that a high electric voltage could be used to give them higher energy. Three ingenious solutions were devised.

> ***First***. In 1928, two physicists, John Cockcroft and Ernest Walton, working at the Cavendish Laboratory in Cambridge, developed a *voltage multiplier*, which could accelerate charged protons up to energies amounting to several hundred thousand electron-volts. With their apparatus they were able, in 1932, to bombard lithium nuclei, and so obtain the first artificial nuclear reaction (p. 182). The machine used was the first to be popularly known as an 'atom smasher'.
>
> ***Second***. In the same years the American physicist Robert Van de Graaff was working at Princeton University. He invented another kind of accelerating machine (called an *electrostatic generator*), which developed very high voltages. Meanwhile, a third type of machine, called a *linear accelerator*, made its appearance: an alternating voltage pushed and pulled the charged particles along successive sections of a line of tubes, driving them along in a rectilinear direction.
>
> ***Third***. Because of their practical limits, linear accelerators were not widely used in those years. Ernest Lawrence at the University of California at Berkeley chose another route: he combined a magnetic field with a high-frequency alternating electric field. The magnetic field drove the particles in a circular path. Each time they completed a half-circle, the electric field accelerated them to higher energies. Lawrence's new device was named a *cyclotron*: it made high-energy particles immediately possible.

Fig. 7.1. Left: the Cockcroft–Walton voltage multiplier in early 1932. (Courtesy The Cavendish Laboratory, University of Cambridge, England.) Right: the first Lawrence cyclotron (its diameter was about 30 centimetres). (Courtesy Lawrence Berkeley National Laboratory.)

Quantum mechanics at work

In the late 1920s and early 1930s quantum mechanics was applied to the study of the properties of atoms, nuclei, particles, molecules and bulk matter, solving problems which previous theories had not been able to resolve.

Particle statistics

Physicists found, with particles like the electron, that their spin could be expressed as half-odd integers ($\frac{1}{2}$, $\frac{3}{2}$ and so on). They also found that particles like photons, or alpha particles, had their spin expressed as zero or as whole numbers (1, 2 and so on): these particles were called *bosons*. Systems of bosons can be dealt with according to a set of rules worked out by Einstein and the Indian physicist Satyendra Nath Bose in the period between the years 1924 and 1925. These rules are now called *Bose–Einstein statistics*. In 1925 Enrico Fermi devised another set of rules that apply to systems of particles with their spin expressed as half-odd integers. Finally, in 1926 Paul Dirac deduced both these statistics from his new quantum mechanics. Fermi's rules are collectively called *Fermi–Dirac statistics*, and particles that obey them are called *fermions*.

Quantum electrodynamics

Dirac, after discovering a new form of quantum mechanics (p. 151), succeeded in applying it to the subject of electromagnetism. In 1927 he published a paper in which he laid down the foundations for a quantum theory of radiation, called *quantum electrodynamics*.[1] In this theory he incorporated Einstein's concept of light quanta (or photons) in a consistent way. He thought of the electromagnetic field not as a continuous entity but as composed of discrete bundles of energy, which he considered as *quanta* of the electromagnetic field itself. He also explained how atoms emit and absorb radiation in the form of these same quanta. Dirac's electrodynamics was completed in 1929–30 by Heisenberg, Pauli and others; and in the 1940s the theory developed into its present modern form.

Dirac's equation

Dirac again! In 1928 he announced a new marvel of quantum mechanics.[2] He proposed a new equation to describe particles with a spin equal to $\frac{1}{2}$; this included the effects of Einstein's special relativity, and accurately predicted some details of the hydrogen spectrum, like its fine structure. Dirac's idea was that the electron could be described by four wave functions (four ψs). Two of these correspond to the electron spin: they are associated with the probability that the electron spin will be *up* or *down*. Thus the *ad hoc* spin hypothesis, suggested three years earlier (p. 141), followed naturally from Dirac's equation.

This equation produced solutions not only for negative electrons, but also for positively charged particles. Dirac first identified these particles with hydrogen nuclei (protons), but the German mathematician Hermann Weyl and the American theoretical physicist J. Robert Oppenheimer, independently of each other, showed that these hypothetical positive particles must have exactly the same mass as the electron. Dirac quickly rectified his error, and in 1931 he predicted the existence of antielectrons – that is, positive particles with the same mass and spin as the electrons. In 1932 a particle of such a kind was in fact discovered in cosmic rays (see p. 183), and was called the *positron* (from *posi*tive elec*tron*). It was the first *antiparticle* to be predicted by theorists of quantum mechanics.

Tunnelling

The Russian-born American physicist George Gamow was not only a successful popular science writer, but also an outstanding nuclear physicist and cosmologist. In the summer of 1928, when he was in Göttingen, he used quantum mechanics to develop a theory that successfully explained the behaviour of radioactive elements when they emit alpha particles (the so-called *alpha decay*). To explain this process, Gamow (and independently Edward Condon and Ronald Gurney at Princeton) devised a new quantum mechanical concept, called *tunnelling*. (Actually, this concept had been discovered some months earlier by J. Robert Oppenheimer, while studying how a weak electric field could separate an electron from its parent atom.)

Let us explain the tunnelling concept with an analogy. In classical physics a ball can bounce over a wall only if its kinetic energy exceeds its gravitational energy at the barrier (which is proportional to the height of the wall). In quantum mechanics, if a particle reaches a region with an energy barrier, but does not possess sufficient energy to pass over it, it has a small, but not zero, probability of passing through; after enough impacts, the odds are that the particle will have been able to actually 'tunnel' through the barrier. (As a matter of interest, this was the first successful application of quantum theory to nuclear phenomena.)

The structure of matter

Molecules and bulk matter are aggregates of atoms held together by electric forces: so the new-born quantum mechanics was used in order to understand their structure. In 1927 Heisenberg applied the new quantum concepts to explain the spectral lines of helium, which had been an unresolved problem since the days of the old quantum theory. In the same way he was able to explain the spectrum of the hydrogen molecule (this was cited in the motivation advanced for his award of the Nobel Prize). In the same year, Max Born and J. Robert Oppenheimer used Schrödinger's equation to obtain the structure of molecules; meanwhile, the American Walter Heitler and the German Fritz London studied the binding forces between atoms in molecules. These and other applications of wave mechanics introduced by a number of scientists greatly helped the creation of *quantum chemistry*.

As it did for chemistry, so quantum mechanics totally transformed the field of *solid-state physics*. Here, once again, Heisenberg succeeded in explaining a particular form of magnetism (ferromagnetism), while Wolfgang Pauli, and independently Arnold Sommerfeld, devised an atomic model that explained the electrical and thermal properties of metals. In the same years Felix Bloch (Nobelist in 1952), a research student of Heisenberg's at Leipzig, used Schrödinger's equation to invent a new theory concerning electrical conduction. He introduced the concept of *energy bands*, which was to become the basis for explaining the behaviour of metals, insulators and semiconductors.

1932

Quantum mechanics honoured

It was in October 1933 that the Swedish Academy of Sciences announced that Werner Heisenberg had been awarded the 1932 Nobel Prize. What had happened inside the Nobel Physics Committee to make its members change their minds? The story is simply this.

During the year the Committee had been flooded with nominations in support of quantum mechanics (Schrödinger alone had received nine nominations, and there had been eight for Heisenberg), so they tried hard to find a way out from the impasse in which they found themselves. Finally they came to a decision, and presented the following recommendation to the Academy (Carl Oseen too was in agreement!):

> The Committee is of the opinion that the point in time has now arisen at which the question of the founders of the new atomic theory should be decided . . .
> The Committee proposes that the Nobel Prize for 1932 be given to Professor Heisenberg (Leipzig) for the presentation of quantum mechanics and applications of it, particularly the discovery of the allotropic forms of hydrogen, and that the Nobel Prize in physics for 1933 be shared between Professor Schrödinger (Berlin) and Professor P. A. M. Dirac (Cambridge) for the discovery of new forms of atomic theory and applications of them.[3]

And now, let us speak a little bit about Heisenberg – the 1932 prizewinner. Who was this genius, this refined theorist and philosopher, who laid the foundations of quantum mechanics? Speaking of his youth and his years at the *Maximilian Gymnasium* in Munich, Heisenberg remembered how his reading of Plato's *Timaeus* had introduced him to a view of atoms which he found poetic and beautiful:

> The smallest particles of matter were said [by Plato] to be right-angled triangles which, after combining in pairs, . . . joined together into the regular bodies of solid geometry: cubes, tetrahedrons, octahedrons and icosahedrons. These four bodies were said to be the building blocks of the four elements, earth, fire, air and water . . . [The] whole thing seemed to be wild speculation . . . Even so, I was enthralled by the idea that the smallest particles of matter must reduce to some mathematical form . . . The most important result of it all, perhaps, was the conviction that, in order to interpret the material world we need to know something about its smallest parts.[4]

After the *gymnasium* (grammar school), Heisenberg studied at the University of Munich, and became a favourite pupil of Sommerfeld's. He loved Munich. There he spent an active and healthy life. Unlike his friend Pauli, who enjoyed visiting cafés and pubs at night, Heinsenberg used his spare time for hiking or bicycling in the countryside.[5]

He then spent the winter semester 1922–3 at the University of Göttingen, working with Max Born. Here fruitful relationships matured, especially those with James Franck. Born was now occupied in attempting to develop a new quantum theory. As Heisenberg remembered: 'it was the mathematical aspect which appealed to Born's mind. Born felt that one should repair physics by introducing a new mathematical tool . . .'.[6]

In July 1923, soon after he had obtained his doctorate in Munich, Heisenberg returned to Göttingen. He was part of a group of young physicists working on atomic structure. And in the spring of 1924 he went to Copenhagen, the Mecca of atomic theory: Bohr, whom he had met two years earlier in Göttingen, had invited him. (Many years later, he described this meeting with Bohr as a 'gift from heaven'.)

The following year Heisenberg was teaching in Göttingen, where he had become a lecturer; and had in the meanwhile 'pursued some work of his own, keeping its idea and purpose somewhat dark and mysterious', as Born recalled. Finally, in June he journeyed to Heligoland (an island in the North Sea), to combat a sudden attack of hay fever. It was there that he made his major breakthrough – the reformulation of atomic theory – 'one of the great jumps – perhaps the greatest – in the development of twentieth-century physics.'[7] Thus he recalled a night – the 'night of Heligoland' – when quantum mechanics was born: 'It was about three o'clock at night when the final result of the calculation lay before me . . . I was so excited that I could not think of sleep. So I left the house . . . and awaited the sunrise on top of a rock.'[8]

Towards the end of June Heisenberg returned to Göttingen and by about the middle of July he had completed his seminal paper. Born recalled the circumstances as follows:

> He came to me with a manuscript and asked me to read it and to decide whether it was worth publishing . . . I remember that I did not read this manuscript at once . . . But when, after a few days, I read it I was fascinated . . . I began to ponder over his symbolic multiplication, and was soon so involved in it that I thought about it the whole day and could hardly sleep at night. For I felt there was something fundamental behind it, the consummation of our endeavours of many years. And one morning . . . I suddenly saw light: Heisenberg's symbolic multiplication was nothing but the matrix calculus, well known to me since my student days from the lectures of [Jakob] Rosanes in Breslau [now Wrocław].[9]

A remarkable year

The year 1932 was a remarkable one for physics, not only because of the Nobel Prize that was awarded to Heisenberg, but also for many important discoveries. Let us follow the chronology of the events.

First, deuterium

In January, the American chemist Harold Urey, then working at Columbia University, New York City, announced the discovery of a heavy isotope of hydrogen, which was named *deuterium*. This is an atom with one external electron and with a mass about twice that of the hydrogen mass (its nucleus contains one proton and one neutron). For his discovery Urey was awarded the 1934 Nobel Prize for chemistry.

Werner Heisenberg (1901–76)

Werner Karl Heisenberg was born in Würzburg, Germany, the son of a university professor of Greek philology. In 1920 he entered the University of Munich, where he studied physics under Arnold Sommerfeld. After obtaining his doctorate in 1923, he went to Göttingen as an assistant to Max Born. Between 1924 and 1925 he spent an academic year at Bohr's institute in Copenhagen.

In 1927, at the age of twenty-six, Heisenberg was appointed Professor of Theoretical Physics at the University of Leipzig, becoming the youngest full professor in the whole of Germany. Under his direction the Leipzig Institute of Physics became a leading research centre for quantum theory. Among his students were Felix Bloch (Nobelist in 1952), Rudolf Peierls, Edward Teller, Victor Weisskopf and Carl Weizsäcker, all of whom would achieve fame in the world of physics.

During the Second World War Heisenberg was a leading figure in the secret German nuclear fission project: in 1942 he moved to Berlin to take over the directorship of the Kaiser Wilhelm Institute of Physics, the principal centre for reactor research at that time in Germany, and a professorship at the University of Berlin. (His role in the project of producing German nuclear weapons has been the subject of heated controversy ever since.) At the end of the war he was captured, together with other German scientists, and they were detained for six months at an English country manor, near Cambridge. Heisenberg then returned to Germany and became director of the Max Planck Institute for Physics and Astrophysics in Göttingen. In 1958 he moved with the institute to Munich, where it was to become a major centre for research in Germany.

After the war Heisenberg sought to establish a prominent role for Germany in the creation of a policy for the development of science and technology, and he devoted himself to re-establishing peaceful international relations. (He considered international cooperation in the field of science a 'main tool to reach understanding between peoples'.) In 1954 he headed the German delegation responsible for the organising of the new European laboratory of CERN, Geneva, Switzerland.

> . . . one of the most marvellous traits of Heisenberg was the almost infallible intuition that he showed in his approach to a problem of physics and the phenomenal way in which the solutions came to him as if out of the blue sky . . . [We] all knew the dreamy expression on his face, even in his complete attention to other matters and in his fullest enjoyment of jokes or play, which indicated that in the inner recesses of the brain he continued his all-important thoughts on physics.
>
> (Felix Bloch)[10]

Second, the neutron

In a letter to the editor of *Nature*, published on 27 February 1932, the English physicist James Chadwick triumphantly announced the discovery of the long-awaited *neutron* (the neutral particle hypothesised by Rutherford in 1920 as a component of atomic nuclei; see p. 122). Chadwick's *experimentum crucis* was carried out at the Cavendish Laboratory. Here, in three steps, is the full story.

1930. The German physicist Walther Bothe and his research student Herbert Becker, working at the PTR in Berlin–Charlottenburg, bombarded beryllium with alpha particles from polonium. They discovered a new form of radiation that was even more penetrating than the gamma rays emitted in the decay of radium. (Bothe's rays could penetrate a metal plate several centimetres thick, without any notable loss of energy.) The newly discovered radiation was assumed to consist of high-energy gamma rays which were understood to be emerging from the target nuclei, and it was called the *radiation of beryllium*.

1931. Irène Curie, the daughter of Pierre and Marie Curie, and her husband, Frédéric Joliot, both working at the *Institut du Radium* in Paris, decided to use their exceptionally strong polonium source (it was the world's most powerful source of alpha particles) for studying the new radiation discovered by Bothe. They found that this radiation could knock protons out of hydrogen atoms contained in a block of paraffin (a substance composed of hydrogen and carbon). They tried to understand this surprising effect, guessing that the radiation produced in the beryllium consisted of energetic photons, which could have struck the protons in a sort of Compton effect. In January 1932 they published their results in the *Comptes Rendus de l'Académie des Sciences* in Paris. (In an ordinary Compton effect, the knocked-out electrons are about two thousand times lighter than protons: this means that if you had wanted to strike a proton out of the paraffin by means of the Joliot-Curies' photons, you would have needed energies even larger than those of the alpha-particle projectiles! The Joliot-Curies in fact soon changed their minds; in a note of February 1932 they wrote that the protons were knocked out not by a sort of Compton effect, but much more probably by a new kind of interaction between gamma rays and protons.)

1932. Soon after reading the Joliots-Curies January paper, Chadwick hurried to his laboratory and repeated their experiment. His apparatus consisted of a source of alpha particles, a beryllium target, a particle detector connected to an amplifing circuit and a cathode-ray oscillograph (see Fig. 7.2). When Chadwick interposed a sheet of paraffin in the path of the 'radiation of beryllium', he observed that the number of deflections recorded on the oscillograph went up significantly: this was due to the arrival in the detector of secondary particles ejected from the paraffin by the mysterious radiation. The secondary particles were readily identified as recoil protons.

Fig. 7.2. Schematic diagram of Chadwick's apparatus. (Source: James Chadwick, *Proceedings of the Royal Society*, A136, 1932, p. 695.)

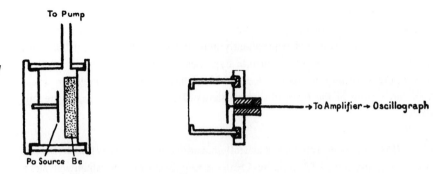

Chadwick went on to explain the 'radiation of beryllium' as consisting not of gamma rays, but rather of massive particles with zero electric charge (they were indeed named *neutrons*). By examining the exchange of mass of a great number of collisions between neutrons and nuclei of different elements, he succeeded in determining their mass. As was to be expected, he found it almost the same as that of the proton.[11]

Fig. 7.3. Schematic diagram of a voltage multiplier. Accelerated protons interact with the atomic nuclei in the target, so producing new particles.

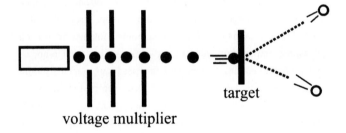

Third, artificial nuclear reactions

The third discovery of the year came in April, and once again the Cavendish Laboratory was responsible – when the first disintegration of light nuclei with artificially accelerated protons was achieved. The experiment was performed by John Cockcroft and Ernest Walton, using their recently invented voltage multiplier (see p. 174). They accelerated protons to energies as high as 150 000 electron-volts, and showed that these interacted with lithium nuclei, which disintegrated, each producing two energetic alpha particles.

The reaction detected by Cockcroft and Walton can be written as

$$\text{proton-1} + \text{lithium-7} = 2\,\text{helium-4}$$

Cockcroft and Walton interpreted their observations as due to the transformation of the lithium isotope of mass 7 (three protons + four neutrons) into two alpha particles. They showed that the observed energy of the emitted alpha particles was in agreement (within twenty per cent) with that computed from the masses of the nuclei appearing in the above reaction (according to Einstein's formula: $E = mc^2$), and also from the energy of the incident proton. This was the first experimental verification of the equivalence of mass and energy, which constitutes the most important prediction of special relativity.[12]

Fourth, the positron

In 1930 the American physicist Carl Anderson, working at Caltech under Robert Millikan, started to set up an experimental apparatus to study the nature of the cosmic radiation discovered by Victor Hess in 1912 (p. 86). The apparatus contained a Wilson cloud chamber equipped with a strong magnet. The chamber rendered visible the tracks created by the passage of charged cosmic-ray particles, which were curved by the magnetic field. It also contained a lead plate through which the observed particles had to pass. Photographs of cosmic-ray tracks were recorded day and night. Upon examination of the photographs, Anderson found not only the tracks of electrons travelling downwards, but also tracks that were moving upwards. One of these events occurred on 2 August 1932 (see Fig. 7.4).

Let us try to interpret the 'August track' in company with Anderson. In the magnetic field a charged particle moves under the influence of the magnetic force in a circular path. Now, it is known that particles with higher speeds have smaller curvature: thus, we are sure that our particle enters the cloud chamber from below and travels upwards. In fact, in crossing the lead plate it loses energy. As a consequence, its speed decreases, and its curvature becomes larger (as you can see in the upper part of Fig. 7.4). Moreover, given the direction of motion and the direction of the magnetic field, any physics student can assure you that such a curvature must belong to a positive particle. Thus, it cannot be an electron. Is it a proton or a heavier nucleus? Both are impossible; such particles would have insufficient energy to cross the plate. Thus, Anderson concluded that the most plausible interpretation was that it must be a 'positive electron' with the same mass as the negative one. The antielectron (it was christened with the name of *positron* in Anderson's paper)[13] that Dirac had predicted (p. 176) had thus eventually been found – despite the fact that Anderson had not been aware of Dirac's prediction.

Quite soon afterwards, Patrick Blackett and his young Italian co-worker, Giuseppe Occhialini, both working at the Cavendish Laboratory, confirmed Anderson's discovery. They had used an improved technique in which a cloud chamber was only triggered after it was certain that a cosmic ray had passed simultaneously through Geiger counters (p. 275) placed above and below the chamber itself. They obtained many photographs of the positrons, and

Fig. 7.4. The 'August track': here you can 'see' the first positron track discovered by Anderson. (Source: Carl Anderson, *Physical Review*, 43, 1933, p. 492.)

early on in 1933 they published their results in the *Proceedings of the Royal Society*.[14]

1933

Schrödinger and Dirac

Continuing our chronicle, it is now the two other creators of quantum mechanics, Erwin Schrödinger and Paul Dirac, who come to claim our attention.

Erwin Schrödinger (1887–1961) was born in Vienna, Austria, the son of the owner of a linoleum factory. He studied at the University of Vienna, where he was awarded his doctorate in 1910. During the First World War he was engaged in military service on the Italian front, as an officer in the Austrian artillery. In 1920 he accepted an academic position in Jena as assistant to Max Wien (Wilhelm Wien's brother), followed by positions at the universities of Stuttgart and Breslau. In 1921 he was appointed to the chair of theoretical physics at the University of Zurich, where he remained until the year 1927. During this period he worked on a variety of subjects in the area of theoretical physics, such as atomic structure, quantum statistics and the specific heat of solids.

After reading de Broglie's doctoral thesis ('I have been intensely concerned these days with Louis de Broglie's ingenious theory. It is extraordinarily exciting . . .', he said at that time), he then wrote, during the Christmas holidays of

1925 in Arosa, in the mountains near Zurich, the first of his four famous papers on wave mechanics (see p. 147). Walter Moore, a biographer of Schrödinger, wrote:

> Erwin wrote to 'an old girlfriend in Vienna' to join him in Arosa, while Anny [his wife] remained in Zurich. Efforts to establish the identity of this woman have so far been unsuccessful . . . Like the dark lady who inspired Shakespeare's sonnets, the lady of Arosa may remain for ever mysterious. We know that she was not Lotte or Irene [or] Felicie . . . Whoever may have been his inspiration, the increase in Erwin's powers was dramatic, and he began a twelve-month period of sustained creative activity that is without parallel in the history of science . . .[15]

(Schrödinger, Moore further commented, '. . . was a passionate man, a poetic man, and the fire of his genius would be kindled by the intellectual tension arising from the desperate situation of the old quantum theory . . . It seems also that psychological stress, particularly that associated with intense love affairs, helped rather than hindered his scientific creativity . . .'.)[16]

In 1927 Schrödinger moved to the University of Berlin to succeed Max Planck in the chair of theoretical physics. In 1933 he decided that he could not continue to live in Nazi Germany. He moved to England, to Oxford, and then back to Austria, to Graz. Later he moved to Belgium, and Italy; and finally in 1939, he went to Dublin, Ireland. There, at the Institute for Advanced Studies, he became director of the School of Theoretical Physics. (During this period he wrote a well-known book entitled *What is Life?*, in which he used quantum physics to explain the physical aspects of the living cell.) He remained in Dublin until 1956, when he returned to his beloved Vienna as Professor Emeritus of Theoretical Physics (a chair created for him).

Schrödinger never liked the generally accepted dual description of atomic physics in terms of waves and particles, as proposed by Bohr, Heisenberg and Born. He tried to set up a theory in terms of waves only. Like Einstein, he sought throughout all his life to unify gravitation and electromagnetism.

Unconventional man

> When he went to the Solvay conferences in Brussels, he would walk from the station to the hotel . . . carrying all his luggage in a rucksack and looking so like a tramp that it needed a great deal of argument at the reception desk before he could claim a room.
>
> (Paul Dirac)[17]

Genius

> His private life seemed strange to bourgeois people like ourselves. But all this does not matter. He was a most lovable person, independent, amusing, temperamental, kind and generous, and he had a most perfect and efficient brain.
>
> (Max Born)[18]

Fig. 7.5. From left to right: Heisenberg's mother, A. Schrödinger, Dirac's mother, P. Dirac, W. Heisenberg and E. Schrödinger. Stockholm, December 1933. (Max-Planck Institute für Physik, courtesy AIP Emilio Segrè Visual Archives.)

Paul Adrien Maurice Dirac (1902–84) was born in Bristol, England, the son of a schoolmaster. He studied electrical engineering at the University of Bristol, where he obtained his degree in 1921. Then he entered St John's College, Cambridge, as a research student in mathematics. He received his Ph.D. in 1926, with a thesis in which he developed his original ideas concerning quantum mechanics (see p. 151). After visits to Copenhagen and Göttingen, Dirac returned to Cambridge, where he was elected a Fellow of St John's College. In 1932 he was appointed Lucasian Professor of Mathematics at the University of Cambridge, the prestigious chair once occupied by Sir Isaac Newton. In 1930 he was elected a Fellow of the Royal Society, and from 1934 on was a member of the Institute for Advanced Study, Princeton, USA. After his retirement in 1968, he left Cambridge and moved to the USA, where he became Professor Emeritus at Florida State University, Tallahassee. His masterpiece entitled *The Principles of Quantum Mechanics*, a work which has been compared in its stature with Newton's *Principia*, has stimulated generations of physicists.

At the Nobel banquet

After formal toasts to the King and to the memory of Nobel, toasts were offered to the prizewinners, who replied at different lengths. [Dirac] . . . said . . . that the cause of the great depression was that people much preferred to collect interest 'through all eternity' . . . Anny thought that his talk was 'a communistic propaganda tirade' . . . Heisenberg spoke very briefly to thank everyone for their hospitality. Schrödinger, perhaps with some inspiration from the champagne, gave the most spirited response . . . [He ended with these words]: '. . . I hope that I may come again . . . It will not be to a celebration in halls bedecked with flags and not with so many formal clothes in my luggage, but with two long skis over my shoulder and a knapsack on my back. And I hope then to learn to know this country that has shown me so much generosity and affection and – if that be possible at all – to learn to love it even more deeply than today.' These words were received with thunderous applause.[19]

Einstein, 1933

Einstein [was] abroad when frivolous conservative intriguers brought Hitler to power on January 30, 1933. [He] was in Pasadena, en route to a semi-annual appointment at the Institute for Advanced Study in Princeton . . . [By] mid-March, he vowed not to return to Germany, to a country of intolerance, unless conditions changed . . . He resigned from the Prussian Academy . . . and in an official communiqué the academy announced Einstein's resignation 'without regret' . . . Einstein's books were burned, his property was confiscated, and in 1934 his German citizenship was revoked.'[20]

1934

Glorious days again

In 1934, although the list of aspiring Nobelists was a long one (it included deferred candidates like Otto Stern, Max Born and Wolfgang Pauli, and new ones like Clinton Davisson, George Paget Thomson, Carl Anderson and Patrick Blackett), the Nobel Physics Committee was not able to identify anyone worthy of an award, so that the Academy backed the Committee's recommendation to wait, and reserve the 1934 prize until the following year. (In 1935 the Academy withheld the 1934 physics prize, and devolved it to a special fund to aid Swedish research.)

But all this did not mean that 1934 was a quiet year for physics research – on the contrary, amazing discoveries were made both in Paris and in Rome.

A family tradition

Pierre and Marie Curie received the Nobel Prize for physics in 1903 for their work on *natural radioactivity* (p. 42). Their daughter Irène and her husband, Frédéric Joliot, discovered *artificial radioactivity* at the *Institut du Radium* in Paris. In a letter to the editor of *Nature*, dated 10 January 1934, they wrote:

> Some months ago we discovered that certain light elements emit positrons [positive beta rays] under the action of alpha particles. Our latest experiments have shown a very striking fact: when an aluminium foil is irradiated on a polonium preparation [alpha rays], the emission of positrons does not cease immediately when the active preparation is removed: the foil remains radioactive and the emission of radiation decays exponentially as for an ordinary radio-element. We observed the same phenomenon with boron and magnesium . . .[21]

Thanks to this one-page paper, Irène and Frédéric were awarded the 1935 Nobel Prize for chemistry. At the ceremony in Stockholm Professor W. Palmaer ended his laudatory speech with these words:

> Madame, 24 years ago, your mother, Madame Marie Skłodowska-Curie, was present at the Nobel festival in order to receive the Chemistry Prize, as a reward for her discovery of radium in the first place; and you, Madame, were also present on that occasion, as a little girl. Her husband, your father Pierre Curie, was already dead; but earlier on, in 1903, he had shared with her half the Physics Prize, an award well deserved for their work on the phenomena of radiation. In collaboration with your husband, you have worthily maintained those brilliant traditions.[22]

Artificial radioactivity, discovered by the Joliot-Curies, can be explained in the following way. An alpha particle (a helium nucleus, two protons + two neutrons) collides with a nucleus of aluminium (thirteen protons + fourteen neutrons). It is absorbed, and then the new system emits a neutron. So the final products are: a nucleus with fifteen (thirteen + two) protons and fifteen (fourteen + one) neutrons plus a free neutron. Now, the new nucleus ($A = 30, Z = 15$) is a radioactive isotope of phosphorus. It decays with a half-life of about three minutes, emitting a positron as it does so.

The lads of Panisperna Road, Rome

Enrico Fermi was appointed in 1927 to the newly created chair of theoretical physics at the University of Rome. The institute of physics where he worked was located on the *Via Panisperna* ('Panisperna Road'). Soon after his arrival, Fermi and Franco Rasetti (a friend from their university days in Pisa) gathered together a small group of first-class students: Edoardo Amaldi, Ettore Majorana, Bruno Pontecorvo and Emilio Segrè. (They were called 'The lads of Via Panisperna'. Judging Fermi infallible, they nicknamed him 'The Pope'.)

Between 1930 and 1934 Fermi began his research in experimental nuclear physics, while continuing his theoretical work. In March 1934 he and his fellow students tried to observe effects similar to those seen by the Joliot-Curies. He decided to use neutrons as bombarding particles because, unlike alpha particles, they are not electrically charged and so do not feel the electric repulsion from the atomic nuclei; therefore, they can easily penetrate inside them. Emilio Segrè thus remembers those days at the institute of physics:

> At first, Fermi started irradiating all the elements he could set his hands on, in order of increasing atomic number . . . [We] . . . discovered about forty new radioactive substances . . . In the spring of 1934 we irradiated the heaviest element then known – uranium . . . Furthermore, we showed by chemical means that none of the [radioactive elements] produced in uranium could be ascribed to elements of atomic number greater than that of lead. We then thought that we had produced

transuranic elements ... In this we were in error, at least in part; while it was true that transuranic elements were formed, ... what we had observed was something quite different.[23]

Work came to a temporary halt for the summer holidays. Fermi went to South America for a series of lectures, and in the autumn experiments were resumed. Segrè continues:

> In the fall of 1934 ... we faced a major surprise. We found ... that neutrons filtered through paraffin were much more effective in producing nuclear reactions than those emerging directly from a radon plus beryllium source ... Within a few hours we were able to check this hypothesis, and on the evening of October 22, 1934, the same day in which we had discovered the effect, we wrote a one-page note ... that firmly established the facts and their interpretation.[24]

Thus, with this second discovery – namely, that slowed-down neutrons produce much larger quantities of artificial radioactivity ('the most important discovery

Fig. 7.6. A historic photograph: from left to right: O. D'Agostino, E. Segrè, E. Amaldi, F. Rasetti, E. Fermi. (Courtesy Edoardo Amaldi Archive, Physics Department, the La Sapienza University in Rome.)

I have made', as Fermi himself was heard to declare), Rome became an important centre of experimental physics at an international level.

> One day, as I came to the laboratory, it occurred to me that I should examine the effect of placing a piece of lead before the incident neutrons . . . [But] I was clearly dissatisfied with something; I tried every excuse to postpone putting the piece of lead in its place. When finally, with some reluctance, I was going to put it in its place . . . I said to myself: 'No, I do not want this piece of lead here; what I want is a piece of paraffin.'
>
> (As reported by Subrahmanyan Chandrasekhar)[25]

> Dear Fermi,
>   I have to thank you for your kindness in sending me an account of your recent experiments in causing temporary radioactivity in a number of elements by means of neutrons. Your results are of great interest, and no doubt later we shall be able to obtain more information as to the actual mechanism of such transformations . . . I congratulate you on your successful escape from the sphere of theoretical physics! You seem to have struck a good line to start with. You may be interested to hear that Professor Dirac also is doing some experiments. This seems be a good augury for the future of theoretical physics!
>
> (From a letter of 23 April 1934 from Rutherford to Fermi)[26]

1935

The neutron

Chadwick's epoch-making discovery of the neutron at the Cavendish Laboratory (see p. 181) merited the 1932 Hughes Medal of the Royal Society of London, which was followed by the 1935 Nobel Prize for physics.

> Apotheosis of the true neutron (Finale)
> 'The Neutron has come to be.
> Loaded with Mass is he.
> Of Charge, forever free.
> Pauli, do you agree?'
>
> (From a parody on Goethe's *Faust*, written and performed in Copenhagen by Bohr's pupils in April 1932)[27]

A persistent search

Let us resume the story of the neutron, which began early in the 1920s. As Chadwick recalled in his Nobel speech:

> The first suggestion of a neutral particle with the properties of the neutron we now know, was made by Rutherford in 1920 [p. 122]. He thought that a proton and an electron might unite in a much more intimate way than they do in the hydrogen atom, and so form a particle of no net charge and with a mass nearly the same as

Fig. 7.7. James Chadwick. (Courtesy The Cavendish Laboratory, University of Cambridge, England.)

that of the hydrogen atom . . . No experimental evidence for the existence of neutral particles could be obtained for years . . . The possibility that neutral particles might exist was, nevertheless, not lost sight of. I myself made several attempts to detect them – in discharge tubes actuated in different ways, in the disintegration of radioactive substances, and in artificial disintegrations produced by [alpha] particles. No doubt similar experiments were made in other laboratories, with the same result . . . The first real step towards the discovery of the neutron was given by a very beautiful experiment of Mme. and M. Joliot-Curie, who were also investigating the properties of [the] beryllium radiation [p. 181].[28]

Chadwick followed up the researches of the Joliot-Curies. As soon as he read their paper of January 1932, he felt that luck had returned to him. As he recollected:

[One morning I read] . . . the communication of the [Joliot-Curies] in the *Comptes Rendus*, in which they reported a still more surprising property of the radiation from beryllium, . . . A little later that morning I told Rutherford . . . As I told him about the [Joliot-Curie] observation and their views on it, I saw his growing amazement; and finally he burst out 'I don't believe it' . . . Of course, Rutherford agreed that one must believe the observations; the explanation was quite another matter. It so happened that I was just ready to begin to experiment, for I had prepared a beautiful source of polonium . . . I started with an open mind, though naturally my thoughts were on the neutron . . . A few days of strenuous work were sufficient to show that these strange effects were due to a neutral particle and to enable me to measure its mass: the neutron . . . had at last revealed itself.[29]

Thus Chadwick made his great discovery. It was not by accident, but as a result of a persistent and thorough search.

James Chadwick (1891–1974) was born in Cheshire, not far from Manchester, England. He studied at the University of Manchester under Rutherford, and obtained his master's degree in 1913. He then went to work with Hans Geiger at the PTR in Berlin-Charlottenburg. From 1919 on he worked at the Cavendish Laboratory, where he studied the disintegration of light elements by bombarding them with alpha particles, and investigated the properties and structure of atomic nuclei. In 1927 he was elected a Fellow of the Royal Society, and in 1935 he became Professor of Physics at the University of Liverpool. From 1943 to 1946 he participated (in the USA) in the Manhattan Project. On his return to England he was elected Master of Gonville and Caius College, Cambridge (1948–59). Chadwick was knighted in 1945, and received the Medal of Merit from the US government in 1946.

Recollections

> I remember vividly the occasion when Chadwick described his experiments to the Kapitza Club [at the Cavendish]. He had been dined and wined well by Kapitza beforehand, in celebration of the discovery, and was in a very mellow mood. The intense excitement of all in the Cavendish, including Rutherford, was already remarkable, for we had heard rumours of Chadwick's results. His account of the experiments was extremely lucid and convincing, and the ovation he received from his audience was spontaneous and warm
>
> (Mark Oliphant)[30]

Particles and forces

The discovery of the neutron at once solved all problems that physicists had had about the structure of the atomic nucleus. In 1932, soon after Chadwick's discovery, Heisenberg explained the atomic nucleus as consisting of protons and neutrons, rather than protons and electrons. Thus every chemical element could be defined simply by two numbers: first, the number of protons in its nucleus, the *atomic number Z*, which coincides with the number of external electrons (it establishes the place of an element in the Mendeleev periodic system); and second, the *mass number A*, which is the number of nucleons (protons and neutrons).

Once the proton–neutron model came to be accepted, physicists realised that there was no explanation for what it is that holds the nucleons tightly packed inside the nucleus: the immense electric repulsion between the electrically charged protons, a hundred million times stronger than the electric attraction of atomic electrons, should cause the nucleus to explode. Clearly, some other type of attractive force, a much stronger one, must be involved.

The idea came once again from Heisenberg. He suggested that in the nucleus protons and neutrons were held together by a kind of 'exchange force', which continually interchanges their natures: a nucleon is first a proton, then a neutron, then a proton, and so on. The new force (now called the *strong nuclear force*) was assumed to be powerfully attractive, but dropping in strength very rapidly with the distance (becoming virtually equal to zero outside the nucleus).

Another major question concerned the nuclear *beta decay* (p. 169). Where did the high-speed electrons emitted by certain radioactive nuclei come from if there were none in the nucleus? The problem was attacked in 1933–4 by Enrico Fermi.[31] He introduced another new force (now known as the *weak nuclear force*), which could account for the conversion in a nucleus of a neutron into a proton with the emission of an electron and a neutrino. Fermi's theory describes the beta decay without assuming that electrons and neutrinos are present in the nucleus from the beginning, but claiming that they are created at the instant of the electron emission, when a nucleon jumps from the neutron quantum state to the proton state.

Finally, in 1935, a Japanese theoretician, of the name of Hideki Yukawa, analysed the problem of nuclear forces thoroughly. He imagined that, inside a nucleus, the nucleons were continually exchanging a massive particle (later on called a *meson*). This particle could either transmit the strong attractive force between two nucleons (the Heisenberg process), or break down into an electron and a neutrino (the Fermi process). Moreover, Yukawa calculated that its mass should be some hundred times greater than that of the electron (p. 229).

1936

Cosmic rays

The term *cosmic rays* was coined by Robert Millikan (Nobel 1923) to indicate 'the radiation from above' which had been discovered by Victor Hess in 1912 (p. 86). Cosmic rays are high-energy subatomic particles that travel all the way through cosmic space and reach right down as far as earth's atmosphere. Most of these particles are protons and come from sources outside the solar system. Cosmic rays were central to the studies of particle physics that were carried out between the 1930s and the early 1950s. In modern astrophysics cosmic rays are carriers of messages from distant regions of our galaxy. Hess made his discovery in August 1912, and he was awarded the Nobel Prize twenty-four years later, in 1936, sharing it with Carl Anderson – the latter for his discovery of the positron (p. 183).

An extraordinary adventure
At the beginning of the twentieth century, physicists found that an electroscope (a device that detected the presence of charged ions produced by radiation in the atmosphere) discharged even if it was kept well away from radioactive sources. The prevailing opinion attributed the source of the radiation which caused the discharging of electroscopes to radioactive materials in the earth's crust. In 1910 it was found that ionisation persisted as one ascended heights of some hundreds

of metres. At that time gamma rays were the most penetrating form of radiation known. If the ionisation had been due to gamma rays originating at the earth's surface, the intensity should have been negligible at such heights. Where did that new radiation come from?

After his balloon ascents of 1912, in which he measured the ionisation of the atmosphere as the altitude increased, Hess made the bold assumption that the mysterious radiation came from somewhere beyond the atmosphere. But this was not generally accepted by the community of physicists, the most sceptical being Millikan himself. Between 1923 and 1926 Millikan undertook a series of experiments, at different mountain altitudes, in order to measure how the intensity of Hess' radiation was absorbed by the atmosphere. The results were so convincing that he changed his mind, and concluded that Hess' radiation did actually come from beyond the earth's atmosphere (he described this radiation as 'ethereal rays of cosmic origin').

But what about the nature of this radiation? Millikan continued to think that it was composed of high-energy gamma rays (the 'birth cry' of atoms – created in cosmic space, he declared). But in 1929 two German physicists, Walther Bothe and Werner Kohlhörster, performed a key experiment that proved that 'the primary cosmic radiation itself consisted of charged particles rather than photons'.

After this, the study of cosmic rays continued in many laboratories. At the University of Florence at Arcetri, Italy, the Italian physicist Bruno Rossi demonstrated that new subatomic particles were being produced when cosmic rays interacted with matter. Again, in 1932, Patrick Blackett and Giuseppe Occhialini, at the Cavendish Laboratory, built the first cloud chamber controlled by Geiger-Müller counters (p. 275). With their apparatus they made observations that marked another milestone in the history of cosmic rays. Not only were they able to confirm Anderson's discovery of the positron, but they also took many pictures showing groups of secondary particles (called *electromagnetic showers*; see Fig. 7.8) that were originating in the walls of the chamber, or in a metal plate placed horizontally across the chamber itself. These showers contained both electrons and positrons.

All these investigations opened up a new world, that of high-energy physics. Here physicists found particles with energies that were a thousand, a million, a billion times greater than those of particles emitted by natural radioactive substances!

Victor Franz Hess (1883–1964) was born in Waldstein Castle, Styria, Austria. He studied at the University of Graz, where in 1910 he received his doctorate in physics. From 1920 to 1937 Hess was professor of physics at the universities of Graz and Innsbruck. Following the Nazi occupation of Austria he was dismissed from his professorship on account of his devotion to Roman Catholicism. He then emigrated to the USA, where he became Professor of Physics at Fordham University in New York City.

Particles and antiparticles

As we have seen, Dirac postulated the positron in 1931 (p. 176), and Carl Anderson established its existence in 1932. Anderson's claim aroused controversy, until his discovery was verified by Blackett and Occhialini at the Cavendish. Besides the positron, these experiments produced other results of crucial importance. Some pictures showed showers where the numbers of electrons and positrons were comparable. Blackett and Occhialini thought of these showers as being produced by high-energy photons in the chamber walls. However, they had to face some puzzling questions: Why did these showers contain electrons and positrons? And why were positrons so rare as to elude observation?

The answers came from Dirac's theory. According to Dirac, photons interacting with matter can produce electrons and positrons, always in pairs. The mass of a particle–antiparticle pair results from a process of 'energy materialisation' (technically called *pair production*). In this process part, or most, of the photon energy is converted into the masses of the produced particles, in agreement with Einstein's mass–energy formula $E = mc^2$. Thus, an electromagnetic shower contains positive and negative electrons.

Concerning the second question – why positrons are so rare – the answer once again came from Dirac. Positrons live forever in empty space. However, they react with the atomic electrons of ordinary matter, and consequently have a very short lifetime. In fact, when a positron meets an electron, the two particles annihilate each other (a process known as *particle–antiparticle annihilation*), and their masses are completely transformed into photons; that is, pure electromagnetic energy.

A new particle

Using his cloud chamber, Anderson and his co-worker, Seth Neddermeyer, carried out experiments at mountain altitudes, and in 1936 discovered a new particle of unit charge but with a mass between those of an electron and a proton – about two hundred times the electron mass. The 'mesotron', as the new particle was called, seemed to be remarkably like the particle predicted by Yukawa in 1935, when explaining the nuclear force existing inside the atomic nucleus (p. 193). It was only ten years later that it was demonstrated that the 'mesotron' (later on renamed *muon*) was not the mediator of the nuclear force, but a kind of more massive electron (see p. 231).

Carl David Anderson (1905–91) was born in New York City. He received his Ph.D. in 1930 from Caltech, where he worked with Robert Millikan. He began his research in cosmic rays there, and in 1932 announced the discovery of the positron. Anderson spent his entire career at Caltech, where he was Professor of Physics from 1939 to 1976. During the Second World War he conducted research on rockets. In 1950, working with his cloud chamber on White Mountain, California, he found many examples of the so-called *strange particles*, thus

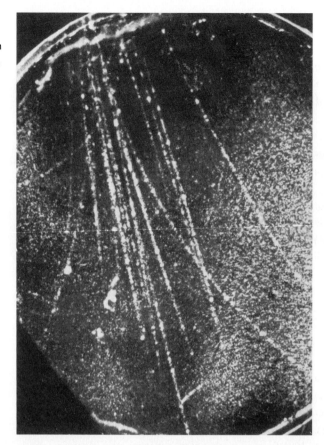

Fig. 7.8. One of the first photographs of a shower with electron and positron tracks in a cloud chamber. (Source: P. M. S. Blackett and G. P. S. Occhialini, *Proceedings of the Royal Society*, A139, 1933, p. 722. Reprinted by permission of the Royal Society, London.)

Fig. 7.9. From left to right: C. Anderson, R. Millikan and S. Neddermeyer. (Courtesy AIP Emilio Segrè Visual Archives.)

confirming a discovery which had been made two years previously in England (see p. 231).

A cold reception

'[When] ... in the fall of 1932 Millikan discussed the positron in a lecture at the Cavendish, various members of the audience coldly suggested that Anderson had doubtless become tangled in some fundamental interpretative error.' ... The situation at the Cavendish Laboratory did not change until Blackett ... obtained results of his own which verified the existence of the new particle. A letter from Ralph Howard Fowler of the Cavendish Laboratory to Millikan in February 1933 indicated that Rutherford was not convinced until Blackett obtained his own evidence.

(Carl Anderson)[32]

Dear Millikan,

I have just had a letter from Rutherford which contains some of Blackett's work which may interest you and Anderson. It is that they have capitulated on the question of positive electrons and agree with Anderson 'that they are present' in large numbers among the tertiary or quaternary (or whatever they are) ionising particles seen in a Wilson photograph of the cosmic ray effects showing 'particles of positive charge and electronic mass' ...

P. S. Viva Caltech and Cav. Lab

(From a letter from Ralph Fowler to Robert Millikan, February 1933)[33]

1937

Electron diffraction

In 1927 Clinton Davisson and Lester Germer carried out their famous experiment, which confirmed de Broglie's hypothesis on matter waves. This was performed at Bell Labs, at that time the world's largest industrial laboratory. Meanwhile, George Paget Thomson and Alexander Reid's experiment on electron diffraction was carried out at a small but well-known university, in Aberdeen, Scotland (see pp. 156–7). The Swedish Academy of Sciences took into consideration the value of both discoveries and awarded the 1937 Nobel Prize to both Davisson and Thomson.

From accident to discovery

One day Davisson and Germer were studying the surface of a piece of metallic nickel by bombarding it with electrons, and observing how many electrons were scattered at various angles. During the experiment an accident occurred. Air entered the vacuum tube containing the nickel target, so that an oxide film formed on its surface. To reduce the oxide on the target the two experimenters heated the nickel to high temperatures. The effect of this was that large crystals were formed in the metal. When they repeated the experiment they obtained

Fig. 7.10. C. Davisson (left) and L. Germer. (Courtesy AIP Emilio Segrè Visual Archives.) '[The] New York experiment was not, at its inception, a test of the wave theory. Only in the summer of 1926, after I had discussed the investigation in England with Richardson, Born, Franck and others, did it take on this character.' (Clinton Davisson)[34]

quite different results. Peaks in the intensity of the reflected electrons occurred at specific angles, in contrast to what they had observed before the accident. They immediately noticed the similarity of this behaviour to the diffraction of X-rays by crystals. It was thus that Davisson and Germer discovered electron waves.

Clinton Joseph Davisson (1881–1958) was born in Bloomington, Illinois, USA. In 1911 he obtained his Ph.D. from Princeton University, and in 1917 he joined the engineering department of the Western Electric Company (later Bell Labs). Here he began a series of investigations into the nature of electron emission from metals, a very important matter in the search to improve the efficiency and power of the vacuum-tube devices then used in telephony. These investigations ultimately led to the discovery that earned him the Nobel Prize. Subsequently, he worked on electron optics, and in the field of crystal physics. After retiring from Bell Labs in 1946, Davisson became a visiting professor at the University of Virginia, in Charlottesville.

George Paget Thomson (1892–1975) was born in Cambridge, England, the son of Sir J. J. Thomson (Nobel 1906). He studied at Trinity College, and worked at the Cavendish Laboratory. From 1922 on he was Professor of Natural Philosophy at the University of Aberdeen, Scotland. In 1930 he became Professor of Physics at the Imperial College of Science and Technology in London; here, he carried out research work in the fields of neutron physics and nuclear fusion. A Fellow of the Royal Society, Thomson was knighted in 1943; nine years later he became Master of Corpus Christi College, Cambridge.

Electron diffraction at work

Electron diffraction ranks among the most important discoveries in physics. Firstly, it demonstrated the reality of the wave nature of electrons, and secondly, it gave rise to a wide number of applications, such as the study of the structure of crystals, the study of surfaces and electron microscopy. Let us devote some words to this last application.

Ordinary-light microscopes are limited to the size of the objects that they can distinguish: they can distinguish (or 'resolve') two dots which are some ten-thousandths of a millimetre apart. To achieve a much greater 'resolving power' physicists thought of using electron waves instead of light waves. The former owe their power to the fact that the de Broglie wavelengths associated with high-speed electrons are far smaller than those of ordinary light, and thus the resolving power obtainable with an *electron microscope* is much greater (it can be about a thousand times greater than that obtainable with an optical microscope).

In an electron microscope a beam of electrons, which are accelerated by a high electric voltage, is focused by using electric and magnetic fields as lenses. Then the beam strikes the object, and forms an image of it, which can be registered on a photographic film, or on a fluorescent screen. The first electron microscope was built in Germany in 1933 by Ernst Ruska (Nobelist in 1986). By 1939 electron microscopes were commercially available; and eventually they reached a magnification of up to one million times.

Rutherford, October 1937

> ... during a meeting commemorating Galvani in Bologna [Italy], word arrived that Rutherford was seriously ill with a hernia. On 19 October 1937, he died. The death was announced at the meeting by Bohr, his voice broken by tears.
>
> (Emilio Segrè)[35]

> Rutherford the individual, and the Cavendish the institution, became fused in one, and together radiated a brilliance rarely matched. At Montreal it had been Rutherford himself, at Manchester Rutherford and his school, and now at Cambridge it was Rutherford, the impersonation of the great Cavendish tradition and part of its glory ... Rutherford was not only one of Cambridge's illustrious Professors; in physics, in the Cavendish, he was 'the' Professor ... The Cavendish

Fig. 7.11. Cavendish research students and academics, June 1932. (Courtesy The Cavendish Laboratory, University of Cambridge, England.) Fourth row (standing): G. Occhialini (second from the right). Second row: Walton (second from the left). First row (sitting); from left to right: P. Kapitza (second), J. Chadwick, J. J. Thomson (fifth), E. Rutherford, C. T. R. Wilson, F. W. Aston, P. M. Blackett (tenth), J. D. Cockcroft.

was Rutherford's domain, his sphere of influence. But one never felt that this stemmed from any formal title; his influence there seemed a wholly natural phenomenon. Benevolent guidance, leadership and intellectual authority flowed from him, and admiration, respect, trust and loyalty were returned. One would no more question his influence on those around him than one would that of the sun on the satellite planets.[36]

1938

The Rome school

We now return to the chronicle of events that we left in 1934 (p. 189). Emilio Segrè is continuing to refer to Enrico Fermi and the Rome School as he relates the story of the Institute of Physics:

> We have reached 1935, the time of the Ethiopian War, the Spanish Civil War, and other forerunners of World War II. It was clear to informed and thinking people that the European situation was approaching a catastrophe. At about this time, for various reasons, the Rome group dispersed: Rasetti, alarmed and disgusted by the situation, came to America; I obtained a physics chair at the University of Palermo in Sicily, relatively distant from Rome; Pontecorvo joined Joliot in Paris; Fermi and Amaldi remained alone in Rome . . . The deteriorating general situation further forced the demise of our group. The enslavement of Italy to Germany by the formation of the Rome–Berlin Axis (Berlin–Rome when seen from Berlin), the promulgation of anti-Semitic laws in Italy, and the pursuit of other foolhardy policies rendered the situation untenable.[37]

Fermi remained in Italy until December 1938, when he won the Nobel Prize. Ruth Moore, a biographer of Niels Bohr, wrote:

> In September 1938, when . . . [Bohr's] institute held its annual seminar, almost none of the German alumni dared attend. But Enrico Fermi, the leading physicist of Italy, was there. Bohr confidentially told him that he [Fermi] was under consideration for the Nobel Prize . . . Fermi then told Bohr that he wanted to leave Italy . . . [He] said that, if he should receive the prize, he would seek a position in the United States.[38]

And Valentine Telegdi, who was associated with Fermi in Chicago in the early 1950s, remembers:

> In 1936, [Fermi] was a visiting professor at Columbia University . . . It was during . . . this visit that the physics department chairman George Pegram first offered him a permanent appointment at Columbia. Fermi was not particularly critical of Mussolini's fascist regime prior to the dictator's promulgation of anti-Jewish laws in 1938. But those new racial laws applied to his wife, Laura. So he decided to leave Italy and discreetly inquired of Pegram whether the position was still available. Fermi's trip to Stockholm to receive the Nobel Prize . . . was a convenient way to emigrate unobtrusively with his whole family.[39]

In November in fact Fermi received the traditional telephone call from Stockholm with the news that he had been awarded the 1938 Nobel Prize for physics (it was for his discoveries on artificial radioactivity and on the properties of slow neutrons). After the Nobel ceremonies the Fermis spent a few days in Copenhagen, cordially received by Bohr and his family. On 24 December they boarded the liner *Franconia* at Southampton, England, and, nine days later, they landed in New York, where Fermi joined the physics department of Columbia University.

The Rome school

> The speed at which it was possible to train a young physicist at the 'Fermi school' was incredible. Naturally a good deal of the success was due to the immense

Fig. 7.12. Enrico Fermi while receiving the Nobel award from the hands of the King of Sweden, December 1938. (National Archives and Records Administration, courtesy AIP Emilio Segrè Visual Archives.)

enthusiasm that had been aroused in the young people – never by exhortations or 'sermons' but by the eloquence of example. After having spent time in the institute in Via Panisperna, one became completely absorbed in physics . . .

(Emilio Segrè)[40]

My greatest impression of Fermi's method in theoretical physics was its simplicity. He was able to analyse into its essential point every problem, however complicated it seemed to be. He stripped it of mathematical complications and of unnecessary formalism. In this way, often in half an hour or less, he could solve the essential physical problem involved . . . His approach was pragmatic . . . He was a master at achieving important results with a minimum of effort and mathematical apparatus

(Hans Bethe)[41]

Enrico Fermi (1901–54)

Enrico Fermi was born in Rome, the son of a railroad official. At the age of seventeen he became a physics student at the elite *Reale Scuola Normale Superiore*, where he obtained his doctorate in 1922. (The *Scuola* is an Italian advanced study and research institution, which was founded by Napoleon in 1810 as a branch of the *Ecole Normale Supérieure* of Paris. It is now associated with the University of Pisa.)

Soon afterwards, thanks to a fellowship, Fermi went first to Germany to study at the University of Göttingen, and then to Leiden in the Netherlands. When he returned to Italy, he began teaching as a lecturer at the University of Florence. In 1925 he discovered statistical laws for describing ensembles of particles, according to Pauli's exclusion principle. In 1927 he was appointed Professor of Theoretical Physics at the University of Rome, the youngest full professor in physics in Italy. The Royal Academy of Italy recognised his scientific work, and in 1929 Fermi became the youngest of its members. In 1933 he devised the first theory of beta-ray emission in radioactivity, which introduced the weak force, one among the four basic forces of nature. In 1934 he turned to experimental nuclear physics. His discoveries of neutron-induced radioactivity, and of nuclear reactions produced by slow neutrons, earned him the Nobel Prize for physics.

In 1938, immediately after the receipt of the Nobel, Fermi emigrated to the USA. Together with other co-researchers he developed the first nuclear reactor (first at Columbia University, then at the University of Chicago). In 1944 he moved to Los Alamos, where he played a leading role in the development of the first nuclear bomb. After the war he returned to the University of Chicago, as a full professor at the Institute for Nuclear Studies (now named the Enrico Fermi Institute). Here students from various countries came to study physics with him. (Among his pupils were Owen Chamberlain, James Cronin, Tsung-Dao Lee, Chen-Ning Yang, Jack Steinberger, and Jerome Friedman, all Nobel laureates.) Recognised as one of the greatest scientists of the twentieth century, Fermi was noted for his vivacious character and the clarity of his lectures. (J. Robert Oppenheimer described him succinctly in these words: 'Not a philosopher. Passion for clarity. He was simply unable to let things be foggy. Since they always are, this kept him pretty active.')

> Fermi devoted a great deal of his time to the graduate students . . . His teaching was exemplary, minutely prepared, clear, with emphasis on simplicity and understanding of the basic ideas, rather than generalities and complications . . . We would knock at his office door, and if free, he would take us in, and then he would be ours until the question was resolved.
>
> (Jack Steinberger)[42]

1939

The cyclotron

The 1939 Nobel Prize for physics went to the American experimentalist Ernest Lawrence, for his invention of the *cyclotron*, the first cycling accelerator (p. 174). (Some people in the community of physicists felt that John Cockcroft and Ernest Walton should have deserved the award before Lawrence, because they had been the first to disintegrate atomic nuclei with artificially accelerated particles (p. 182). But Manne Siegbahn (Nobel 1924), as a member of the Nobel Physics Committee, strongly backed Lawrence's candidacy, and he was able to gather support in favour of Lawrence both in the Physics Section and in the Academy. It was not until 1951 that Cockcroft and Walton's work was recognised with a Nobel Prize.)

Ernest Orlando Lawrence (1901–58) was born in Canton, South Dakota, USA, the son of a school superintendent. He received his master's degree from the University of Minnesota, and his Ph.D. in physics from Yale University, Connecticut, in 1925. Three years later he joined the faculty of the University of California at Berkeley as Associate Professor of Physics, and in 1930 he became a full professor there.

Lawrence first thought of the idea of the cyclotron in 1929, and proposed it in a note published in *Science*.[43] In the summer of 1930 he suggested the matter to one of his graduate students, Stanley Livingston, who undertook the project and succeeded in building a device that accelerated molecular hydrogen ions to an energy of 13 000 electron-volts. During the year 1931, Lawrence and Livingston then built a second cyclotron, which accelerated protons to an energy of 1.2 million electron-volts.[44] (This was the first time that subatomic particles had been accelerated to such high energies.) In 1936 Lawrence founded the Radiation Laboratory at Berkeley, where more and more powerful cyclotrons were built. One of these cyclotrons produced *technetium* (which was discovered in 1937 by Emilio Segrè, Nobelist in 1959: this was the first element to be discovered which does not occur naturally, and so has to be made artificially). With the cyclotron Lawrence also produced radioactive isotopes for medical purposes.

During the Second World War Lawrence worked on the Manhattan Project, and developed a process for the separation of uranium isotopes. In 1957 he received the Fermi Award from the US Atomic Energy Commission. In his honour the Berkeley laboratory was later named the Lawrence Berkeley National Laboratory, and the element with mass number 103 was called *lawrencium*.

The principle of the cyclotron

In a cyclotron a particle-source is located at the centre of an evacuated cylindrical metal chamber, placed between the poles of a strong electromagnet, which creates a uniform magnetic field perpendicular to the chamber's flat faces. A charged

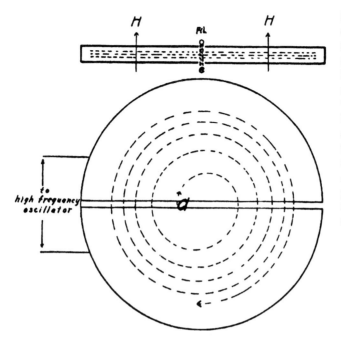

Fig. 7.13. Schematic diagram of the cyclotron. (Source: Ernest O. Lawrence and M. Stanley Livingston, *Physical Review*, 40, 1932, p. 23 © 1932 by the American Physical Society.) The source of the particles is located near the centre (*a*), and the magnetic field *H* is perpendicular to the plane of the Ds. The spiral represents the trajectory of the particles.

particle, emitted from the source, describes a circle in the field; and the time taken by it to make this circle is the same at any speed (as long as the speed of the particle is much less than the speed of light).

An oscillating electric voltage is applied to a pair of semicircular electrodes called 'Dees', so named from their shape, each half of the cylindrical chamber being a D-shaped conductor (see Fig. 7.13). The source of the voltage is a circuit, similar to a radio transmitter, which operates at a frequency equal to that of the revolutions of the particles in the field. This voltage creates an oscillating electric field in the gap between the Ds, so that each time a particle crosses the gap, the electric field pushes or pulls it, thus speeding it up. As a consequence, the particle continually gains kinetic energy (and speed), and its path becomes a spiral-like series of semicircles of increasing radius. The resulting particle beam (bunches of particles) is thus extracted at the maximum radius (and maximum energy), and it is directed at a target, where the accelerated particles can interact with the target material.

Fortune in an article

One evening early in 1929 as I was glancing over current periodicals in the University library, I came across an article in a German electrical engineering journal . . . on the multiple acceleration of positive ions. Not being able to read German easily, I merely looked at the diagrams and photographs of . . . [the] apparatus and . . . was able to determine [the] general approach to the problem – i.e. the multiple acceleration of the positive ions by appropriate application of

Fig. 7.14. D. Cooksey (left) and E. Lawrence beside one of the Berkeley cyclotrons. (Courtesy Lawrence Berkeley National Laboratory.)

radio-frequency oscillating voltages to a series of cylindrical electrodes in line ... I asked myself the question, instead of using a large number of cylindrical electrodes in line, might it not be possible to use two electrodes over and over again by sending the positive ions back and forth through the electrodes by some sort of appropriate magnetic field arrangement.

(Ernest Lawrence)[45]

Splitting the nucleus

Let us go back for a moment to the autumn–winter of 1938. The protagonists are Otto Hahn, Lise Meitner, Otto Frisch and Fritz Strassmann, and the places Berlin, Stockholm and Copenhagen.

Otto Hahn was a German chemist. In the early 1900s he had worked with Sir William Ramsay (Nobel for chemistry 1904) at University College, London, and later on with Ernest Rutherford at McGill University, Montreal, Canada. In 1938 he was head of the department of radiochemistry at the Kaiser Wilhelm Institute for Chemistry in Berlin–Dahlem, where he had been since 1913, and where he had devoted many years to radiochemistry. In 1934, following Fermi's discoveries (p. 188), he began to investigate the radioactive products formed when uranium was bombarded by neutrons.

Lise Meitner was born in Vienna, the third of eight children of a Jewish family. After receiving her doctorate in physics in 1906, she went to Berlin to attend Max Planck's lectures. Later on she joined Hahn in research on radioactivity: her collaboration with him lasted some thirty years. She and Hahn were the first to isolate an isotope of the radioactive element protactinium. After the *Anschluss* (the Nazi annexation of Austria, in March 1938), Lise had to leave Germany. She fled to Sweden in the summer of 1938, aided by Niels Bohr, and went to work at the Nobel Institute of Physics in Stockholm, directed by Manne Siegbahn (Nobel 1924).

Otto Frisch was a nephew of Lise Meitner's. He was born in Vienna and, studied at the same university as his aunt. He received his doctorate in physics in 1926. In 1933 Frisch, with Otto Stern and Immanuel Estermann, measured the magnetic moment of the proton at the University of Hamburg (see Nobel 1943). In 1938 he fled to Copenhagen and worked under Niels Bohr.

Fritz Strassmann was a German analytical chemist who had studied at the Technical University of Hannover. He received his doctorate in 1929, and in 1934 he joined Hahn and Lise Meitner in their scientific work.

And now for a story
In 1934 Fermi and his co-workers, while working in Rome, produced radioactive isotopes by neutron bombardment, and came upon the following puzzle: were there any *transuranic elements* (elements with an atomic number greater than that of uranium) among the products after uranium had been bombarded by neutrons?

Hahn and Meitner, assisted by the young Strassmann, started to probe the possibilities. They bombarded solutions of uranium salts with neutrons, and found, by chemical analysis, that a number of new radioactive elements were present among the products. At first their results seemed to be in agreement with those found by Fermi's group in Rome, but it became increasingly difficult to understand fully the whole question of transuranic elements.

While Meitner was escaping to Sweden, the puzzle was actually becoming more and more complicated. She corresponded with Hahn, while he was trying to identify his new elements. During autumn 1938 Hahn and Strassmann obtained a surprising result: one of the products from uranium was a radioactive form of a much lighter element that they thought must be radium. They performed further tests and found that the uranium product was in fact radioactive barium, not radium.

Hahn wrote to Meitner in December giving her the result. At that time Frisch happened to be staying near by for the holidays. ('. . . I went to visit her at

Christmas. There, in a small hotel in Kungälv near Göteborg I found her at breakfast brooding over a letter from Hahn. . . . We walked up and down in the snow, I on skis and she on foot . . . and gradually the idea took shape . . .'.)[46]

Meitner and Frisch realised that in Hahn's experiment the uranium nuclei had split into two smaller fragments. They named the splitting of the atomic nuclei *fission*, after the equivalent cell division in biology. Meitner and Frisch then used the *liquid-drop model* of the nucleus, which had been first proposed by Niels Bohr in 1936. (Bohr thought of the nucleus as a liquid drop, in which the individual nucleons – protons and neutrons – are analogous to molecules in a liquid, held together by strong forces and surface-tension effects.) Taking into account the effects of the nuclear and electric forces, and using Einstein's formula for mass–energy equivalence ($E = mc^2$), they were able to calculate the large amount of energy released in the reaction (200 million electron-volts, about ten times larger than in any nuclear reaction that had up to then been produced).

At the beginning of January, Frisch rushed back to Copenhagen in great excitement, and told Bohr of Hahn's and Meitner's discoveries. (Frisch remembers that, soon after he had started to speak, Bohr exclaimed: 'Oh, what fools we have been! We ought to have seen that before.')[47] Then, on 13 January 1939, using an ionisation chamber (this produced a pulse of elecric current when a fission fragment ionised the gas contained in it, p. 100), Frisch carried out a crucial experiment in Bohr's laboratory in Copenhagen, and fully confirmed the Berlin results.

Hahn and Strassmann published their discovery in the German scientific magazine *Die Naturwissenschaften* on 6 January 1939 (their paper was received by the editor on 22 December 1938). Meitner and Frisch explained nuclear fission in a paper that was submitted to *Nature* on 16 January, and published on 11 February. These results were immediately confirmed all around the world. In 1946 Otto Hahn was awarded the 1944 Nobel Prize for chemistry 'for his discovery of the fission of heavy nuclei', but no mention was made of Lise Meitner. (When the war was over Bohr proposed both Meitner and Frisch for a Nobel Prize in physics – but in the event, neither of them received the award.) To conclude the story, we find ourselves asking a question. Why did not Fermi and his fellow students discover nuclear fission? Here is an explanation, as given by Emilio Segrè:

> We had performed an experiment that came extremely close to a demonstration of fission. In order to see possible short-life alpha [radioactive nuclei] produced by neutron bombardment in uranium, we assumed that such activities would give rise to energetic alpha particles. We covered our sample with an aluminium foil to stop lower energy alpha particles emitted by uranium even without any bombardment, and as a result we did not observe anything. If we had removed the aluminum foil, we would have seen the huge ionisation pulses produced by the fission fragments.[48]

Einstein to Roosevelt

Let us resume the events of 1939. On 16 January Bohr arrived in the USA for a series of conferences, so it was then that Fermi and other physicists first heard of the discovery of nuclear fission. Soon Fermi's co-workers rushed into their laboratory at Columbia University and repeated Frisch's experiment. They obtained the same results, which they presented at a conference held in Washington, DC, at the end of January. In March, experiments performed by Frédéric Joliot in Paris, by Fermi's group and by the Hungarian physicist Leo Szilard, both at Columbia, showed that free neutrons were ejected in the fission process. Thus physicists thought that a self-sustaining chain reaction would be possible, and that it could be exploited to build a controlled fission reactor, or a bomb.

In August 1939 Einstein, persuaded by his fellow scientists (in particular Szilard and Eugene Wigner), presented the military potential of nuclear fission to the President of the USA, Franklin Delano Roosevelt. On 1 September Germany invaded Poland and the Second World War broke out. Here are some excerpts from Einstein's letter to Roosevelt:

> Some recent work by E. Fermi and L. Szilard . . . leads me to expect that the element uranium may be turned into a new and important source of energy in the immediate future . . . In the course of the last four months it has been made probable – through the work of Joliot in France as well as Fermi and Szilard in America – that it may become possible to set up a nuclear chain reaction . . . by which vast amounts of power . . . would be generated . . . This new phenomenon would also lead to the construction of bombs, and it is conceivable – though much less certain – that extremely powerful bombs of a new type may thus be constructed . . . In view of this situation, you may think it desirable to have some permanent contact maintained between the Administration and the group of physicists working on chain reaction in America.[49]

1940

War time

The Nobel Prize for physics was not awarded in 1940, 1941, and 1942, because of the war. In many countries – but principally in the UK, the USA and Germany – physicists were mobilised for war research, particularly in the fields of nuclear energy and radar.

Nuclear energy

In natural uranium one atom in about 140 (0.7 per cent) is the isotope uranium-235 (92 protons and 143 neutrons in its nucleus). The rest is uranium-238 (92 protons and 146 neutrons). In February 1939 Bohr predicted that uranium-235 would undergo fission when bombarded by slow neutrons, while uranium-238 would do

Fig. 7.15. A chain reaction. One neutron splits a uranium nucleus, thus releasing two neutrons that can produce two fissions that release four neutrons, and so on. For each fission the energy released is 20 million times more than that liberated in each atom of carbon burned as coal!

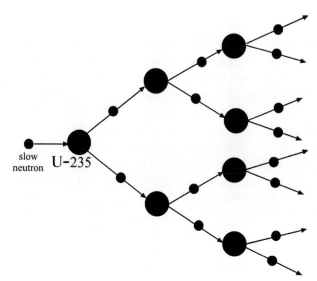

the same with fast neutrons. (Bohr's conjecture was confirmed by experiments carried out about one year later.) As neutrons released in fission reactions would not have had enough energy to cause additional fission in uranium-238, a chain reaction sustained by this isotope alone would not have been feasible. So physicists began to attack the problem of how to obtain an enrichment of the uranium-235 content in a mixture of the two isotopes. Early in 1940 the US government made funds available for research in this field, which was soon initiated in universities and industrial laboratories.

The refugees

The 1930s saw a convulsion in Europe which the scientists did not anticipate, and which then disrupted many of their lives. They were forced into the greatest emigration of intellectuals since the collapse of Byzantium, and one far more dramatic and influential than that. Einstein, himself homeless, had to help make provision for Jewish scientists now deprived . . . Bohr arranged for Copenhagen to be a staging post . . . The Göttingen faculty was broken up. Born found himself in Edinburgh; others were scattered round American or British universities . . . Hans Bethe arrived at Cornell . . . Towards the end of the decade, just before the beginning of the war, a recent power in the scientific world, the Physics School in Rome, had to cross the Atlantic. Their leader, Enrico Fermi, was not himself Jewish, but had a Jewish wife . . . The United States, because it was rich and because it was the most secure refuge . . . received a high proportion of the Jewish scientists. It was the most significant influx of ability of which there is any record. America, of course, was already producing its native-born Nobel Prize winners. The refugees made it, in a very short time, the world's dominant force in pure science. They also helped create what was soon called the Jewish explosion,

Fig. 7.16. The Institute of Physics of the University of Rome in Via Panisperna. The building was erected in about the year 1880 on land that had once belonged to an old monastic order. This became the seat of the research activity of Enrico Fermi and his group during the late 1920s and the 1930s, and the address was immortalised in giving its name ('the boys of *Via Panisperna*') to the elite band of Italian researchers who worked there. (Photo: courtesy Edoardo Amaldi Archive, Physics Department, the *La Sapienza* University in Rome.)

a burst of creativity in all fields, not only science ... The refugees gave the explosion a new dynamic.

(C.P. Snow)[50]

Segrè's memories

The physics building was perfectly adequate for scientific work in the 1920s, and it compared favorably with other major European laboratories. The equipment was fair, including mainly instruments for optical spectroscopy and some adequate subsidiary apparatus ... The location of the building in a small park on a hill near the central part of Rome was convenient and beautiful at the same time. The garden, landscaped with palm trees and bamboo thickets, with its prevailing silence (except at dusk, when gatherings of sparrows populated the greenery), made the institute a most peaceful and attractive centre of study. I believe that everybody who ever worked there kept an affectionate regard for the old place, and had a poetic feeling about it.

(Emilio Segrè)[51]

Chapter 8
The nuclear age

Up to the previous decade, with some notable exceptions, physics had been predominantly a European affair. But this was set to change in a few years' time with the coming to power in Germany and Austria of the Nazis. The German scientific community, one of the greatest in the world, fought an unequal battle with the anti-Semitic and anti-intellectual atmosphere which now prevailed. Many prominent scientists fled abroad, and took refuge in other nations – especially in the USA. Refugees also came from other European countries, notably from Italy, where, in the late 1930s the Fascist regime too began to implement racial laws. American physics had already begun to grow rapidly – but now it benefited greatly from a colossal infusion of intellectual power and creativity. So it is the USA which, displacing Europe, rapidly came to occupy centre stage, a position which, due also to its economic hegemony, it has held without serious rivalry ever since.

Nazism and Fascism brought with them the Second World War. Among the major events for science and technology during this period, and soon after, are the *nuclear reactor* and the *nuclear bomb*, so marking the birth of the nuclear age, and the development of *radar* and the first *digital computer*. War research produced not only new technology, but also much fundamental insight into physics, especially in the fields of nuclear and electromagnetic phenomena.

As the war finished, scientists returned to their universities and laboratories. A massive boom in science and technology soon followed. Physics in particular increasingly became a symbol and a stimulus for economic growth and national strength. Large collaborative efforts came to dominate many fields of physics, heralding an increase in size and complexity. Fresh generations of physicists made their debut. Significant advances were made in all sectors of physics: the *transistor* was invented; *nuclear magnetic resonance*, fine features in atomic spectra and new subatomic particles were discovered. On the theoretical side, a modern *quantum electrodynamics* was created, which accurately described the interaction of photons and electrons.

Seven out of the eight Nobel laureates of the period 1943–50 (in the years 1941 and 1942 the Nobel Prize was not awarded) were honoured for achievements that date right back to pre-war years.

1941

The Manhattan Project

Early in 1941, scientists at the University of California at Berkeley discovered a new isotope, named *plutonium-239*, which undergoes fission under slow neutron bombardment, and can be used instead of uranium as a fissionable fuel. On 6 December President Roosevelt authorised the organisation of a secret programme, known as the 'Manhattan Project'. This programme included work on chain reactions, and research to develop ways of producing plutonium-239 and to find methods of obtaining uranium-235 in high concentrations. The final goal was the construction of a nuclear bomb. (The very next day, the Japanese attacked Pearl Harbor, and the USA was at war!)

Many physicists and chemists participated in the project, and a number of them were Nobel laureates, including Hans Bethe, Niels Bohr, Arthur Compton, Enrico Fermi, Richard Feynman, James Franck, Ernest Lawrence, Emilio Segrè, Harold Urey, and Eugene Wigner.

Radar

Radar (short for *RAdio Detecting And Ranging*) is used to locate a distant object, and determine the direction in which it is moving; this depends on the time lapse between the emission of a radio pulse and its return, when reflected back by the object itself. The first radar system began to be tested during the 1920s. In the mid-1930s, British scientists and engineers began to build a network of ground-based radar to detect aircraft. It was fully operational towards the end of the decade, and it became a vital instrument in detecting oncoming Nazi planes despite darkness and fog during the famous Battle of Britain (1940).

In 1939 a new device, called the *magnetron*, was developed: it generated high-frequency radio pulses, and made possible the development of the microwave radar. In the same year a laboratory, called the Radiation Laboratory, was established in the USA, at MIT in Cambridge, Massachusetts, in order to develop radar research. It was headed by the American physicist Isidor Rabi (Nobel 1944). Here, many future Nobelists (such as Luis Alvarez, Polykarp Kusch, Willis Lamb, Edward Purcell, Norman Ramsey and Julian Schwinger) worked in the development of radar technology, which was soon to become the basis for anti-ship, anti-submarine and air-surveillance systems during the Second World War. (It is said that radar helped win the war, whereas the nuclear bomb helped end it!)

1942

The first chain reaction

A fission chain-reaction experiment was first performed at Columbia University in 1941, under the leadership of Enrico Fermi. The operation was then transferred to the University of Chicago. Here, towards the end of the year 1942, Fermi and his team finally succeeded in making the world's first nuclear reactor.

2 December 1942

The first nuclear reactor, built at the University of Chicago, contained 52 tons of uranium. On 2 December 1942 a group of scientists and technicians (about forty persons) gathered near the reactor. Enrico Fermi, imperturbable as usual, directed the experiment. At noon, as was his custom, he stopped for lunch. Operations were resumed at 2:00 p.m., and about two hours later, a self-sustaining fission reaction was under way. Here is an account of that historic moment:

> The clicks [of the counter] came more and more rapidly, and after a while they began to merge into a roar; the counter couldn't follow any more. That was the moment to switch to the chart recorder. Everyone watched in the sudden silence the mounting deflection of the recorder's pen. It was an awesome silence. Everyone realised the significance of that switch; we were in the high intensity regime. Again and again, the scale of the recorder had to be changed to accommodate the neutron intensity which was increasing more and more rapidly. Suddenly Fermi raised his hand. 'The pile has gone critical,' he announced. No one present had any doubt about it.[1]

Eugene Wigner (Nobelist in 1963) presented Fermi with a bottle of *Chianti* to mark the event; and Arthur Compton (Nobel 1927) made a long-distance call announcing the success. 'Jim', he said, 'the Italian navigator has just entered the new world.' It was in this way that humanity entered the *nuclear age*.

Most of the research was then concentrated in a new laboratory that had been created in a secret location, Los Alamos in New Mexico. Its scientific director was J. Robert Oppenheimer. By 1945 enough uranium-235 and plutonium-239 were available for the construction of nuclear-fission bombs. The first test was performed at 5:30 a.m. on 16 July 1945, at a site on the Alamogordo air base, 193 kilometres south of Albuquerque, New Mexico. The bomb used plutonium and had an explosive power of about 20 000 tons of trinitrotoluene. The first nuclear bomb to be used in warfare was dropped on Hiroshima, Japan, on 6 August 1945, and a second one was used against Nagasaki three days later.

1943

Proton magnetism

After the gap in the annual succession of Nobel Prizes, due to the Second World War, it was not until the autumn of 1944 that the Swedish Academy of Sciences announced that for 1943 the prize would be awarded to the German-American physicist Otto Stern, and that for 1944 to the American Isidor Rabi.

Stern had been living in the USA since 1933, when the Nazis had risen to power in Germany, and he, being Jewish, had had to leave his country. He became a professor of physics at the Carnegie Institute of Technology in Pittsburgh. The Nobel Prize was granted to him, 'for his contribution to the development of the molecular ray method and his discovery of the magnetic moment of the proton'.

Otto Stern (1888–1969) was born in Sorau, Upper Silesia, Germany (now Zory, Poland). He studied physical chemistry at the University of Breslau, and obtained his doctorate in 1912. He then went to the University of Prague as an assistant to Einstein, and later followed him to the Zurich polytechnic (ETH). He taught at the universities of Frankfurt (1914–21) and Rostoc (1921–2). In 1923 he was appointed Professor of Physical Chemistry at the University of Hamburg. In the late 1910s he developed his *method of molecular beams*, which he used to measure the velocities of gas molecules, so proving the assumptions of the kinetic theory of gases. In Frankfurt, Stern and Walther Gerlach used the molecular-beam method to perform a historic experiment on the deflection of silver atoms by a non-uniform magnetic field (see p. 129); and in 1933 the same method was used to measure the magnetic moment of the proton. Stern, one of the foremost experimental physicists of his time, also performed experiments similar to those of Davisson and Germer (see Nobel 1937): in 1930, by using helium atoms and hydrogen molecules, he succeeded in demonstrating the wave properties of matter.

The proton magnetic moment
In the late 1920s physicists thought that atomic nuclei too should have a magnetic moment. Thus the proton, the nucleus of the hydrogen atom, should have had a magnetic moment about 1840 times smaller than that of the electron. (This is because the formula expressing the magnetic moment of a spinning particle contains its mass in the denominator. Since the mass of a proton is about 1840 times that of the electron, its magnetic moment should be 1/1840 of the magnetic moment of the electron – this last being used as a unit for the magnetic moments of particles, and named the *Bohr magneton*.) In 1933 Stern was at the University of Hamburg; Wolfgang Pauli too was there as an assistant professor of theoretical physics. Emilio Segrè reports that 'Pauli told Stern that if he enjoyed doing difficult experiments, he could do them [on the proton magnetic moment], but that it was a waste of time and effort because the result was already known [just 1/1840 of the magnetic moment of the electron].'[2] Stern carried out the experiment and, to his surprise, he found that the magnetic moment of the proton was actually about 2.8 times more than had been theoretically anticipated, that is, only about 1/660 of the electron magnetic moment.

The arrogance of theorists
> A good example of the arrogance of theorists . . . is the following story: There was a seminar held . . . in Göttingen, and Otto Stern came down to Göttingen from Hamburg and gave a talk on the measurements he was about to finish of the

Fig. 8.1. Otto Stern. (Courtesy Emilio Segrè Visual Archives, Segrè Collection.) In 1913, after Bohr's theory of the atom, Stern and Max von Laue declared that they would give up physics if 'there was anything to this Bohrian nonsense'. In 1922, after his famous experiment with Gerlach, Stern sent a postcard to Bohr, in which he congratulated him 'on the confirmation of your theory' (see Fig. 6.2).

magnetic moment of the proton. He explained his apparatus, but he did not tell us the result. He took a piece of paper and went to each of us saying, 'What is your prediction of the magnetic moment of the proton?' Each theoretical physicist from Max Born down to Victor Weisskopf said, 'Well, of course, the great thing about the Dirac equation is that it predicts . . . [the] magnetic moment . . . for a particle of spin one-half.' Then he asked us to write down the predictions; everybody wrote 'one [nuclear magneton, equal to 1/1840 of a Bohr magneton].' Then, two months later, he came back again to give a talk about the finished experiment, which showed that the value was 2.8. He projected the paper with our predictions on the screen. It was a sobering experience.

(Victor Weisskopf)[3]

1944

Magnetic resonance

Isidor Rabi, at Columbia University, New York City, made a decisive step forward in the application of Stern's molecular-beam method. Between 1936 and 1937 he invented the so-called *molecular-beam magnetic resonance*, with which he was able to measure the magnetic properties of nuclei, atoms and molecules with very high precision. For his work he received the 1944 Nobel Prize for physics.

Isidor Isaac Rabi (1898–1988) was born in Raymanov, Austria (now in Poland). His parents were Jewish, and had emigrated to the USA at the end of the nineteenth

Fig. 8.2. Isidor Rabi. (Courtesy AIP Emilio Segrè Visual Archives, Bainbridge Collection.) Rabi set up and developed one of the world's finest physics departments at Columbia University; it was to produce several Nobel winners. His magnetic resonance method made possible numerous practical applications like the atomic clock, the maser, as well as magnetic resonance imaging.

century, where they settled in New York City. Rabi graduated in chemistry at Cornell University, Ithaca, New York, and received his Ph.D. from Columbia University in 1927. He then spent two years in Europe, working with Sommerfeld, Bohr, Pauli, Stern and Heisenberg. In 1929 he returned to Columbia University, where he became Professor of Physics in 1937. During the Second World War he was director of the Radiation Laboratory at MIT (p. 214). He was then chairman of the General Advisory Committee of the Atomic Energy Commission (1952–6), and one of the founders of the Brookhaven National Laboratory, Upton, New York.

The Rabi method

The molecular-beam magnetic resonance method invented by Rabi exploited Stern's molecular-beam technique, and the behaviour of a spinning atomic object, such as an atom or a molecule, in a magnetic field. Imagine, for example, an atom possessing a magnetic moment (like a minuscule compass needle), while

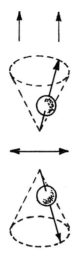

Fig. 8.3. The atomic magnetic moment describes a cone around the static magnetic field (vertical arrows). The horizontal oscillating magnetic field (horizontal arrow) acts to 'flip' the dipole either up or down when its oscillating frequency exactly equals the *precession frequency* (the number of revolutions per second of the magnetic dipole around the vertical field). This frequency is related to the two energy levels, E_1 (magnetic moment up) and E_2 (magnetic moment down) by Bohr's formula $v = (E_2 - E_1)/h$.

spinning in a magnetic field. It experiences magnetic forces, so that its magnetic moment (represented by the arrow in Fig. 8.3) describes a cone about the direction of the magnetic field. This motion is called *precession* (it is analogous to that of a spinning top in the earth's gravity field). Classical physics predicts that the orientation angle between the magnetic moment itself and the direction of the field can change continuously. But according to quantum mechanics, it can assume only certain definite values; that is, the atom can find itself only in specific quantum states. This ultimately means that the atom possesses a discrete set of energy levels.

Let us see briefly how Rabi used these properties. When he applied an oscillating magnetic field to the magnetic dipole (Fig. 8.3) exactly at the precession frequency (a condition known as 'resonance'), this field caused the dipole to 'flip' down, so determining a transition between the corresponding energy levels. And by measuring this resonance frequency (the differences between the energy levels are so small that it is in the range of radio frequencies) one can obtain the value of the magnetic moment. Rabi's method soon became a way for measuring the magnetic moments of atomic objects with a fantastic precision. It revealed the internal magnetic properties of atoms and molecules, and it provided a wealth of information on spins and magnetic moments in atomic nuclei. It ultimately led to the new research field of *radio-frequency spectroscopy*.

1945

The Nobel to Pauli

Princeton, New Jersey, USA – 10 December 1945. At the Institute for Advanced Study a formal-dress dinner is taking place to honour Wolfgang Pauli, who has been awarded the 1945 Nobel Prize for physics. At the end of the dinner, toasts are offered to the prizewinner. Among others, Einstein addresses Pauli, who recalls the occasion some time later: 'Never will I forget the speech that he gave about me . . . It was like a king who abdicates and appoints me as a sort of "son elect" as successor.'[4]

Pauli had received nominations for eight years ever since 1933, but the Nobel Physics Committee had always passed him over. (It was again Carl Oseen who considered Pauli's *exclusion principle* not worthy of a Nobel; he claimed it was merely a 'philosophical construction'.) In 1945 Einstein sent a telegram to the Committee in which he wrote: 'Nominate Wolfgang Pauli for physics prize . . . [His] contributions to modern quantum theory consisting in the so-called Pauli or exclusion principle became a fundamental part of modern quantum physics . . .'.[5] On 15 November 1945, the Committee decided to award the prize to Pauli precisely for his discovery of the exclusion principle (p. 142). (Oseen had died in 1944, and had been replaced by a pupil of his, of the name of Ivar Waller – a theoretical physicist who had worked with Pauli himself, and with Paul Dirac and Lawrence Bragg.)

On 13 December 1946, in his Nobel lecture, Pauli thus recalled how he had discovered his principle:

> The history of the discovery of the 'exclusion principle' . . . goes back to my student days . . . [It] was at the University of Munich that I was introduced by Sommerfeld to the structure of the atom – somewhat strange from the point of view of classical physics . . . A new phase of my scientific life began when I met Niels Bohr personally for the first time. This was in 1922, when he gave a series of guest lectures at Göttingen, in which he reported on his theoretical investigations on the Periodic System of the Elements . . . Following Bohr's invitation, I went to Copenhagen in the autumn of 1922, where I made a serious effort to explain the so-called 'anomalous Zeeman effect' . . .[6]

Pauli's investigations into this effect, and the alkali spectra, played a great part in leading him to the formulation of his principle:

> On the basis of my earlier results on the classification of spectral terms in a strong magnetic field the general formulation of the exclusion principle became clear to me . . . The exposition of [it] . . . was made in Hamburg in the spring of 1925, after I was able to verify some additional conclusions . . . during a visit to Tübingen with the help of the spectroscopic material assembled there.[7]

Enfant prodige

Wolfgang Pauli (1900–58) was born in Vienna, at that time the capital of the Austro-Hungarian Empire, and the centre of a flourishing culture. His father was a professor of biochemistry at the university, and a close friend of the philosopher Ernst Mach. Pauli considered the contacts that he had had with Mach during his youth to have been 'the most important event' in his intellectual life. He remembered:

> Among my books there is a somewhat dusty case; it contains a silver cup made in the *art nouveau* style (*Jugendstil*) and in it there is a card . . . written in old-fashioned, adorned letters: 'Dr. E. Mach, Professor at the University of Vienna.' . . . [He] had kindly agreed to assume the role of my godfather . . . [and] the result seems to be that I was baptised in this way to [become more] anti-metaphysical than Catholic . . . [That] cup . . . remained the symbol of '*aqua permanens*' [holy water], which drives away the evil metaphysical spirits.[8]

Pauli studied under Arnold Sommerfeld at the University of Munich, and obtained his doctorate in 1921. In the same year he wrote a masterly review article on relativity for the German *Encyclopaedia of Mathematics*, which Einstein praised with words of admiration ('No one studying this mature, grandly conceived work would believe that the author is a man of twenty-one.').[9] He taught physics at the universities of Göttingen (1921–2), Copenhagen (1922–3) and Hamburg (1923–8), and then became professor of theoretical physics at the Zurich polytechnic (ETH), a post that he held for the rest of his life. He was visiting professor at the Institute for Advanced Study at Princeton for many periods between 1935 and 1954.

Pauli's most famous contributions to modern physics were the exclusion principle, for which he received the Nobel Prize, and the neutrino hypothesis (see p. 169). Other major contributions dealt with the electric properties of metals (leading to the quantum theory of electrons in metals), and with the relation between spin and statistics for elementary particles.

Pauli was noted for his wit, and for his sharp critical tongue. He was known as the 'conscience of twentieth-century physics'. During his 'life-crisis' (1930–1), he entered into contact with the famous psychologist Carl Gustav Jung. This relationship lasted until 1958 and led Pauli to be involved with psychological problems. (He wrote a paper on the personality of Johannes Kepler.) He was also interested in mysticism, and developed a long friendship with Gershon Scholem – the great scholar of Jewish mysticism and the kabalah.[10]

Weisskopf's recollections

> . . . I knocked, and no answer. I knocked again, and no answer. After about five minutes he said, rather roughly, 'Who is it? Come in.' I opened the door, and here was Pauli . . . He said, 'Who is this? First I must finish calculating . . .' Again he let me wait for about five minutes and then: 'Who is that?' 'I am Weisskopf'. 'Uhh,

Fig. 8.4. Wolfgang Pauli receiving a cake to celebrate his forty-fifth birthday, Princeton 1945. (Courtesy CERN – Pauli Archive.)

Weisskopf, ja, you are my new assistant.' Then he looked at me and said, 'Now, you see I wanted to take Bethe, but Bethe works now on the solid state . . . Solid state I don't like . . . This is why I took you.' . . . He gave me a problem, some calculations, and then he said, '. . . go and work'. So I went, and after 10 days or so, he came and said, 'Well, show me what you have done.' And I showed him. He looked at it and exclaimed: 'I should have taken Bethe!' He was probably right . . .

One of our joint works was a paper that turned out to be the beginning of meson theory . . . In the beginning . . . I tried to tell Pauli the problem. He was in a bad humour. (In spite of his occasional rough manners, he was one of the kindest and warmest persons I ever met.) . . . [He] said, 'Dummheit, nonsense, go away.' . . . Then I became angry and told him a quotation 'Oh, master, why so much excitement and so little calm; I think that your judgement would be more mature if you listened more carefully.' He then looked at me and said, 'What is that?' and I said, 'That is from Richard Wagner, Die Meister-singer.' . . . 'Wagner? . . . I don't like him at all.' And then, of course, it was over, I had to go. Two days later, I again said, 'Look, there is an interesting problem here,' and he said 'Uh, why didn't you tell me this right away?' Then a wonderful collaboration began . . .

(Victor Weisskopf)[11]

1946

High-pressure physics

The first Nobel Prize after the end of the Second World War, in 1945, was awarded for research work carried out in the field of high-pressure physics. The prizewinner was Percy Bridgman, an American professor of physics at Harvard University in Cambridge, Massachusetts.

Percy Williams Bridgman (1882–1961) was born in Cambridge, Massachusetts, the son of a newspaper reporter. His scientific and academic life was entirely associated with Harvard University. He received his bachelor's and master's degrees there; and his Ph.D. in 1908. He was appointed assistant professor in 1919, and became Hollis professor of mathematics and natural philosophy in 1926, and Higgins university professor in 1950. His researches in the field of high-pressure physics, rewarded with the Nobel Prize, began in 1908, and continued throughout his whole career.

Bridgman's work

Whilst studying for his doctorate, Bridgman happened to be investigating certain optical phenomena under high pressure. In the course of an experiment he was carrying out, the apparatus exploded. He was then able to devise new, stronger materials, and also to construct a new apparatus with which he eventually succeeded in producing pressures of 20 000 atmospheres. In his forty years of research, he was able to extend the pressure range up to more than 400 000 atmospheres. (Under these pressures, atoms and molecules of a substance are forced into a more compact structure.)

Bridgman carried out extensive investigations into the properties of matter at high pressures, including studies on electric and thermal conductivity, on the viscosity of fluids and on the elastic properties of solids. As the range of pressure was extended, new and unexpected phenomena appeared; for example, he was able to convert a form of phosphorus, a substance that normally does not conduct electricity, into a conducting form. Using high pressures, he also produced new types of ice – forms of water that become solid at temperatures far higher than water's ordinary freezing temperature (this being zero degrees Celsius, at normal atmospheric pressure). Finally, in 1955 his pioneering work on high-pressure physics became the basis for the production of synthetic diamonds by scientists at the General Electric Company. The results of Bridgman's work had important implications in several fields of science, including solid-state physics and geophysics.

1947

The ionosphere

In 1947 Sir Edward Appleton, professor of natural philosophy at the University of Cambridge, England, received the Nobel Prize, 'for his investigations into the physics of the upper atmosphere, and especially for the discovery of the so-called Appleton layer'.

Edward Victor Appleton (1892–1965) was born in Bradford, Yorkshire, England. In 1914 he graduated at St John's College, Cambridge, and then studied under J. J. Thomson at the Cavendish Laboratory. In 1924 he was appointed Professor of Physics at the University of London, and in 1936 Professor of Natural Philosophy at the University of Cambridge. In 1939 he became secretary of the Department of Scientific and Industrial Research, where he was responsible for many laboratories, including those involved in radar and atomic energy research.

Appleton's discovery
When Guglielmo Marconi succeeded in 1901 in sending radio signals across the Atlantic, scientists were stunned. Marconi's achievement seemed to defy any explanation – for radio waves, they maintained, travel in straight lines. How, they asked, can these waves possibly propagate around the curved surface of the earth? The answer came from a British physicist, Oliver Heaviside, and independently from an American radio engineer, Arthur Kennelly. They suggested that the radio waves might have been reflected back to earth by a layer, high up in the atmosphere, that was capable of conducting electricity. The *Kennelly–Heaviside layer*, as it was called, was finally located late in 1924 by Appleton: by measuring the strength of radio waves of different frequencies reflected back from the upper atmosphere, he was able to determine that there really was a conducting layer some 100 kilometres up.

In 1926 Appleton found that at night radio waves of shorter wavelength were not reflected back by that layer, but were reflected from a still higher layer at a height of 230 kilometres, which was later named the *Appleton layer*. The whole region discovered by Appleton was called the *ionosphere*. Here, atoms, electrically charged ions and electrons are present, due to the ionising effects of the ultraviolet rays of the sun or other forms of radiation. The methods developed by Appleton can be looked upon as precursors of radar (p. 214). Apart from the field of radio communication, they were also to become of immense importance in other sciences, such as astronomy, geophysics and meteorology.

1948

Nuclei and particles

The last three Nobel Prizes in the first half of the twentieth century (1948, 1949 and 1950) concerned the new-born fields of nuclear and particle physics. The first of these prizes was awarded to a Cavendish man, Patrick Blackett, for his important achievements in both these fields.

Patrick Maynard Blackett (1897–1974) was born in London. After graduating in 1921 at Cambridge University, he remained there to carry out research work at the Cavendish Laboratory. In 1933 he was appointed Professor of Physics at the

University of London, and in 1937 he succeeded Lawrence Bragg at Manchester, in the chair formerly held by Rutherford. Here he gathered around him a cosmopolitan group of cosmic-ray scientists, and he also stimulated the development of new research lines in the field of radio astronomy (this led to the building of the first radio telescope at Jodrell Bank). During the Second World War Blackett was appointed director of Naval Operational Research at the Admiralty in Britain; here, he headed a research group engaged in studying means of coping with the German U-boat menace. In 1953 he was appointed Professor and Head of the Physics Department of Imperial College, London; and in 1965 he was named president of the Royal Society.

Blackett's work

How Blackett began to investigate nuclear physics is described in the following extract from his Nobel lecture:

> In 1919 Sir Ernest Rutherford ... found that the nuclei of certain elements, of which nitrogen was a conspicuous example, could be disintegrated by the impact of fast alpha particles from radioactive sources ... Rutherford [looked] to the Wilson cloud method to reveal the finer details of this newly discovered process. The research worker chosen to carry out this work was a Japanese physicist Shimizu ... [who] built a small cloud chamber and camera to take a large number of photographs of the tracks of alpha particles in nitrogen ... Unfortunately Shimizu had to return unexpectedly to Japan with the work hardly started. Rutherford's choice of someone to continue Shimizu's work fell on me ...[12]

Thus it was that in 1924 Blackett was able to perform an experiment with Shimizu's cloud chamber that resulted in obtaining the first photographs of the nuclear transmutation of nitrogen into oxygen (see an example in Fig. 5.20).

In 1931, together with Giuseppe Occhialini (Beppo to his friends), Blackett designed the counter-controlled cloud chamber, an invention by which cosmic rays could 'take their own photographs':

> [Walther] Bothe and [Bruno] Rossi had shown that two Geiger counters placed near each other gave a considerable number of simultaneous discharges, called coincidences, which indicated in general the passage of a single cosmic ray through both counters. Rossi devised a neat valve circuit by which such coincidences could easily be recorded. Occhialini and I decided to place Geiger counters above and below a vertical cloud chamber, so that any ray passing through the two counters would also pass through the chamber.[13]

With their improved chamber, Blackett and Occhialini succeeded in obtaining at least one cosmic-ray track in each picture, instead of one track in some tens of pictures, as had been the case before with a chamber triggered at random. (Occhialini thus recalled their initial success: 'Blackett emerged from the darkroom with four dripping photographic plates in his hands exclaiming for all the

Fig. 8.5. Patrick Blackett started his career as a cadet in the Royal Navy. He took part, during the First World War, in the battle of the Falkland Islands. At the end of the war he resigned from the navy, and went to the Cavendish to study physics. He was created a life peer in 1969, with the name of Baron Blackett of Chelsea. (Photo: © The Royal Society, London.)

lab to hear, "one on each, Beppo, one on each!" ')[14] As we have seen (p. 183), our two heroes, using their innovative technique, were able to confirm Anderson's discovery of the positron; and, still more, they succeeded in demonstrating that gamma rays can transform themselves into pairs of electrons and positrons (the phenomenon called 'pair production', p. 195).

Discoveries and inventions

John Bardeen and Walter Brattain (Nobelists in 1956) at Bell Labs invented the first transistor, the *point-contact transistor*, in December 1947. Some months later, their team leader, William Shockley (Nobelist in 1956), invented the *junction transistor*. On 30 June 1948 the Director of Research for Bell Labs held a press conference in which he announced the invention of this epoch-making device. Here is an excerpt from that press conference.

> We have called it the Transistor, T-R-A-N-S-I-S-T-O-R, because it is a resistor or semiconductor device which can amplify electrical signals as they are transferred through it from input to output terminals. It is, if you will, the electrical equivalent

Fig. 8.6. The first transistor, December 1947. (Courtesy Lucent Technologies/Bell Labs.)

of a vacuum tube amplifier. But there the similarity ceases. It has no vacuum, no filament, and no glass tube. It is composed entirely of cold solid substances. This tiny cylindrical object which I am holding up is a Transistor. Although it is a 'little bitty' thing it can . . . do just about everything a vacuum tube can do and some unique things which a vacuum tube cannot do . . .[15]

The invention was reported – in amongst other news items – in an article published in the *New York Times*.

A device called a transistor, which has several applications in radio where a vacuum tube ordinarily is employed, was demonstrated for the first time yesterday at Bell Telephone Laboratories, 461 West Street [New York City], where it was invented. The device was demonstrated in a radio receiver, which contained none of the conventional tubes . . .

Fig. 8.7. From left to right: Abraham Pais, Mrs Yukawa and Hideki Yukawa. (Courtesy AIP Emilio Segrè Visual Archives, Uhlenbeck Collection.) In November 1949 the *New York Times* published the news that Yukawa had been awarded the Nobel Prize. According to the newspaper, Yukawa had turned to theoretical physics as a result of his inability to master the art of glass-blowing in his student days. Happy choice, for he became the first Japanese scientist to receive a Nobel Prize for physics!

1949

Nuclear forces

As we have seen (p. 193), in 1935 the young Japanese theorist Hideki Yukawa had predicted that a new massive particle could be the 'mediator' of the forces that act between nucleons inside the atomic nucleus. This particle was actually discovered in cosmic rays in 1947 (see Nobel 1950). Two years later the Swedish Academy of Sciences decided to reward Yukawa with the 1949 Nobel Prize for physics.

Hideki Yukawa (1907–81) was born in Tokyo, Japan, the son of a university professor of geology. He graduated from Kyoto University in 1929, and in 1938 he earned his doctorate from Osaka University. He then returned to Kyoto University as a professor of theoretical physics; here in 1953 he became director of the Research Institute for Fundamental Physics, and in 1970 he was granted the title of emeritus. During his scientific career Yukawa was principally engaged in investigations concerning the theory of elementary particles. He was invited as visiting professor to the Institute for Advanced Study in Princeton (1948–9), and to Columbia University (1949–53).

Yukawa's theory

As we have seen, soon after Chadwick's discovery of the neutron in 1932 (p. 181), physicists realised that a new force must exist in nature, which would have the task of binding protons and neutrons inside the atomic nuclei. Meanwhile, Eugene Wigner (Nobelist in 1963), then working at Princeton University, had pointed out that this nuclear force must have a very short range, corresponding roughly to the dimension of an atomic nucleus. Thus it was that in 1934 Yukawa came to consider the problem of the nature of this new force. He thought that the concept of electromagnetic field could be modified so as to give a force that might have a short range. He imagined a field of force as associated with some kind of quanta or mediators (the so-called *mesons*) that transmit the force itself:

> In order to obtain exchange forces, we must assume that these mesons have the electric charge $+e$ or $-e$ [e is the elementary charge], and that a positive (negative) meson is emitted (absorbed) when the nucleon jumps from the proton state to the neutron state, whereas a negative (positive) meson is emitted (absorbed) when the nucleon jumps from the neutron to the proton. Thus a neutron and a proton can interact with each other by exchanging mesons just as two charged particles interact by exchanging photons.[16]

And what about the mass of these mesons? He recalled years later:

> I was trying to sleep in a small room in the back of the house, but as usual, I was thinking; my insomnia was back again. Beside my bed lay a notebook, so that if I had an idea, I could write it down. That went on for several days. The crucial point came to me one night in October. The nuclear force is effective at extremely small distances, of the order of [two-hundred-thousand-billionth, of a metre]. That much I knew already. My new insight was the realisation that this distance and the mass of the new particle that I was seeking are inversely related to each other. Why had I not noticed that before?[17]

Yukawa used Heisenberg's uncertainty principle to obtain a mathematical relation between the mass of the meson and the range of the force:

$$\text{mass} = h/(\text{range} \times c)$$

(The symbol h always means Planck's constant, and c the speed of light in empty space.) He found that the force mediators would have to be at least 200 times heavier than the electron. He thus concluded his reminiscences:

> Not long afterwards, in November, I presented the new theory to the Osaka branch of the Physico-Mathematical Society of Japan . . . I finished a paper in English by the end of November and sent it to the society for publication . . . I felt like a traveller who rests himself at a small tea shop at the top of a mountain slope. At that time I was not thinking about whether there were any more mountains ahead.[18]

Fig. 8.8. Cecil Powell. (Courtesy Department of Physics, University of Bristol, England.)

1950

Mesons

The first half of the century ended with the 1950 Nobel Prize awarded to another old student of the Cavendish Laboratory – Cecil Powell. He received the much sought-after accolade for his discovery of the mediators of the nuclear force.

Cecil Frank Powell (1903–69) was born in Tonbridge, Kent, England. He studied at Cambridge University, where he graduated in 1925. He then worked at the Cavendish Laboratory, under C. T. R. Wilson and Rutherford; and in 1927 he obtained his Ph.D. there. He then moved to the University of Bristol, where he was first a research assistant, then a lecturer and in 1948 a professor of physics. In 1964 he was appointed director of the physics laboratory at Bristol. During the 1940s Powell developed new photographic techniques for directly recording the tracks of subatomic particles. He then exposed photographic plates at the tops of high mountains, or sent them up in high-altitude balloons, so as to be able to 'photograph' cosmic rays.

A new discovery

The Wilson cloud chamber enabled physicists to see the tracks left by subatomic particles in the gas contained in its interior. It was a truly wonderful detector, extensively used between the 1920s and the 1950s. In the middle 1940s, however, physicists succeeded in perfecting another technique, which had been widely used in studies on the radiation emitted by radioactive substances in the early years of the twentieth century. It was the *photographic emulsion technique*, and it was brought back into use by a group of physicists at the University of Bristol under Powell's leadership. Working in collaboration with scientists of the Ilford Laboratories, they prepared new photographic emulsions which were much more sensitive to fast-moving charged particles. Briefly, they worked as follows: after a charged particle had passed through a photographic emulsion, the silver bromide grains of the emulsion were developed. Under a microscope, the trajectories of individual particles appeared as lines of dark specks. By measuring quantities such as the density of these specks and the maximum distance traversed by the particle, it was possible to determine the mass of the particle itself. Nuclear emulsions enabled the whole sequence of interactions and decays to be studied in a single photographic image.

In 1947 Powell and Giuseppe Occhialini (an old collaborator of Blackett's, p. 226) exposed nuclear emulsions to cosmic rays at mountain altitudes (at the *Pic du Midi* astronomical observatory, in the French Pyrenees). In these plates they found several pictures in which a particle, with a mass of a few hundred times the mass of the electron, came to rest, and decayed into a lighter one (see Fig. 8.9). They came to the conclusion that the parent particle was the mediator of the nuclear force; that is, the *meson* of Hideki Yukawa's theory (see Nobel 1949). Moreover, they concluded that the lighter daughter particle should be identified with the previously known *muon*. Their discovery was reported in a paper published in the 24 May 1947 issue of *Nature*,[19] and the parent particle became known as the *pi-meson* or *pion*.

Pions are produced in high-energy nuclear interactions; they have a mass of 273 electron masses, and their average lifetime is 2.5 hundred-millionths of a second (that is, about a hundred times shorter than that of muons). Moreover, their decay products are muons and neutrinos. By this time physicists had gathered evidence that muons did not interact strongly with protons and neutrons (the first experiment revealing this fact was performed at the University of Rome by the Italian physicists Marcello Conversi, Ettore Pancini and Oreste Piccioni, in 1945). Muons have a mass of 207 electron masses, and a mean lifetime of 2.2 millionths of a second. Their decay products are electrons (positrons for positive muons), neutrinos and antineutrinos.

To complete this chronicle of the second half of the 1940s, we should mention that just a few months after the paper announcing the discovery of the pion appeared, two physicists, George Rochester and Clifford Butler, then working with Patrick Blackett at the University of Manchester, published two remarkable cloud-chamber photographs of totally unexpected particles that they had discovered in cosmic rays. This discovery was subsequently confirmed by experiments carried out at mountain altitudes by Butler and Rochester themselves – and by Carl Anderson and Eugene Cowan in the USA (p. 195). The new particles (named *K*-mesons) belong to a new class of strong interacting particles, which were later called *strange particles*.

Fig. 8.9. The photograph of the first pion discovered by Powell's group. The pion (track π) had decayed into a muon (track *AB*). (Reprinted by permission from C. M. G. Lattes, H. Muirhead, G. P. S. Occhialini, and C. F. Powell, *Nature*, 159, 1947, p. 695 © 1947 Macmillan Publishers.)

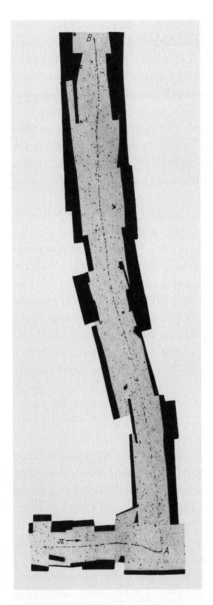

Summarizing the story

At the end of the 1920s only three fundamental particles were known to physicists: atoms and nuclei were thought to consist solely of *protons* and *electrons*; and the massless *photon* represented the quantum of the electromagnetic field. In 1932 two new particles were added to the list: in two beautiful experiments, James Chadwick discovered the *neutron* (see Nobel 1935); and Carl Anderson photographed the first tracks of the *positron* (see Nobel 1936). In the same period, Wolfgang Pauli decided that a new kind of neutral massless particle was needed to

Fig. 8.10. The entrance of the Pupin Physics Laboratory at Columbia University, New York City, circa 1950. Here in 1939 Enrico Fermi, Leo Szilard and other scientists began work on developing a nuclear chain reaction. (Courtesy Columbia University Archives – Columbiana Library.)

Fig. 8.11. The Berkeley Radiation Laboratory, Berkeley, California: 'Old Town', the site of the 184-inch cyclotron, soon after the Second World War. (Courtesy Lawrence Berkeley National Laboratory.)

preserve energy conservation in the radioactive beta decays, the *neutrino* (p. 169). Thus in 1935 the number of known particles and antiparticles (including those whose existence was proved only years later) stood at nine.

In 1936, Carl Anderson and Seth Neddermeyer discovered the *muon* in cosmic rays (p. 195): this was later proved not to be Yukawa's mediator of the nuclear forces, but to be endowed with characteristics similar to those of a heavy electron. Yukawa's particle, the *pion*, was instead discovered in 1947 by Cecil Powell and colleagues. In the same year, new cosmic-ray particles appeared in cloud chamber experiments. They were called *strange particles*, with mass values between the mass of a pion and that of a proton. Finally, in 1950 the *neutral pion* was discovered in experiments with a high-energy accelerator performed at the University of California at Berkeley. Thus, the 1950 particle list totalled fourteen items, taking account of both particles and antiparticles – plus a few strange particles.

Part III
New frontiers (1951–2003)

Fig. 9.1. View of Harvard College (circa 1767), drawn by Joseph Chadwick and engraved by Paul Revere. (Courtesy of the Harvard University Archives.)

Harvard University, the oldest institution of higher learning in the United States, was founded in 1636, 'To advance Learning and perpetuate it to Posterity . . .'.

Chapter 9
Wave of inventions

The *transistor*, which first saw the light of day at the end of 1947, would have a greater impact on twentieth century science and technology than most other inventions of the time. In the 1950s a multitude of new devices appeared, the most remarkable being the *maser*, the *laser* and the *integrated circuit*. These all generated, in the following years, a flood of new technologies which transformed not only science but many aspects of our everyday life.

The first nuclear chain reaction – born in 1942 under the football stadium at the University of Chicago – put at our disposal the energy of the atomic nucleus; in fact, after less than fifteen years, the first nuclear reactors for the production of electric power came into existence. During the same period, physicists extended their investigations, principally in order to delve deeper into the internal structure of the nucleus itself.

The development of radar during the war brought about new achievements in the fields of atomic and subatomic physics. Among these achievements, nuclear magnetic resonance, which was discovered round about the middle of the 1940s, developed steadily into a multi-purpose tool, with uses ranging very widely: from the study of bulk matter, to the chemistry of living organisms and, most popularly, to medical diagnostics.

Towards the end of the 1950s a breakthrough came in condensed-matter physics: at long last, more than four decades after its discovery, a successful microscopic theory was worked out to explain the phenomenon of *superconductivity*.

And then again, new subatomic particles were discovered; the existence of the elusive neutrino was proved in 1956; old conservation laws broke up and fresh ones were elaborated. A new generation of particle accelerators advanced, reaching higher and higher energies, and a new device, the *bubble chamber*, replaced the cloud chamber as a major research apparatus. And so it was that particle research shifted from cosmic rays to accelerator laboratories, and all this means that the era of high-energy particle physics had arrived.

In this decade (1951–60), twelve of the twenty Nobel laureates received their prizes for discoveries and inventions originating quite soon after the Second World War.

1951

Old days at the Cavendish

Back at the Cavendish Laboratory in Cambridge, in the early 1930s, John Cockcroft and Ernest Walton first succeeded in accelerating charged particles to high energies, thus producing the first nuclear reaction using an accelerator (p. 174). Twenty years later, they shared the 1951 Nobel Prize for physics, precisely for that pioneering work on the transmutation, under human control, of atomic nuclei.

It so happened that, late in 1928, George Gamow was at Bohr's institute in Copenhagen, and decided to visit the Cavendish. He had shown that there was some probability that alpha particles would escape from radioactive atomic nuclei, even if their energy was not enough to overcome the energy barrier (see p. 177). Cockcroft thought that if a particle could *tunnel out* of a nucleus then it must also be able to *tunnel into* the same nucleus from outside. He wrote in his Nobel lecture:

> I discussed with him [Gamow] the converse problem – the energy which would be required for a proton accelerated by high voltages to penetrate the nuclei of the light elements. As a result of these talks I prepared a memorandum which I sent to Rutherford showing that there was a quite high probability for the boron nucleus to be penetrated by a proton of only 300 kilovolts energy whilst conditions for lithium were even more favourable. Rutherford then agreed to my beginning work on this problem and I was soon joined by Dr. Walton.[1]

Thus in 1932 Cockcroft and Walton, using their newly invented voltage multiplier, accelerated protons to high energies and looked for alpha particles produced in the disintegration of lithium. Success was immediate. Thus Walton remembers the events of Thursday, 13 April:

> When the voltage and the current of protons reached a reasonably high value, I decided to have a look for scintillations . . . Immediately I saw scintillations on the screen. I then went back . . . and switched off the power to the proton source. On returning to the hut, no scintillations could be seen. After a few more repetitions of this kind of thing, I became convinced that the effect was genuine. Incidentally, these were the first alpha-particle scintillations I had ever seen and they fitted in with what I had read about them. I then 'phoned Cockcroft who came immediately. He had a look at the scintillations and after repeating my observations, he also was convinced of their genuine character. He then rang up Rutherford who arrived shortly afterwards. With some difficulty we manoeuvred him into the rather small hut [see Fig. 7.1 left] and he had a look at the scintillations. He shouted out instructions such as, 'Switch off the proton current'; 'Increase the accelerating voltage', etc., but he said little or nothing about what he saw. He ultimately came out of the hut, sat down on a stool and said something like this: 'Those scintillations look mighty like alpha particle ones. I should know

an alpha-particle scintillation when I see one for I was in at the birth of the alpha particle and I have been observing them ever since.'[2]

In this way Cockcroft and Walton obtained the first nuclear reaction induced by artificially produced particles. The British novelist C. P. Snow thus described Cockcroft's enthusiasm: 'In about the only magniloquent gesture of a singularly modest and self-effacing life, Cockcroft walked with soft-footed games player's tread through the streets of Cambridge and announced to strangers, "We've split the atom. We've split the atom."'[3]

Cockcroft and Walton's discovery gave impetus to the development of new machines capable of accelerating particles to high energies, and so opening the door to the age of accelerators in the new field of nuclear physics; it also represented the first direct proof for Einstein's formula expressing how mass is converted into energy (p. 183).

John Douglas Cockcroft (1897–1967) was born in Todmorden, Yorkshire, England, the son of an entrepreneur in the weaving industry of the Calder Valley. He studied at the University of Manchester and at St John's College, Cambridge. He joined the Cavendish Laboratory in 1924, and first worked with the Russian physicist Peter Kapitza (Nobelist in 1978) on the production of intense magnetic fields. In 1928 he began research on the acceleration of protons by high-voltage equipment. In 1939 Cockcroft was appointed Professor of Natural Philosophy at Cambridge University. During the Second World War he was leader of the Canadian Atomic Energy project. Sir John, as he later became, was also a leading scientist in the development of radar. In 1946 he returned to England and organised research in the field of nuclear power reactors; in 1959 he was elected Master of Churchill College, Cambridge.

Ernest Thomas Walton (1903–95) was born in Dungarvan, County Waterford, Ireland, the son of a Methodist minister. In 1927 he received his master's degree from Trinity College, Dublin. He then went to the Cavendish Laboratory to work under Rutherford; he received his Ph.D. there in 1931. Three years later he returned to Trinity College, where, in 1946, he became professor of natural and experimental philosophy. He also worked on hydrodynamics and on microwaves. For their discovery, which was rewarded with the Nobel Prize, Walton and Sir John Cockcroft in 1938 also earned the Hughes Medal from the Royal Society of London.

1952

Nuclear magnetic resonance

Nuclear magnetism was first studied in the 1930s by Otto Stern and Isidor Rabi (Nobelists in 1943 and in 1944). (When measuring the magnetic moments of atomic nuclei, Rabi exploited both Stern's molecular-beam method and resonance with radio waves).

Soon after the Second World War a new technique, called *nuclear magnetic resonance* or NMR, came out. It made possible the investigation of the nuclear magnetism of solid, liquid and gaseous substances. It was discovered independently by two physicists, the Swiss-born American Felix Bloch, working at Stanford University, and the American Edward Purcell from Harvard University. For their discovery they were awarded the 1952 Nobel Prize for physics.

Felix Bloch (1905–83) was born in Zurich, Switzerland. In 1924 he entered the Swiss Federal Institute of Technology (ETH) in that city, and here attended courses held by the physical chemist Peter Debye, the mathematician Hermann Weyl and Erwin Schrödinger. He received his doctorate in 1928 at the University of Leipzig, Germany, studying under Heisenberg (he was Heisenberg's first graduate student). He wrote a thesis dealing with the application of quantum mechanics to the motions of electrons in crystals, and to electrical conduction in metals (p. 00). This led in turn to the modern quantum theory of solids, and eventually to modern electronics. In 1934 Bloch, who was Jewish, left Germany and emigrated to the USA, joining Stanford University in Palo Alto, California. His interests and studies shifted to experimental physics, and he soon proposed a method for splitting a beam of neutrons in a magnetic field. In 1939 he and the American physicist Luis Alvarez (Nobel 1968) used this method at the Berkeley cyclotron, and they were there able to measure, for the first time ever, the magnetic moment of the neutron.

During the Second World War Bloch worked at Los Alamos on the Manhattan Project, and on radar techniques at Harvard University, thus acquiring extensive skills in modern electronics. After the war he returned to Stanford where he developed, together with his co-workers, the nuclear magnetic resonance principle, which we will soon describe in greater detail (p. 241).

Edward Mills Purcell (1912–97) was born in Taylorville, Illinois, USA. He graduated in electrical engineering from Purdue University (Indiana) in 1933, and received his Ph.D. from Harvard University in 1938. During the Second World War Purcell headed a group which developed radar microwave techniques at the MIT Radiation Laboratory in Cambridge, Massachusetts. After the war he returned to Harvard as Associate Professor of Physics. Here he discovered nuclear magnetic resonance in liquids and solids, which earned him the Nobel Prize. In 1952 Purcell detected 21-centimetre-wavelength radio waves emitted by hydrogen atoms in interstellar space. These waves had been predicted in the 1940s (they are due to transitions between very closely spaced energy levels in the hydrogen atom – the so-called 'hyperfine structure', p. 251), and their study enabled astronomers to determine the location of hydrogen clouds in our galaxy. Purcell became full professor at Harvard in 1949, and emeritus in 1980.

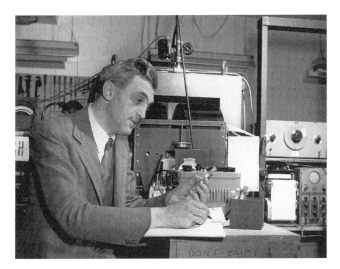

Fig. 9.2. Felix Bloch in his laboratory. (Courtesy AIP Emilio Segrè Visual Archives, Bainbridge Collection.)

Fig. 9.3. Edward Purcell. (Courtesy AIP Emilio Segrè Visual Archives, Bainbridge Collection.) 'I remember, in the winter of our first experiments, . . . looking on snow with new eyes. There the snow lay around my doorstep – great heaps of protons quietly precessing in the earth's magnetic field. To see the world for a moment as something rich and strange is the private reward of many a discovery.' (Edward Purcell)[4]

The principle of NMR

In two papers published in the *Physical Review*, Bloch and Purcell reported in 1946, almost simultaneously, that they had independently observed the nuclear magnetic resonance due to protons. Let us try to grasp the principle of this phenomenon.

Consider a sample of atoms, for example those contained in a drop of water. In their nuclei there are protons (some thousand billion billion), acting as microscopic magnetic dipoles. They are randomly distributed in space. If one applies a strong magnetic field (for example, one some ten thousand times stronger than that around the earth), a fraction of the proton dipoles (some billions) become oriented parallel to the field. At the same time, each dipole starts to revolve (or to carry out a *precession*) about the field direction (see Rabi's magnetic resonance method, p. 219), with a precession frequency which is in the radio-frequency range. Now, imagine stimulating the drop with radio waves of exactly the same frequency (as when you synchronise your receiver on the frequency of a radio station). In this situation, called 'resonance', the dipoles will absorb energy from the radio waves, and will flip over (again as in Rabi's method). If now the radio waves are switched off, each dipole re-orients itself and starts to re-emit radio waves of the same frequency. These can easily be detected (as your radio detects a broadcasting station). This, in fact, was the method used by Bloch, as he noted in his Nobel lecture: 'The essential fact of the magnetic resonance consists in the change of orientation of nuclear moments . . . The acquaintance with radio techniques during the war suggested to me . . . [detection of their] reorientation . . . through the normal methods of radio reception . . . I believe that this is the most general and distinctive feature of our discovery . . .'[5]

Purcell used a method that was different, but based on the same principle. Soon after Bloch and Purcell's discovery, physicists observed NMR in practically all the atomic nuclei with nuclear magnetic moments.

NMR then became an extremely useful tool for the study of bulk matter. But the most spectacular use of NMR is in medicine. Here it has become a powerful diagnostic technique, called *magnetic resonance imaging* or MRI, which is used to generate thin-slice images of any part of the body. (Our body is in fact filled with microscopic magnetic dipoles, the most abundant of which are protons, which are present in the body's water and fat molecules.) MRI distinguishes between normal tissues and diseased or damaged ones. This technique is used to examine the brain, the kidneys and the spinal cord, as well as other organs, and researchers using it were awarded the 2003 Nobel Prize for medicine.

1953

A new microscope

About forty years after the award to Heike Kamerlingh Onnes, another Dutch person, the fourth, was crowned a Nobel laureate. He was Frits Zernike, who won the 1953 Nobel Prize in physics for his invention of the *phase-contrast microscope*. Here, the principle involved belongs to wave optics, the branch of physics which concerns optical phenomena explained by the wave theory of light. This was one of the few prizes to be awarded to classical physics. As the Swedish Academy noted:

> When . . . a Nobel Prize is awarded for contributions in classical physics, the fact is so remarkable that we must go back to the very earliest Nobel Prizes to find a

counterpart. All later Nobel Prizes, with the exception of a couple of awards where the stress was rather upon the technical aspect, have been awarded for discoveries in atomic and nuclear physics, the physics of this century.[6]

Frits Zernike (1888–1966) was born in Amsterdam, the Netherlands. His father and mother were mathematics teachers at local schools. He obtained his doctorate from the University of Amsterdam in 1915, after becoming assistant at the University of Groningen. Here he was full professor of mathematical physics from 1920 to 1958. His early interest in optics began when he was still a boy, and was focused mainly on astronomical telescopes, as reported by his grandnephew Gerardus 't Hooft (Nobelist in 1999):

> My grandmother, Zernike's sister, used to tell us anecdotes about her brother when they were young. One day, for instance, he had purchased a telescope at a local market. That night, the police came to their door to warn her parents that there were 'zinc thieves on their roof'; it was Frits however, trying out his new telescope and studying the heavens.[7]

In the early 1930s, while studying optical irregularities in diffraction gratings, Zernike discovered the phase-contrast principle. Thus he observed in his Nobel lecture:

> On looking back to this event, I am impressed by the great limitations of the human mind. How quick are we to learn, that is, to imitate what others have done or thought before. And how slow to understand, that is, to see the deeper connections. Slowest of all, however, are we in inventing new connections or even in applying old ideas in a new field.[8]

Zernike was able to distinguish light rays passing through different transparent materials. Thus a few years later he constructed the first phase-contrast microscope, 'with which he had stunned biologists by showing them moving images of a living cell.'[9]

Seeing minute biological objects

Conventional light microscopes use lenses to form the image. They can magnify up to 1000 times the dimensions of the object they observe, giving a resolution (the minimum dimension of the object seen) of up to a twenty-thousandths of a millimetre.

Microscopic biological objects, however, such as living cells or microorganisms, consist of transparent and colourless structures. They are difficult to distinguish from the surrounding medium, so one cannot observe their structures when using a conventional microscope. However, since the transparency of the structure varies slightly from the surroundings, a light wave propagating through such an object will shift its waveform slightly (a quarter of a wavelength) relative to the wave passing around the structure itself. Zernike was able to use

Fig. 9.4. Frits Zernike and his phase-contrast microscope. (Courtesy AIP Emilio Segrè Visual Archives, W. F. Meggers Gallery of Nobel Laureates.) In the beginning Zernike's invention was neglected by industry (he had tried to interest the famous German company Zeiss; but without success). However, during the Second World War, the Germans took up all inventions which might serve the war effort, and so they developed Zernike's microscope. After the war various firms produced such microscopes, which soon became a powerful tool for biological research.

this shift to convert variations in the transparency of the object into visible contrasts in the intensity of the image (technically, the intensity distribution is related to the phase changes produced by the structure of the object itself). In this way different kinds of structures can be observed in high contrast, even though the specimens themselves are completely transparent.

X-rays, sixty years later

Since Röntgen's discovery in 1895, X-rays had become vital to research and to numerous practical applications. X-rays revolutionised medical diagnostics by introducing a new science, radiography. With X-ray spectroscopy chemists identified new elements and compounds, and with X-ray diffraction the structure of crystals could be studied: it was also applied with success to bio-molecular structures, such as proteins. The method was used for one of the most important discoveries in twentieth-century biology: the establishing of the structure of deoxyribose nucleic acid (DNA), the molecule that contains the genetic blueprint for living organisms. This happened precisely in 1953.

At that time the American biologist James Watson and the English biophysicist Francis Crick were working at the Cavendish Laboratory, which was under the direction of Sir Lawrence Bragg (Nobelist in 1915). ('For almost forty years Bragg . . . had been watching X-ray diffraction methods solve structures of ever-increasing difficulty . . . Thus in the immediate post-war years he was especially keen about the possibility of solving the structures of proteins, the most complicated of all molecules.')[10] Using experimental X-ray diffraction studies of DNA made by the British biophysicist Maurice Wilkins and the physical chemist Rosalind Franklin (both working at King's College, London), they discovered the famous double-helix structure of the molecule. For their discovery Watson, Crick and Wilkins shared the 1962 Nobel Prize for physiology or medicine – 'a prize for which Franklin would have been considered but for her untimely death'.

Here is an excerpt from Watson's book, *The Double Helix*, where he describes the last days of their discovery:

> Sir Lawrence was shown the paper . . . [He] enthusiastically expressed his willingness to post it to *Nature* with a strong covering letter. The solution to the structure was bringing genuine happiness to Bragg. That the result came out of the Cavendish and not Pasadena [Caltech] was obviously a factor. More important was . . . the fact that the X-ray method he had developed forty years before was at the heart of a profound insight into the nature of life itself.
>
> The final version was ready to be typed on the last week-end of March [1953]. Our Cavendish typist was not on hand, and the brief job was given to my sister . . . Francis and I stood over her as she typed the nine-hundred-word article that began, 'We wish to suggest a structure for the salt of deoxyribose nucleic acid (DNA). This structure has novel features which are of considerable biological interest.' On Tuesday the manuscript was sent up to Bragg's office and on Wednesday, 2nd April, it went off to the editors of *Nature*.[11]

1954

Two veterans

Max Born, the last pioneer of quantum mechanics to be honoured with a Nobel Prize, was rewarded by the Swedish Academy of Sciences some three decades after the golden age that had witnessed the birth of the new theory. (Born's crucial role in the development of quantum mechanics was not recognised when the 1932 and 1933 Nobel Prizes for physics came to be awarded, and this aroused no little criticism. In point of fact, Heisenberg himself wrote to Born to express surprise and regret that they were not going to share the prize.)

Born, then in his seventies, shared the 1954 Nobel Prize with Walther Bothe, another veteran of pre-war German physics. The discoveries of Born and Bothe were made at Göttingen (Born) and at the PTR in Berlin-Charlottenburg (Bothe) in the mid-1920s. Here are the citations of the Academy: Born received the prize 'for his fundamental research in quantum mechanics, especially for his statistical interpretation of the wavefunction': and Bothe 'for the coincidence method and his discoveries made therewith'.

Quantum probability

As we related in our chronicle of the year 1925 (p. 143), in July young Heisenberg completed his epoch-making paper which marked the birth of quantum mechanics. Two months later, Born and his assistant, Pascual Jordan, used abstract mathematical devices (matrices) to interpret Heisenberg's findings. In November, Born, Jordan and Heisenberg himself outlined the foundations of *matrix mechanics* in their famous 'three man paper'.

The following year, Erwin Schrödinger found a new way to formulate quantum mechanics, the so-called *wave mechanics* (p. 147). He suggested that waves, represented by the wave function ψ, were the only reality: '. . . he hoped that wave mechanics would turn out to be a branch of classical physics – a new branch, to be sure, yet as classical as the theory of vibrating strings or drums or balls.'[12]

But what did the wave function ψ mean? Nobody knew. It was up to Born to provide a physical interpretation, in a paper published in 1926. Born proposed that ψ would be related to the probability of finding a particle at a certain point in space (p. 00). This was a radical break with the older conception. Einstein and Schrödinger, however, were the strongest opponents of this view. ('I wrote about it to Schrödinger, and it made him furious, because he didn't want that', Born remembered later on.)[13] In December 1926 Einstein wrote a famous letter to Born; in it he noted: 'Quantum mechanics is certainly imposing. But an inner voice tells me that it is not yet the real thing. The theory says a lot but does not really bring us any closer to the secret of the "old one." I, at any rate, am convinced that *He* is not playing dice.'[14] However, in spite of this, Born's ideas were generally accepted by other physicists.

The beginning of matrix mechanics

In July 1925 Born delivered a lecture at the University of Tübingen. Alfred Landé, then Professor of Physics there, later on remembered Born's speech with these words:

> . . . He told [us] . . . that . . . [the Göttingen theoreticians] had a completely new approach to quantum mechanics, and everything [was] dominated by multiplication, [in which] A times B differs from BA. I did not understand a single word about it, and I do not think that Born and the whole group in Göttingen 'understood' much more than the mere formulae.[15]

Max Born (1882–1970) was born in Breslau, Germany (now Wrocław, Poland). His father was Professor of Anatomy at the university ('My father was a scientist . . . My mother's family was Silesian, a very old Jewish family there . . . [Her]

Fig. 9.5. J. Franck (left) and M. Born, Göttingen, 1924. (Courtesy AIP Emilio Segrè Archives, W. F. Meggers Gallery of Nobel Laureates.)

parents were great industrialists – textiles.') Born studied in different universities (Breslau, Heidelberg, Zurich, Göttingen and Cambridge), under the best teachers of the time (Klein, Hilbert, Minkowski, Schwarzschild, J. J. Thomson). He received his doctorate from the University of Göttingen in 1907, and from 1908 to 1909 he worked at the University of Breslau.

In 1915 he was appointed Professor of Physics at the University of Berlin; then at the universities of Frankfurt, and in 1921 Göttingen. Here he became head of the Institute of Theoretical Physics. Under him the institute became one of the most important centres for quantum theory; among his collaborators and students were physicists who became famous later (p. 170). But Hitler's laws against Jews forced Born to leave Germany. He went to Cambridge, and to the Institute of Science in Bangalore, India, for a year. Then seventeen years followed as Professor of Natural Philosophy at the University of Edinburgh, Scotland. In 1953 he retired and went to live in Bad Pyrmont near Göttingen.

The coincidence method

The idea of the *coincidence method*, which earned Walther Bothe the Nobel Prize, was to use two counters in such a way that they would register only particles which they had detected simultaneously.

Bothe, together with Hans Geiger (Rutherford's erstwhile assistant in Manchester from 1907 to 1912), when working in 1924 at the PTR in Berlin-Charlottenburg, invented the coincidence method in order to investigate the Compton effect (p. 135). They used two particle counters, called *needle counters* (these had been developed by Geiger himself while working in Manchester), which they arranged in such a way that they were able to detect both the scattered photon and the recoil electron when produced in a Compton process. Bothe and Geiger succeeded in detecting 'temporal coincidences' between the two counters within a very short time interval (about a thousandth of a second). This demonstrated that the scattered photon and the recoil electron were generated 'simultaneously'. As Bothe wrote in his Nobel lecture:

> The 'question to Nature' which the experiment was designed to answer could therefore be formulated as follows: is it exactly a scatter quantum [X-ray photon] and a recoil electron that are simultaneously emitted in the elementary [Compton] process, or is there merely a statistical relationship between the two? ... The final result we obtained was ... that, in each elementary Compton process, a scatter quantum and a recoil electron are generated *simultaneously*.[16]

Bothe and Geiger's experiment definitely established the particle nature of radiation, and also proved that energy was in fact conserved even in elementary processes such as the collision of a photon with an electron. (Some months earlier Niels Bohr, Hendrik Kramers and the young American physicist John Slater had published a paper in which they tried to explain such elementary processes without the photon hypothesis; they also suggested that in each of such elementary processes energy could not be conserved.)

Bothe's coincidence method was greatly improved upon in subsequent years. In the late 1920s Bruno Rossi invented the vacuum-tube coincidence circuit, which greatly reduced the number of chance coincidences recorded by the counters, so improving the accuracy of the experiments (in 1954 when Bothe received the Nobel Prize the resolution coincidence time was already reduced to about ten-millionths of a second). Bothe also used the coincidence method in other fields, such as those of cosmic rays and nuclear reactions. It soon became one of the most important tools in the study of nuclear and particle physics.

Walther Bothe (1891–1957) was born in Oranienburg, near Berlin. He obtained his doctorate from the University of Berlin in 1914. During the First World War he was taken prisoner and spent a year in Siberia. When he returned to Germany in 1920, he went to work with Hans Geiger at the PTR. He taught at the Universities of Berlin (1920–31), Giessen (1931–4) and Heidelberg (1934–57). During the Second World War he became one of the foremost scientists involved in Germany's nuclear energy project.

Travelling with Enrico Fermi
On 28 November 1954, Enrico Fermi died at the age of fifty-three. Here is an excerpt from a tribute to Fermi that the American physicist Samuel K. Allison, a collaborator of Fermi's during the 1940s, gave at the University of Chicago.

Professor A. H. Compton [Nobelist in 1927], Enrico Fermi, and I were travelling together . . . on the train . . . The hours seemed to drag crossing the mountains, and Enrico, who always disliked travelling, was restless and bored. After some long silences, Mr Compton said: 'Enrico, when I was in the Andes mountains on my cosmic-ray trips, I noticed that at very high altitudes my watch didn't keep good time. I thought about this considerably and finally came to an explanation which satisfied me. Let's hear you discourse on this subject.'

Enrico's eyes flashed. A problem! A challenge! Something to work on! Having been in several such situations before, I relaxed and prepared to enjoy the fireworks that would surely follow. He found a scrap of paper and took from his pocket the small slide rule he always carried. During the next five minutes he wrote down the mathematical equations for the entrainment of air in the balance wheel of the watch, the effect on the period of the wheel, and the change in this effect at the low pressures of high altitudes. He came out with a figure which checked accurately with Mr Compton's memory of the deficiencies of his timepiece in the Andes. Mr Compton acknowledged the correctness of the calculation, and I shall not forget the expression of wonder on his face. It is with such a man that we in the Institute could consult daily, and it is such a man that we have lost.[17]

1955

Splitting spectral lines

Two discoveries, both made at Columbia University in the second half of the 1940s, were rewarded with the 1955 Nobel Prize. The laureates were two Americans: Willis Lamb and Polykarp Kusch.

Lamb's findings concerned the spectrum of the hydrogen atom, whereas Kusch measured anomalies in the magnetic moment of the electron. They discovered truly minute deviations from the quantum theory of that time, which described the interaction of electrons and electromagnetic radiation. The results of their experiments were published in 1947 in two distinct papers in the same volume of the *Physical Review*. Their discoveries were astonishing and far-reaching, and ultimately led to a new formulation of quantum electrodynamics.

Let us start with Lamb. Here are the first sentences from his Nobel lecture:

> In order to determine the properties of elementary particles experimentally it is necessary to subject them to external forces or to allow them to interact with each other. The hydrogen atom which is the union of the first known elementary particles, electron and proton, has been studied for many years, and its spectrum has taught us much about the electron.[18]

So Lamb began to study the spectral lines of the hydrogen atom when stimulated by radio waves, and discovered subtle new features of its spectrum which earned him the Nobel Prize.

The Lamb shift
Dirac's 1928 relativistic theory of the electron (p. 176) predicted the structure of the hydrogen atom in great detail, including the *fine structure* of its spectrum.

Now, we are going to try to explain to you this splitting of the spectral lines, and we shall do it in the following way. First of all, remember how the spinning electron in the hydrogen atom acts as a microscopic bar magnet. Next, we want you to imagine yourself moving while you are seated on the electron itself. As you ride along, you will see the positively charged nucleus (the proton) revolving around yourself – just as, to you on earth, the sun seems to be orbiting our planet. This circulating charged nucleus creates a magnetic field at your location, so that the electron feels it. The resulting interaction between this magnetic field and the electron's magnetic moment causes very small differences in the energy levels; as a consequence the spectral lines are split into multiple lines that lie close together.

This kind of splitting is known as the *fine structure* of the spectrum: it was first discovered by Albert Michelson in the hydrogen Balmer-alpha line (p. 65); and was first explained in 1916–17 by Arnold Sommerfeld (p. 110). (The Balmer-alpha line has a frequency of about 450 000 billion cycles per second, which corresponds to the colour red at a wavelength of 656.3 nanometres. It is emitted when a hydrogen atom changes from the energy level with quantum number $n = 3$ to that with $n = 2$.)

Of particular interest is the fine structure of the first excited state of the hydrogen atom (corresponding to $n = 2$). According to Dirac, it should split as in Fig. 9.6: two states with two different energy levels (labelled $2P$), corresponding to an angular momentum quantum number $l = 1$; and one state $2S$, corresponding to $l = 0$, which has energy identical to the lower of the 2P states (they differ, however, in the spatial distribution of the electron around the nucleus). In the 1930s several investigators used optical spectroscopy to test Dirac's theory. Some of them had reported small discrepancies between the measured optical spectra and the predictions of the theory.

At the end of the war Lamb, with a graduate student, Robert Retherford, began an experiment to investigate the structure of those quantum states, $2S$ and $2P$. They did not attempt to 'see' the fine splitting of the red Balmer line

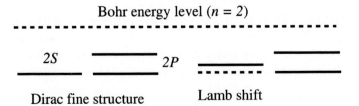

Fig. 9.6. The energy level 2P is split into two components called the *fine structure* of the levels. The *Lamb shift* further displaces the 2S level upwards with respect to the lower of the two 2P levels.

by analysing the light emitted. Instead they employed radio waves to directly stimulate a possible transition between the levels, using Isidor Rabi's atomic-beam resonance technique (p. 218). As Lamb remembers it:

> During the war, at the Columbia Radiation Laboratory, I received some first-hand acquaintance with microwave radar and vacuum-tube construction techniques . . . and my interest was started in what was to become the very active post-war field of microwave spectroscopy. In teaching a summer session class in atomic physics in 1945 using a textbook by Herzberg, I found references to some attempts made in 1932–5 to detect absorption of short-wavelength radio waves in a gas discharge of atomic hydrogen. At first it seemed to me that these experiments had failed because of inadequate microwave techniques. I thought of repeating them with the greatly improved facilities developed during the war.[19]

Lamb and Retherford shone radio waves with a frequency of about one billion cycles per second on a beam of hydrogen gas. They measured a strong absorption of energy, meaning that the hydrogen atoms had been stimulated to shift their $2S$ level upward with respect to the lower of the two $2P$ levels by a factor of little more than 2 millionths (1/450 000). The effect (now known as the *Lamb shift*) showed thereby that Dirac's theory was incomplete.

Electron magnetism

And now let us move on to Kusch, still starting from Dirac's electron theory, and from an experiment which had been performed in Rabi's laboratory at Columbia University.

In his theory, Dirac also predicted that the magnetic moment of the electron should be equal to a certain value, usually named a *Bohr magneton* (p. 216). Then, at the beginning of 1947, Rabi and his co-workers, using his molecular-beam resonance method, carried out an experiment in which they demonstrated a further splitting of the energy levels of the hydrogen atom. This splitting, which is called the *hyperfine structure* of the levels, is the result of the interaction of the magnetic moment of the spinning electron and the magnetic moment of the hydrogen nucleus (the proton), and is even smaller than that which gives the fine structure (about ten times smaller). Moreover, Rabi's results showed that the hyperfine interval of the hydrogen ground level emerged as being a little larger than that predicted by Dirac (about two-thousandths larger). Some physicists proposed that the reason could be found in the magnetic moment of the electron: it could be somewhat different from the value of exactly one Bohr magneton, as predicted by Dirac's theory. As Kusch recalls in his Nobel lecture:

> After the war, Nafe and Nelson, working with Rabi, made the first measurements of the hyperfine structure splitting of hydrogen in the ground state . . . However, a discrepancy of about one quarter per cent was noted between the observed and the predicted magnitude of the [hyperfine] splitting . . . The discrepancy led Breit to

Fig. 9.7. Willis Lamb (left) while conversing with Alfred Kastler (Nobel 1966). (Photograph by Speck, courtesy AIP Emilio Segrè Visual Archives.)

suggest that the electron may possess an intrinsic magnetic moment greater than [a Bohr magneton] . . . The question of the existence of an anomalous magnetic moment was then investigated in detail by Foley and me.[20]

Kusch and his co-worker, Henry Foley, began their experiment in the same year as Lamb (1947). They used Rabi's molecular-beam resonance technique, and measured the hyperfine structure of an excited level of the element gallium. They were able in a few months to discover that the magnetic moment of the electron was indeed slightly larger (about 1/1000) than that predicted by Dirac.

Instead of the electron magnetic moment, physicists prefer to speak of another quantity, called the *gyromagnetic ratio* (symbol: g). The g-factor is the ratio of the magnetic moment of a spinning particle and its intrinsic spin angular momentum. For the electron, following Dirac's theory, it should be exactly equal to 2. Kusch and Foley instead obtained $g = 2 + 0.002\,29$.

Lamb and Kusch's experiments posed two questions. What is the origin of the Lamb shift? And how could one explain the magnetic moment anomaly of the electron? Quantum electrodynamics of the 1930s could not answer these puzzles.

The two Columbians
Willis Eugene Lamb, Jr., was born in 1913 in Los Angeles, California, the son of a telephone engineer. He studied physics under J. Robert Oppenheimer at the University of California, Berkeley, where he graduated in 1934, and received a Ph.D. in 1938. Lamb then went to Columbia University, and began to work in

Rabi's laboratory. During the Second World War he contributed to the development of microwave radar and vacuum-tube techniques there. In 1947 he started research on atomic physics, and made the discovery that was rewarded with the Nobel Prize. In 1951 he went to Stanford University in California, as full professor of physics. Here he devised microwave techniques for examining the hyperfine structure of the spectral lines of helium. From 1956 to 1962 Lamb was a professor at Oxford University, then at Yale, and from 1974 onwards at the University of Arizona.

Polykarp Kusch (1911–93) was born in Blankenburg, Germany, the son of a clergyman. His family emigrated to the United States in 1912; here he studied physics at the Case Institute of Technology, Cleveland, Ohio, and obtained his Ph.D. from the University of Illinois at Urbana–Champaign in 1936. The following year he went to Columbia University, and worked there with Rabi on atomic-beam resonance. During the Second World War Kusch was engaged in military research on the application of vacuum tubes and microwaves at the Westinghouse Electric Corporation, and at Bell Labs. After the war he returned to Columbia, and became a full professor of physics. In 1972 he moved to the University of Texas, Dallas.

A new electrodynamics

In June 1947 a group of leading physicists (among them, Rabi, Lamb, Hans Bethe, Richard Feynman and Julian Schwinger) met on Shelter Island, Long Island, New York, to discuss pressing problems in the field of physics. Rabi described his measurements of the hyperfine structure of the ground level of the hydrogen atom, and spoke about the anomaly that Kusch had found regarding the electron magnetic moment. Lamb reported his discovery of a new component in the fine structure of the hydrogen spectrum. All the results left no doubt: Dirac's theory had to be reshaped. Bethe was able within a week to calculate an approximate value of the Lamb shift, while Feynman and Schwinger were able within two years to completely revise Dirac's theory. They created a modern quantum electrodynamics (for convenience: QED, see p. 297). Similar ideas had been published a few years earlier by the Japanese theorist Sin-Itiro Tomonaga.

A central point of this new QED was that the electron observed in experiments was to be considered different from a hypothetical isolated electron, which had been imagined to exist in space. The differences, regarding properties such as its mass and charge, had previously been ignored. However, it was demonstrated that these differences can alter the properties of the electron when it is in an atom. In the new QED the Lamb shift arises from the fact that the electron in a hydrogen atom jitters about, so that it perceives the charge of the nucleus as being somewhat diffused in space. The result is an apparently subtle alteration of the electric interaction (the Coulomb force) between the two particles, the

Fig. 9.8. Particle–antiparticle pairs continuously appear and disappear in the vacuum.

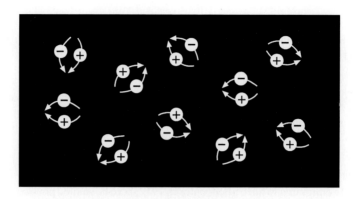

electron and the proton, when they are close together, so giving rise to a minute perturbation in the energy levels of the atom.

Another effect taken into account by modern QED is the interaction of the electron with the space surrounding it (this is called the *vacuum*). Physicists do not see the vacuum as a completely empty space, but as being a complicated medium populated by crowds of ghost-like particle–antiparticle pairs. These pairs are continuously created, and they continuously annihilate each other (their ephemeral life lasts less than billionths and billionths of a second); in such a vacuum there are also of course photons. (All these particles are 'virtual', meaning that they cannot be observed directly.) Physicists accounted for the effect of this crowded vacuum on the electron, and obtained an increase in the g-factor to slightly more than two, which agreed perfectly with Kusch's result.

18 April 1955

In March of that year, Einstein had occasion to remember three old friends. He wrote to Kurt Blumenfeld, 'I thank you belatedly for having made me conscious of my Jewish soul'. He wrote his last autobiographical sketch . . . In this note, he mentioned 'the need to express at least once in my life my gratitude to Marcel Grossmann', the friend whose notebooks he had used as a student, who had helped him to get a job at the patent office, to whom he had dedicated his doctoral thesis, and with whom he had written his first paper on the tensor theory of general relativity. In the same month Michele Besso died, another trusted friend from his student days, later his colleague at the patent office, and his sounding board in the days of special relativity. In a letter to the Besso family, Einstein wrote, 'Now he has gone a little ahead of me in departing from this curious world.'

. . . On the morning of Wednesday, April 13, the Israeli consul called on Einstein at his home in order to discuss the draft of a statement Einstein intended to make on television and radio on the occasion of the forthcoming anniversary of Israel's independence . . . That afternoon Einstein collapsed at home . . . On Friday he was moved to Princeton Hospital. That evening a call was made to his son Hans Albert in Berkeley, who immediately left for Princeton and arrived on Saturday afternoon . . . Alberta Rozsel, a night nurse at the hospital, was the last person to see Einstein alive.

(Abraham Pais)[21]

1956

Transistor breakthrough

One of the most revolutionary technological inventions of the twentieth century, the transistor, was effected at Bell Labs in December 1947. The news of its birth (p. 226) was not received with acclaim; for some time it was grossly undervalued. And then, about ten years later, the transistor burst on to the stage of modern electronics, where it has held a prime place ever since. For this the three inventors, John Bardeen, Walter Brattain and William Shockley, shared the 1956 Nobel Prize for physics. Let us summarise briefly the story of this invention.

At the end of the Second World War vacuum tubes were essential components of radio and telephone equipment. They were also used in early televisions and computers, and in many other electronic circuits. However, vacuum tubes used up too much power and produced too much heat. They were fragile, bulky and short-lived.

Mervin Kelly, during the period that he was executive director of Bell Labs, recognised that better switches and better amplifiers were needed for telephone equipment, and he felt that the answer might lie in semiconductors:

> Towards the end of 1945 . . . Bell Laboratories' management anticipated a rapid expansion of communications in the post-war era. However, they also saw several areas of major concern that would pose serious limitations on future telecommunications systems. Near the top of Mervin Kelly's list of limitations was the slow speed of the electro-mechanical relays and the high power-dissipation and poor reliability of vacuum-tube amplifiers. Kelly . . . established a group in the summer of 1945 to focus on understanding semiconductors.[22]

Experts in solid-state physics, physical chemistry and electronics were summoned to operate under the leadership of Shockley, a very competitive and sometimes combative man. He was a brilliant theoretician, thirty-five years old, an expert in solid-state physics, who had joined Bell Labs in 1936 to work with Clinton Davisson (Nobelist in 1937). The second person was Brattain, a very skilled experimenter. He was forty-three years old, and had been working at the Bell Labs since 1929. His research work mainly concerned the surface properties of semiconductors. The third man in the team, John Bardeen ('whispering John' to his friends), was thirty-seven years old, and a versatile theoretician (the only person to be awarded the Nobel Prize for physics twice). He was a true expert in quantum atomic theory; he had been taken on by Shockley in the autumn of 1945.

The miracle month

Before the Second World War Shockley had already foreseen the possibility of using semiconductors for amplifying electric signals. He had also tried to build an amplifying device, but he had been unsuccessful. Soon after the war he returned to this problem and together with Brattain and Bardeen began studying amplification in semiconducting crystals; they tried to exploit a physical effect, called the *field-effect mechanism*, which had been known since the 1930s. But again the idea did not work out.

So Bardeen went on to analyse the reason for the failure. During the year 1946, he succeeded in explaining it, and published his idea in the *Physical Review*. Immediately afterwards he went on to work with Brattain, in an effort to make an electronic amplifier based on his own understanding of the behaviour of the charge carriers in a semiconducting crystal. Shockley supervised their research occasionally, but he preferred to work alone on certain properties of solids. Bardeen and Brattain were also working very hard, and finally on 16 December 1947, the miracle occurred: they achieved the amplification effect; that is, they observed that an electrical signal, applied to metal contacts on a sliver of germanium, gave an output larger than the input. It was thus that the 'transistor effect' was discovered.

On 23 December the device (called the *point-contact transistor*) was demonstrated to Bell Labs' managers, who soon realised that it was a big breakthrough. But Shockley felt left out, because he had not been directly involved. This was how he lived out those momentous days:

> [He] was torn by conflicting emotions. Although he recognised that Bardeen and Brattain's invention had been a 'magnificent Christmas present' to Bell Labs, he was chagrined that he had not had a direct role to play in this obviously crucial breakthrough. 'My elation with the group's success was tempered by not being one of the inventors,' he recalled a quarter century later . . . 'I experienced frustration that my personal efforts, started more than eight years before, had not resulted in a significant inventive contribution of my own.'[23]

Shockley, employing his knowledge of the quantum physics of semiconductors, searched hard for an explanation of the physical effect that had been discovered, and over a few weeks he managed to develop another amplifying device, which was more practical and easier to manufacture. It was a kind of sandwich made up of three regions of a semiconducting material containing different concentrations of impurities (see p. 259), together with two boundaries joining these regions, called 'junctions': it came in fact to be known as the *junction transistor*. In February 1948 Bell Labs applied for four patents on semiconductor amplifiers, including the point-contact transistor, and in June they also applied for a patent on Shockley's junction transistor. On 30 June the Bell Labs management announced the invention of the transistor in a historic press conference (p. 226).

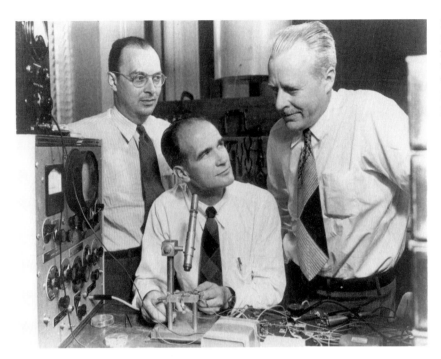

Fig. 9.9. A historic photograph. From left to right: J. Bardeen, W. Shockley and W. Brattain. (Courtesy Lucent Technologies/Bell Labs.)

The Bell Labs trio

John Bardeen (1908–91) was born in Madison, Wisconsin, USA, the son of a professor of anatomy at the local university. He graduated in electrical engineering there, and became a research assistant (1928–30). During the period 1933–8 he worked with Eugene Wigner, John Van Vleck and Percy Bridgman (all Nobel laureates), and in 1936 he obtained a Ph.D. from Princeton University. He taught at the University of Minnesota, Minneapolis, from 1938 to 1941. After the Second World War Bardeen joined Bell Labs, and in 1951 he moved to the University of Illinois at Urbana-Champaign, where he was Professor of Physics and Electrical Engineering till 1975. Here Bardeen resumed research on superconductivity, and together with Leon Cooper and Robert Schrieffer came up with the first successful interpretation of superconductivity, for which he was awarded his second Nobel Prize in 1972 (see p. 321).

Walter Houser Brattain (1902–87) was born in Amoy, China, the son of a science and mathematics teacher. In 1929 he received his Ph.D. from the University of Minnesota, and in the same year he became a research physicist at Bell Labs. He was a highly proficient experimenter; his chief field of research involved the surface properties of solids. After leaving Bell Labs in 1967, Brattain became a professor at Whitman College, Walla Walla, Washington State.

Fig. 9.10. The first point-contact transistor for the telephone, circa 1948. (Courtesy Lucent Technologies/Bell Labs.)

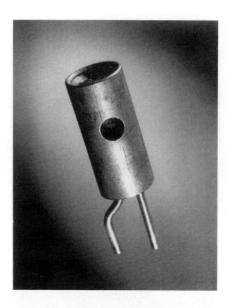

William Bradford Shockley (1910–89) was born in London, the son of an American mining engineer. He studied physics at Caltech, Pasadena, and then at MIT, where in 1936 he obtained his Ph.D., with a thesis under John Slater. That same year he joined Bell Labs, where he began research on semiconductors. 'He was a systematic inventor, with a wide knowledge of physics, whose principles, especially as applied to materials, he understood as well as anybody.'[24] He was granted more than fifty patents for his inventions. After leaving Bell Labs, Shockley established his own firm in 1955, the Shockley Semiconductor Laboratory at Mountain View, near Palo Alto, California. This sparked off an explosion of activity that was to grow into the phenomenon later known as 'Silicon Valley'. In 1963 he became Professor of Engineering Science at Stanford University. In later life Shockley became a figure of public controversy because of his beliefs concerning racial differences in human intelligence.

Hunting the elusive neutrino

In his theory of nuclear beta decay (p. 193), Enrico Fermi was the first person to use the concept of the *neutrino* – the particle that Wolfgang Pauli had first postulated in 1930 (p. 169). From then on physicists had always assumed this particle in their theories, despite the fact that it had never been actually observed. This is because capturing a neutrino is an extremely rare event: a neutrino, to interact with matter, must travel on average through some 10 000 billion kilometres of lead! (Neutrinos have no electric charge, and they interact with matter only through the weak nuclear force.) So it was that physicists did not think seriously about hunting for them, until they were able to employ very intense fluxes of neutrinos.

The transistor

In metals some of the atomic electrons are free to roam through the material. When an electric voltage is applied to a metal, an electric current flows. In insulators, on the contrary, there are no free electrons, and electric charges cannot move through the material. Semiconductors are intermediate in their properties between good conductors and good insulators. In a semiconductor, the conduction of electric charges is partly by electrons and partly by *holes* (these are sites of missing electrons and act like positive charges).

Semiconductors are made from elements such as germanium or silicon. In these materials the number of charge carriers, either electrons or holes, can be increased by adding impurities, and their electrical properties will then become very sensitive even to extremely small concentrations of such impurities. The idea of two types of carriers was put forward in the 1930s, at a time when important uses were beginning to be found for crystal rectifiers. Attempts were made to control the interplay of electrons and holes in those crystals without success. Finally, the discovery of the transistor effect constituted the key to both the control mechanism and the amplification effect.

The term *transistor* was coined by John Pierce, who was a Bell Labs engineer, and a science-fiction writer in his spare time, in order to combine the words *trans*fer and res*istor*. It is made up of different layers of a semiconductor containing different impurities (for example boron and phosphorus). These impurities change the way an electric current flows through the semiconductor, so that the current flow can be controlled, and the device can act either as a conductor or as an insulator. The transistor has two basic functions: it can be used either to *amplify* weak electrical signals (like radio signals in an antenna) or to work like a *switch* (such as in computer operations).

The transistor soon proved to be as effective as the vacuum tube but with reduced power requirements, and with higher reliability. To help speed progress, Bell Labs offered to share what had been discovered with other companies by giving them the opportunity to take advantage freely of their transistor patent-rights. Many firms now shifted their production from vacuum tubes to transistors, so marking the beginning of modern electronics.

In 1954 the first transistorised radio was introduced. In the following years, silicon supplanted germanium in solid state devices. In 1958 the first *integrated circuit* was invented, and the first transistorized computer was developed. Henceforward, transistors began to be used in a wide array of electronic equipment, ranging from radio, television, electronic computers and scientific instruments, right up to communication satellites.

This happened in the early 1950s, when the first nuclear fission reactors were built.

In 1953 the American physicists Clyde Cowan and Frederick Reines decided to use a nuclear reactor as a source of neutrinos. Reines later remembered that they '. . . initially considered the use of a nuclear bomb test as the source of neutrinos, but soon decided that [a] reactor . . . would be better.'[25] After a preliminary experiment at Hanford, Washington State, they moved to the Savannah River Plant in South Carolina, where a more intense neutrino flux was possible. Cowan and Reines' apparatus consisted of large tanks containing 400 litres of water and cadmium chloride, placed between scintillation counters. These they erected near the reactor, where radioactive fission fragments furnished floods of antineutrinos. These were then made to bombard the protons in the atomic nuclei of the liquid. The two physicists planned to detect the particles produced by protons in the capture of antineutrinos. It happened in the following way. When a neutron in the fission fragments broke down, it yielded a proton, an electron and an antineutrino (the reaction involved is: neutron $\rightarrow$ proton + electron + electron antineutrino). Now, if the antineutrino existed in the free state, its absorption by a proton would have to produce the reverse reaction, that is: electron antineutrino + proton $\rightarrow$ neutron + positron. So there were two particles to be looked for: a positron and a neutron.

In June 1956 Cowan and Reines succeeded in identifying these two particles in their detectors; hence, they were able to be certain, without any kind of doubt, that they had caught antineutrinos. So it was that they triumphantly sent off a telegram to Pauli informing him that the neutrino (strictly speaking the antineutrino) that he had predicted more than 25 years earlier had now been found! The telegram read: 'We are happy to inform you that we have definitely detected neutrinos from fission fragments by observing inverse beta decay of protons.'.[26] Pauli then replied to Reines: 'Thanks for message. Everything comes to him who knows how to wait.'[27]

1957

Broken symmetries

Until 1956 physicists believed that *right–left* or *mirror symmetry* was a fundamental principle of nature. It was assumed that there was no absolute distinction between a real physical event and its mirror image. It happened that a paper published in the *Physical Review* issue of 1 October 1956 caused much astonishment to the community of physicists. The authors were two imaginative and inventive Chinese-American physicists, Tsung Dao Lee of Columbia University, New York City, and Chen Ning Yang of the Institute for Advanced Study, Princeton, New Jersey. They claimed that phenomena involving radioactivity could in fact distinguish between right and left! The astonishment was even greater when, a few months later, it was reported in the same journal that experiments clearly confirmed Lee and Yang's predictions. Less than a year later the Swedish Academy of Sciences decided to reward that dramatic discovery with the 1957 Nobel Prize for physics.

Mirror symmetry

Since ancient times, philosophers have debated at great length whether nature exhibits mirror symmetry. Here is how T. D. Lee describes it:

> The concept that nature (i.e., physical law) is symmetrical with respect to right and left dates back to the early history of physics. Of course, in our daily life left and right are quite distinct from each other. Our hearts, for example, are usually on our left sides. However, such asymmetry in daily life is attributed to either the accidental asymmetry of our environment or the initial condition in organic life.[28]

According to pre-1956 physics, nature does not distinguish between right and left. Its physical laws are symmetrical for mirror reflection. Your right hand and your left hand are surely different, but they are precise counterparts; looking at your right hand in a mirror, you cannot tell that you are not looking at your left hand. So it is only a matter of convention when we indicate the direction *right* and its opposite, *left*; and consequently, we could never communicate, for instance, to a Martian which side our heart is on!

In their technical language, particle physicists speak of *parity* as a mathematical concept related to mirror reflection. For example, they say: 'you are doing a parity operation', instead of, 'you are reflecting yourself in a mirror'. More subtle is the mathematical meaning they give to parity. It concerns the so-called wave function ψ by which quantum mechanics describes the wave characteristics of a particle, and depends on its position in space (p. 150). Parity is a property of such a wave function, characterised by the behaviour of its sign when the particle it describes is reflected in a mirror (or, more precisely, is completely inverted in space): the wave function either remains unchanged (*even* parity) or changes in sign (*odd* parity). (The concept of parity was first introduced in atomic physics by Eugene Wigner, Nobelist in 1963)

Until 1956 physicists had assumed that parity, like energy or electric charge, could never change. There was a conservation law which stated that, 'in any reaction parity is conserved' (in general, every conservation law in physics has its origin in an elementary symmetry of nature). That is to say, when particles either interact to form new particles or break down (decay) into other particles, the parity on both sides of the equation must balance, just as energy or charge does.

Now let us turn our attention to what was happening in the mid-1950s. It was found that certain charged *K*-mesons (discovered in the late 1940s, see p. 231) decayed sometimes into two pions and sometimes into three pions. Physicists carefully analysed the decay processes of these particles in order to obtain the overall parity of the products, and hence the parity of the parent *K*-mesons. To their surprise they found that, in order to conserve parity in those decays, the

K-mesons (called *theta*-mesons) which decayed into two pions must have *even* parity; whereas odd parity should be given to the *K*-mesons (called *tau*-mesons) which decayed into three pions. In conclusion, it seemed that in nature there existed two different types of *K*-mesons, each with its own parity! However, these two particles seemed in all other respects identical: same mass, same lifetime, and so on. It was hard to believe that they were different. This baffling paradox was called the *theta–tau puzzle*.

Particle physicists of the mid-1950s had evidently reached an impasse: the question now was, 'how to solve the puzzle?' Lee and Yang entered the picture by suggesting that perhaps *theta* and *tau* were actually the same particle, and that maybe instead there was something wrong with the principle of parity conservation.

Breaking the law

In the spring of 1956 Lee and Yang began investigating the foundation of parity conservation. They examined previous experiments concerning particle decays, and found that the assumption of parity conservation in weak interactions (which give rise to beta radioactivity and the decay of particles like pions, muons and *K*-mesons) had no experimental support. Yang explained this in his Nobel lecture:

> The situation that the physicist found himself in at that time has been likened to a man in a dark room groping for an outlet. He is aware of the fact that in some direction there must be a door which would lead him out of his predicament. But in which direction? That direction turned out to lie in the faultiness of the law of parity conservation for the weak interactions. But to uproot an accepted concept one must first demonstrate why the previous evidence in its favour was insufficient.[29]

Lee and Yang examined the problem in detail, and came to the conclusion that either past experiments had not been able to reveal a lack of right–left symmetry in weak interactions, or specific experiments to test this symmetry had never been carried out in the past. Yang went on:

> Once these points were understood it was easy to point out what were the experiments that would unambiguously test the previously untested assumption of parity conservation in the weak interactions. Dr. Lee and I proposed in the summer of 1956 a number of these tests concerning [beta decay], [pion–muon, muon–electron] and strange-particle decays. The basic principles involved in these experiments are all the same: *One constructs two sets of experimental arrangements which are mirror images of each other, and which contain weak interactions. One then examines whether the two arrangements always give the same results in terms of the readings of their meters* (or counters). If the results are not the same, one would have an unequivocal proof that right–left symmetry, as we usually understand it, breaks down.[30]

One of the experiments suggested by Lee and Yang to test mirror symmetry in beta radioactivity was carried out by another Chinese-American physicist, Chien Shiung Wu (known as Mrs Wu, she happened to be a colleague of Lee's at Columbia University), in collaboration with a group of physicists at the National Bureau of Standards in Washington, DC.

Very schematically, Wu's classic experiment functioned like this: atoms of a certain radioactive isotope of cobalt were exposed to a strong magnetic field; since the cobalt atoms function just like microscopic bar magnets, they lined up in the direction of the field, exactly like compass needles. To make orderly alignment possible, thermal oscillations of the atoms were reduced to a minimum by cooling the system to a very low temperature (about a hundredth of a degree above absolute zero). It was then necessary to observe how the electrons emitted by the lined-up cobalt atoms emerged. Within a few weeks Mrs Wu and her co-workers were able to give the answer: the electrons were indeed emitted asymmetrically, most of them coming out in the opposite direction with respect to the magnetic field (the numbers of counts registered by the two counters in Fig. 9.11 were

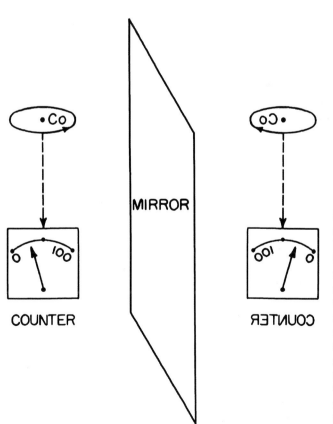

Fig. 9.11. Mrs Wu's experiment and its mirror image: if mirror symmetry had been valid, the mirror-image arrangement (right) would have had to have registered the same number of electrons as those recorded in the cobalt decay on the left. (Source: Chen Ning Yang's Nobel lecture, © The Nobel Foundation, Stockholm.)[31]

completely different). This meant that conservation of parity did break down in weak interactions governing beta radioactivity.

Two other experiments, performed during the same months by two teams, one at the University of Chicago led by Valentine Telegdi, and the other at Columbia University led by Leon Lederman (Nobelist in 1988), confirmed the overthrow of parity in weak decays of particles such as pions and muons. Thus physicists could conclude that the *theta* and the *tau* were effectively one and the same particle, which broke down with odd parity in some cases, and with even parity in others.

In this way one of the universally accepted laws about the behaviour of subatomic particles was upset; nature does in fact distinguish between right and left! We can now communicate to the Martian which side our heart is on. As Richard Feynman suggests, we can tell him: 'Listen, build yourself a magnet, and put the coils in, and put the current on, and then take some cobalt and lower the temperature. Arrange the experiment so the electrons go from the foot to the head, then the direction in which the current goes through the coils is the direction that goes in on what we call the right – and comes out on the left.'[32]

Lee and Yang's discovery is a fine example of how the scientific method functions with physics (remember, 'Observation, reason, and experiment make up what we call the scientific method', p. 8). It was well described by T. D. Lee in his Nobel lecture:

> The progress of science has always been the result of a close interplay between our concepts of the universe and our observations of nature. The former can only evolve out of the latter and yet the latter is also conditioned greatly by the former. Thus in our exploration of nature, the interplay between our concepts and our observations may sometimes lead to totally unexpected aspects among already familiar phenomena. As in the present case, these hidden properties are usually revealed only through a fundamental change in our basic concept concerning the principles that underlie natural phenomena. While all this is well-known, it is nevertheless an extremely rich and memorable experience to be able to watch at close proximity in a single instance the mutual influence and the subsequent growth of these two factors – the concept and the observation.[33]

Tsung Dao Lee (T. D. to his friends) was born in Shanghai, China, in 1926, the son of a businessman. He studied at the National Southwest Associated University in Kunming, Yunnan. In 1946, thanks to a Chinese government scholarship, he went to the University of Chicago as a graduate student; in 1950 he obtained his Ph.D. in physics there. After working for one year at the University of California at Berkeley, in 1951 Lee became a member of the Institute for Advanced Study, Princeton, where he started working with Yang. In 1953 he went to Columbia University, New York City, as an assistant professor of physics, becoming full professor in 1956 at the age of twenty-nine (the youngest professor in the University).

Fig. 9.12. T. D. Lee (left) and C. N. Yang. (Alan W. Richards, courtesy AIP Emilio Segrè Visual Archives.)

Lee is a well-known theorist. As well as his work on non-conservation of parity, he is well known for his studies in statistical mechanics, nuclear and particle physics, astrophysics and turbulence. J. Robert Oppenheimer considered Lee 'one of the most brilliant theoretical physicists then known, whose work was characterised by a "remarkable freshness, versatility, and style." '[34] Being awarded the Nobel Prize at the youthful age of thirty-one, Lee is the second youngest scientist ever to receive a Nobel for physics (the youngest was Sir Lawrence Bragg in 1915).

Chen Ning Yang (also known as Frank) was born in Hofei, in the province of Anhwei, China, in 1922, the son of a professor of mathematics at Tsinghua University, near Peking. Like T. D. Lee, Yang studied at the Southwest Associated University in Kunming; then he went to Tsinghua University, where he received his master's degree in 1944. The next year, on a fellowship, he went to the USA. ('There being no commercial passenger traffic between China and the United States at that time, I had to wait several months in Calcutta for a berth in a troop transport', he later recounted.)[35] On arrival at Chicago, he enrolled as a graduate student at the university, and in 1948 took his Ph.D. in nuclear physics, under the supervision of the Hungarian-American physicist Edward Teller.

Yang remained in Chicago as an assistant to Enrico Fermi, who was probably the most influential person in his scientific development. In 1949 he became associated with the Institute for Advanced Study in Princeton, and was appointed full professor there in 1955. Yang's research work deals with a variety of subjects, chiefly regarding particle and statistical physics. (His work with Robert Mills on the so-called *gauge theories* is considered to be among the most important of all contributions to twentieth-century theoretical physics, p. 355).

Fig. 9.13. Mrs Wu and Wolfgang Pauli. (Courtesy CERN – Pauli Archive.) Chien Shiung Wu (1912–97) graduated at the University of Nanking, China, in 1934, and obtained a Ph.D. from the University of California at Berkeley, in 1940. In 1958 she became a professor of physics at Columbia University. In 1975 she received the US Medal of Science, and in 1978 she was awarded the Israeli Wolf Prize in physics.

Recollections

Pauli's bet. In a famous letter to Victor Weisskopf dated 17 January 1957 (before Wu's results were published in the *Physical Review*), Pauli wrote:

> I do not believe that the Lord is a weak left-hander, and I am ready to bet a very high sum that the experiment will give symmetric angular distribution of the electrons. I do not see any logical connection between the strength of an interaction and its mirror invariance.[36]

The θ–τ puzzle

> Various participants of the Sixth Rochester Conference (April 1956) felt and expressed doubts about the universal validity of parity conservation in view of the experimentally-found near equality in mass and, above all, in the lifetime of the charged theta- and tau-mesons . . . [On] the train back from Rochester to New York, Professor Yang and the present author each bet Professor John Wheeler that the theta- and tau-mesons were distinct particles; and . . . Professor Wheeler has since collected two dollars.
>
> (Abraham Pais)[37]

Discoveries and inventions

Soon after the discovery of parity violation, theoreticians began to think of some mechanism which could explain why the weak interactions are able to distinguish right from left. A theory worked out by Robert Marshak and George Sudarshan, and soon afterwards by Richard Feynman and Murray Gell-Mann (Nobelists in 1965 and in 1969), attributed this ability to the neutrino, the elusive particle first detected by Clyde Cowan and Frederick Reines (p. 260) in the same year as Lee and Yang's discovery. In the new theory (a reformulation of Fermi's 1934 theory on beta decay) the experimental results obtained up to that time could be explained only with 'left-handed' neutrinos (neutrinos with their spins pointing back along their direction of motion), and antineutrinos only in a 'right-handed' form (with their spins pointing forwards). An experiment (described by C. N. Yang as 'devilishly ingenious') was carried out in December 1957 by the American scientist Maurice Goldhaber – one of the most capable experimenters of the time in the field of nuclear physics – and his co-workers, at the Brookhaven National Laboratory. It demonstrated conclusively that in fact the structure of the weak interactions implied only left-handed neutrinos and right-handed antineutrinos. (For their respective discoveries Goldhaber and Valentine Telegdi shared the 1982 Israeli Wolf Prize for physics.)

The dawn of the space age

On 4 October 1957 Soviet scientists launched the first man-made satellite, the *Sputnik I*, into an orbit around the earth. Thus the direct exploration of outer space was born. This is an excerpt taken from *Physics Today*:

> Shortly after midnight on October 5 at precisely 0015 GMT, the British Brodcasting Corporation's listening station at Tatsfield, just south of London, recorded the now famous 'beep-beep' of Sputnik I on its very first transit. News of this detection and the official Russian announcement of the satellite launching were flashed almost simultaneously to a startled world.[38]

1958

First Russian Nobelists

In 1958 the Soviet Union earned the Nobel Prize for physics for the first time. The three prizewinners were Pavel Cherenkov, Ilja Frank and Igor Tamm, all from the Lebedev Institute of Physics, Moscow. The Nobel was awarded for the discovery of an optical phenomenon (called the *Cherenkov effect*), made by Cherenkov himself in experiments carried out in 1934. Three years later, two colleagues of his, Frank and Tamm, analysed the effect, and gave a simple theoretical explanation based on classical electromagnetism.

The Cherenkov effect

Let us now describe the mechanism of this phenomenon. When a charged particle, such as an electron, travels through a transparent medium (for example, water) at a speed greater than that of light in the medium, though not greater than that of light in empty space (as prescribed by special relativity), it leaves behind it a faint bluish glow. This radiation forms a wave analogous to the shock wave (the sonic boom) which is produced in the air by a jet aircraft when travelling faster than the speed of sound.

The Cherenkov radiation appears, for example, as a bluish glow in the pools of water shielding certain nuclear reactors (the radiation is caused by electrons from the reactor travelling at speeds greater than the speed of light in water). The discovery acquired importance during the 1950s, when the Cherenkov effect found application in the development of new counters to track high-speed subatomic particles at accelerators (for instance, they played a decisive role in the discovery of the antiproton, p. 269). (Huge Cherenkov detectors, containing distilled water, are now used in underground laboratories to detect solar and cosmic neutrinos, see Nobel 2002.)

Pavel Aleksojevic Cherenkov (1904–90) was born in Novaya Chigla, in the Voronezh Region, Russia, the son of peasants. He graduated from Voronezh State University in 1928. Two years later, he went as a research student to the Lebedev Institute of Physics of the Academy of Sciences of the Soviet Union, where he received his doctorate in 1940. His career advanced at the Lebedev Institute, where he became Professor of Experimental Physics in 1959. He worked principally on the development of electron accelerators and in the field of nuclear physics.

Ilja Michajlovich Frank (1908–90) was born in St Petersburg, Russia, the son of a professor of mathematics. He graduated from Moscow State University in 1930. From 1931 to 1934 he was a senior scientist at the State Optical Institute in Leningrad (St Petersburg). He then moved to the Lebedev Institute in Moscow, where he became head of the Atomic Nucleus Laboratory. Frank was elected a member of the Soviet Union Academy of Sciences in 1946. He was interested principally in the study of gamma rays and neutrons.

Igor Eugenevich Tamm (1895–1971) was born in Vladivostok, Siberia, Russia, the son of an engineer. He graduated from Moscow State University in 1918 and taught there from 1924 to 1937. In 1934 he became head of the theoretical division at the Lebedev Institute. His early studies on crystalline solids had important applications in semiconductor physics. A member of the Soviet Union Academy of Sciences from 1933 on, Tamm spent the last decades of his career working on thermonuclear fusion.

Discoveries and inventions
Van Allen belts
The US satellites *Explorer I* and *Explorer III*, launched in the first months of 1958, carried Geiger counters to measure both radiation at very high altitudes, and more especially, the intensity of cosmic rays. Up to heights of about 1000

kilometres the counters detected the expected number of particles, above 2000 kilometres the count fell virtually to zero (the counters had apparently become 'jammed').

The American physicist James Van Allen at the University of Iowa, who was in charge of this research programme, explored all the possible causes of this extraordinary fact. He came up with an explanation: the counters had blanked out, not because there was no radiation at those heights, but because there was too much. The satellites had penetrated regions of intense concentration of radiation. He announced his discovery at a meeting held in Washington, DC, on 1 May 1958, suggesting that the radiation must consist of charged particles (electrons and protons) originating from the sun, which remain trapped by the earth's magnetic field. Satellites launched towards the end of the 1958 showed that there are two regions of such a radiation surrounding the earth. They are named the *Van Allen radiation belts*, and the whole region is called the *magnetosphere*.

The integrated circuit

In 1958 Jack Kilby (Nobelist in 2000), working at Texas Instruments, Dallas, USA, invented the *integrated circuit* (or *chip*) – a single circuit, containing transistors and other electronic components, made out of a single crystal of silicon. Some months later Robert Noyce, working at a company named Fairchild Semiconductors in Palo Alto, California, independently conceived the same idea. Integrated circuits very soon replaced transistors in electronic equipment; and in the late 1960s what was called *large-scale integration* (thousands of chips on the same wafer of silicon, with only a few millimetres on each side) began. This completely revolutionised the solid-state electronic industry.

1959

Antiprotons

Two of Fermi's old pupils were the 1959 Nobel prizewinners: Emilio Segrè and Owen Chamberlain, both from the University of California at Berkeley. In 1955, together with Clyde Wiegand and Thomas Ypsilantis, they had discovered the *antiproton*, the second antiparticle to be found (the first was the positron in 1932, see Nobel 1936). The discovery was made at the Bevatron, the big proton accelerator that had been recently constructed at the Berkeley Radiation Laboratory.

A not unexpected discovery

The antiproton is the proton antiparticle: it is like the proton, except that it has a negative instead of a positive electric charge (the same relation existing between the positron and the electron). Its existence had been predicted, like that of the positron, by Dirac's theory of antiparticles (p. 176). As in the case of an electron–positron pair, an antiproton can be created only in pair with a proton. But the energy needed to produce such a pair is about 2000 times greater (about two billion electron-volts). As Chamberlain noted in his Nobel lecture:

> [It] was ... known that high-energy collisions were necessary ... A likely process might be one in which a very fast proton struck a stationary neutron. The proton would have to have about 6 billion electron-volts of energy in order that 2 billion electron-volts might be available to make new particles ... I am told that the men responsible for the early design of the Bevatron ... were well aware that about 6 billion electron-volts were needed to produce antiprotons, and that this fact was taken into consideration when the Bevatron energy, just 6 billion electron-volts, was decided upon. Professor Edwin McMillan and the late Professor Ernest Lawrence were among the men who selected the energy of the Bevatron ...[39]

Segrè and Chamberlain used those high-energy protons to bombard a copper target. They estimated that an antiproton might be produced once in each million collisions between the protons and the copper nuclei. After running the experiment for hours, they were successful in catching some sixty antiprotons (one every fifteen minutes on average). It was far from easy to identify them. For every antiproton produced, 50 000 particles of other types came into existence. But their system of counters was so designed that only antiprotons were selected. So the particle could be recognised without doubt.

Professor E. Hulthén, chairman of the Nobel Physics Committee, had this to say in presenting the award-winners:

> It has been said of the *Bevatron*, the great proton accelerator at Berkeley University in California, that it was constructed chiefly with a view to the production of antiprotons. This is perhaps an exaggeration, but it is correct in so far as its peak achievement, 6 [billion] electron-volts, was set with a view to the energy required for the pair formation of protons–antiprotons ... But even if antiproton research was thus first made possible through this technologically very impressive machine, the actual discovery and investigation of the antiproton was chiefly the merit of Chamberlain and Segrè.[40]

Emilio Gino Segrè (1905–89) was born in Tivoli, near Rome, the son of an industrialist. He studied at the University of Rome under Fermi, and received his doctor's degree in physics in 1928. In 1930 he worked under Otto Stern at Hamburg, and under Pieter Zeeman at Amsterdam. Four years later he participated in the pioneering neutron experiments directed by Fermi (p. 188). In 1936 he became professor of physics at the University of Palermo, where, together with the Italian mineralogist Carlo Perrier, they discovered *technetium*, the first man-made element not to be found in nature.

In 1938 Segrè left his Italian university (he was Jewish), and emigrated to the USA, where he became a research associate at the Berkeley Radiation Laboratory. Here he discovered the element *astatine*; and in 1941, together with Glenn Seaborg

(Nobel for chemistry 1951) and other colleagues, he participated in the discovery of the isotope *plutonium-239*. From 1943 to 1946 he was a group leader of the Manhattan Project at the Los Alamos Laboratory. In 1946 he became full professor of physics at the University of California, Berkeley; and in 1974 he was appointed Professor of Nuclear Physics at the University of Rome, Italy. Segrè is well known for his books on nuclear and particle physics, and also on the history of physics.

Emilio Segrè had been Fermi's first assistant in Rome. In his Nobel lecture he wrote:[41] 'Of Enrico Fermi I would only say, quoting Dante as he himself might have done,

Tu se' lo mio maestro e il mio autore;
Tu se' solo colui da cui io tolsi
Lo bello stilo che mi ha fatto onore.'

[Thou art my master and my author;
thou alone art he from whom I took
the good style that hath done me honor.]

Owen Chamberlain was born in 1920 in San Francisco, California, the son of a radiologist. He graduated from Dartmouth College in 1941. He then went to the University of California, Berkeley, but his studies were interrupted at the outbreak of the Second World War. From 1942 to 1946 Chamberlain worked on the Manhattan Project under Segrè, both in Berkeley and in Los Alamos. Here he studied nuclear interactions with neutrons and spontaneous fission of heavy elements. After the war he completed his Ph.D. thesis at the University of Chicago, under Fermi, working at the Argonne National Laboratory, Illinois. In 1948 he began teaching at the University of California, Berkeley, where he was appointed full professor in 1958. After the discovery of the antiproton, Chamberlain continued to work with Segrè, and in 1956 they confirmed the existence of the antineutron – the antiparticle of the neutron.[42] He then undertook research on alpha-particle decays, neutron diffraction, and phenomena in the field of high-energy interactions produced by protons.

Particle accelerators

In the early 1930s John Cockcroft and Ernest Walton used their voltage multiplier to accelerate protons to energies as high as hundreds of thousands of electron-volts (p. 174). Later, towards the end of the decade, the University of California at Berkeley had a cyclotron (invented by Ernest Lawrence, p. 174) capable of reaching energies up to some 20 million electron-volts.

After the end of the Second World War the science of accelerating particles developed greatly. The cyclotron then reached its limit. Particles were travelling so fast that their mass increased with energy (as predicted by Einstein's theory

Fig. 9.14. From left to right: E. Segrè, C. Wiegand, E. Lofgren, O. Chamberlain and T. Ypsilantis. (Courtesy Lawrence Berkeley National Laboratory.)

of special relativity), and the particles failed to keep pace with the electric kicks. Physicists got round this limitation by synchronising the particle increase in energy with an alternating electric field. Successively, new accelerating devices were designed, which balanced the increasing energy with an electric field of increasing strength; they were called *synchrotrons*. In a synchrotron particles are kept in a single circular path instead of spiralling outwards.

In the early 1950s the Brookhaven National Laboratory (a consortium of American universities, at Upton, Long Island, in New York) built a proton-accelerating synchrotron that reached 3 billion electron-volts (it was called the *Cosmotron*). In 1954 the University of California completed its *Bevatron*, capable of accelerating protons up to 6 billion electron-volts (p. 269). Then, in 1957, the Soviet Union announced that it had actually constructed one of 10 billion electron-volts.

The advance continued. Magnets producing magnetic fields of different shape were introduced, which kept the particles focused in a narrow beam. This further increased the energy levels attained. In November 1959 CERN (Geneva) completed a proton synchrotron (in short PS) which reached 28 billion electron-volts. Its diameter was 200 metres, and it produced large bursts of protons (10 billion) every three seconds. CERN PS's sister-machine was the Brookhaven Alternating Gradient Synchrotron (AGS), which started operations in July 1960, and accelerated protons to an energy of 33 billion electron-volts. The AGS was used for

Fig. 9.15. The CERN Proton Synchrotron ring, from a photograph of 5 November 1959. (Courtesy CERN.)

important discoveries in the field of subatomic particles (scientists working with it have been awarded no less than three Nobel Prizes for physics).

1960

A new particle detector

The glorious old Wilson cloud chamber saw its last years in the 1950s, when a new generation of accelerators began to produce high-energy particles (the last experiment carried out using a cloud chamber was performed by Leon Lederman (Nobelist in 1988) and his group at the Brookhaven Cosmotron accelerator in the late 1950s). A new device, called a *bubble chamber*, then came out and began to dominate particle physics. It was invented in 1952 by Donald Glaser, a young experimental physicist, then at the University of Michigan, Ann Arbor. He was granted the 1960 Nobel Prize in physics for this invention.

The bubble chamber

A bubble chamber uses a pressure-tight vessel containing a suitable liquid (usually hydrogen) which is heated far above its usual boiling point, but is maintained under high pressure to prevent it boiling. When the pressure is suddenly reduced, the liquid becomes overheated, so that high-speed charged particles traversing the chamber form trails of tiny bubbles along their paths (like the stream left against the blue sky by a jet aircraft). Photographs of these bubbles are then taken through windows, so that one can see precise details of the interactions

Fig. 9.16. One of the first glass bubble chambers, used by Glaser (its volume was only three cubic centimetres). You can see a track due to a cosmic-ray particle. (Source: Donald Glaser's Nobel lecture, 1960; © The Nobel Foundation, Stockholm.)[43]

caused by the particles with the nuclei of the liquid atoms, and hence study these interactions more effectively.

Donald Arthur Glaser was born in Cleveland, Ohio, USA, in 1926, the son of a businessman. In 1946 he graduated in physics and mathematics at the Case Institute of Technology, Cleveland; soon afterwards he went to Caltech, Pasadena, as a graduate student. After receiving a Ph.D. in 1949, he moved to the University of Michigan, where he began his teaching and research career, becoming in 1957 Professor of Physics. In 1959 Glaser moved to the University of California at Berkeley; from 1964 on he served there as Professor of Physics and Molecular Biology.

Tracking particles

Discoveries in nuclear and particle physics have always closely paralleled the development of instrumentation designed for detecting and identifying subatomic particles and radiation.

Rutherford was the first to 'see' a subatomic particle. In his pioneering experiments on alpha-particle scattering (p. 80), he and his assistants counted the scattered particles by observing the tiny light flashes, or 'scintillations', produced when they hit a fluorescent screen. (They first had to sit in the dark for thirty

Fig. 9.17. Installation of a large bubble chamber at CERN in the 1970s. (Courtesy CERN.)

minutes in order to sensitise their eyes!) This was the first example of a *scintillation counter* (fortunately, today light flashes of modern scintillation counters are converted into electric pulses which are counted electronically!)

In the early 1910s C. T. R. Wilson invented the cloud chamber, in which particle tracks in a gas could be photographed. It was one of the most widely used tools of experimental physics in the first half of the twentieth century. Rutherford and his 'boys' used this instrument for studying nuclear reactions during the 1920s (p. 118). Subsequently cloud chambers were used in cosmic-ray studies (remember C. Anderson and P. M. Blackett's epoch-making experiments which led to the discovery of the positron, p. 183). In the late 1920s Hans Geiger and a student of his, Walther Müller, developed the so-called *Geiger–Müller counter*: this could count individual charged particles or high-energy photons, and it was mainly used in cosmic-ray studies.

In the mid-1940s physicists succeeded in perfecting *photographic emulsion detectors*. Thanks to their advantages (they were simple, compact and could be

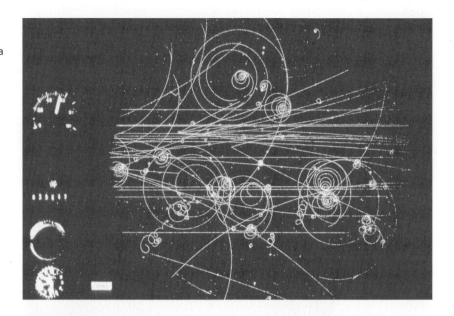

Fig. 9.18. A spray of particles created in a high-energy collision in a liquid-hydrogen bubble chamber. (Courtesy CERN.) Physicists of the 1960s discovered new exotic states of matter while analysing bubble tracks like these.

carried anywhere), they were used in both cosmic-ray and accelerator experiments. This new method was instrumental in the discovery of new subatomic particles (like the pion, p. 231). In the same period the *scintillation counter* and the *solid-state counter* were invented. (In a scintillation counter high-speed particles traversing certain solids or liquids produce flashes of light, which are then converted into electric pulses and detected.) These counters, together with the *Cherenkov counter*, have been used right up until today in high-energy experiments. Then along came the *bubble chamber*. It was better suited to high-energy accelerators than the cloud chamber. It used a liquid as a medium – so increasing the number of interactions produced. It was amenable to magnetic fields, and not least it recovered faster after each exposure. It became the work-horse of particle physics in the 1960s and 1970s.

The first laser

In 1957 Charles Townes was a professor of physics at Columbia University and a consultant to Bell Labs. His former assistant, Arthur Schawlow, was also at the Bell Labs as a research physicist. They decided to work together, studying how the principle of the *maser* could be extended to visible light. (A maser is a device that amplifies and produces an intense and pure beam of microwave radiation; it had been conceived and built four years earlier by Townes himself, see Nobel 1964.) One year later Townes and Schawlow published a seminal paper, in which they had worked out a theory for amplifying visible and infrared radiation. They then patented their idea; and later on the device they had projected was called a *laser*, an acronym for *light amplification by stimulated emission of radiation*. Soon researchers began investigating ways of constructing a practical laser device.

Fig. 9.19. Theodore Maiman holding his first laser.[44]

It came as a surprise when a little-known young physicist, whose name was Theodore Maiman (he was a researcher at Hughes Research Laboratories in Malibu, California), announced that he had successfully constructed an operative laser. On 15 May 1960 he happened to use a ruby bar, some few centimetres long, with silvered-end faces to serve as mirrors. He then inserted the bar into a spring-shaped flash-lamp. When he turned the lamp on, the ruby bar emitted a bright pulse of deep red light – the first laser beam. The device was small enough to fit into Maiman's hand. He announced his invention at a press conference on 7 July 1960; his initial paper, amazingly enough, was rejected by the *Physical Review Letters*, but it was soon published in the 6 August issue of *Nature*.

Fig. 9.20. An aerial view of the Bell Labs, Murray Hill, circa 1960. (Courtesy Lucent Technologies/Bell Labs.)

The Bell Labs spirit

In 1925 the American company AT&T founded its own research laboratories, the Bell Telephone Laboratories (universally known as 'Bell Labs'). In-house basic research soon became the heart and soul of the new laboratories, diversifying over different fields. 'By the mid-1930's the problems, approaches and atmospheres of fundamental research at Bell Labs were remarkably similar to those of university laboratories.'[45] New research teams were created. One of these went on to contribute enormously to the advancement of solid-state physics: their research culminated in the late 1940s in that epic invention of the transistor (p. 255). The 'Bell Labs spirit' of the mid-twentieth-century is well expressed by Oliver Buckley, Bell Labs' president at the time, in a letter he wrote to the *New York Times* (August 1949):

> One sure way to defeat the scientific spirit is to attempt to direct inquiry from above. All successful industrial research directors know this, and have learned from experience that one thing a 'director of research' must never do is to direct research . . . Successful research goes in the direction in which some inquiring mind finds itself impelled . . . The director of research does his part by building teams and seeing that they are supplied with facilities and given freedom to pursue their inquiry. He also . . . protects them from interference or diversion arising from demands of immediate operating needs . . .[46]

Chapter 10
New vistas on the cosmos

The scientific boom of the 1950s continued during the 1960s. The launch in 1957 of the first artificial satellite – the *Sputnik I* – by Russian scientists, had given a boost to space conquest. Artificial satellites travelled to Mars and Saturn, and on 21 July 1969 the US astronaut Neil Armstrong became the first human being to walk on the moon.

The period following on the early 1950s saw huge advances in our knowledge of the cosmos. The emergence of new technologies, like radar technology, and the development of space science opened up new vistas on the cosmos, extending astronomical observations to the whole of the electromagnetic spectrum. Astronomers started using new techniques such as radio waves, infrared and ultraviolet radiation, X-rays and gamma rays; these methods, combined with the more traditional optical ones, led to the flowering of a new astrophysics. Amazing discoveries soon made world headlines. Notable examples are: the quasars (early 1960s); the first X-ray cosmic source (1962); the cosmic microwave background radiation (1964); and the first pulsar (1967). And not long afterwards the detection of solar neutrinos furnished a new tool for investigating the working mechanisms inside stars.

In the field of particle physics, recently built accelerators gave access to a wider range of energies, so that a flood of unexpected new particles emerged. Meanwhile, theoretical speculations flourished; physicists invented elegant symmetry schemes, they then postulated the existence of new particles (the *quarks*) as the ultimate building blocks of hadrons. Finally, towards the end of the decade a new theory was successfully worked out, capable of combining the apparently different electromagnetic and weak nuclear forces into a single mathematical framework. All this introduced a new order, and a better understanding of the world of subatomic particles. At the same time semiconductors and superconductors revealed novel characteristics, so opening up new horizons to condensed-matter physics.

In this decade (1961–70), nearly half of the laureates received their Nobel Prizes for discoveries in the fields of nuclear and particle physics – discoveries which date back to the period between the late 1940s and early 1960s. The Nobel Physics Committee, however, promptly recognised the inventions of the maser and the laser; and at last, in 1967, the first Nobel award arrived to reward astrophysics.

1961
Nuclear physics

It was nuclear physics that won the contest for the 1961 Nobel Prize. The award was shared by the American physicist Robert Hofstadter and the German Rudolf Mössbauer. Hofstadter was honoured for his studies on the distribution of electric charge in atomic nuclei, which he had carried out at Stanford University in the 1950s, Mössbauer for his discovery of a new effect involving the emission and absorption of gamma rays by atomic nuclei.

A picture of the nucleus
You will remember how Rutherford discovered the atomic nucleus in 1911, while studying Geiger–Marsden's scattering of alpha particles by atoms of matter (p. 80). He also showed that the dimension of a nucleus was nearly 10 000 times smaller than that of an atom.

In the early 1950s, when Hofstadter began his work at Stanford University, physicists thought of the nucleus as a sphere, with a positive electric charge (the charge of protons) uniformly distributed inside. This spherical nucleus had a radius of about one-million-billionth of a metre, slowly increasing with its mass.

In 1953 Hofstadter began a systematic study of the size of nuclei, and succeeded in obtaining new results. Particularly interesting were those concerning the structure of nucleons (the proton and the neutron). He used an electron linear accelerator that had been constructed quite recently on the campus of Stanford University.

His method was similar to that of Rutherford's. He shot high-energy electrons at a target of matter (Hans Geiger and Ernest Marsden, by contrast, had employed alpha particles for this task). The electrons were bounced off by the atomic nuclei, and deviated by strong electric forces due to the nuclear charge. The highest energy of the electrons used by Hofstadter was about one billion electron-volts, so that their de Broglie wavelength, which decreases as the energy increases (p. 139), was reduced to the size of the nuclei themselves. (And now here is a very important principle of nature that follows from quantum mechanics: when you want to 'see' smaller and smaller objects, you need higher- and higher-energy radiation in order to probe them, so Hofstadter's accelerator acted as an electron microscope, some thousand times more powerful than an ordinary electron microscope.)

In his high-energy *electron scattering* experiments, Hofstadter first separated, with ingenious instruments, the scattered electrons of different energies. He then measured the number of electrons which had been deviated in each particular direction. Thus he obtained a detailed map of the distribution of the electric charge in the nuclei; and he was also able to give indications of the structure of the proton and the neutron. Thus he concluded his Nobel lecture:

... it may be appropriate ... to raise the question ... of the deeper, and possibly philosophical, meaning of the term 'elementary' particle. As we have seen, the proton and the neutron, which were once thought to be elementary particles are now seen to be highly complex bodies. It is almost certain that physicists will subsequently investigate the constituent parts of the proton and the neutron ... What will happen from that point on? One can only guess at future problems and future progress, but my personal conviction is that the search for ever-smaller and ever-more fundamental particles will go on as long as Man retains the curiosity he has always demonstrated.[1]

Hofstadter's method of using electron beams for the mapping of charge and magnetic nuclear structure was extended in precision during the following years, and became a powerful source of information about atomic nuclei.

The Mössbauer effect

The Mössbauer effect concerns the emission and absorption of gamma rays by certain radioactive atomic nuclei. Let us describe this phenomenon briefly.

These nuclei, like atoms, exist in discrete energy states. A gamma-ray photon is emitted each time a nucleus drops from a higher energy level to a lower one (the energies involved here are a million times greater than those of atoms). It thus happens that, when a gamma ray is emitted from a nucleus, the nucleus itself experiences a recoil (like a rifle when it fires a shot); so that the gamma photon loses an amount of energy which is given to the recoiling nucleus. On the other hand, for a gamma ray to be absorbed by a similar nucleus, its energy must be exactly equal to the difference between the two energy levels (the excited level and the lower one); this phenomenon is called *resonance absorption*. (Resonance occurs, for example, when incoming radio waves are received only if the receiver is tuned to exactly the same frequency as the transmitter.) Thus, when a gamma ray is ejected from a *free* nucleus (the transmitter), it cannot be absorbed by another similar nucleus (the receiver), because, due to the recoil of the emitting nucleus, its energy is less than the 'resonance energy'.

Mössbauer discovered that when atomic nuclei are embedded in the lattice of crystalline solids, it happens that, at low temperatures, their recoils are absorbed by the crystal lattice as a whole. As a consequence, the resulting energies of the emitted gamma rays are of the right value, and resonance absorption may take place. Mössbauer made his discovery in the course of experiments that he was carrying out in preparation for his doctoral thesis.

The discovery of the Mössbauer effect made it possible to produce gamma rays of extremely precise energies (or frequencies). This proved to be a useful tool for several applications, such as measuring the magnetic fields that surround atomic nuclei in magnetic materials, or studying important properties of solids. A new spectroscopy, based on the Mössbauer effect, also became available to the chemist for the study of molecular structure and chemical bonds.

Another use of the Mössbauer effect was to verify Einstein's gravitational red shift, one of the historic tests of the general theory of relativity (p. 108). A high-precision experiment was carried out by two American physicists, Robert Pound and Glen Rebka, in the early 1960s, at Harvard University. They made use of gamma rays emitted in the decay of a certain radioactive isotope of iron, which were embedded in a crystal. They then measured the frequency shift of these gamma rays at a vertical separation of 22.5 metres. The result was a brilliant confirmation of one of the predictions of general relativity.

Robert Hofstadter (1915–90) was born in New York City. After graduating at the City College of New York, Hofstadter went to Princeton University to study physics; he took a master's degree there, and in 1938 a Ph.D. in physics. Later, he worked as a researcher first at Princeton and then at the University of Pennsylvania. During the Second World War he continued his research work at the National Bureau of Standards, and later at the Norden Laboratory Corporation of New York. In 1946 he left industry and became Assistant Professor of Physics at Princeton University, where he performed research on the Compton effect, on photoconductivity and on crystal and scintillation counters. In 1950 he left Princeton and joined Stanford University, where he became a professor of physics. From then on Hofstadter's principal scientific interest was focused on nuclear physics, and he began his researches on electron scattering by atomic nuclei.

Rudolf Ludwig Mössbauer was born in Munich, Germany, in 1929. He studied at the local Technical University, and from 1955 to 1957 he carried out research for his doctoral thesis in the physics department of the Max Planck Institute for Medical Research in Heidelberg. In the course of this work he observed the phenomenon which now bears his name and earned him the Nobel Prize. In 1961 he went to Caltech, Pasadena, California, where he remained for three years as a visiting professor. In 1964 he returned to Munich and became a full professor of physics at the Institute for Technical Physics. Mössbauer has worked mainly on problems of nuclear and solid-state physics.

1962

Superfluidity

The 1962 Nobel Prize for physics was the second one to be awarded to Russian scientists, the first being the awards that went to Cherenkov, Frank and Tamm in 1958. The laureate was Lev Landau, Professor of Theoretical Physics at the University of Moscow; and the prize was awarded for his investigations into condensed matter (that is, matter in the solid and liquid states), and in particular for his pioneering theories concerning helium superfluidity. Here, in brief, is his story.

Lev Davidovich Landau (1908–68) was born in Baku in Azerbaijan, Russia, the son of Jewish parents (his father was an engineer working in the local oil industry, and his mother a physician). He studied physics at Baku University and in Leningrad (now St Petersburg). After graduating in 1927, he began his research work at the Leningrad Physico-technical Institute. In 1929 he was given the opportunity to go abroad, something that was extremely difficult at that time for a Soviet citizen. He first went to Göttingen and Leipzig, then to Copenhagen to Bohr's institute. Later still he went to Cambridge University to work with Paul Dirac, and lastly to Zurich to work with Wolfgang Pauli. After his return in 1932, Landau became the head of the theoretical division of the Ukrainian Physico-technical Institute in Kharkov. Here he established the first school of theoretical physics in the Soviet Union, which soon acquired a world-wide reputation. With his former fellow student, Evgenii Lifshitz, he wrote a set of ten volumes, the magnificent *Course of Theoretical Physics*, which has taught generations of physicists. In 1937 Peter Kapitza (Nobelist in 1978) – then director of the Institute for Physical Problems in Moscow – persuaded him to head the theory division of the institute.

Landau ('an ardent communist, very proud of his revolutionary roots', as was recalled by Hendrik Casimir, a physicist who met him in Copenhagen)[2] was imprisoned during Stalin's Terror (1938). Only a personal intervention by the influential Kapitza saved him. (Kapitza wrote to Joseph Stalin, claiming that only Landau could explain superfluidity, which Kapitza himself had just discovered.) Besides superfluidity, Landau worked in many other areas of theoretical physics:

Fig. 10.1. Lev Landau. (Courtesy AIP Emilio Segrè Visual Archives, Physics Today Collection.) 'His admirers saw [Landau] as an ivory tower theorist – bold, impudent and charming but detached from the humdrum of everyday existence. They ignored two political aspects of his life: his year in Joseph Stalin's prisons in the late 1930s, and his contributions to the dictator's nuclear bomb a decade later.'[3]

he made remarkable contributions to the theory of plasma; he also worked in the fields of atomic and nuclear physics, superconductivity, stellar theory, properties of metals and particle physics. As well as the Nobel Prize, Landau received many other honours. He was a member of the Soviet Union Academy of Sciences, and a foreign member of both the Royal Society of London and the US National Academy of Sciences.

A theory for superfluid helium

In superfluid helium, viscosity (which represents its resistance to flow) is greatly reduced (less than one-thousandth of that of helium in its non-superfluid state). This means that superfluid helium flows through extremely small holes, which no other liquid can do. This phenomenon is observed in the liquid state of common helium (helium-4) at temperatures below 2.17 kelvins (about -271 °C). It was discovered in the period 1937–8 by the Russian experimentalist Peter Kapitza in Moscow, and by two Canadian physicists, Jack Allen and Donald Misener, in Cambridge, England (see Nobel 1978). (The characteristic temperature of 2.17 kelvins is known as the *lambda point*.) In brief, this is the phenomenon.

At normal atmospheric pressure, common helium-4 (with two protons and two neutrons in its nucleus, and two electrons around it) liquefies at a temperature of 4.2 kelvins (see Nobel 1913). From this temperature down to the lambda point helium-4 behaves as a normal liquid, and is called *helium-I*. Below the lambda point it loses almost all of its viscosity (it acts like a mixture of a perfect superfluid and a normal viscous fluid), and it is called *helium-II*. Besides superfluidity, helium-II presents two other unusual characteristics: its thermal conductivity (which represents its capacity to conduct heat) is very high (some three million times higher than that of helium-I); and, unlike helium-I, it flows spontaneously from a cold region to a hotter one.

Another important substance is the rare isotope of helium-4, called *helium-3* (in nature we find about one atom of helium-3 among ten million atoms of helium). It has two protons and one neutron in its nucleus, and it also exhibits superfluid properties, but only at temperatures lower than about 0.003 of a kelvin. Superfluidity in helium-3 was first observed by the American physicists David Lee, Douglas Osheroff and Robert Richardson, in 1971, and all three won a Nobel Prize for this in 1996. (Many of the properties of superfluid helium-3 were later explained by another researcher, Anthony Leggett, Nobelist in 2003.)

The first decisive theoretical breakthrough concerning superfluid helium-4 was made by Landau in 1941, when he published a classical paper which constituted a landmark in the history of superfluidity, and the basis for his Nobel Prize. Landau thought of superfluid helium-4 as a *quantum liquid*, in which the atoms, being composed of an even number of particles, each with a spin equal to $\frac{1}{2}$, are considered as Bose particles, or *bosons* (p. 175), with a spin equal to zero. He then extended quantum mechanics rules over the entire liquid at the temperature of absolute zero, so that its atoms, being bosons, behave as if they were in the same

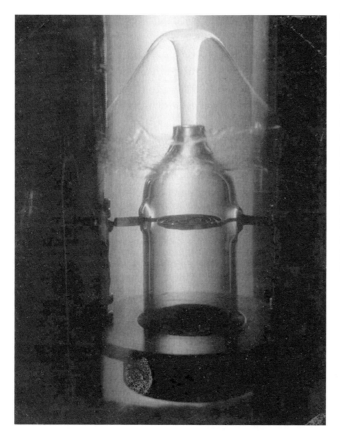

Fig. 10.2. The fountain effect – the most visually exciting manifestation of superfluidity: superfluid helium-II spouts out of a vessel through small holes when heat is applied to the inside. The effect was photographed in the 1970s by Jack Allen, one of the discoverers of superfluidity. (Allen was the first to observe this spectacular phenomenon in 1938.) (Photo: courtesy of Professor Russell J. Donnelly, University of Oregon in Eugene, USA.)

quantum state. As a result, their motions are highly correlated, so reinforcing the physical properties of the liquid in its superfluid state.

To explain the properties of such a quantum liquid, Landau also introduced the notion of *quasiparticles*. These are quanta of energy which behave in many respects like ordinary particles, and are carriers of motion in the liquid. (The concept of the 'quasiparticle' is a very important one, and it has been used to explain many other phenomena in condensed-matter physics.)

After the Second World War many of the predictions contained in Landau's theory were confirmed by experiments. The theoretical picture of superfluid helium was successively developed by other physicists, notably Richard Feynman (Nobelist in 1965), Fritz London and Lars Onsager (Nobel for chemistry 1968).

The two-neutrino experiment

Soon after the neutrino had been observed by Clyde Cowan and Frederick Reines in 1956 (p. 260), physicists began to study this most elusive among the subatomic particles, in an effort to determine its properties and its role in the particle physics world.

In 1961–2 three physicists from Columbia University, Leon Lederman, Melvin Schwartz and Jack Steinberger, carried out an experiment at the Brookhaven Alternating Gradient Synchrotron (AGS), which had recently begun operating (see p. 272). They discovered that two kinds of neutrinos exist in nature, one related to the electron (named the *electron neutrino*), the other one related to the muon (named the *muon neutrino*). Stated more clearly, they discovered that the neutrino emitted with a muon in the breakdown of a pion is not identical to the neutrino emitted with an electron in a radioactive beta decay of an atomic nucleus. What is the difference? Let us see it as exemplified in the so-called *two-neutrino experiment*.

The 'Columbia team', as the three were called, fired a beam of protons, provided by the AGS at an energy around fifteen billion electron-volts, against a fixed target of beryllium. The protons collided with the atomic nuclei in the target, and created a large number of pions. These were allowed to travel a free-flight distance of about twenty metres. Since pions are unstable, some of them decayed as they proceeded to that distance, each giving rise to a muon and a neutrino. At the end of the path there was an absorbing shield some thirteen metres thick (a wall of steel, taken from the battleship *USS Missouri*, which was being cut up for scrap). Pions, muons and other unwanted accompanying particles were absorbed in the shield, but the neutrinos were not, so an intense and clean beam of neutrinos (hundreds of billions) emerged. Behind the shielding wall there was a measuring device, which consisted of a ten-ton apparatus with particle detectors called *spark chambers*. These were large and thick aluminium plates that were piled up, separated by gaps and filled with a gas. A voltage was applied between each pair of plates, so that a charged particle traversing the plates ionised the gas, and created a trail of luminous sparks along its path; these were then photographed.

The experiment started in September 1961 and lasted eight months. During this period 100 000 billion neutrinos traversed the spark chambers, yielding about 5000 spark photographs. Among these, fifty-one clearly showed a long straight track characteristic of a muon produced by a neutrino interaction; only very few pictures showed events that could have been electrons. The Columbia team then concluded that only muons were produced by the neutrinos, not electrons. If the neutrinos coming from the pions were identical to those coming from common beta radioactivity, they should have produced as many charged electrons as muons in their interactions. The experimental result proved, however, otherwise: a new type of neutrino exists, and this one is associated with the muon, whereas the electron is associated with its own neutrino.[4]

Some months later, physicists at CERN near Geneva repeated the Brookhaven experiment. They detected hundreds of muons, produced by neutrino interactions, and no electrons, so that the Brookhaven discovery was fully confirmed.

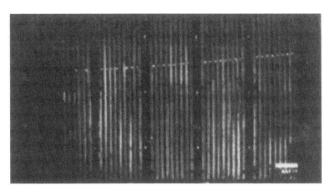

Fig. 10.3. A track left in the spark chambers by a muon produced in an interaction of a muon neutrino in the Brookhaven detector. (Source: G. Danby et al., *Physical Review Letters*, 9, 1962, p. 40.[4] © 1962 by the American Physical Society.)

Niels Bohr, November 1962

On 18 November 1962 Niels Bohr died at his home, from heart failure. He was seventy-seven years old. Abraham Pais concluded his book entitled *Niels Bohr's Times, in Physics, Philosophy, and Polity*, quoting the following brief note by Bohr himself in memory of Rutherford, and saying: 'I know of no better way of concluding this book than by applying these words to Niels Bohr himself.'[5]

> Those of us who had the good fortune to come into contact with ... [him] will always treasure the memory of his noble and generous character. In his life all honours imaginable for a man of science came to him, but yet he remained quite simple in all his ways ... [He] was, ... and always remained, open to listen to what a young man had on his mind. This, together with the kind interest he took in the welfare of his pupils, was indeed the reason for the spirit of affection he created around him wherever he worked ... The thought of him will always be to us an invaluable source of encouragement and fortitude.'[6]

1963

Inside the nucleus

Maria Goeppert Mayer was the second woman in history to win a Nobel Prize for physics, the first being Marie Curie in 1903. She shared a part of the 1963 prize with Hans Jensen, for their discoveries in nuclear physics. Eugene Wigner, the third prizewinner, was awarded the other part of the prize for his theoretical studies in the field of nuclear and particle physics.

Eugene Paul Wigner (1902–95) was born in Budapest, Hungary, the son of Jewish parents who managed a leather tannery. Wigner received his doctorate in chemical engineering from the Technical University of Berlin in 1925. He then returned to Budapest: 'From a distance he ecstatically witnessed the birth of quantum mechanics, through the papers of Heisenberg, Born, Jordan, and Schrödinger. To be nearer to the action in physics he returned to his old school in

Berlin in 1926, and took a job as an assistant in crystallography.'[7] Wigner taught at the universities of Göttingen and Berlin from 1927 to 1930, before leaving for the USA (he was an early refugee from Nazi Germany). He was Professor of Physics at the University of Wisconsin at Madison until 1938. He then moved to Princeton University, where he remained for the rest of his career.

Wigner's work

Wigner, truly one of the greatest theoreticians of the twentieth century, contributed to many fields of physics. When at Göttingen and Berlin he published, among others, three seminal papers with a friend of his, the Hungarian John von Neumann, and a book, the celebrated *Group Theory and Its Application to the Quantum Mechanics of Atomic Spectra*, in which he developed the concept of parity, and explained important features of the atomic spectra. Then, in the 1930s, he established many of the foundations of the modern theory of solids.

At Princeton Wigner discovered that nuclear forces, which bind neutrons and protons together in a nucleus, are necessarily short-range and independent of the electric charge of the nucleons. He also applied symmetry principles to explain the energy levels of the atomic nuclei. As noted by the Nobel Committee: '[Wigner's] investigations of the symmetry principles in physics are important far beyond nuclear physics proper. His methods and results have become an indispensable guide for the interpretation of the rich and complicated picture which has emerged from . . . experimental research on elementary particles.'[8]

Fig. 10.4. Eugene Wigner with his daughter. (Courtesy AIP Emilio Segrè Visual Archives, Wigner Collection.) 'From his earliest years Wigner knew his vocation was science . . . As an aspiring young physicist he "decided that physics had a duty to provide a living picture of our world, to uncover relations between natural events, and to offer us the full unity, beauty and natural grandeur of the physical world."'[9]

During the Second World War Wigner collaborated with Enrico Fermi in the construction of the first nuclear reactor (p. 214). And after the war, he participated in the design of nuclear reactors, but soon returned to Princeton to pursue 'the deep questions that had first drawn us to physics'.

The nuclear shell model

Maria Goeppert Mayer and Hans Jensen proposed the nuclear shell model independently of each other in 1948–9. The model describes atomic nuclei by analogy with the atomic shell structure (p. 143).

In their model, the nuclei are thought of not as random aggregates of neutrons and protons, but rather as a structure of spherical concentric layers (called *shells*), each of which is occupied by neutrons and protons. Their distribution among these shells produces the stability of each type of nucleus. Like atoms, each shell corresponds to an energy level, in which nucleons are paired, neutron with neutron and proton with proton. (Nuclear energy levels are typically separated by millions of electron-volts, some million times larger than for atomic levels.) A nuclear shell is filled (or closed) when the number of neutrons or the number of protons equals 2, 8, 20, 28, 50, 82 (and 126 for neutrons): these are called 'magic numbers' (the name seems to have been proposed by Wigner). Nuclei with a magic number have both proton and neutron shells closed. They are tightly bound and especially stable. (They are like atoms with a completely filled outermost electron shell, which are very stable.) The nuclear shell model accurately predicted certain properties of atomic nuclei, such as their angular momentum. But for nuclei in highly unstable states, the model was no longer adequate and had to be replaced by others.

Maria Goeppert Mayer (1906–72) was born in Kattowitz, Germany (now Katowice in Poland), the daughter of a university professor of paediatrics. Maria, 'the beauty of Göttingen', studied physics at that university, under the guidance of Max Born. Here, in 1930, she obtained her doctorate in theoretical physics. Soon afterwards she moved to the USA, to Johns Hopkins University, in Baltimore, Maryland, and in 1939 to Columbia University, New York City (where she worked on the separation of uranium isotopes). After the war she transferred to the University of Chicago, and in 1960 to the University of California at San Diego, where she was a professor of physics.

Johannes Jensen (1907–73) was born in Hamburg, Germany. He studied at the universities of Freiburg, and Hamburg, where he obtained his doctorate in 1932. Jensen worked at Hamburg (1932–41), and at the Institute of Technology in Hannover. In 1949 he became Professor of Physics at the University of Heidelberg. In the 1950s and 1960s, Jensen went to the USA as a guest professor in various universities and research institutions.

Fig. 10.5. V. Weisskopf (left), M. Mayer and M. Born, riding bicycles in Göttingen, circa 1929. (Courtesy AIP Emilio Segrè Visual Archives.)

1964

The maser–laser

If you replace the initial letter '*l*' of *light* by the letter '*m*' of *microwave* in the word *laser*, you obtain the acronym standing for *maser*. This marks the beginning of the following story, the heroes involved being one American, Charles Townes, and two Russians, Nikolai Basov and Alexander Prokhorov. These three physicists together shared the 1964 Nobel Prize in physics for their inventions.

The story

The physical process at the heart of the maser–laser operation is *stimulated emission*. Albert Einstein suggested this in 1916–17 (p. 111), by pointing out that photons of just the right frequency can stimulate excited atoms or molecules to emit photons of exactly the same frequency. These photons can be multiplied in number and used to obtain an intense, narrow, beam of monochromatic radiation (see p. 293).

For many years physicists had thought that, under normal conditions, absorption of radiation by atoms would always predominate over stimulated emission, thus not permitting the amplification of the radiation. The question then arose: how to create the opposite, that is, stimulated emission dominating absorption? The answer came many years later, in the 1950s in fact.

During the Second World War Charles Townes worked at Bell Labs, designing radar systems; as a consequence he became very experienced in microwave techniques. He then began to work in a new field which was at that time emerging; this was microwave spectroscopy (he examined molecules by means of the microwaves emitted or absorbed). In 1951 Townes was a professor of physics at Columbia University. He and his two co-workers, the graduate student James Gordon and the postdoctoral fellow Herbert Zeiger, thought of using ammonia molecules to obtain *coherent* electromagnetic waves by stimulated emission.

About the same period, Nikolai Basov and Alexander Prokhorov were also working on microwave spectroscopy at the Lebedev Institute of Physics in Moscow. They had the same idea of amplifying electromagnetic waves, all in phase and with the same wavelength. They presented their maser idea during a conference on radio spectroscopy in 1952, and published it in a Russian scientific journal in 1954. (Similar concepts occurred in the same years to Joseph Weber of the University of Maryland.) But let us go back again to Townes and his ammonia maser. One day in the spring of 1951 Townes was in Washington, DC, to attend a meeting. As he himself remembers:

> By coincidence, I was in a hotel room with my friend and colleague Arthur L. Schawlow, later to be involved with the laser. I awoke early in the morning and, in order not to disturb him, went out and sat on a nearby park bench to puzzle over [how it might be possible to generate waves shorter than those produced by radar techniques] . . . Perhaps it was the fresh morning air that made me suddenly see that this was possible: in a few minutes I sketched out and calculated requirements for a molecular-beam system . . .[10]

Townes went back to Columbia and immediately started to work on his new idea, together with Gordon and Zeiger. But his maser project did not get much encouragement at the physics department. He recalls: 'One day, . . . Rabi [Nobel 1944] and Kusch [Nobelist in 1955], the former and current chairmen of the department, . . . came into my office and sat down. They were worried. Their research depended on support from the same source as did mine. "Look," they said, "you should stop the work you are doing. You're wasting money. Just stop!"'[11]

Finally, towards the end of the year 1953, Townes and his co-workers were able to build the first maser and to produce the first beam of very pure microwave radiation (with a frequency near 24 billions of oscillations per second, and a wavelength of 1.25 centimetres). As Townes again recalls: 'The first maser had been born. This was about three months after Poly Kusch had insisted it would not work. But when it worked, he was gracious about it, commenting that he should have realised that I probably knew more about what I was doing than he did.'[12]

Four years later Townes, who was then also a consultant to Bell Labs, and Arthur Schawlow (Nobelist in 1981), who had been working on condensed matter at Bell Labs since 1951, began to think of how the principles of the maser could be extended to shorter wavelengths, that is, to the visible and infrared regions of the electromagnetic spectrum. This project is recalled by Schawlow with these words:

> ... I was beginning to think seriously about the possibility of extending the maser principle from the microwave region to shorter wavelengths, such as the infrared region of the spectrum. It turned out that he [Townes] was also thinking about this problem, so we decided to look at the problem together ... Without interrupting our other duties, over the next few months we worked on this in odd moments.[13]

Townes and Schawlow published their laser idea in a paper which was received on 26 August 1958, and published in the 15 December issue of the *Physical Review*.[14] The paper soon spurred many other researchers on to making a laser. As we have seen (p. 277), less than two years later, Theodore Maiman succeeded in constructing the very first laser.

Charles Hard Townes was born in Greenville, South Carolina, USA, in 1915, the son of an attorney. He studied at Caltech in Pasadena, where he received his Ph.D. in 1939. In the same year he joined Bell Labs, where he worked until 1948. He then moved to Columbia University, where he became Professor of Physics in 1950. From 1959 to 1961 he was Director of Research for the Institute for Defense Analyses in Washington, DC. He was then appointed Professor of

Fig. 10.6. The three prizewinners. From left to right: A. Prokhorov, C. Townes and N. Basov. (Sources for the History of Lasers; courtesy AIP Emilio Segrè Visual Archives.)

Physics and provost at MIT, Cambridge, Massachusetts, where he continued his research work, principally in the fields of quantum electronics and astronomy. In 1967 Townes moved to the University of California at Berkeley, where he was a university professor of physics, becoming emeritus in 1986.

The laser

As you will certainly remember, quantum mechanics limits atoms to specific energy levels. To go to an excited level an atom must absorb energy; for instance, it can absorb a photon whose frequency corresponds to the energy difference between the excited and the ground levels (given by Bohr's formula $v = (E_2 - E_1)/h$, see p. 93). Left alone, the atom, in the excited level, can *spontaneously* drop (or decay) to the ground level, releasing a photon of the same frequency.

In a laser there is another process which plays an important role. It is called *stimulated emission*: light of the right frequency can stimulate (or trigger) an atom to drop from a higher energy level to a lower one. However, there is competition between absorption and stimulated emission. Which of the two processes is going to dominate? It depends on whether there are more atoms in the ground level, capable of undergoing absorption, or more in the upper level, capable of stimulated emission. Physicists have invented methods for 'pushing' most of the atoms to the upper energy level, so that the dropping from a higher level to a lower one will predominate. (In laser terminology, the process of energising the atoms of a laser medium to an excited level is known as 'pumping'; when it is achieved by irradiating the medium with an intense beam of light, it is called 'optical pumping'.)

Thus, an incoming photon of the right frequency triggers one atom which falls from a higher to a lower energy level. In such a process this atom does not gain energy, and, in addition, a second photon is released. The two photons can now strike two atoms, so that two more photons are released, and so on. The result is that few photons enter the material, but a flood of them emerges, all having exactly the same frequency, and moving in exactly the same direction. The incoming light in this way becomes greatly amplified.

Laser light is said to be *coherent*, meaning that the peaks and the valleys in its waves coincide perfectly. This is very different from ordinary light, which is not so organised. An ordinary light source emits photons in many different colours, and they point in various directions. Laser light, on the contrary, is highly 'monochromatic' (of a very pure single colour, or frequency). Furthermore, laser beams can be extremely intense, and can travel over long distances without spreading outwards.

Laser action has been obtained from solids, from liquids, from columns of gas and from semiconductors. A variety of lasers and masers now produce coherent radiation in large intervals of wavelengths, ranging from microwaves, through the infrared and visible, and so on, up to X-ray regions.

Nikolai Gennadievich Basov (1922–2001) was born at Usman near Voronezh, Russia, the son of a professor at a local university. He graduated from the Moscow Institute of Physics and Engineering in 1950, and then went to the Lebedev Institute of Physics in Moscow, where he obtained his doctorate in 1956. From 1961 on Basov carried out theoretical and experimental research in the field of powerful lasers, and a few years later he and his co-workers developed semiconductor lasers. Basov became Deputy Director of the Lebedev Institute in 1958, a member of the Soviet Union Academy of Sciences in 1962 and Professor of Physics at Moscow Institute of Physics and Engineering in 1963.

Alexander Mikhailovich Prokhorov (1916–2002) was born at Atherton, Queensland, Australia. His parents moved to the Soviet Union in 1923. After obtaining his doctorate from the University of Leningrad (now St Petersburg) in 1946, he went on to the Lebedev Institute of Physics in Moscow, as a senior associate, where he worked on microwave spectroscopy and quantum electronics. In 1954 Prokhorov became head of the Oscillation Laboratory there, and later professor of physics at Moscow State University. Subsequently he worked on lasers emitting infrared radiation, and on nonlinear optics.

Fig. 10.7. A laser beam travelling through a hollow fibre filled with a noble gas. These fibres are used to compress laser pulses down to a few million-billionths of a second.[15] (Photo: courtesy Professor Orazio Svelto, The Polytechnic of Milan, Italy.)

Laser applications

Since the mid-1970s there has been an explosive growth of laser applications, so that the laser has become a powerful tool in research, in industry and in countless fields of everyday life.

Lasers come in many different forms, from tiny semiconductor lasers, no bigger than a grain of salt, to monsters as long as whole buildings. For many years the most common lasers were helium–neon gas lasers (like those used to read bar codes at the tills of supermarkets). Today the most widely used are semiconductor lasers (they account for more than fifty per cent of the world market for lasers). These are small chips of semiconductors, and are used in handling and processing information. They also play the music in compact disc players, and are used in telephone links, sending thousands of telephone calls through many kilometres of optical fibres. Laser applications in the field of image-recording include computer printers, copiers and fax machines.

Engineers also make use of lasers to cast straight lines when constructing tunnels, or when planning irrigation projects, and geologists are able to measure small motions of the earth's crust plates. Lasers are widely used in industry for soldering and welding, or for cutting and drilling metals, plastic and other materials. Medical lasers are used in ophthalmology and surgery, as well as in cosmetic procedures. In research laboratories chemical lasers help scientists to analyse the structure of atoms and molecules, and the composition of materials. High-power lasers (which consume as much electrical energy as a small town) are used in laboratories to study fusion reactions.

Mysterious radio signals

Arno Penzias and Robert Wilson were two radio astronomers working at Bell Labs in Holmdel, New Jersey, a few miles from Princeton University. Near the laboratory, Bell Labs engineers had built a radio antenna. It had been used in the early 1960s to transmit and receive microwave signals to and from Telstar satellites. When the antenna became available for research, Penzias and Wilson decided to use it to see whether the Milky Way's halo (a vast spherical region surrounding the disc of our galaxy) was made up of glowing radio waves.

Before starting the experiment, they had to calibrate their instrument; in particular, they wanted to measure its 'noise' (that is, the random signals created by the instrument itself). They decided to set it up for a wavelength of 7.35 centimetres (a frequency equal to 4080 megacycles per second), a wavelength at which the Milky Way's halo should have been practically invisible – hence they expected that the intensity of radio waves of such a wavelength would have been quite low. The waves captured by the antenna were to be detected by a very sensitive solid-state maser.

Penzias and Wilson began their measurements in May 1964, and immediately, to their surprise, they detected a microwave noise higher than they had expected.

They suspected at first that this could be radio signals originating either in the atmosphere or in the antenna itself, or issuing from man-made sources. However, thanks to careful tests, made during the following months, they showed that the radio waves were coming from outer space, and that their intensity was the same in all directions. They were thus forced to come to the amazing conclusion that the universe is filled uniformly with microwaves, with a wavelength of about 7.35 centimetres. (The excess radiation that they had detected at such a wavelength would in fact be produced by a perfect radiator, a black body, at a temperature of about 3.5 degrees above absolute zero.)

What they had discovered was the afterglow of the *Big Bang*, that titanic explosion from which the universe was born (p. 253). This was one of the greatest cosmological discoveries of the twentieth century, well worthy of winning a Nobel Prize (see Nobel 1978).

1965

Quantum electrodynamics

The 1965 Nobel Prize for physics went to three theoretical physicists. Two of them were Americans, Richard Feynman and Julian Schwinger. The third was a Japanese professor, Sin-Itiro Tomonaga. These three scientists had made fundamental contributions to the development of modern quantum electrodynamics, QED for short, the most precise theory of electrons and electromagnetic radiation. Here, in a few words, is a behind-the-scenes account of what happened:

> Feynman, Schwinger and Tomonaga . . . could possibly have been awarded that prize several years earlier, but Niels Bohr . . . suspected the new theory, and his negative attitude deterred the Nobel Committee from acknowledging it. The decision to give the Nobel Prize to the authors of QED was made only after Bohr's death. When it was published, R. Oppenheimer ('the father' of the atomic bomb) was in Israel, as Yuval Ne'eman's guest. He sent Feynman a telegram from Tel Aviv with only one word: '*Enfin*' [French for 'At long last!'].[16]

A long march

Let us summarise briefly, starting from the old classical theory of electrodynamics. This was born in the 1860s when James Clerk Maxwell formulated his monumental equations, which describe electric and magnetic fields created by charged bodies and electric currents (p. 19). The theory was beautifully verified by experiment when Heinrich Hertz, in the late 1880s, demonstrated the existence of radio waves (p. 20). However, classical electrodynamics failed to describe the world of atoms. In particular, it did not explain the characteristic atomic line spectra, and black-body radiation (p. 25). Then quantum theory arrived. Here are the main steps.

In 1900 Planck put forward his quantum of action h to explain the nature of black-body radiation; in 1905 Einstein came out with his revolutionary idea of light-quanta. In 1913 Bohr explained how light of sharp characteristic colours is emitted from atoms. He was followed in 1923–4 by Louis de Broglie with his particle–wave hypothesis. Finally, a definite quantum mechanics was developed by Heisenberg, Born, Schrödinger, Dirac and Bohr again between 1925 and 1927. Quantum mechanics was applied not only to matter but also to radiation. A few years later, Dirac, Heisenberg, Pauli and others (p. 176) laid the foundations of a theory which was called *quantum electrodynamics*, and which described the electromagnetic field as being composed of distinct entities or radiation quanta. Despite the successes scored, the theory contained within itself serious contradictions, which hindered its development for some twenty years. These difficulties were solved in the 1940s by Feynman, Schwinger, Tomonaga and others (important contributions also came from Hans Bethe and Freeman Dyson), who reformulated and extended the old theory into a new one.

Modern QED

Modern QED is a relativistic quantum theory of electromagnetism, which describes how electrically charged particles, such as electrons, interact with electromagnetic radiation, and how these same particles interact with one another due to their electric charge.

The theory builds on the idea that charged particles are surrounded by a cloud of photons. So two such particles interact by the mutual exchange of these photons, which are considered as acting like *messengers*, occupied in carrying the electromagnetic force between the particles themselves. The two interacting particles change their speed and the direction of their motion as they emit or absorb such photons; this means that a force is exerted on each of them.

The photons carrying the force are *virtual*, that is, their existence can only be deduced. They are continuously emitted and re-absorbed by their source, unless they meet another charged particle which absorbs them, but they exist for a time span too short to be anything other than conjectured (this is why physicists use the term 'virtual'). (Photons, such as those of sunlight or radio waves, are in contrast real; consequently, they are emitted in a free state and can actually be observed.) Virtual photons, like real photons, have zero mass. This means that the range of the electromagnetic force is infinity; its strength gradually declines with increasing distance without falling off abruptly.

Modern QED explains the interaction of two charged particles, such as two electrons, as occurring in a series of processes of increasing complexity. In the simplest process, only one virtual photon is involved; in more complex processes, there are two, three (and so on) virtual photons. Each process can be represented by means of geometrical graphics in the Minkowski space-time (p. 71). These were invented by Feynman, and are known as *Feynman diagrams*.[17]

They are extremely useful in prescribing a method of calculating the physical quantities involved in the interaction.

A consequence of the internal contradictions within the old 1930s theory was that when physicists tried to calculate such physical quantities as the charge and the mass of the electron, which ought to be finite, these turned out to be actually infinite (without any physical significance!). This is well known as the 'problem of the so-called infinities'.

Attempts had already been made in those years to overcome these contradictions, but without any success. The experimental results presented in 1947 at the famous Shelter Island conference by Isidor Rabi and Willis Lamb (p. 253), gave new impetus to such attempts. A group of theorists, including our three heroes, developed a mathematical method in which quantities, like the mass and the charge of the electron, were 'redefined' (or, as theorists like to say, *renormalised*) with quantities obtained from experiments (these quantities, having a finite value, naturally have more physical significance). The corrected theory thus provided a way of calculating, with whatever degree of accuracy was desired, any phenomenon resulting from the interaction of electrons with the electromagnetic radiation.

Modern QED thus became one of the most precise and successful of theoretical constructs. Today's particle physicists consider it as a model for the elaboration of their ambitious theories of all the particles and forces of nature. Here are some proofs of QED's spectacular success.

The most recent measurement of the Lamb shift gives a difference between theoretical predictions and experimental results that is less than two-millionths.[18] The best experimental value that has been reported for the electron g-factor (p. 252) is $g/2 = 1.001\,159\,652\,188$, to be compared with the theoretical value, $1.001\,159\,652\,140$. The agreement between theory and experiment is impressive![19] In the 1960s and 1970s a series of experiments on the magnetic moment of the muon were carried out at CERN near Geneva.[20] It emerged that there was no difference between the muon and the electron as regards their anomalous magnetic moment. The most recent results (in 2004) have been obtained in an experiment carried out at the Brookhaven National Laboratory.[21] They have confirmed the muon anomaly to a precision of about 0.5 parts per million.

The QED triumvirae

Richard Phillip Feynman (1918–88), one of the most brilliant theoretical physicists of the twentieth century, was born in New York City, the descendant of Russian and Polish Jews. He studied physics at MIT and Princeton, where he obtained his Ph.D. in 1942. During the Second World War he worked on the Manhattan Project at Princeton, and at Los Alamos. After the war years he was appointed Professor of Physics at Cornell University from 1945 to 1950, when he moved to Caltech, Pasadena; here, in 1959, he became Richard Chace Tolman Professor of Theoretical Physics. Apart from QED, Feynman made far-reaching contributions to other fields of physics, ranging from low-temperature physics, to superfluidity, to quantum mechanics and to high-energy particle physics. Together with Murray Gell-Mann (Nobel 1969) he devised a new theory on weak interactions, and in 1968 he introduced the concept of *partons*, which led to the understanding of quarks, the modern building blocks of hadrons.

> The physical universe presented him with a fascinating series of puzzles and challenges, and so did his social environment. A lifelong prankster, he treated authority and the academic establishment with the same sort of disrespect he showed for stuffy mathematical formalism . . . His autobiographical writings contain amusing stories of Feynman outwitting the atom-bomb security services during the war, Feynman cracking safes, Feynman disarming women with outrageous behaviour. He treated his Nobel Prize . . . in a similar take-it-or-leave-it manner.
>
> (Paul Davies)[22]

Julian Seymour Schwinger (1918–94) was born in New York City, the son of a Jewish family. An *enfant prodige*, Schwinger published his first scientific paper while a freshman at City College, New York, and received his Ph.D. from Columbia University at the tender age of twenty-one. He then worked under J. Robert Oppenheimer at the University of California, Berkeley. During the Second World War he joined the Radiation Laboratory at MIT, where he worked on the theory of radar devices. In 1946 he was appointed Associate Professor

Fig. 10.8. A historic photo. From left to right: W. Lamb, A. Pais, J. Wheeler, R. Feynman, H. Feshbach and J. Schwinger; Shelter Island Conference, 1947. (Courtesy AIP Emilio Segrè Visual Archives.)

Fig. 10.9. Sin-Itiro Tomonaga. (Courtesy AIP Emilio Segrè Visual Archives.)

at Harvard University, and one year later he became one of the youngest full professors in Harvard's history. From 1972 onwards Schwinger was Professor of Physics at the University of California, Los Angeles.

Sin-Itiro Tomonaga (1906–79) was born in Tokyo, Japan, the son of a university professor of philosophy. Tomonaga graduated from Kyoto University, where he studied with Hideki Yukawa (Nobel 1949). He then became a research associate at the University of Tokyo. From 1937 to 1939 he was at the University of Leipzig, Germany, working with Werner Heisenberg. In 1941 Tomonaga became professor of physics at Tokyo University of Education, and in 1943 he published (in Japanese) his fundamental paper on QED. After the war he became chairman of the Japanese Science Council.

Schwinger and Tomonaga

Schwinger and Tomonaga's contributions are examples of the momentum of internal factors in the advances of physics. That so remarkably similar solutions were advanced by Tomonaga and by Schwinger working in complete isolation from one another, in two remarkably dissimilar cultures under radically differing working conditions – war-ravaged Tokyo and affluent Harvard – is surely proof that physicists advance explanatory schemes that transcend national boundaries and national styles. Schwinger and Tomonaga's contributions to QED also illustrate the impressive efficacy and instrumental power of modern physical theories.

(Silvan S. Schweber)[23]

1966

Studying atomic energy levels

When the 1966 Nobel Prize for physics was awarded to Alfred Kastler, France obtained, after a gap of thirty-seven years, its eighth Nobel laureate. Kastler won the Nobel Prize for having developed new and sophisticated experimental methods for studying the energy levels of atoms.

The first method developed by Kastler towards the end of the 1940s was called *double resonance*. Atoms of a gas, contained in a cell, were illuminated by a beam of suitable polarised light; the luminous vibrations of this light ran along a fixed direction, and were of the right wavelength to cause the atoms to jump from the ground or lowest level to one of the excited levels. These atoms were then stimulated by radio-frequency waves, which caused transitions from one of the populated levels of the excited state to nearby empty sublevels. Then the same atoms re-emitted light waves with a different state of polarisation, which were detected by sensitive devices and carefully analysed. Thanks to these observations Kastler and his colleague and former student, Jean Brossel, were able to interpret the very finest details of the structure of the excited atomic levels.

The second method, proposed in 1950, was called *optical pumping*. Using light of the right frequency to raise atoms of a substance from lower to higher energy levels, Kastler was able to explain how light can be used to stimulate atoms so that they attain well-defined excited levels. Kastler's techniques also permitted the construction of very accurate atomic clocks, and of sensitive devices for measuring extremely feeble magnetic fields.

Alfred Kastler (1902–84) was born in Guebwiller, France. He studied physics at the *Ecole Normale Supérieure* in Paris, and did his doctoral thesis at the University of Bordeaux. After working at the universities of Bordeaux and Clermont-Ferrand in France, and of Louvain in Belgium, he returned in 1941 to the *Ecole Normale*, where he was a professor of physics until 1968. During his long career he trained an entire generation of French physicists in the field of atomic physics (among these was Claude Cohen-Tannoudji, Nobelist in 1997). From 1968 to 1972 he was director of research at the National Centre of Scientific Research (CNRS) in Paris. Kastler was a leading scientist and an internationally recognised authority in the field of atomic spectroscopy. He was also active in peace movements, and in organisations protesting against nuclear proliferation.

1967

How do stars shine?

On 30 October 1967 the Swedish Academy of Sciences awarded the year's Nobel Prize for physics to the German-born theorist Hans Bethe, then Professor of physics at Cornell University in Ithaca, New York. The prize was awarded for '... his contributions to the theory of nuclear reactions, especially his discoveries concerning energy production in stars'.

Bethe's award was the first in astrophysics, the branch of science dealing with the physics of the stars, galaxies and all the phenomena regarding the largest objects in the universe. The Nobel Physics Committee's attitude towards fields which they had considered peripheral to physics, such as astronomy, or 'cosmical physics', had remained negative for many years. ('Cosmical physics' was the name which, earlier in the 1900s, embodied both astrophysics and geophysics.) As an example of this ostracism, in 1923, when the American astrophysicist George Ellery Hale and the French astronomer and meteorologist Henri Deslandres were brought to the Committee's attention for a Nobel award, most of the Committee members 'doubted whether their contributions belonged to physics'. Even the influential Svante Arrhenius, who previously had been a supporter of cosmical physics, prepared a report in which he maintained that astrophysics was not part of physics. He claimed '... that astrophysics had expanded so rapidly since 1900, when it had been eligible for the physics prize, that it now encompassed all of astronomy. With a rhetorical sleight of hand, [he] then declared that astronomy had therefore become astrophysics. But since Alfred Nobel clearly had not intended astronomy to be eligible for his prizes, astrophysics no longer could be eligible.'[24] The Committee's attitude was so unyielding that leading astrophysicists such as Arthur Eddington, Edwin Hubble and Henry Norris Russell, who had been nominated during the preceding decades, were persistently ignored and in fact never qualified for a Nobel Prize. However, the Committee's composition changed over the years, and so did its attitude towards astrophysics. (However, in 1947, with the Nobel Prize to Edward Appleton, doors had already been opened to geophysics.)

Let us read the last passages of the presentation speech, at the Nobel ceremony (in December 1967). This was delivered by Oskar Klein, a theoretical physicist in Stockholm and a member of the Swedish Academy of Sciences:

> Professor Bethe. You may have been astonished that among your many contributions to physics, several of which have been proposed for the Nobel Prize, we have chosen one which contains less fundamental physics than many of the others and which has taken only a short part of your long time in science. This, however, ... does not imply that we are not highly impressed by the role you have played in so many parts of the development of physics ever since you started doing research some forty years ago. On the other hand your solution of the energy source of stars is one of the most important applications of fundamental physics in our days, having led to a deep-going evolution of our knowledge of the universe around us.[25]

Energy from the stars

What makes the sun and all the other stars shine? How could the sun have been shining for hundreds of millions of years, throughout the whole time that the earth has existed? The sun is pouring out immense amounts of energy in the form of light and heat radiation: about 400 billion billion megawatts. Where does all this energy come from? For centuries all these questions remained a mystery.

In the mid-1800s Hermann von Helmholtz and Lord Kelvin suggested that the sun should gradually contract under the tremendous crushing weight of its outer layers, so that its interior fluids would become compressed. These would therefore become hot enough to radiate energy out into space. On this basis, however, the sun would have liberated enough energy for only a few tens of million years or so. This would not have sufficed. It was thus clear that the energy of the sun could not come from the usual familiar sources; some unknown process must be at work in its interior. Only after the discovery of radioactivity, and that of the structure of the atomic nucleus, did physicists recognise the identity of the new source that they were looking for.

In the 1920s Arthur Eddington (the British astronomer who had observed the famous solar eclipse in 1919) considered the core of the sun as consisting of a hot gas, and showed that the temperature there must be extraordinarily high – in the range of 15–20 million kelvins (at such temperatures, atoms are dissolved into electrons and nuclei). Another astronomer from Britain, Robert Atkinson, and the Austrian physicist Fritz Houtermans, had the idea that collisions between hydrogen nuclei (protons) and those of other elements could liberate an amount of energy enough to account for the observed radiation of the sun, but at that time very little was known about nuclear reactions. And so it was only some ten years later, when sufficient information had been accumulated, that the source of the energy of the sun could be positively identified.

In 1938 Bethe advanced his theory. He proposed that solar energy is produced by the *nuclear fusion* of protons into helium nuclei. Consequently, it was easy to calculate that the sun can, fortunately, continue to radiate energy for billions of years to come! Bethe then worked out two possible ways of explaining the origin of solar energy, which involved the fusion of four hydrogen nuclei into one helium nucleus, and the conversion of a tiny amount of mass into a large amount of energy, according to Einstein's famous formula $E = mc^2$.

Proton–proton chain

The first way, called the *proton–proton chain*, was proposed by Bethe and his co-worker Charles Critchfield, and involves the direct conversion of hydrogen into helium. (They used both Fermi's theory of nuclear beta decay and Gamow's concept of tunnelling.) The chain reaction begins with two protons colliding and forming a nucleus of *deuterium* (an isotope of hydrogen with one proton and one neutron bound together) causing the release of a positron and a neutrino. After capturing a third proton, the deuterium nucleus becomes absorbed within the low-mass isotope of helium, called helium-3 (two protons and one neutron). In the final step, two nuclei of helium-3 combine to produce a nucleus of ordinary helium (helium-4), plus two protons (see Fig. 10.10). The quantity of energy released is about 600 000 billion joules per kilogram of hydrogen, in perfect

Fig. 10.10. The proton–proton chain.

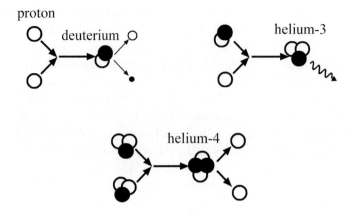

agreement with the observed value (it is about twenty million times greater than that released by burning the same quantity of coal).

CNO cycle

In the same year, 1938, Bethe developed a second process, one that is a little more complicated. This process involves carbon as an intermediate in the transformation; in fact it proceeds through successive addition of three protons to the nucleus of the isotope carbon-12 (six protons and six neutrons), in the end obtaining a nucleus of nitrogen-15 (seven protons and eight neutrons). The largest part of the energy is then produced in the final and fusion of a fourth proton with nitrogen-15, giving as a result a helium nucleus and the original carbon-12. Within the sun the energy released in this process (it is called the *CNO cycle*, CNO standing for carbon–nitrogen–oxygen) is about one hundredth of the energy released in the proton–proton chain which is so predominant. The energies produced by the two processes are both in perfect agreement with the observed values pertaining to the sun. The two processes possess different sensitivities to temperature; in stars which are hotter than the sun (whose central temperature is about 16 million kelvins), the carbon cycle predominates, whereas for stars where the central temperature does not exceed 16 million kelvins, hydrogen burning occurs predominantly through the proton–proton chain.

Hans Albrecht Bethe was born in Strasbourg, Alsace-Lorraine (then in Germany), in 1906. He studied physics first at the University of Frankfurt, then at Munich, where he obtained his doctorate in 1928, writing a thesis under the supervision of Arnold Sommerfeld. He then became an instructor in physics, first at Frankfurt and then at Stuttgart. Between 1930 and 1932 he worked with Rutherford at Cambridge, and with Fermi in Rome. He then taught as a lecturer at the University of Tübingen until 1933, when he was dismissed for political reasons (his mother was Jewish). Bethe first emigrated to Britain, and taught at the universities of Manchester and Bristol. In 1935 he went to the USA, and was

Fig. 10.11. Hans Bethe. (Courtesy AIP Emilio Segrè Visual Archives, Physics Today Collection.)

> It was in Rome [1930–1932] . . . that Bethe wrote his famous article 'One- and Two-Electron Systems' for the 'Handbuch der Physik'. I remember that he stood at a high table, and without books wrote page after page of this treatise without interruptions, and recalculated all results of previous authors to check their correctness.
>
> (Emilio Segrè)[26]

appointed Assistant Professor of Physics at Cornell University, where he became full professor in 1937, and emeritus in 1975. During the Second World War Bethe worked at MIT on the development of radar, and headed the theoretical physics division of the Manhattan Project in Los Alamos. After the war he became famous as a crusader in the cause for the peaceful use of nuclear energy and the reduction of armaments.

Bethe, in a very long life, has made many fundamental contributions to other fields of physics, such as atomic physics, crystals, solid state and metal physics. He is a member of the Royal Society of London, as well as of the US National Academy of Sciences.

Looking deep into the sun

As we have just seen, the fusion reactions in Bethe's proton–proton chain involve four protons which fuse together and make one nucleus of helium. Two of the original protons are transformed into neutrons through the weak nuclear force, and emit two positrons and two electron neutrinos. Other branches of the chain also produce neutrinos. Therefore nuclear fusion reactions create a prodigious outflow of neutrinos from the sun's centre: every second hundred billion billion billion billion neutrinos are released, and every second about ten billion of them pass through every square centimetre of your body here on earth. So detecting neutrinos is a means of looking deep into the sun's core and discovering what kind of physics is at work there.

During the 1960s physicists began to think seriously about the possibility of observing solar neutrinos. Because these interact very weakly with matter (99.9999 per cent of neutrinos pass through the earth as though it were not there), intercepting them is a real challenge. So physicists thought they would place very massive detectors deep below ground level: thus screened from cosmic rays, only neutrinos would be able to reach the detectors.

The first *solar neutrino experiment* was carried out by Raymond Davis of the Brookhaven National Laboratory. He and his co-workers built a neutrino detector and placed it in a mine in South Dakota, USA. Davis started his experiment in the mid-1960s and, in a few years, he managed to detect a significant number of solar neutrinos, so confirming that the sun actually shines by nuclear fusion reactions. The rate observed, however, was lower than that predicted. Other experiments, carried out in the following years in underground laboratories have also indicated a solar neutrino rate lower than expected. (One of these was carried out in a mine in Japan, by a group headed by the Japanese physicist Masatoshi Koshiba.) At this point an obvious question arises: does the sun perhaps not produce enough neutrinos? We will consider this problem when speaking of the 2002 Nobel Prize for physics, which was awarded in fact to Davis and Koshiba.

Nuclear-fusion fire

Fusion reactions in the sun are the ultimate source of all forms of energy on earth. Physicists have dreamed for half a century of the possibility of creating such reactions in their laboratories – making them proceed at a controlled pace – and thereby producing huge amounts of energy. In addition, fusion power is more attractive than fission power: in fact, fusion produces a smaller amount of possibly dangerous radioactive nuclear waste than fission does. Therefore fusion power represents a potential source of efficient and less harmful energy which could solve all mankind's energy problems. Nevertheless, obtaining useful power from controlled nuclear fusion constitutes a very difficult challenge. Starting in the early 1950s physicists and engineers began to develop fusion power technology. They used the nuclei of the two heavy isotopes of hydrogen: *deuterium* (one proton and one neutron) and *tritium* (one proton and two neutrons). This is the fusion reaction producing energy:

$$\text{deuterium} + \text{tritium} = \text{helium-4} + \text{neutron} + \text{energy}$$

In order to make the nuclei of deuterium and tritium fuse together, these must be made to approach each other at very high speed so as to overcome the repulsive electric force between them and attain a sufficiently small separation to cause the strong nuclear force to dominate. In a gas at the extremely high temperatures characteristic of stars the average energy of nuclei is high enough to allow them to fuse. (The reactions are called *thermonuclear fusion reactions*). A temperature of hundreds of millions of kelvins is needed for deuterium–tritium fusions to occur at a sufficient rate. At such temperatures the fuel changes its state from a neutral gas to a *plasma* (an ionised gas consisting of negatively charged electrons and positive nuclei, p. 315).

Since a plasma consists of charged particles, magnetic fields are needed to confine them in order to produce enough fusion reactions (this method is called *magnetic confinement*). A second approach to heating the plasma is to focus powerful laser or particle beams on a tiny solid pellet made of deuterium and tritium. The pellet is compressed and heated by the pressure of the incident beams, and the resultant high temperature produces fusion reactions (this method is called *inertial confinement*). In these last decades significant progress has been made in controlled nuclear fusion. So, as we all hope, the goal of lighting the nuclear-fusion fire may not be all that very far away.

1968

Bubble chamber at work

On being awarded the 1968 Nobel Prize, Luis Alvarez, in his lecture at Stockholm, commenced by explaining vividly the excitement he found in his work as an experimental physicist. This is what he said: 'Most of us who become experimental physicists do so for two reasons; we love the tools of physics because to us they have intrinsic beauty, and we dream of finding new secrets of nature as important and as exciting as those uncovered by our scientific heroes.'[27] This is exactly what Alvarez succeeded in doing in his work at the Berkeley Radiation Laboratory, when he discovered new groups of subatomic particles, called *resonance particles*, by using a hydrogen bubble chamber.

Alvarez then proceeded to describe the beginning of his adventure in the world of particle physics, and the exploits which finally brought him to Stockholm:

> In late April of 1953 I paid my annual visit to Washington, to attend the meeting of the American Physical Society. At lunch on the first day, I found myself seated at a large table in the garden of the Shoreham Hotel . . . A young chap . . . was seated at my left, and we were soon talking of our interests in physics. He expressed concern that no one would hear his 10-min contributed paper, because it was scheduled as the final paper of the Saturday afternoon session . . . I admitted that I wouldn't be there, and asked him to tell me what he would be reporting. And that

is how I heard first hand from Donald Glaser [Nobel 1960] how he had invented the bubble chamber ... He showed me photographs of bubble tracks in a small glass bulb ... I was greatly impressed by his work, and it immediately occurred to me that this could be the 'big idea' I felt was needed in particle physics.[28]

Alvarez returned to Berkeley and, together with his co-workers, started to develop a bubble chamber containing liquid hydrogen. In February 1954 they succeeded in photographing the first particle tracks. Alvarez' group then constructed large-scale bubble chambers (containing hundreds of litres of liquid hydrogen), and used them during the 1960s to study high-energy interactions of protons with hydrogen at the Bevatron accelerator. They produced millions of photographs of such interactions, and developed high-speed devices to automatically measure and analyse the particle tracks produced. So they discovered the new states of matter called 'resonance particles'.

Luis Walter Alvarez (1911–88) was born in San Francisco, California. He studied physics at the University of Chicago, where he obtained his Ph.D. in 1936. He then joined the University of California, Berkeley, where he became professor of physics in 1945. In 1939 Alvarez and Felix Bloch (Nobelist in 1952) made the first measurement of the magnetic moment of the neutron. During the Second World War Alvarez worked on microwave radar research at MIT, and participated in the Manhattan Project at the University of Chicago and in Los Alamos. After the war he conceived a plan for the first linear accelerator for protons, and subsequently led a group which built it at Berkeley. Early on in the 1950s he began research work in high-energy particle physics; he used the Bevatron accelerator at the Berkeley Radiation Laboratory, and developed the first large liquid hydrogen bubble chambers.

Resonance particles

The term 'resonance' is commonly used in physics when a system absorbs energy with a maximum degree of efficiency. In high-energy physics, a 'resonance particle' means a system of particles which are grouped together for an ultra-short time span (of the order of one- hundred-thousand-billion-billionth of a second), due to the effect of the strong nuclear force (which is so powerful that it takes this very short time to be transmitted across the resonance itself). Then the resonance breaks down into particles, owing to the fact that the phenomenon is possible from an energy point of view. In spite of its ultra-short lifetime, a resonance has a mass, a spin and other quantum numbers, just as all particles do, so as to permit physicists to treat it as a real individual entity.

Due to its extremely short lifetime, there is no way of observing a resonance directly. The distance traversed by such a particle system between the point at

Fig. 10.12. Luis Alvarez (left) and his son Walter. (Courtesy Lawrence Berkeley National Laboratory.) Alvarez was a versatile scientist. Early in his career he worked in the fields of optics, cosmic rays and nuclear physics. After working in particle physics, he looked for hidden chambers in Egyptian pyramids by measuring the intensity of cosmic-ray muons. In 1980, he and his son Walter, a geologist, suggested a theory which ascribed the extinction of the dinosaurs, some 65 million years previously, to a massive meteorite that struck the earth and caused climatic changes.

which it is created and the point at which it breaks down is too short (some hundred-billionths of a millimetre), so that its track cannot be recorded in any detector. Physicists have then used alternative techniques for studying a resonance particle. By counting and analysing its breakdown products the existence of a resonance can be deduced, and its properties revealed. Another way is that of increasing the energy of the interacting particles; a resonance occurs when the energy is just enough to produce its mass. (At this particular value of the energy, a sharp increase in the frequency of particle interactions is clearly apparent.)

During the 1960s dozens of resonances had been discovered. How could they fit into the list of particles which were already known? At first physicists tried to explain most of them as excited states of low-energy hadrons. Later, the American theorist Murray Gell-Mann (Nobel 1969) proposed the 'quark model' (see p. 314). In this way a totally new light was shed on resonances.

A jungle of particles

In the 1950s, experiments carried out in the field of cosmic rays, and at the newly operating accelerators, produced handfuls of fresh subatomic particles. The number of *strange particles*, which were discovered at the end of the 1940s (p. 231), actually increased to nearly ten. They were divided into two groups: those with masses about one half the proton mass (the so-called K-mesons) in the first group; and those with masses greater than that of the neutron in the second. These particles are produced in strong interactions, and were named 'strange' because they presented several puzzles. How to account for all their properties? As we shall see, the answer came once again from Gell-Mann (p. 312).

While we are busy with the 1950s we should add other discoveries: the *antiproton* and the *antineutron* (p. 269); the *electron neutrino* (p. 260); and the *muon neutrino* (p. 286). In the 1960s physicists focused their activities on the new short-lived resonances, so that the particle list expanded into a veritable 'jungle'. All known particles (excluding the photon) were divided into two groups: the first group includes *hadrons*, which 'feel' the strong nuclear force; and the second group includes *leptons*, which do not 'feel' the strong force. Hadrons were further divided into two sub-groups: hadrons which are fermions were called *baryons*; and hadrons which are bosons were called *mesons*.

1969

Order and harmony

It was the American theorist Murray Gell-Mann who in the 1950s and 1960s was responsible for major contributions to our knowledge of hadrons, and devised new methods for their classification. The Swedish Academy of Sciences recognised his original contributions to particle theory, and awarded him the 1969 Nobel Prize for physics.

Murray Gell-Mann was born in New York City in 1929, the son of a Jewish Austrian-born immigrant. Another *enfant prodige* to make his debut in modern physics, Gell-Mann entered Yale University at the age of fifteen, and graduated at the age of nineteen. He then went to MIT, where he obtained his Ph.D. when he was twenty-two, having as a supervisor Victor Weisskopf. (Gell-Mann '... would later state that Weisskopf had convinced him of the importance of "prizing agreement with the evidence above mathematical sophistication," of always searching for simplicity, and of "avoiding cant and pomposity." '[29])

In 1952 Gell-Mann spent a year as a postdoctoral fellow at the Institute for Advanced Study in Princeton, under the leadership of J. Robert Oppenheimer. He then moved to the Institute for Nuclear Studies of the University of Chicago, where he became associate professor in 1954. The following year he was invited to Caltech, and became a full professor of theoretical physics there.

Gell-Mann ranks among the greatest theoretical physicists of the last fifty years; he has developed many of the major theories of particle physics. He has received prestigious awards and *honoris-causa* degrees from several universities for his scientific work. Gell-Mann is a member of the US National Academy of Sciences, and is at present affiliated with the Santa Fe Institute for the study of complex systems. All his biographers recall Gell-Mann's passion for the history of languages, for collecting archaeological artefacts and for bird study. He has also become involved with problems of the environment (in 1969 he participated in the organisation of an environmental study programme sponsored by the US National Academy of Sciences).

Three of Gell-Mann's major contributions were cited by the Swedish Academy of Sciences in awarding him the Nobel Prize: *strangeness* (a property which explained the newly discovered strange particles); the *eightfold way* (a new classification scheme in which hadrons form families); and *quarks* (fractionally charged particles which physicists think of as the fundamental building blocks of hadrons).

Playing cat and mouse

> When [Weinberg] . . . made his Caltech debut, both Feynman and Gell-Mann needled him mercilessly. Weinberg [Nobelist in 1979], feeling a little traumatized, thought Feynman was just playing cat-and-mouse: he loved making people squirm. Murray's aggressiveness at least seemed motivated by a genuine interest in the physics . . . Gell-Mann and Weinberg developed a grudging, prickly respect for each other. But from then on, Weinberg felt gun-shy about lecturing at the school. Physicists started to talk about a Caltech style of doing physics, more brutal even than Oppenheimer's relentless grilling or Pauli's acerbic wit. Ideas were attacked with a viciousness that some found shocking, others exhilarating . . . Particle physics is so intensely competitive that even close

Fig. 10.13. Murray Gell-Mann during a lecture. (Courtesy AIP Emilio Segrè Visual Archives, Physics Today Collection.)

collaborators like Yang and Lee found their friendship could not survive... But no school was more combative than Caltech. Visiting speakers learned to prepare themselves for what might develop into a no-holds-barred fight if either Murray or Dick was in the audience.[30]

Strangeness

As we have mentioned (p. 310), strange particles are created in strong interactions (for example, in high-energy collisions of protons), and they break down to produce other hadrons (such as pions, or even other strange particles), which feel the strong force. One might then guess that strange particles should decay by means of the same force, and so have lifetimes of some hundred-thousand-billion-billionths of a second. Experiments, however, had shown that the lifetimes of strange particles are actually some ten thousand billion times longer, which is typical of the decays produced by the slower weak nuclear force. Furthermore, it had already been pointed out that strange particles must be produced in groups of two, and never as single particles. This fact was soon proved by experiments carried out at accelerators.

Physicists then asked themselves: Why do these particles have such a long lifetime? Why do we never observe reactions that produce single strange particles? Could it be that new conservation laws exist, which forbid single production, and at the same time account for the particles' slow decay? (Atomic and subatomic physics are ruled by conservation laws, which state what 'cannot' occur. When processes are found which do not contradict any of the known conservation laws, and have never yet been observed, physicists suspect that hitherto unknown conservation laws must have been in existence which forbid those processes from taking place.)

Finally, in 1953 Gell-Mann solved the mystery. In a paper published in the *Physical Review*[31] he explained that particles possess an additional property (a quantum number), which he named *strangeness*. He ascribed zero strangeness to all hadrons except the strange ones, and stated that strangeness must be conserved only in strong and electromagnetic interactions (here is the basic rule: the total strangeness after a reaction must be equal to the strangeness prior to the reaction). However, in weak interactions (and weak decays) strangeness can change. It was thus that all strange-particle puzzles were resolved. (Something of the sort was independently proposed by the Japanese physicist Kazuhiko Nishijima.)

The eightfold way

The discovery of strangeness opened a way through the jungle of hadrons. These were first organised into groups, called *charge multiplets*. Each multiplet contained hadrons with approximately the same mass, and identical in all other properties, except for the electric charge. For example, the proton and the neutron together form a doublet: they can be seen as two charge states $(+1, 0)$ of a single particle, the nucleon; pions form a triplet, with three charge states $(+1, 0, -1)$.

When only strong interactions are involved, all members of a charge multiplet are equivalent, they are 'indistinguishable', as if they were a single particle. However, weak and electromagnetic interactions 'break' this symmetry, so that each member acquires its own identity, characterised by the values of strangeness and of electric charge.

In 1961 Gell-Mann, and, independently, the Israeli physicist Yuval Ne'eman, had an idea. They invented a more complete scheme for classifying hadrons in which the old charge multiplets were united in larger families, called *supermultiplets*, on the basis of eight quantum numbers.[32] In this new scheme, the members of a particular supermultiplet can be regarded as excited states of basic members. For example, mesons with spin zero and baryons with spin $\frac{1}{2}$ form octets (families of eight, see Fig. 10.14); baryons with a spin of $\frac{3}{2}$ form decuplets (families of ten). All the hadrons known up to that time did actually fit into such supermultiplets (mesons into singlets and octets, and baryons into singlets, octets and decuplets), and all resonances then known were capable of being fitted into other supermultiplets. Formulae were also found which described various characteristics for the particles in a supermultiplet.

Gell-Mann called this scheme the *eightfold way*, both because of the eight quantum numbers which are at the basis of the scheme itself, and also because the supermultiplet to which the nucleon belongs is an octet (rather than a triplet as was thought earlier). (As Gell-Mann himself wrote, the name eightfold way carries echoes of a teaching attributed to the Buddha: 'Now this, O monks, is noble truth that leads to the cessation of pain: this is the noble *Eightfold Way*: namely, right views, right intention, right speech, right action, right living, right effort, right mindfulness, right concentration.'[33])

The definitive victory of the eightfold way was signalled by the discovery in 1964 of the omega minus – the missing particle in a decuplet of baryons. (Its existence was proved by a photo – the only one out of about 100 000 – of particle tracks in a bubble chamber at the Brookhaven AGS accelerator.

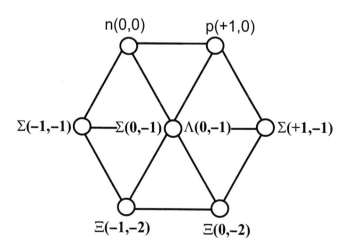

Fig. 10.14. The eightfold way: an octet of baryons, each with spin equal to $\frac{1}{2}$: n and p represent the neutron and the proton respectively. In parentheses: the first number represents the electric charge in units of the elementary charge; the second one indicates the quantum number, *strangeness*.

Quarks

Despite the success of the eightfold way in classifying baryons and mesons, physicists still strove hard to find an even simpler and more fundamental structure for all the hadrons.

It was once again Gell-Mann who showed, in a paper published in 1964, that such a simpler structure could be devised.[34] According to him, all of the hadrons then known could be considered as composed of only three even more fundamental bits of matter. (The same idea came independently to another theorist, George Zweig, also from Caltech.) The three building blocks which Gell-Mann named *quarks*, a fanciful term from a line of James Joyce's novel *Finnegans Wake*: 'Three quarks for Muster Mark!', were designated with the arbitrary labels *up*, *down* and *strange*. The new particles have the astonishing property of having a *fractional electric charge*.

If we just use the proton charge as a unit, *up* quark has a charge equal to $+\frac{2}{3}$; and both *down* quark and *strange* quark have a charge equal to $-\frac{1}{3}$. The *strange* quark is needed only to make up strange particles (the *up* and *down* quarks both have zero strangeness, whereas the *strange* quark has strangeness equal to -1; and the corresponding antiquarks have opposite charge and strangeness). Here are two simple cooking rules to make up hadrons: a *meson* is made up of one quark and one antiquark, whereas a *baryon* is made up of three quarks.

To obtain the charge or the strangeness of a particle, you have only to sum up those of the constituent quarks and antiquarks. Just two of the three fundamental quarks, *up* and *down*, suffice to explain the structure of the hadrons of all the matter which composes us and the world around us: a proton in an atomic nucleus is formed of two *up* quarks and one *down* (charge $= +\frac{2}{3} + \frac{2}{3} - \frac{1}{3} = +1$); a neutron is made up of two *down* quarks and one *up* (charge $= -\frac{1}{3} - \frac{1}{3} + \frac{2}{3} = 0$).

The quark model brought order into the jungle of hadrons, and predicted the results of interactions between them. It explained the eightfold-way pattern; it also gave a new meaning to strangeness and other quantum numbers, and to their conservation laws. In the following years an impressive body of experimental evidence arrived to support the correctness of the quark hypothesis. This led to a better understanding of the different aspects of the force that binds the quarks together inside hadrons.

Recollections

> In 1963, when I assigned the name 'quark' to the fundamental constituents of the nucleon, I had the sound first, without the spelling, which could have been 'kwork'. Then, in one of my occasional perusals of *Finnegans Wake*, by James Joyce, I came across the word 'quark' in the phrase 'Three quarks for Muster Mark.' Since 'quark' (meaning, for one thing, the cry of a gull) was clearly intended to rhyme with 'Mark', as well as 'bark' and other such words, I had to find an excuse to pronounce it as 'kwork' ... From time to time, phrases occur in the book that are partially determined by calls for drinks at the bar. I argued, therefore, that perhaps one of the multiple sources of the cry 'Three quarks for

Muster Mark' might be 'Three quarts for Mister Mark', in which case the pronunciation 'kwork' would not be totally unjustified. In any case, the number three fitted perfectly the way quarks occur in nature.

(Murray Gell-Mann)[35]

1970

Plasma and magnetism

After forty-six years Sweden received its third Nobel Prize for physics, but for France this was its seventh. The 1970 prize was in fact shared between the Swedish physicist Hannes Alfvén (Royal Institute of Technology in Stockholm), and the French Louis Néel (University of Grenoble). They had investigated two different forms of matter: Alfvén the so-called *plasma*, and Néel, magnetic materials.

Hannes Olof Alfvén (1908–95) was born in Norrköping, Sweden. His parents were both physicians. Alfvén studied physics at Uppsala University, where he obtained his doctorate in 1934. In 1940 he was appointed Professor of Theoretical Electricity at the Royal Institute of Technology in Stockholm, where he later became professor of plasma physics. After 1967 he was also associated with the University of California at San Diego.

A founder of plasma physics

When atoms in a gas are ionised (that is, split into negatively charged electrons and positive ions), the gas becomes a *plasma*, sometimes named the *fourth state of matter* (fourth in addition to the solid, liquid and gaseous states). Plasmas occur in the sun, in stars and in interstellar space; aurorae and lightning are also plasmas. The earth itself is struck by a tenuous plasma, a breeze of charged particles blowing from the sun, which is called the *solar wind*; it is also surrounded by layers of a dense plasma, which is the material that forms the ionosphere (p. 224). Plasmas are additionally produced in laboratories by heating gases to such extremely high temperatures that they become ionised; a similar process occurs in thermonuclear fusion reactors (p. 307).

Alfvén is considered a pioneer in the study of the behaviour of plasmas in magnetic fields – the science which is called *magneto-hydrodynamics*. He devised new and powerful mathematical methods to describe the motion of the charged particles composing a plasma when they are moving in electric and magnetic fields. These methods enabled physicists to interpret many plasma phenomena, both in the laboratory and in cosmic space. So Alfvén succeeded in explaining the origin of the magnetic field in the interstellar medium, the cause of sunspots and magnetic storms, the origin of the aurorae, the propagation of cosmic rays in space, and the motion of charged particles trapped in the magnetosphere. The method introduced by Alfvén in the study of plasmas became later one of the fundamental approaches to controlled thermonuclear fusion.

Louis Eugène Néel (1904–2000) was born in Lyons, France. He studied at the *Ecole Normale Supérieure* in Paris; and, in 1932, he obtained his doctorate at the University of Strasbourg. In 1937 he was appointed professor of physics at the same university, and in 1945 at the University of Grenoble, where he became director of the laboratory for electrostatics and the physics of metals, and also of the Centre for Nuclear Studies, where he developed extensive research programmes in neutron diffraction and crystal growth. A member of the French Academy of Sciences, Néel created a school that was internationally known as one of the leading research centres in the field of magnetism.

New forms of magnetism

Materials which can become permanent magnets are called *ferromagnetic* materials (iron is the most familiar example). Within a ferromagnetic material, each atom acts as a tiny bar magnet, with adjacent atoms tending to be locked together and lined up in clusters called 'domains' (each about a tenth of a millimetre across and just visible under a microscope). When these materials are not made magnetic, their domains are randomly oriented, though within each domain the tiny bar magnets are aligned along the same direction. But when an external magnetic field is applied, these tiny magnets turn to line up with the field, and the piece of material becomes a permanent magnet; it manifests itself in its capacity to attract other ferrous materials. Above a certain critical temperature, which is called the *Curie point*, however, the ferromagnetic material loses its special magnetic properties.

In papers published from 1932 to 1936 Néel described the characteristics of special materials, later on called 'antiferromagnetic' materials, but also known as *antiferromagnets*. Unlike ferromagnetic materials, in which the atomic magnetic dipoles line up like compass needles, all parallel to each other, in an antiferromagnet pairs of adjacent atomic magnets are oriented in the opposite direction. At high temperatures this new magnetic state disappears; the transition temperature is known as the *Néel temperature*.

In 1948 Néel made another fundamental discovery. He was able to explain theoretically the magnetism found in certain ceramic materials called *ferrites*. These are not ferromagnetic – in Néel's terminology they are named *ferrimagnetics*. These materials have at least two different kinds of atomic magnets, which are oriented in opposite directions. Ferrites have important technical applications, such as in computer memories and in microwave electronics.

Other important subjects approached by Néel during his long life include rock magnetism, magnetic anisotropy, magnetism of fine particles, and thin films. Néel's influential views were based on an effective mixture of sound physics and simple ideas. His works exemplify great lucidity, and at the same time, simplicity of style.

Fig. 10.15. California Institute of Technology (Caltech): Richard Feynman (after the announcement of his award of the 1965 Nobel Prize for physics) in front of Throop Hall, gazing at a banner exclaiming 'Win Big RF'. (Courtesy of the Archives, California Institute of Technology.)

Fig. 10.16. Massachusetts Institute of Technology (MIT): The Great Dome at Killian Court. (Courtesy Donna Coveney/MIT.)

Chapter 11
The small, the large – the complex

The decade of the 1970s saw exciting developments in our knowledge of the subatomic world – let us call these advances steps into the *small*. Experiments at particle accelerators uncovered a succession of unexpected phenomena: the evidence for the existence of quarks within protons and neutrons (early 1970s); the discovery in 1974 of a new particle containing a new type of quark (the 'charm quark'). And again, the discovery a short time later of a fifth member of the lepton family (the 'tau lepton'), and a fifth quark (the 'bottom quark') in 1977. Finally, experiments carried out at CERN and at Fermilab in 1973–4 produced the first proof of the validity of the electroweak theory.

Not content with fascinating advances in the world of the small, we also witnessed amazing strides forward in the context of the *large*. Since the 1960s general relativity has been experiencing a renaissance – and has in fact become one of the most active and exciting branches of physics. New experimental tests now succeed in verifying the validity of the theory up to remarkable levels of precision. Astronomical discoveries such as the cosmic microwave background, radiation pulsars, black hole candidates (this last in the early 1970s), and the first binary pulsar in 1974, all demonstrate that general relativity would have important applications in the fields of astrophysics and cosmology.

Turning now to the *complex*, we find ourselves in the field of condensed-matter physics. Physicists are able to use the quantum theory of bulk matter – which had begun to evolve around 1930, shortly after the development of quantum mechanics – to explain many phenomena such as the flow of electric current in metals, and the properties of magnetic materials, insulators, and semiconductors. Starting from the 1940s and continuing through the 1950s, scientists had been able to explain superfluid helium-4, and superconductivity in metals, and in the early 1960s, tunnelling effects in semiconductors and superconductors. These accomplishments were recognised with two Nobel laureates in the 1960s and no less than ten in the 1970s. Lastly, during the period we are considering (1971–80), two extraordinary phenomena were discovered, both of which would have far-reaching influence: they are superfluidity in helium-3 in the early 1970s, and, no less important, the *quantum Hall effect* in 1980.

1971

Holography

The 1971 Nobel Prize for physics was awarded to the Hungarian-English engineer Dennis Gabor, from the Imperial College of Science and Technology, London, for his invention of *holography*, an ingenious method of obtaining three-dimensional photographic images.

Dennis Gabor (1900–79) was born in Budapest, Hungary. After receiving his doctorate in electrical engineering in Berlin in 1927, he joined the company Siemens and Halske. In 1933 he fled Nazi Germany and went to England to work at the Thomson–Houston Company. In 1949 he moved to Imperial College in London, where he became Professor of Applied Electron Physics. Besides holography, Gabor's research included communication theory, high-speed oscilloscopes, electron devices and television. (He was awarded more than a hundred patents.)

The principle of holography
In an ordinary photograph, the image reproduces the variations in the 'intensity' of light reflected from different points of the object. It produces brighter spots where more light is reflected and darker spots where less light is reflected. This method does not exploit, however, another important characteristic of light waves, the so-called 'phase'; as a consequence, ordinary photography loses the three-dimensional geometry of the object.

Holography was developed by Gabor in the period 1947–8. (It is said that Gabor 'received a sudden vision' of the principle of holography during the Easter holiday of 1947, while waiting his turn for a game of tennis.) Unlike ordinary photography, holography reproduces not only the intensity of light waves but also their phase distribution. Gabor, in fact, called the photograph obtained with this method a *hologram* (from the Greek *holos* for 'whole' and *gramma* for 'letter'), which means in effect 'complete', since the resulting photograph contains all of the information.

In a hologram an interference pattern is stored on a photographic film when monochromatic light, scattered from an object, interferes on the film with a direct light beam produced by a laser (this second beam is called the 'reference beam'). When the hologram is then illuminated only by the reference beam, the original shape of the object is reconstructed, and the result is a truly three-dimensional image. We can find holography applied in the storage of digital data, in microscopy, in particle-size analysis, in high-speed photography, as well as in a wide variety of other technological applications.

1972

Explaining superconductivity

On 10 December 1956, John Bardeen, Walter Brattain and William Shockley received the Nobel medal and diploma from the hands of the King of Sweden, Gustav VI, for their invention of the transistor (p. 255). The chronicle of the day reports:

> Bardeen [had] brought only one of his three children to Stockholm so as not to disrupt the other two sons' studies at Harvard. King Gustav scolded [him] . . . about leaving his family behind on such an important occasion. [So Bardeen] assured the King that the next time he would bring all his children.[1]

Bardeen maintained his promise to the King. Sixteen years later, in December 1972, he went to Stockholm to receive his second Nobel Prize, once again for physics, but this time bringing all three of his children with him! The 1972 Nobel Prize was also shared between three people: Bardeen and two younger American physicists, Leon Cooper and Robert Schrieffer, 'for their jointly developed theory of superconductivity, usually called BCS theory'.

A three-man team

Let us start our chronicle from those glorious days at Bell Labs, when Bardeen, Brattain and Shockley had finally concluded their prodigious transistor adventure:

> After the invention of the transistor, the mood in the lab took a turn for the worse. Shockley resented the fact that he missed the invention . . . Relationships fell apart completely when Shockley blocked Bardeen from working on things that interested him. By 1951 Bardeen had started looking for a new job. When [a friend of his] . . . convinced the University of Illinois to make Bardeen an offer of $10,000 a year, he left Bell Labs with little regrets. In a memo to Mervin Kelly [the executive director of Bell Labs] he wrote: 'My difficulties stem from the invention of the transistor. Before that there was an excellent research atmosphere here.'[2]

So Bardeen left Bell Labs and went to the University of Illinois at Urbana-Champaign. Here he returned to his first (scientific) love: superconductivity. He had been working on this subject since the 1930s. But, as he reported in his Nobel lecture: 'It was not until 1950 . . . that I again began to become interested in superconductivity, and shortly after moved to the University of Illinois.'[3]

And now let us turn to his two companions. Schrieffer, who at the time was a research student of Bardeen's, recalled:

> In 1955, stimulated by writing a review article on the status of the theory of superconductivity, John Bardeen decided to renew the attack on the problem. He invited Leon Cooper, whose background was in elementary particle physics and

who was at that time working with C. N. Yang [Nobelist in 1957] at the Institute for Advanced Study, to join in the effort, starting in the fall of 1955. I had the good fortune to be a graduate student of Bardeen's at that time, and, having finished my graduate preliminary work, I was delighted to accept an invitation to join them.[4]

It was thus that the three-man team, a professor, a research associate and a graduate student, embarked on the study of something that had been waiting for an explanation for more than forty-five years; that is, the phenomenon of superconductivity, originally discovered in 1911 by Heike Kamerlingh Onnes (Nobel 1913).

The breakthrough came in the eighteen months between February 1956 and July 1957. The first successful step was made by Cooper, when he developed, between February and March 1956, the idea of *electron pairs* (from then on named *Cooper pairs*) – a new ordered state for the conduction electrons in a superconductor (it is these that form the 'superconducting electric current'). After he had published this important concept in 1956, the team sought to explain how these electron pairs interacted with the crystalline lattice. What was needed was a wave function (the ψ of Schrödinger), which was supposed to conduct the parade in the superconducting crystal of a troop of some thousand billion billion *Cooper pairs*, all moving in a perfectly synchronised fashion (they all behave in exactly the same way, because they are all forced to stay in the same quantum state – a situation somewhat similar to that of bosons in a *Bose–Einstein condensate*, see Nobel 2001).

The Nobel Prize awards are celebrated in December. Bardeen, in 1956, was due to receive his first Nobel Prize. But, before leaving, he urged Schrieffer to continue his work for another month. And Schrieffer in an intuitive leap was actually able to discover a manageable form for the electron-pair wave function. When Bardeen came back from Stockholm with all his honours, the team soon set to work on an intensive study. In the space of a few months they were able, by properly managing the wave function devised by Schrieffer, to work out a complete theory which explained all the experimentally established aspects of superconductivity then known. They described their findings in a paper, which was published in 1957 in the *Physical Review*.[5] It was thus that Bardeen, Cooper and Schrieffer unlocked the mysteries of superconductivity (and Schrieffer was at last able to complete his doctoral thesis!)

The BCS theory

For many years after Kamerlingh Onnes' discovery, physicists had believed that, except for the abrupt disappearance of resistance to an electric current, superconductors had the same properties as other ordinary materials. But in 1933 another basic property of metallic superconductors was discovered; that is, that they are strongly repelled by a magnetic field (physicists say that they are *diamagnetic*). This phenomenon is called the *Meissner effect*, after one of the two physicists who discovered it. Theories developed in the following years attempted to explain the observed phenomena, but they were unable to solve the riddle of the true origin of superconductivity.

At last, in 1957, Bardeen, Cooper and Schrieffer developed their theory, which satisfactorily explained what goes on inside a superconductor at a microscopic level. It was named the *BCS theory* in their honour (BCS is short for Bardeen–Cooper–Schrieffer). Let me do my best to explain this sophisticated theory in a few simple words.

In a normal metal one or more electrons per atom, the so-called conduction electrons, roam freely through the entire crystal. When an electric voltage is applied to the metal these freely moving electrons slowly drift in the direction opposite to the electric field, so that an electric current starts to flow. While flowing, the electrons are also scattered by obstacles in the crystal lattice (impurity atoms, defects or lattice vibrations). As a consequence the current decreases in intensity; a resistance is created to the flow of the current through the material. Cooper used quantum mechanics to study how the conduction electrons interact with the lattice vibrations of a superconducting material, when the latter is below its *transition temperature* (the temperature at which the material becomes a superconductor). He showed that if two such electrons have equal and opposite momentum and spin, they will tend to attract each other and bind together to form a pair (something like a diatomic molecule), despite the fact that they both have a negative electric charge, and normally repel each other.

These electron pairs move in a mutually correlated way, and a single wave function (like that invented by Schrieffer) can describe the entire assembly of pairs. This means that the pairs are prevented from becoming scattered by the lattice obstacles, so that, when a voltage is applied, they start to move and to create a current which flows with no resistance at all. And when the voltage is removed, the current continues to flow indefinitely because the pairs encounter practically no opposition. Here again, as a superconductor is warmed up, electron pairs separate into individual electrons, and the material becomes normal from the electrical point of view.

The BCS theory successfully described many other unusual features of superconductivity. It was enthusiastically received by the physics community, and acted as a powerful stimulus for both theoretical and experimental studies. The theory had a great impact on other, different fields, such as astrophysics, nuclear structure and the behaviour of superfluid helium. During the 1970s novel superconducting materials were discovered, which required new developments of the BCS theory in order to explain their properties. High-temperature superconductors discovered in the mid-1980s (see Nobel 1987) have opened new ways in the field of superconductivity; the question of the mechanism by which superconductivity is induced in such materials continues to be a hotly debated topic in solid-state physics circles.

Fig. 11.1. The Meissner effect: a magnet levitating above a superconductor. (Courtesy Lawrence Berkeley National Laboratory.)

The creators of the BCS theory

John Bardeen (see p. 257). From 1951 to 1975 Bardeen was a full professor at the University of Illinois, where he was engaged in the Department of Physics and Electrical Engineering. He served also as a member of scientific governmental committees, such as the US President's Science Advisory Committee (1959–62), and the White House Science Council (1982–3). The most celebrated solid-state theorist of his time, Bardeen received seventeen honorary degrees from different universities, and about twenty awards (among them, two Nobel Prizes, and the US National Medal of Science). He was a member of several academies, among them the US National Academy of Sciences and the Royal Society of London.

Leon N Cooper was born in New York City in 1930. He studied at Columbia University, where in 1954 he received his Ph.D. After being a member of the Institute for Advanced Study, Princeton, he became a postdoctoral research associate at the University of Illinois (1955–7). In 1958 he moved to Brown University, in Providence, where he is now the Thomas J. Watson Senior Professor of Science and the Director of the Institute for Brain and Neural Systems and the Brain Science Program. Cooper specialises in theoretical physics, including low-temperature physics, and in modelling neural networks, which are networks of nerve cells. His scientific work has been recognised with prizes and honorary degrees from many scientific institutions. He is a member of scientific academies, among them the US National Academy of Sciences and the American Academy of Arts and Sciences. Cooper is also chairman of Sention, Inc., a company specialising in memory storage, as well as a member of advisory boards of two other high-tech companies in the fields of lasers and neural networks.

John Robert Schrieffer was born in Oak Park, Illinois, in 1931. After graduation in physics at MIT, he went to the University of Illinois to work as a research

Fig. 11.2. From left to right: J. Bardeen, L. Cooper and R. Schrieffer. (Courtesy Professor Leon Cooper.)

student under Bardeen. He received his Ph.D. there in 1957, and then spent an academic year at the University of Birmingham, England, and at the Niels Bohr Institute in Copenhagen, where he continued his research work into superconductivity. In 1964 he was appointed professor of physics at the University of Pennsylvania in Philadelphia; in 1980 he moved to the University of California, Santa Barbara, where he became Director of the Institute for Theoretical Physics (1984–9). In 1992 Schrieffer transferred to the Florida State University, in Tallahassee, where he became Chief Scientist of the National High Magnetic Field Laboratory. Like his two colleagues, he is a member the US National Academy of Sciences.

1973

Quantum tunnelling

It was again solid-state physics that received the Nobel Prize in 1973. This award recognised the importance of discoveries in the fields of semiconductors and superconductors, discoveries which were made in Japan, at the Sony Corporation, Tokyo, in the USA, at the General Electric Company, and in England, at the Cavendish Laboratory, Cambridge. The prize was shared among three scientists: the Japanese physicist Leo Esaki, the Norwegian-born experimenter Ivar Giaever, and Brian Josephson, a British theorist from Cambridge University. All three made their discoveries in the late 1950s and early 1960s, while they were still graduate students.

Tunnelling in semiconductors

Leo Esaki was born in Osaka, Japan, in 1925. He studied at the University of Tokyo, and then, in 1956 went to work at the Sony Corporation. Here he carried out his pioneering research on electron tunnelling in semiconductors, which resulted in the invention of the first quantum electronic device, the so-called *tunnel* (or *Esaki*) *diode*. It was for this work that he was rewarded with the Nobel Prize. After receiving his Ph.D. in 1959 he moved to the USA on an IBM fellowship, and went to work at the IBM Thomas J. Watson Research Center in Yorktown Heights, New York. Since the late 1960s Esaki and his co-workers have pioneered designs of semiconductor quantum structures (which are called *superlattices*), thereby exploring a quite new frontier of solid-state physics.

The tunnel diode

To understand the tunnel diode invented by Esaki, we have to consider two simple concepts: the concept of the *junction diode* and that of the quantum phenomenon of *tunnelling*.

The junction diode is a semiconductor crystal with two regions, the two regions having different concentrations of positive and negative charge carriers (just as in a junction transistor, p. 256). Moving on instead to the concept of tunnelling, you will doubtless remember that it was introduced by George Gamow and others in the late 1920s to explain how alpha particles have some small chance of tunnelling through the barrier due to the forces which bind those particles in the atomic nuclei. The same concept of tunnelling applies to other particles, such as electrons in semiconductors.

In 1958 Esaki was in fact able to observe electrons tunnelling across a barrier (an electric potential difference) that was formed through a very narrow region (less than fifteen nanometres) in the junction of a germanium diode. This discovery led to the tunnel diode, which has had important applications in many fields of modern electronics, such as high-speed computers and space communication systems. Moreover, Esaki's discovery opened the way to new studies involving the tunnelling effect in several different types of materials.

Tunnelling in superconductors

Ivar Giæver was born in Bergen, Norway, in 1929. He studied at the Norwegian Institute of Technology, where in 1952 he graduated in mechanical engineering. In 1954 he emigrated to Canada, and went to work at the General Electric Company. Two years later he moved to the USA, and joined the General Electric Research and Development Center in Schenectady, New York. He then decided to study physics at the Rensselaer Polytechnic Institute in Troy, New York, where in 1964 he obtained his Ph.D. From 1958 to 1969 Giaever worked in the fields of thin films, tunnelling and superconductivity. In 1988 he left General Electric and became a professor both at the Rensselaer Polytechnic Institute, and at the University of Oslo, Norway.

Giaever's work

The BCS theory on superconductivity (p. 323) predicts, among other things, that in a superconductor none of the superconducting electrons can have energies within the range of a small forbidden gap. So that, if one tried to inject electrons into a superconductor with these forbidden energies, they would be rejected. It was this *energy gap* that Giaever tried to observe and study.

Giaever performed his experiment in 1960, while he was still a graduate student at the Rensselaer Polytechnic Institute, and was working at General Electric on thin films (material sheets with a thickness of the order of about fifty atomic layers). Using a special technique, he produced two metal layers, one made of aluminium and the other of lead, separated by a very thin insulating layer. Then he cooled down this tri-layer sandwich to less than 4 kelvins (-269 °C), using liquid helium; at this temperature the lead, but not the aluminium, became a superconductor. Next, Giaever applied an electric voltage across the junction (this acted as a barrier to the flow of electrons from the metal into the superconductor). At very low voltages almost no current was measured (electron energies were in the forbidden gap). But when the voltage reached a certain value the current began to grow (electron energies were then in the allowed range). Giaever's experiment gave direct evidence for the existence of the forbidden energy gap predicted by the BCS theory, and opened a new way to understanding superconductors and tunnelling.

The Josephson effect

Brian David Josephson was born in 1940 in Cardiff, Wales, Great Britain. He studied at the University of Cambridge, where in 1960 he obtained his bachelor's degree. He then became a graduate student at the Cavendish Laboratory, and here, in 1964, he obtained his Ph.D. (He had been elected a Fellow of Trinity College in 1962.) In 1965–6 he went to the USA as a research assistant professor at the University of Illinois. He then returned to Cambridge as assistant director of research, and in 1974 he was appointed Professor of Physics at Cambridge University. From 1970 on he has been a Fellow of the Royal Society. In the 1970s Josephson, who had always been interested in the brain, broadened his intellectual and scientific interests to include 'Eastern mysticism'. He is now directing the Mind–Matter Unification Project in the Theory of Condensed Matter Group at the Cavendish.

Josephson's discovery

In 1962 Josephson was a graduate student at the Cavendish under Brian Pippard (a leading figure in the field of condensed-matter physics, and later on seventh Cavendish Professor, in succession to Nevill Mott). Being interested in superconductivity, he began to explore whether Cooper pairs could 'tunnel' through a non-superconducting barrier from one superconductor to another. He based his studies mainly on the results obtained by Giaever; on the BCS theory, as further

Fig. 11.3. Brian Josephson. (Courtesy The Cavendish Laboratory, University of Cambridge, England.)

developed by the Russian physicist Lev Gor'kov; and on work produced by the American theorists Philip Anderson (Nobelist in 1977), Morrel Cohen and others. Josephson was able in a comparatively short time to work out a new theory, and to predict very peculiar characteristics concerning the tunnelling of electron pairs between two superconductors. (From then on this phenomenon became known as the *Josephson effect.*)

He demonstrated firstly, that when two pieces of a superconducting material separated by a very thin insulating layer are below their critical transition temperature, an electric current can flow between the two superconductors – without any electric voltage source having to be applied (no battery being in fact connected). Secondly, if a voltage is maintained across the junction, the current starts to oscillate in time at a high frequency. Josephson published his discovery in the July 1962 issue of *Physics Letters.*[6] After some perplexity his theory was fully accepted by the world of physics, and Josephson made a considerable name for himself. (The most important person who disagreed with Josephson's theory when it was published was the Nobelist John Bardeen, the grand creator of the BCS theory.)

Soon after Josephson's paper appeared, experiments started in many laboratories, in an effort to witness at first hand the new and surprising effect. One of the first experiments was performed at Bell Labs in Murray Hill, New Jersey, USA. Philip Anderson, who was a Bell Labs research scientist (p. 345), after returning home from the Cavendish (he had taught for a year there; Josephson had attended his lectures, and they had many discussions about his ideas on tunnelling in superconductors), in the autumn of 1962 collaborated in the experiment with John Rowell. The latter was a skilled experimenter at Bell Labs; they were the first people to actually 'see' the Josephson effect.

The most successful application of the Josephson effect is an ultra-sensitive device called the *SQUID*, short for *superconducting quantum interference device*. SQUIDs are capable of detecting even the weakest magnetic fields. They have been employed for measuring tiny magnetic field fluctuations, ranging from those produced by neurones in the human brain to seismic tremors of the earth. The Josephson effect was also used to determine a physical constant, e/h (e is the elementary electric charge, and h the Planck constant); it is called the *Josephson constant*, and is used to define the quantum standard of electric voltage in the modern system of measurement.

An historic debate

> After Josephson's talk, Bardeen rose to describe his theory of single-particle tunnelling, including his previously-published comment that pairing does not extend into the barrier. As Bardeen spoke, Josephson interrupted him. The exchanges went back and forth several times, with Josephson answering each criticism of his theory. The scene was quite civil, because both men were soft spoken, not given to the bluster of verbal combat, even though, as history would show, a Nobel Prize hung in the balance ... It was youth versus maturity, daring spirit versus depth of experience, and mathematics versus intuition ... After [Anderson and Rowell's] experimental confirmation [of the Josephson effect], Bardeen graciously withdrew his objections to Josephson's theory.[7]

1974

Celestial radio waves

The 1974 Nobel Prize for physics was the second to be given to astronomy and astrophysics (the first being the one awarded to Hans Bethe in 1967). The Swedish Academy of Sciences rewarded two British scientists, both from Cambridge University: Martin Ryle, and Antony Hewish. They earned the prize for their pioneering work in radio astronomy, more specifically, Ryle for having developed revolutionary radio telescopes, and Hewish for his discovery of the first *pulsar*.

Radio astronomy

In the early 1930s Karl Jansky, a young American engineer, was working at Bell Labs in Holmdel, New Jersey. While investigating the sources of static that might interfere with radio voice transmissions, he detected a very faint radio noise which he suspected could not be coming from any of the usual terrestrial sources. After making day-by-day sweeps of the sky, he discovered that strong radio signals were coming from the central area of our galaxy – the Milky Way. Jansky hence concluded that he was detecting radio signals from some astronomical source. And it was thus that *radio astronomy* was born.

During the Second World War, British radar scientists discovered that bursts of radio noise coming from the sun interfered with their radar operations. This aroused their interest in radio astronomy, so that shortly after the war they pioneered the building of large antennae to produce radio maps of the sky, and so locate cosmic objects emitting radio waves. In the early 1950s the first really large radio telescope (a seventy-five-metre dish) was built at Jodrell Bank in Cheshire, England, under the supervision of Sir Bernard Lowell.

Astronomers discovered a number of radio sources within the Milky Way, other than the one discovered by Jansky: one of these proved to be the famous *Crab Nebula*, a source located at a distance of 6300 light years (about 60 million billion kilometres) from us, in the constellation Taurus. It was the first radio source to be associated with an actual visible cosmic object. These discoveries strengthened the opinion that cosmic radio signals arose primarily from extended regions of turbulent gas.

As radio telescopes became more powerful, it seemed that radio waves were also being emitted by individual compact objects. These were puzzling objects which, though looking like individual stars, did not seem to be quite like those that had been observed up to then. Their distance was estimated to be some billion light years away, and their extraordinary luminosity appeared to be hundreds of times more intense than that of a normal galaxy. They were called *quasi-stellar* (meaning 'star-resembling') *radio sources*, shortened to *quasars*. (The first quasar was discovered in the early 1960s by the American astronomer Allan Sandage.) But let us return to the person of Martin Ryle, and his contributions to radio astronomy.

Martin Ryle (1918–84) was born in Brighton, England. He was the son of a professor of social medicine at Oxford University, and the nephew of the well-known philosopher Gilbert Ryle. He studied at Bradfield College and Oxford, where he graduated in 1939. During the Second World War he worked on radar for the Royal Air Force. He then joined the Cavendish Laboratory at Cambridge, where he began to work on radio astronomy. From 1948 on Ryle was a lecturer at Cambridge University, and he was appointed the first professor of radio astronomy in 1959. In this period he became director of the Mullard Radio Astronomy Observatory of Cambridge University, and started developing advanced radio telescope systems. He recalled:

> I think that the event which, more than anything else, led me to the search for ways of making more powerful radio telescopes, was the recognition, in 1952, that the intense source in the constellation of Cygnus was a distant galaxy – 1000 million light years away. This discovery showed that some galaxies were capable of producing radio emission about a million times more intense than that from our

own galaxy or the Andromeda nebula . . . It was not until 1958 that it could be
shown with some certainty that most of the sources were indeed powerful
extra-galactic objects, but the possibilities were so exciting even in 1952 that my
colleagues and I set about the task of designing instruments capable of extending
the observations to weaker and weaker sources, and of exploring their internal
structure.[8]

For studying distant radio sources such as quasars, Ryle developed a technique called *aperture synthesis*. He used a number of relatively small antennae whose reciprocal positions could be changed – within a distance of some kilometres – into a much larger array. He thus boosted the effective aperture of the telescope system, and greatly increased its power of resolution (the ability of the system to distinguish the fine details of the source).

Ryle was elected a Fellow of the Royal Society in 1952; he was knighted in 1966, and was appointed Astronomer Royal of Great Britain in 1972. He headed the Cambridge radio astronomy group, which in the late 1950s published one of the most famous lists of radio sources. It was entitled *Third Cambridge Catalogue*, and it listed some four hundred radio sources. (Research into the distribution in space of these radio sources reinforced acceptance of the hypothesis of an expanding universe, as opposed to the rival 'steady state' theory of a static universe.)

The first pulsar

In February 1968 Antony Hewish, the second Nobel laureate, announced, in a paper published in *Nature*, a sensational discovery: the existence of a new type of star which emitted periodic bursts of radio waves. The new object was nicknamed a *pulsar*, a shortened form for 'pulsating radio star'. Briefly, here is the fascinating story as related by Hewish himself:

In 1965 I drew up plans for a radio telescope with which I intended to carry out a
large-scale survey of more than 1000 radio galaxies . . . The final design was an
array containing 2048 dipole antennas. Later that year I was joined by a new
graduate student, Jocelyn Bell . . . The radio telescope was complete, and tested,
by July 1967 and we immediately commenced a survey of the sky.[9]

It was Miss Bell who managed the instruments, which were able to record very fine details of the time structure of the radio signals coming from the sky. Less than one month later, something strange and mysterious appeared on the recordings. It indicated bursts of radio waves from a place between the constellations Vega and Altair. Hewish thus continued:

One day around the middle of August 1967 Jocelyn showed me a record
indicating fluctuating signals . . . [We] first thought that the signals might be
electrical interference . . . [It] was not until November 28th that we obtained the
first evidence that our mysterious source was emitting regular pulses of radiation

> at intervals of just greater than one second. I could not believe that any natural source would radiate in this fashion and I immediately consulted astronomical colleagues at other observatories to enquire whether they had any equipment in operation which might possibly generate electrical interference . . .[10]

Hewish received negative answers; so he continued to pursue his prey:

> Still sceptical, I arranged a device to display accurate time marks at one second intervals . . . To my astonishment the readings fell in a regular pattern, to within the observational uncertainty of 0.1 [of a second], showing that the pulsed source kept time to better than 1 part in [a million, and that] the duration of each pulse . . . was approximately 16 [thousandths of a second]. Having found no satisfactory terrestrial explanation for the pulses we now began to believe that they could only be generated by some source far beyond the solar system . . . The months that followed the announcement of our discovery were busy ones for observers and theoreticians alike, as radio telescopes all over the world turned towards the first pulsars and information flooded in at a phenomenal rate.[11]

Antony Hewish was born in Fowey, Cornwall, England, in 1924. He studied at Cambridge University, and from 1943 to 1946 he was engaged in war service, working on radar with Martin Ryle. After the war he returned to Cambridge, where he graduated in 1948, and joined Ryle's group. After obtaining his Ph.D. in 1952, Hewish became a research fellow, and in 1971 he was appointed Professor of Radio Astronomy at Cambridge University. In 1968 he was elected Fellow of the Royal Society, and from 1982 to 1988 he was director of the Mullard Radio Astronomy Observatory.

Fig. 11.4. A. Hewish (left) and M. Ryle. (Courtesy AIP Emilio Segrè Visual Archives, Physics Today Collection.)

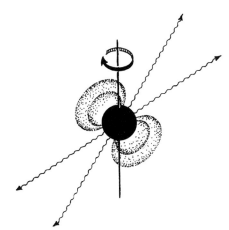

Fig. 11.5. A schematic view of a pulsar. The beam of radio waves sweeps around the pulsar's rotational axis, in the same way as the beam from a lighthouse, so astronomers receive only radio pulses as the pulsar's radiation beam sweeps over the earth.

Pulsars

Pulsars are rapidly spinning *neutron stars*, very dense stars composed almost entirely of neutrons, which are left behind by *supernovae* – explosions occurring at the end of the lives of certain stars. Their spinning periods vary from a few seconds to thousandths of a second.

Neutron stars have diameters of some 20 kilometres, and masses of more than 1.4 times the mass of the sun; as a consequence they are extremely dense. Astrophysicists have calculated the upper limit to their mass; it worked out as between two and three solar masses. A neutron star also has an enormously intense magnetic field around it (millions of times stronger than any magnetic field produced in our laboratories), and a large induced electric field. Charged particles moving at nearly the speed of light stream out of the pulsar's magnetic poles; they are then accelerated along the lines of force of the magnetic field, and produce powerful beams of radiation. (Since the news of the first pulsar, astronomers have discovered over a hundred others, and estimate that there must be at least a million active pulsars in our galaxy.)

1975

Nuclear structure

The Swedish Academy of Sciences awarded the 1975 Nobel Prize for physics to three scientists for their studies regarding the structure of atomic nuclei. They were two theoreticians working in Denmark, Aage Bohr (Niels Bohr's son) and the American-born Ben Mottelson, and an American physicist, James Rainwater. They had developed, in the 1950s, a theory which incorporated aspects of both the nuclear shell model developed by Maria Goeppert Mayer and Hans Jensen

Fig. 11.6. The Crab Nebula. (Courtesy European Southern Observatory.) It is the remnant of a supernova: in its centre there is a pulsar, which was discovered in late 1968; it emits radio waves, light, and also X-ray and gamma-ray pulses. In the chronicles of the Chinese Sung Dynasty, the imperial astronomer Yang Wie-T'e thus described the unusual celestial event which occurred in 1054:

> Prostrating myself, I have observed the appearance of a guest star; on the star there was a slightly iridescent yellow colour. Respectfully, according to the dispositions for Emperors, I have prognosticated, and the result said: The guest star does not infringe upon Aldebaran; this shows that a Plentiful One is Lord, and that the country has a Great Worth. I request that this be given to the Bureau of Historiography to be preserved.[12]

(see Nobel 1963), and the liquid-drop model (this last having been proposed by Niels Bohr in 1936, p. 208).

The nuclear collective model

Goeppert Mayer and Jensen's shell model best explained the properties of those atomic nuclei with a number of protons or neutrons close to the magic numbers (p. 289). As more experimental data became available, the nucleus revealed unexpected facets which could not be explained by the old theories. First of all, its shape. The fine details on the electric charge distribution observed in certain nuclei indicated that they were not spherical (the most extreme case was that of the so-called rare earths). As spheroids, they seemed to be deformed; they were lemon-shaped, in fact, and this had certainly not been predicted by the nuclear shell model. At the time, no one could explain this phenomenon.

But a solution did appear in the summer of 1950, when James Rainwater, then at Columbia University, came out with a new idea that he then published in the *Physical Review*. He pointed out that the motion of individual nucleons, placed in the outside shells, exerted an internal pressure on the surface of a nucleus, and led to its distorted shape. This paper constituted a crucial element in Rainwater's Nobel award, and was responsible for further developments later on in the field.

It was precisely at this time that Aage Bohr came on to the scene. He was visiting Columbia on a one-year research fellowship, and he shared an office with Rainwater at the Pupin Physics Laboratory. They often had discussions on the subject, so much so that Bohr was stimulated to work out a new theory. He thus recalls how he was drawn into the matter:

> James Rainwater had been thinking about the origin of the [nuclear asymmetric shape] . . . and conceived an idea that was to play a crucial role in the following development . . . On my return to Copenhagen in the autumn of 1950, I took up the problem of incorporating the . . . [idea] suggested by Rainwater into a consistent dynamical system describing the motion of a particle in a deformable core.[13]

Aage Bohr published an initial paper in which he proposed a new theory concerning the structure of the nucleus. In his model there is an interplay between two basic motions: the motion of individual nucleons and what was named the *collective motion* (that is, the motion of the nucleus as a whole). It is this latter motion which can distort the shape of the nucleus. (Speaking through an analogy which has often been used, let us think of a swarm of bees: 'the motion of each bee seems rapid and erratic, but the swarm will move slowly as a unit'.)

The development of these ideas and the comparison of the predictions with actual experiments was made by Bohr together with Ben Mottelson, a young American physicist, then working at Niels Bohr's institute in Copenhagen on a fellowship from Harvard. ('Soon, I was joined by Ben Mottelson in pursuing the consequences of the interplay of individual-particle and collective motion for the great variety of nuclear phenomena that was then coming within the range of experimental studies.'[14]) They wrote three papers, published in the years 1952–3, in which they presented a formidable collection of experimental data supporting their *nuclear collective model*.

The conclusion of their analysis was that: 'One is . . . led to describe the nucleus as a shell structure capable of performing oscillations in shape and size.'; that is, they had unified the shell model and the liquid-drop model. A remarkable discovery made by Bohr and Mottelson was that the rotational motion of a deformed nucleus, which is governed by the laws of quantum mechanics, gives energy levels

Fig. 11.7. A. Bohr (right) with his father Niels. (Courtesy Niels Bohr Archive, Copenhagen.)

corresponding to the rotational excited states of the nucleus. All these results were in perfect accord with experiment.

Aage Niels Bohr was born in Copenhagen in 1922, the fourth son of Niels and Margrethe Bohr. In his Nobel autobiography he wrote:

> I began studying physics at the University of Copenhagen in 1940 . . . In October 1943, my father had to flee Denmark to avoid arrest by the Nazis, . . . and I followed after him . . . [We] travelled together spending extensive periods in London, Washington, and Los Alamos. I was acting as his assistant and secretary and had the opportunity daily to share in his work and thoughts.[15]

On his return to Copenhagen Aage Bohr resumed his studies and in 1954 he received his doctorate from that university. In 1948 he was a member of the Institute for Advanced Study in Princeton, USA, and during the period 1949–50 he was associated with Columbia University, New York City. In 1956 Aage Bohr was appointed Professor of Physics at the University of Copenhagen. After the death of his father in 1962 he became, until 1970, director of the Niels Bohr Institute. In 1975 he was appointed head of the Nordita (Nordic Institute for Theoretical Atomic Physics), which had been founded in 1957 by his father.

> During my early childhood, my parents lived at the Institute for Theoretical Physics (now the Niels Bohr Institute), and the remarkable generation of scientists who came to join my father in his work became, for us children, Uncle Kramers, Uncle Klein, Uncle Nishina, Uncle Heisenberg, Uncle Pauli, etc. When I was about ten years old, my parents moved to the mansion of Carlsberg, where they were hosts for widening circles of scholars, artists, and persons in public life.
>
> (Aage Bohr)[16]

Ben Roy Mottelson was born in Chicago in 1926. He studied at Purdue University and at Harvard, where he obtained his Ph.D. in 1950 with a thesis under Julian Schwinger (Nobelist in 1965). Between 1950 and 1953 he worked at the Niels Bohr Institute, 'where so much of modern physics had been created and where there were such special traditions for international co-operation.'[17] And from 1953 to 1957 he worked in the CERN study group that was formed in Copenhagen. In 1957 he was appointed Professor of Physics at the Nordita.

Leo James Rainwater (1917–86) was born in Council, Idaho, USA. He graduated in physics at Caltech in 1939, and then began graduate study at Columbia University, under Isidor Rabi (Nobel 1944) and Enrico Fermi (Nobel 1938). During the Second World War he worked on the Manhattan Project. In 1946 he received his Ph.D., and then continued at Columbia, first as an instructor, and from 1952 on as a full professor of physics. His experimental researches included neutron spectroscopy, pion interactions and atomic X-ray studies.

1976

Charmed particles

The 1976 Nobel Prize for physics was shared equally between two American experimentalists, both working in the field of high-energy particle physics: Burton Richter, from Stanford University, and Samuel Ting, from MIT. They had discovered, independently (Richter at the SLAC electron–positron collider, and Ting at the Brookhaven AGS accelerator) a completely new class of particles. These had properties which required a fourth quark (besides the three proposed by Gell-Mann, p. 314); this was thought of as providing the fourth building block of hadrons. The first of the new particles that they had discovered was then named 'J/ψ', and the new quark was called *charmed quark*.

The SLAC and BNL experiments
On the West Coast

Richter's team was made up of physicists from SLAC and from the Lawrence Berkeley Laboratory. They used the three-kilometre-long accelerator at SLAC (Stanford) to generate high-energy electrons and positrons, which were further injected into a collider (about eighty metres in diameter), named SPEAR (Stanford Positron–Electron Asymmetric Ring). In the magnetic fields of SPEAR the electrons (negatively charged) and the positrons (positively charged) circulated in opposite directions. The two beams collided at a point along the ring, with the result that electrons and positrons at this point annihilated each other. When an annihilation occurred, all the energy of the system – electron plus positron (about 8 billion electron-volts) was made available for the creation of new states of matter. Detectors surrounded the interaction point to record and analyse the products of the interactions.

Richter's team started the experiment by increasing the energy of the colliding beams in small steps. Over the weekend of 9 and 10 November 1974 they observed a sudden peak in the number of annihilations that were giving rise to new particles. This 'resonance' (as the peak is technically named) took place at a particular value of the beam energy of 3.1 billion electron-volts (an equivalent mass about three times the mass of a proton). When the energy was increased further, the number of detected annihilations dropped off again (see Fig. 11.8). It was thus that Richter's team discovered a new class of particles: the first member was christened the ψ-*particle* (ten days later they found a second member of the ψ family).

On the East Coast

At that time, at the Brookhaven National Laboratory (BNL), Ting's team (whose members were from the MIT and the BNL itself) were performing an experiment at the AGS accelerator. In August they had had evidence of an another unexpected peak, and Ting had checked and rechecked until November in order to have definite proof of the new effect.

The Brookhaven experiment was quite different. They used high-energy protons (with an energy of about 29 billion electron-volts) supplied by the AGS, and these they fired off to strike a fixed target of beryllium. They looked for heavy hadrons, which would break up into electron–positron pairs (the inverse of the SLAC process); in the case of a resonance the number of detected pairs would show a sharp peak at some value of the energy corresponding to the mass of the parent particle.

Ting's experiment presented one great difficulty, which was that of picking out a significant event from a forest of the millions of non-significant particles that were produced in each interaction. It really reminded one of the proverbial 'hunting for a needle in a haystack'! To reduce the *background* (the technical word for the 'haystack' in our dictum) of unwanted particles, the apparatus was shielded by tons of shielding material. In time it became clear that a new resonance was being formed in the collisions, at exactly the same energy as in the SLAC experiment. Ting's group christened the new-born resonance particle '*J*' which is a letter similar to the Chinese character for his name 'Ting'.

11 November

It was on this day that Ting and Richter met at SLAC, in the office of Wolfgang Panofsky, the director of the laboratory; imagine their astonishment when what each had discovered separately proved to be the identical particle! An announcement appeared at once, and their two papers followed within a week in the same issue of the *Physical Review Letters*.[18] A few days later, the results of the discovery were confirmed at the ADONE collider in Frascati, Italy;[19] and later on

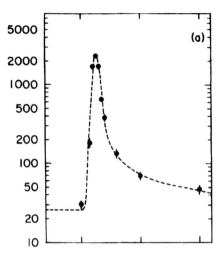

Fig. 11.8. The resonance peak found in Richter's diagram: it represents the proof of the existence of the charm particle 'ψ'. (Source: J.-E. Augustin et al., *Physical Review Letters*, 33, 1974, p. 1407. © 1974 by the American Physical Society.)

still at the electron–positron storage ring at DESY (Deutsches Electronen Synchrotron) in Hamburg, Germany. (Within the community of physicists the SLAC-BNL discovery is called the 'November revolution'!)

After Richter and Ting's discoveries were announced, theorists started to ask themselves what the J/ψ particle might be. The leading candidate appeared to be a hadron made up of a new massive quark bound to the corresponding antiquark by the same strong force which binds three quarks into the proton and the neutron.

The theory that explained the new quark had been proposed by a number of theorists some years earlier. In order to construct a parallel to the four known leptons, they had added, to the three Gell–Mann original quarks (up, down and strange), a fourth quark, which needed a new characteristic (a new quantum number) in order to distinguish it from the known quarks. The theorist Sheldon Glashow (Nobelist in 1979) and his colleague B. J. Bjorken had by 1964 already given the new quantum number the name *charm*, and the new quark had naturally been called the *charmed quark*. Finally, in 1970, the charm quark was given an important role in particle physics through the work of Glashow himself and two of his co-workers, the Greek John Iliopoulos and the Italian Luciano Maiani (all three men were then working at Harvard University).

The two team leaders

Burton Richter was born in 1931 in Brooklyn, New York City. He studied physics at MIT, where he graduated in 1952, and obtained his Ph.D. in 1956. He then became a research associate at Stanford University. Here, at the High Energy Physics Laboratory, he began, together with other physicists, the construction of a new kind of accelerator, called an *electron storage ring*. In this machine two beams of electrons circled in opposite directions in two 'storage rings' which had a cross-section in common; electrons collided head-on with each other here. In

1965, using this machine to study electron–electron scattering, he established that QED was correct to distances less than one-million-billionth of a metre. In 1972, in collaboration with John Rees, Richter completed (at SLAC) the construction of the electron–positron collider, SPEAR. In 1967 he became full professor of physics at Stanford University, and from 1984 to 1999 he served as director of SLAC. In the 1980s and 1990s Richter headed a project for the construction of an electron–positron linear collider.

Samuel Ting was born in Ann Arbor, Michigan, USA, in 1936, where his father, a Chinese professor of engineering, was studying at the University of Michigan. His early adolescence was spent in Taiwan, and when he was twenty years old he returned to the USA to study at the same university as his father. In 1959 he graduated in both mathematics and physics, and he obtained his Ph.D. in 1962, writing his thesis under Martin Perl (Nobelist in 1995). The following year Ting went to CERN, Geneva, as a postdoctoral fellow, where he worked at the Proton Synchrotron with the Italian physicist Giuseppe Cocconi. He returned to the USA in 1965 and went to teach at Columbia University; one year later he went to DESY in Hamburg, Germany, to perform experiments on quantum electrodynamics. In 1967 he went back to the USA again, and joined MIT, where he was appointed full professor of physics in 1969. Ting is now leading a group engaged in carrying out an experiment on a space station; the experiment is called the Alpha Magnetic Spectrometer (AMS), and it has been designed in an effort to search for dark matter, missing matter and antimatter in the cosmos.

1977

Highlights in condensed matter

In the year 1977 condensed matter came under review for the eleventh time by the Nobel Physics Committee. This resulted in a total of eighteen prizewinners having been elected in this field since the beginning of the Nobel era, so highlighting how very important the science of solids and liquids had become. The prizewinners in 1977 were three theorists: two of them were Americans, Philip Anderson, from Bell Labs, and John Van Vleck, from Harvard University, while the third scientist was Sir Nevill Mott from the Cavendish Laboratory, England.

In awarding the prize, the Swedish Academy of Sciences emphasised the fundamental contributions of the three to the specific fields of magnetic and disordered materials. But the most noteworthy aspect of the prize, which differentiated it with respect to all preceding ones, is that it was awarded not so much for one specific discovery, as for the leading role that the three prizewinners had exerted, over the years, in the field of condensed-matter physics. Van Vleck and Mott were among the brilliant young physicists who in the mid-1920s had entered the then new-born field of quantum mechanics. They succeeded in laying the foundations

Fig. 11.9. Left: S. Ting is showing his diagram with the famous peak which proved the existence of the charmed particle 'J'. (Courtesy Brookhaven National Laboratory.) Right: B. Richter (seated on the right), in conversation with his co-workers at SLAC. (Courtesy Stanford Linear Accelerator.)

for the understanding of the properties of matter in bulk, and they soon became pioneers of the innovative science of *solid state*.

Anderson, a generation younger, has provided, during the last half century, many of the most influential theoretical ideas underlying the modern science of condensed matter. He is considered one of the founders of what is called *many-body theory* – the use of advanced quantum techniques to solve condensed-matter problems.

Let us commence by taking a closer look at our three Nobelists, and at their principal contributions.

John Van Vleck

John Hasbrouck Van Vleck (1899–1980), a tenth-generation American, was born in Middletown, Connecticut, USA, where his father was a celebrated university professor of mathematics. He studied at the University of Wisconsin in Madison, where he graduated in 1920. Later he went to Harvard, where in 1922 he obtained a Ph.D. in theoretical physics (the first Ph.D. on quantum theory awarded in the USA). One year later he went to teach at the University of Minnesota, Minneapolis, where he remained for the following five years. During this period Van Vleck worked on the application to atomic spectroscopy of the old Bohr–Sommerfeld quantum theory, and in 1926 he published his first major work, a very successful book on this subject. These were also the years of quantum revolution. Van Vleck soon realised its significance, and rapidly managing to master the radical new concepts, he became one of the first and most distinguished of American quantum theorists.

Father of modern magnetism

In 1928 Van Vleck moved to the University of Wisconsin as a professor of theoretical physics, where he remained for the next six years. Here he continued his research work on magnetism, which he had started at Madison; he used the new quantum mechanics and developed a microscopic theory of magnetism. In 1932 he published a remarkable book entitled *The Theory of Electric and Magnetic Susceptibilities*, which is considered the first treatise to describe the application of quantum mechanics to solid-state physics, and has come to be regarded as the 'bible' of modern magnetism by generations of students.

Magnetism was Van Vleck's main area of research for about fifty years; it earned him an international reputation as an authority in the field, and the title of 'father of modern magnetism'. He recalled in his Nobel autobiography: 'I was . . . lucky in choosing the theory of magnetism as my principal research interest, as this is a field which has continued to be of interest over the years, with new ramifications continuing to make their appearance (magnetic resonance . . . microwave devices, etc). So often a particular field loses general interest after a span of time. My last paper dealing with magnetism was published fifty years after my first one.'[20]

Return to Harvard

In 1934 Van Vleck returned to Harvard, where he remained until his retirement in 1969. He served as Director of the Physics Department there (1945–9), and first dean of the Harvard School of Applied Science (1951–7). From 1951 on he was the Hollis Professor of Mathematical and Natural Philosophy there.

Among Van Vleck's major achievements, cited by the Swedish Academy of Sciences, we should mention his theory on the behaviour of an atom or ion in a crystal or in a cluster in solution – the so-called *ligand field theory*. Very briefly, supposing you imagine the introduction of a foreign atom or an ion into a crystal, you will find that its electrons feel the electric field from the neighbouring atoms or ions clustering about it (technically these are called *ligands*). Van Vleck used quantum mechanics to describe how the energy levels of the system are modified by the *ligand field*, and, as a consequence, how magnetic, electrical and optical properties of the system itself are subject to changes. This theory was to be widely used in the chemistry of magnetic ions, with applications to molecular biology and medicine (an example is the study of the chemical behaviour of iron in blood). The theory was used, too, in the field of solid-state spectroscopy, with applications in solid-state lasers and semiconductor devices. Another important contribution of Van Vleck, singled out by the Academy, was his emphasis on the importance of what he called *electron correlation*; that is, the interaction between the motions of electrons in atoms or ions, which may lead to the formation in metals of local magnetic dipoles (these will act as 'minimagnets').

Magnetism, though the most important, was by no means the only topic covered by Van Vleck's research work. Besides the related subject of ligand fields, his interests also included certain electric properties of insulators, and molecular spectra. His theoretical studies on molecular spectral lines influenced the work of chemists and astrophysicists, as well as that of solid-state physicists. Van Vleck's long and prestigious career was marked by many honours and awards, among them the US National Medal of Science; and memberships of the US National Academy of Sciences and the Royal Society of London.

Nevill Mott

Nevill Francis Mott (1905–96) was born in Leeds, England. His parents had worked under J. J. Thomson at the Cavendish Laboratory, and one of his great-grandfathers was Sir John Richardson, the famous Arctic explorer. Mott studied at Cambridge (St John's College), where he received his bachelor's degree in 1927. During the following year he went to Copenhagen to work under Niels Bohr; then he went to Göttingen under Max Born, and to Manchester with Lawrence Bragg. On his return to Cambridge he obtained his master's degree in 1930, and remained at the Cavendish Laboratory for three years. During this period he developed a quantum theory of atomic and nuclear collisions, and wrote a highly successful book on the subject.

Founder of the Bristol School

In 1933 Mott became Professor of Theoretical Physics at the University of Bristol, where he found an active team working in the field of solid state. He recalled: 'I was fascinated to learn that quantum mechanics could be applied to problems of such practical importance as metallic alloys and it was this as much as anything else that turned my interest to the problem of electrons in solids.'[21] It was thus that Mott began to work on the properties of metals, alloys and semiconductors. Under his leadership, Bristol saw the flourishing of the first and most renowned school of solid-state physics in Britain. (In those years Bristol attracted many outstanding physicists who had fled from Nazi Germany, including Hans Bethe, Nobel 1967.)

During his Bristol period, Mott and his co-workers worked on many subjects: ferromagnetism, transition-metal physics, rectification of currents (how it is that the electric current through contacts between solids passes much more easily in one direction than in the other), to mention only a few. He also collaborated with industry, and developed a theoretical model of the photographic latent image, which had great influence on research into photographic processes.

During the Second World War he carried out military research in London. After the war he returned to Bristol, where he became head of the physics department. Towards the end of the 1940s he investigated the electrical behaviour of certain crystalline materials when they change from a metallic to an insulating state. He explained this alteration by taking into account the detailed interactions between electrons. He showed that when the distances between atoms of the material in a metallic state increase over a certain critical value, a form of transition from the metallic to an insulating state occurs: this phenomenon is known as the *Mott transition*. (Mott used this idea to explain why nickel oxide, which, according to the quantum energy-band theory of crystals, ought to be a conductor, is actually an insulator.)

Cavendish Professor

In 1954 Mott returned to Cambridge. He had been appointed sixth Cavendish Professor of Physics and director of the Cavendish Laboratory, as successor to Sir Lawrence Bragg. He held this post until 1971 when he retired.

In Cambridge he continued his research study, which he had started at Bristol soon after the end of the war, on dislocations (imperfections in the lattice structure of crystals), and he also worked on the mechanical strengths of metals. But he soon turned to studying the electronic properties of disordered materials, such as glasses, alloys and semiconductors containing impurities. He contributed greatly to the understanding of their electrical behaviour, introducing new fundamental concepts. These include the concept of *impurity conduction* (by adding impurities in a semiconductor, a transition can occur, whereby the semiconductor is transformed into a fully conducting state), and the concepts of *minimum metallic conductivity* and *electron hopping*. (These concepts were to play a central role in all quantum transport phenomena.) Let us try to grasp something of their meaning.

Take an amorphous material (a non-crystalline system, whose atoms are put together not in a regular lattice, but in a disordered way), where the density of conduction electrons is sufficient to make it behave as a conductor. Now, if you manage in some way to lower this density, its conductivity (a measure of the ability of an electric current to flow through the material) falls till it reaches a value which Mott called the 'minimum metallic conductivity': this represents the smallest value at which the material can still behave as a conductor. This quantity is proportional to e^2/h, where e is the elementary charge and h the Planck constant, and plays an important role in the quantum Hall effect (see Nobel Prizes 1985 and 1998).

If the density is lowered still further, a transition occurs, which Mott called an *Anderson transition*; the electrons can no longer roam through the whole material, but they become 'localised' in space (this is what the concept of *electron localisation* developed by Philip Anderson is all about, p. 347). These electrons can only 'hop' from one place to another. When such a transition occurs the material changes its electrical behaviour: its conductivity falls below the minimum metallic conductivity value, and tends towards zero as the temperature decreases (this is a typical characteristic of insulators).

Fig. 11.10. N. Mott (left) and J. Van Vleck. (Courtesy AIP Emilio Segrè Visual Archives, Ramsey & Muspratt (Mott), Weber Collection (Van Vleck).) In November of this same year (1977), Pierre-Gilles de Gennes (Nobel 1991) wrote an article for the French Magazine *La Recherche*[22] placing the three prizewinners in two different categories: these were 'tennis players' and 'golf players'. Mott and Anderson, because they lived continuously exchanging ideas with colleagues and pupils, were tennis players; whereas Van Vleck was a golf player, as, persistent and alone, he played the ball from hole to hole!

It was thus that metal–insulator transitions became a favourite subject in research, and the studies of Mott and Anderson greatly contributed to our understanding of amorphous materials, which began to be used in the many technological applications to be found in electronic devices.

Mott habitually summarized his research in any given period of his scientific life by writing a book (altogether he wrote a total of 13 books). He was elected Master of Gonville and Caius College, Cambridge, in 1959, and was knighted in 1962. In addition, he was a member of several societies, including the Royal Society of London and the US National Academy of Sciences.

Philip Anderson

Philip Warren Anderson was born in 1923 in Indianapolis, Indiana, USA, the son of a university professor of plant pathology. During the wartime years 1943–5 he was sent to the Naval Research Laboratory in Washington, DC, where he worked on the design of antennae. He then went on to study at Harvard University and obtained his Ph.D. in 1949, writing his dissertation under Van Vleck. From 1949 to 1984 he was a research scientist at Bell Labs. In the periods 1961–2 and

1967–75 Anderson was a tenured visiting professor at the Cavendish Laboratory, Cambridge. When he returned to the USA, he continued his work at Bell Labs with the role of consulting director at the physical research division, while also teaching at Princeton University. Here, in 1978, he was promoted to the Joseph Henry Chair of Physics, a professorship that in 1997 became emeritus.

Anderson's research in the field of condensed matter has covered a vast number of subjects. He has made fundamental contributions to the theoretical understanding of the electrical properties of crystalline and amorphous materials, ferromagnetism and antiferromagnetism, magnetic resonance, superconductivity and superfluidity. His work, undertaken in an industrial laboratory (Bell Labs), has made important technical developments possible, and has created an exceptional atmosphere for a large group of experimenters. Let us pick out some of Anderson's outstanding contributions, especially those cited by the Swedish Academy of Sciences.

Theory of magnetism

In 1959 Anderson developed a theory to explain the coupling (called *super-exchange*) of the magnetic moments of two atoms in an insulating magnetic crystal, through their interaction with a non-magnetic atom occupying a position between them. He showed how two magnetic moments can acquire opposite orientations, so explaining the properties of many antiferromagnetic materials. (The phenomenon of antiferromagnetism was discovered in the 1930s by Louis Néel, Nobelist in 1970, but it was only thanks to Anderson's work that it came to be understood in greater detail.)

Two years later Anderson devised a quantum model, widely known as the *Anderson model*, to describe the behaviour of impurity atoms with a magnetic moment in non-magnetic metals. Building on Van Vleck's original work on electron correlation, and on his own work on super-exchange, he showed how these atoms behave in the metal, and when they can form local magnetic moments (or 'minimagnets'). This rendered possible a microscopic explanation of certain magnetic properties of metals (for example, these magnetic moments can occur when iron atoms are dissolved in copper, which in pure form is a non-magnetic metal). Thanks to his model, Anderson contributed to a clarification of one of the most subtle effects regarding magnetism, the so-called *Kondo effect* (an anomalous minimum value of the electrical resistance in diluted magnetic alloys at low temperatures).

Disordered materials

Anderson has greatly advanced our understanding of the electronic properties of disordered materials. In 1958 he wrote a seminal paper regarding the propagation of electron waves in a disordered medium. He was able to show that, if the atoms are put together in a high disorder, these electron waves are confined to precise locations in space, so that the electrons themselves remain localised in small areas; they are never free to roam through the entire volume of the material, as they do in a crystalline metal. This phenomenon has taken on the name of *Anderson localisation*, and has become fundamental to our understanding of the physical nature of disordered materials (as we have seen, it is a cornerstone of some of Mott's metal–insulator transitions). In the late 1970s, Anderson and his co-workers further developed the concept of localisation in a more complete theory. (This concept shows up in many physical phenomena, for example it is an important ingredient in the theory of the *fractional quantum Hall effect*, see Nobel 1998.)

Anderson's interests also focused on low-temperature properties of glasses, and on the so-called *spin glasses*. These are disordered magnetic materials, with atomic magnetic moments fixed in random orientations. (To explain spin glasses physicists have developed a new form of statistical mechanics, which has come to be applied in other fields of science, such as computer science and neural networks.)

Superconductivity, superfluidity

In the 1960s Anderson concentrated particularly on superconductivity and superfluidity, working with his young pupil, the French Pierre Morel. (They pointed out, among other matters, the nature of possible superfluid states of helium-3, which were actually discovered in 1971, see Nobel 1996.) In 1962 Anderson and his colleague, John Rowell, carried out at Bell Labs a pioneering experiment that demonstrated for the first time the Josephson effect (see p. 328). After the discovery of high-temperature superconductivity in 1986, Anderson threw himself into the study of the new phenomenon, and contributed a great many interesting ideas.

Anderson was among the first to realise the significance of *broken symmetry*, a fundamental concept which is applicable in many fields, ranging from condensed-matter physics, and successively on to particle physics. Such symmetry breaking occurs, for example, in a ferromagnet: at high temperatures the atomic magnetic moments point in any direction, and the physical laws governing them are perfectly symmetrical with regard to directions in space. Below a specific temperature all the magnetic moments point in some specific direction, so breaking the symmetry among a range of different directions. (Incidentally, the concept of 'change of symmetry', had already been introduced in the late 1930s by Lev Landau in his theory of phase transitions.)

Emergence versus reductionism

In 1993 Silvan S. Schweber wrote in *Physics Today*:

> Traditionally, physics has been highly reductionist, analysing nature in terms of smaller and smaller building blocks and revealing underlying, unifying fundamental laws . . . Now, however, the reductionist approach . . . is being superseded by the investigation of emergent phenomena, the study of the

Fig. 11.11. Philip Anderson. (Lucent Technologies' Bell Laboratories, courtesy AIP Emilio Segrè Visual Archives, *Physics Today* Collection.) As well as the Nobel Prize, Anderson received the US National Medal of Science, and memberships in the Royal Society of London, the *Accademia dei Lincei* (Rome), and the Japanese Academy of Sciences. Since 1986 he has been involved with the Santa Fe Institute, an interdisciplinary institution dedicated to the sciences of complexity.

> properties of complexes whose 'elementary' constituents and their interactions are known . . . These conceptual developments in fundamental physics have revealed a hierarchical structure of the physical world. Each layer of the hierarchy is successfully represented while remaining largely decoupled from other layers.[23]

Reductionism, the philosophical position held by the majority of particle physicists (a believer in the importance of reductionism is Steven Weinberg, Nobelist in 1979), has been strongly challenged, for more than thirty years, by Philip Anderson, whose philosophical credo is instead, 'emergence at every level'. In an article entitled 'More is different', published in 1972 in *Science*, Anderson wrote:

> The ability to reduce everything to simple fundamental laws does not imply the ability to start from those laws and reconstruct the universe. In fact, the more the elementary particle physicists tell us about the nature of the fundamental laws, the less relevance they seem to have to the very real problems of the rest of science . . . The constructionist hypothesis breaks down when confronted with the twin difficulties of scale and complexity. The behaviour of large and complex aggregates of elementary particles, it turns out, is not to be understood in terms

of a simple extrapolation of the properties of a few particles. Instead, at each level of complexity entirely new properties appear, and the understanding of the new behaviours requires research which I think is as fundamental in its nature as any other.[24]

1978

Russian and American Nobelists

The 1978 Nobel Prize for physics was shared in two equal parts. One half went to the Russian academician Peter Kapitza. The other half was to be shared equally by the radio astronomers Arno Penzias and Robert Wilson of Bell Labs in Holmdel, New Jersey, USA.

Kapitza was rewarded for his inventions and discoveries which, according to the Swedish Academy, 'have been basic to the modern expansion of the science of low-temperature physics'. Penzias and Wilson worked in a totally unrelated field, and won the prize for their discovery of the *cosmic microwave background radiation* (p. 296).

Peter Leonidovich Kapitza (1894–1984) was born in Kronstadt, near St Petersburg, Russia, the son of a general in the Czar's engineer corps. He studied at the Petrograd (St Petersburg) Polytechnical Institute, where he graduated in 1918 and subsequently lectured there for three years. In 1921 Kapitza went to the Cavendish Laboratory to work with Ernest Rutherford. At Cambridge he developed magnets which produced very strong magnetic fields (they were not surpassed for more than thirty years). He became a Fellow both of Trinity College, and of the Royal Society (the first foreigner to become a Fellow of the Society in two hundred years!). He then became Messel Professor of the Royal Society while at Cambridge University, and director of the Mond Laboratory (1930–4). In the early 1930s he turned his attention to low-temperature physics, and developed a new and original method for liquefying helium.

In 1934, during a visit to the Soviet Union, Kapitza was forced to remain there. He then became director of the Institute for Physical Problems in Moscow, where he continued his research on strong magnetic fields and low-temperature physics. He carried out experiments that led to the discovery of superfluidity in helium-4, and late in the 1940s he turned his attention to microwave techniques and plasma physics. Kapitza received many honorary degrees. Among other distinctions, he was a member both of the Soviet Union Academy of Sciences and of the US National Academy of Sciences.

The discovery of superfluid helium

In 1935, the Dutch physicist Willem Keesom, and his daughter Annie, were carrying out experiments at the famous Kamerlingh Onnes Laboratory in Leiden:

Fig. 11.12. James Chadwick (right) at his wedding, accompanied by his best man, Kapitza. (Courtesy the Cavendish Laboratory, University of Cambridge, England.)

Kapitza arrived in England at age twenty-seven – thin, unhappy, unknown, looking like 'a tragic Russian prince' . . . [He] began to put down roots in Cambridge . . . and made some deep, enduring friendships . . . Lady Cockcroft related . . . her first impression of Kapitza [in these words]: 'A wild kind of character, untidy, . . . bursting with energy, his words tumbling out . . . He drove a high-powered Lagonda, the sporting car of the day." '[25]

they found that liquid helium-4 at temperatures below 2.17 kelvins (the *lambda point*, p. 284) conducted heat almost instantaneously (some hundred times more rapidly than copper, at that time the best heat conductor known). Two years later, Peter Kapitza, while in Moscow, demonstrated why helium conducts heat so well. At the same time, Jack Allen and Donald Misener, at the Royal Society Laboratories in Cambridge, found that at temperatures below the lambda point helium flows remarkably well (it is known as helium-II, p. 284). Early in 1938, all these three scientists reported their findings in two papers published in *Nature*.

In his paper Kapitza reported that the viscosity (or internal mobility) of helium-II was only a few thousandths that of gaseous hydrogen (which was at the time the least viscous fluid known). He wrote: 'the helium below [the lambda] point enters a special state which might be called a "superfluid"'. Soon after Kapitza's and Allen and Misener's papers were published, the so-called fountain-effect (see Fig. 10.2) was discovered. Kapitza continued to carry out experiments which indicated that helium-II is a quantum fluid with a perfect atomic order. Super-fluid helium-II was explained by a colleague of Kapitza's, a leading theorist by the name of Lev Landau, who you will remember meeting when we spoke about the 1962 Nobel Prize.

Cosmic microwave background

Let us return to the events surrounding the discovery made by the two other prizewinners, Arno Penzias and Robert Wilson; that is, the discovery of the *cosmic microwave background radiation* (see p. 296).

After they had detected mysterious microwave signals from the sky in May 1964, Penzias and Wilson continued for almost a year to look for any possible source of that faint background noise. Meantime, Penzias happened to discuss his observations with a colleague, who suggested calling in Robert Dicke, a professor of physics at Princeton University (he was the man who in the 1960s measured the equivalence between the inertial mass and the gravitational mass, p. 67).

Dicke and James Peebles (one of his postdoctoral fellows) had made detailed calculations about the radiation that should have filled the universe immediately after the Big Bang. They had deduced that, due to the expansion of the universe, the primordial short-wavelength photons must have become long-wavelength photons, in the microwave region of the electromagnetic spectrum. These photons now form a radiation field, pervading the whole universe, as though they were emitted by a black body, the temperature of which can be obtained by Planck's radiation law. At Princeton in those years David Wilkinson and Peter Roll (two other young collaborators of Dicke's) had just assembled a radio antenna expressly to investigate this microwave radiation.

Penzias telephoned Dicke, and the realisation then dawned on him that he and Wilson had discovered the cooled-down cosmic background radiation left over from the blazing hot Big Bang. The two groups decided to announce the discovery in the *Astrophysical Journal*. In their paper Penzias and Wilson wrote:

> Measurements of the . . . noise temperature of the 20-foot horn-reflector antenna . . . at the Crawford Hill Laboratory, Holmdel, New Jersey, at 4080 [megacycles per second] have yielded a value of about 3.5 [kelvins] higher than expected . . . A possible explanation for the observed excess noise temperature is the one given by Dicke, Peebles, Roll and Wilkinson . . . in the companion letter in this issue.[26]

Dicke and his collaborators meanwhile proposed their view, which was that the radiation that Penzias and Wilson had studied was the remnant of the Big Bang:

Fig. 11.13. R. Wilson (left) and A. Penzias with their antenna. (Photo by Robert Isear, courtesy AIP Emilio Segrè Visual Archives, Physics Today Collection.)

> ... we recently learned that Penzias and Wilson ... of the Bell Telephone Laboratories have observed background radiation at 7.3-cm wavelength. In attempting to eliminate (or account for) every contribution to the noise seen at the output of the receiver, they ended with a residual of [3.5 kelvins]. Apparently this could only be due to radiation of unknown origin entering the antenna ... While all the data are not yet in hand we propose to present here the possible conclusions to be drawn if we tentatively assume that the measurements of Penzias and Wilson ... do indicate black-body radiation at 3.5 [kelvins].[27]

Penzias and Wilson's pioneering observations were soon followed by many other measurements of the intensity of the cosmic background radiation at a variety of frequencies: these were carried out by different groups all over the world. These measurements showed that the radiation definitely followed a black-body spectrum, so proving conclusively that it was the frozen remnant, at 2.73 kelvin, of the Big Bang itself.

Arno Allan Penzias was born in Munich, Germany, in 1933. His parents, being Jewish, fled Nazi Germany in 1940 and emigrated to the USA. He attended the City College of New York and, after serving in the US Army as a radar officer, continued to study at Columbia University, obtaining his Ph.D. in 1962, under the supervision of Charles Townes (Nobelist in 1964). Penzias worked at Bell Labs, first as a staff member (1961–72), then as head of the Radio Physics Research Department. From 1981 to 1995 he was vice-president of research for Bell Labs. He has also held several academic positions: at Princeton, at the National Radio Astronomical Observatory in West Virginia, and at Stanford University in California.

Robert Woodrow Wilson was born in Houston, Texas, in 1936, the son of a chemist. He studied at Rice University, Houston, and at Caltech, Pasadena, where he received his Ph.D. in 1962. From 1963 to 1976 he worked at the Bell Labs at Holmdel, and then became the head of Bell's Radio Physics Research Department. As well as his research on cosmic microwave background, Wilson has also worked on subjects such as galactic radiation and millimetre-wave measurements of interstellar molecules.

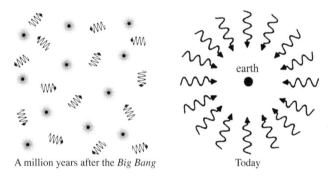

Fig. 11.14. A million years after the Big Bang protons and electrons began combining to form hydrogen atoms, and the universe became transparent to electromagnetic radiation. The black-body radiation therefore survived, cooling to its present 2.73 kelvin as the universe expanded and cooled down.

The Big Bang

In 1948, George Gamow, Hans Bethe (Nobel 1967), and Ralph Alpher (a student of Gamow's) wrote a paper where they discussed how the chemical elements could have been produced in a hot, dense universe, which was supposed to have originated from a violent explosion. (Later on the British astronomer Fred Hoyle called this explosion the *Big Bang*.) In the same year, Alpher joined forces with Robert Herman, and worked out the calculations more rigorously, including the idea of an expanding universe. They pointed out that, if the Big Bang had taken place, the radiation accompanying it should have lost energy as the universe expanded, and it should still be around today in the form of radio waves. (At microwave wavelengths this *cosmic microwave background radiation* should be characteristic of a perfect radiator, a black body heated to a temperature of about 2.73 kelvins.)

This idea was carried further by Robert Dicke in the early 1960s at Princeton. Thus Penzias and Wilson's discovery is considered to be the conclusive evidence in favour of the very hot Big Bang model. It is now generally accepted that the Big Bang did take place – as recently as 13 billion or so years ago.

Living in an expanding universe

Starting nearly 13 billion years ago. Scientists have used particle physics, nuclear and atomic physics, and cosmology to trace the sequence of events as they occurred in the universe, beginning right from its birth. Our story starts at an infinitesimal fraction of time after the Big Bang – the explosive event which marked the beginning of the expansion of the universe. Before this time (called the *Planck time*), about 10^{-43} (less than ten-million-billion-billion-billion-billionth) of a second, all four forces of nature are unified.

- ***At 10^{-43} of a second*** after the Big Bang, the energy of the universe is concentrated at an unimaginably high density, and the temperature is as high as 10^{32} (100 000 billion billion billion) kelvins. Gravity emerges, functioning separately from the strong, electromagnetic and weak forces, these three still being indistinguishable. Quarks and leptons freely transform one into the other.
- ***At 10^{-35} of a second*** the temperature of the universe has decreased to some 10^{27} kelvins. A sudden expansion of space lasts for a time less than a million-billion-billionth of a second; the strong force separates from the electroweak force, and the universe becomes a soup of quarks, leptons, gluons, weak bosons and photons; a minutely small quantity of particles exceeds the quantity of antiparticles, so leading to a dominance of matter over antimatter.
- ***At 10^{-12} of a second*** the temperature is 10^{15} kelvins, and the electromagnetic and weak forces separate from each other.
- ***At 10^{-6} of a second*** (the time of *confinement*) the temperature reaches 10^{13} kelvins, so that quarks can bind together to form individual protons and neutrons.
- ***At 100 seconds*** after the Big Bang, the universe expands further and its temperature drops down to some billion kelvin. Protons and neutrons bind together to form the first atomic nuclei; fusion reactions continuously produce and transform nuclear matter.
- ***At 1 million years*** the temperature is 3000 kelvins. Hydrogen and helium atoms begin to form, and the universe becomes transparent enough for photons to be able to travel along straight lines.
- ***At 300 million years*** the gravitational attraction pulls atoms together, so that stars and galaxies emerge.
- ***At 13 billion years or so*** after the Big Bang, the universe is progressively cooled to about 2.73 kelvins. Chemical processes link atoms together to form molecules, and macro-molecules: and so it is that man is able to emerge from the dust of stars!

1979

A unifying theory

The 1979 Nobel Prize was awarded for one of the crowning achievements of twentieth-century physics: the merger of two of the four fundamental forces of nature, the electromagnetic and weak nuclear forces, into a single theory, called the *electroweak theory*. This successful synthesis constituted a milestone in that age-old dream of physicists – to describe the structure of our universe in terms of a minimal number of fundamental particles and forces.

The prizewinners, the architects of the new theory, were three theoretical physicists: the Americans Sheldon Glashow and Steven Weinberg, and the Pakistani scientist Abdus Salam.

The path to unification
First step
Quantum electrodynamics is the prototype of certain theories called *gauge field theories*, which embody a symmetry principle, known as *gauge invariance* (the curious word 'gauge' is a historical legacy which has got rooted in the jargon of particle physics). In the case of QED, this technical expression has the following meaning : you can make certain special changes in the quantities describing the electromagnetic (or photon) field, and in those describing charged particles (such as electrons), and still the form of the equations governing the electromagnetic interactions between those particles will remain unchanged. This symmetry principle is intimately related to the fact that the photon, the carrier of the electromagnetic force, possesses a mass exactly equal to zero.

In 1954 Chen Ning Yang (Nobelist in 1957) and his younger co-worker Robert Mills, then both at the Brookhaven National Laboratory, began to study the possibility of extending a similar symmetry principle to theories describing other fundamental forces. They discovered that their abstract symmetry affirmed the existence of force messengers with a zero mass and a spin equal to one (a kind of a generalised version of the photon). Although the Yang–Mills theory had a great appeal, it presented serious difficulties: for instance, some of these massless messengers would have had to carry an electric charge; but it turned out that particles of this type were not to be found.

Second step
Our chronicle now shifts to Harvard University. Here, in the mid-1950s, Sheldon Glashow was carrying out his doctoral thesis work under Julian Schwinger (Nobelist in 1965). He was working on developing an idea, suggested by his supervisor, which involved precisely the problem of devising a theory in which the electromagnetic and weak forces could be combined into a unique mathematical formalism. After obtaining his Ph.D., Glashow went to Copenhagen to the Niels Bohr Institute and to CERN, Geneva, on a research fellowship. Here,

in 1960–1, he extended Schwinger's idea, which was based on the work of Yang and Mills, and worked out a gauge theory of the weak and electromagnetic interactions. The news of his discovery he then published in the scientific journal *Nuclear Physics*.[28]

Glashow's theory required four messengers for transmitting the two united forces: the massless photon, as carrier of the electromagnetic force, and three heavy particles, called *weak bosons*, as carriers of the weak force. He also predicted the existence of a new sort of weak interaction in which the interacting particles did not change their electric charge: it was called the *neutral-current interaction*. However, this theory presented some problems. In particular, it did not allow the weak bosons to have a mass, whereas the short range of the weak force requires heavy messengers. (Glashow added the three boson masses *ad hoc*, leaving the photon massless; thus he 'broke the symmetry' between the four messengers, which was embodied in the equations of the theory itself.) Moreover, Glashow's theory concerned only leptons: it did not include hadrons, which can also feel the electromagnetic and weak forces. (Similar ideas had been suggested in 1964 by Abdus Salam and his co-worker John Ward.)

Third step

It was only ten years after the Yang-Mills paper that physicists began to understand how to give the Yang-Mills messengers their masses. In 1964, two theoreticians, Robert Brout and François Englert, who were working at the *Université Libre de Brussel* (Belgium), and shortly afterwards Peter Higgs from Edinburgh University (Scotland), published in their scientific papers a theoretical construct, named *spontaneous symmetry breaking*, which succeeded in endowing those messengers with a mass. (Analogous phenomena had been discussed one year before by Philip Anderson (Nobelist in 1977) in the context of condensed-matter physics.) Let us try to sketch, in plain words, their ingenious construct for which all the three were awarded the 2004 Israeli Wolf Prize for physics.

In synthesis, the underlying idea is that the vacuum, as imagined by physicists, is pervaded by an elusive field (a kind of cosmic 'molasses'), whose quanta, (which have by now become universally known as *Higgs particles*) have large masses and a spin of zero. Messenger particles (such as the weak bosons) that interact with this 'molasses' slow down and acquire a mass, just as happens with a transparent medium that slows down the photons of a beam of light as they traverse it. On the contrary, particles that do not interact with the 'molasses' remain without a mass. As a consequence, the symmetry remains intact, even if 'hidden' in the basic theory; in other words, the symmetry is said to be *spontaneously*, rather than *dynamically*, *broken* (as it was in Glashow's theory). (Because of their huge mass, to date no one has yet discovered a 'Higgs particle'; one of the chief tasks of the future accelerators, like the LHC at CERN, is to search for them.)

Fourth step

In 1967–8, Weinberg from MIT, and Salam from Imperial College in London, independently developed a more complete theory to unite the two forces in a single framework.[29] Like Glashow they thought of four messengers, one massless (the photon) and three with a mass (two of them, designated W^+ and W^-, carrying an electric charge; the third, designated Z, being electrically neutral). But Weinberg and Salam, thanks to the 'Brout–Englert–Higgs mechanism', were able to explain how the masses of the weak messengers W and Z could be generated. Their model still had, however, some remaining difficulties. It presented the same kind of internal contradictions as had the old quantum electrodynamics (Glashow's theory, too, suffered the same defect) – that is, the so-called *problem of infinities*. Moreover, this model, too, was only suitable for leptons; hadrons were not included.

Fifth step

Now it was Glashow who came to the rescue. As we have seen (p. 339), in 1970 he himself, together with John Iliopoulos and Luciano Maiani, reconsidered the idea of a fourth quark, the charm quark,[30] which was actually discovered four years later in experiments by Burton Richter and Samuel Ting (see Nobel 1976).

The introduction of the charm quark permitted the classification of quarks and leptons into two families of four, each family being composed of two quarks and two leptons: the first family, the up and down quarks, together with the electron and its neutrino, provided all the matter constituting the whole world around us; the second group, the strange and charm quarks, together with the muon and its neutrino, were able to explain certain weak processes of strange particles, then known that were, however, difficult to understand. Thus Glashow's idea was successful in showing how Weinberg and Salam's electroweak theory could be extended to include the weak and electromagnetic interactions of hadrons as well as of leptons.

The solution to the *problem of infinities* came instead from two Dutch theorists, namely Gerardus 't Hooft and Martinus Veltman. They published some papers in 1971–2 in which they showed how it was possible to handle calculations capable of eliminating the infinities in theories like that of Weinberg and Salam (see Nobel 1999). Thus, the Glashow–Salam–Weinberg (G–S–W) model, which united the electromagnetic and weak interactions in a unique mathematical framework, became a complete, and later a widely accepted, theory.

Sixth step

Finally, in 1973, the first piece of experimental evidence of the G–S–W electroweak theory was produced at CERN, near Geneva. Here, physicists discovered the neutral-current interaction, a by-product of the electroweak unification.

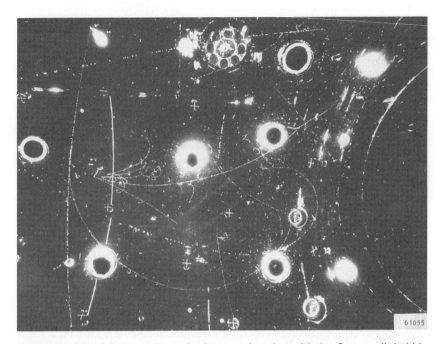

Fig. 11.15. One of the most dramatic photographs taken with the *Gargamelle* bubble chamber. (Courtesy CERN.) A neutrino (invisible) enters from the left and interacts with an atomic nucleus of the heavy liquid. All the recorded emerging tracks were identified as strongly interacting particles; no tracks due to charged leptons were visible, so no charge change has occurred in the interaction. The event was thus interpreted as a weak neutral–current interaction.

They directed a high-energy beam of neutrinos at a giant bubble chamber (it was called *Gargamelle* after the mother of the giant Gargantua in the satirical writings of François Rabelais; it was ten cubic metres in volume, and was filled with eighteen tons of liquid freon). Occasional neutrino interactions with the freon atoms produced visible tracks of charged particles. It was then possible to photograph these tracks through windows situated around the chamber. Early in that same year, the CERN team found photographs showing events in which a neutrino had been scattered off target particles, without changing its identity (see Fig. 11.15). These events were interpreted as examples of neutral-current interactions, mediated precisely by the neutral Z boson. Identical results were obtained some months later at Fermilab, near Chicago, with a different experimental method. Thus, the Swedish Academy of Sciences was able to award the 1979 Nobel Prize with these words: ' . . . for their contributions to the theory of the unified weak and electromagnetic interaction between elementary particles, including, inter alia, the prediction of the weak neutral current'.

> ### The electroweak theory
>
> In the Glashow–Salam–Weinberg *electroweak theory*, there are four messengers of the united forces: the three weak bosons (W^+, W^-, Z) and the *photon*. At very high energies, and therefore at very short distances, all these four messengers appear as being without a mass; the electromagnetic and weak interactions have the same strength, and act over an infinitesimal range. Physicists describe all this by saying that at very high energies the mathematical description of the electroweak interaction embodies a *gauge symmetry*, or that its equations are *gauge invariant* (this means, for example, that you can interchange the quantities describing particles of different electric charge, such as the electron and the neutrino, in the equations, without changing their form).
>
> At the lower energies of the experiments in our laboratories, however, the unification is no longer evident: the symmetry is *spontaneously broken*, so that the two forces acquire their own identities, and they differ in strength, range and other characteristics. Meanwhile, the three weak messengers pick up a mass through the so-called 'Brout–Englert–Higgs mechanism', and carry the short-range weak force, while the photon remains without mass, and carries the much stronger infinite-range electromagnetic force.
>
> According to the Glashow–Salam–Weinberg theory, the masses of the three weak bosons should be in the range of 80–90 billion electron-volts. Their discovery came to represent the best experimental evidence of the theory. But, at that time, no existing accelerators were powerful enough to produce these particles. Physicists had to wait until 1982–3, when experiments carried out at CERN finally revealed their existence (see Nobel 1984).

Glashow, Weinberg and Salam

Sheldon Lee Glashow was born in 1932 in New York City, the son of Jewish immigrants from Russia. He recalls his parents with these words:

> My parents . . . immigrated to New York City from Bobruisk in the early years of this century. Here they found the freedom and opportunity denied to Jews in Czarist Russia. After years of struggle, my father became a successful plumber, and his family could then enjoy the comforts of the middle class. While my parents never had the time or money to secure a university education themselves, they were adamant that their children should. In comfort and in love, we were taught the joys of knowledge and of work well done. I only regret that neither my mother nor my father could live to see the day I would accept the Nobel Prize.[31]

He also recalled his high-school studies with Weinberg: 'Among my chums at the Bronx High School of Science were Gary Feinberg and Steven Weinberg.

Fig. 11.16. From left to right: S. Weinberg, S. Glashow (the two friends who rendered famous the Bronx High School of Science of New York) and P. Frampton. (Photo by Will Owens, courtesy AIP Emilio Segrè Visual Archives, Physics Today Collection.)

We spurred one another to learn physics while commuting on the New York subway.'[32] After graduation in 1950, Glashow went on to Cornell University; here he obtained a bachelor's degree in 1954. He then went to Harvard, where he earned his master's degree in 1955 and his Ph.D. in 1959. After a period at the University of California, Berkeley, he returned in 1966 to Harvard as full professor of physics. A member of the US National Academy of Sciences, Glashow was in 1979 nominated Eugene Higgins Professor of Physics, and in 1990 he became emeritus, at Harvard.

Steven Weinberg was born in 1933 in New York City. He studied physics together with Glashow at Cornell University. After graduating in 1954, he went for a year to the Niels Bohr Institute in Copenhagen, where he started his research career in physics. He then returned to the USA, and received his Ph.D. in 1957 from Princeton University. Weinberg worked first at Columbia University and then at the University of California at Berkeley (1959–66). His research work during this period included a wide variety of subjects: high-energy physics, weak interactions and muon physics.

From 1966 to 1969 he taught at Harvard and MIT, where he then became a full professor of physics. In 1973 he moved to Harvard, where he succeeded Julian Schwinger in his chair as Eugene Higgins Professor of physics; in the same year he was also appointed senior scientist at the Smithsonian Astrophysical Observatory. During the 1970s Weinberg worked on the electroweak theory and on quantum chromodynamics, the theory of strong interactions, so moving towards the unification of all the three forces of nature. In 1982 he transferred to Austin as Josey Regental Professor of Science at the University of Texas. He has received many awards and honours, among them the US National Medal of Science, and memberships of the US National Academy of Sciences and of the Royal Society of London.

Abdus Salam (1926–96) was the first Pakistani, and the first Muslim scientist, to win a Nobel Prize. He was born in Jhang, a small town now in Pakistan, the son of an official in the local department of education. We quote from Salam's Nobel biography:

> His family [had] a long tradition of piety and learning. When he cycled home from Lahore, at the age of 14, after gaining the highest marks ever recorded for the Matriculation Examination at the University of the Punjab, the whole town turned out to welcome him. He won a scholarship to Government College, University of the Punjab, and took his MA [Master of Arts] in 1946. In the same year he was awarded a scholarship to St John's College, Cambridge . . .[33]

Salam graduated at Cambridge University in 1949, and obtained a Ph.D. in 1951. He then returned to Pakistan to teach at Government College in Lahore. In 1952 he became Head of the Mathematics Department of Punjab University. Two years later he went back to Cambridge, and in 1957 he became professor of theoretical physics at Imperial College of Science and Technology, London. In 1964 Salam founded the International Centre for Theoretical Physics in Trieste, Italy, a scientific institution which provides support to young physicists from developing countries.

> The creation of physics is the shared heritage of all mankind. East and West, North and South have equally participated in it. In the Holy Book of Islam, Allah says . . . 'Thou seest not, in the creation of the All-merciful any imperfection. Return Thy gaze, seest thou any fissure. Then Return thy gaze, again and again. Thy gaze, Comes back to thee dazzled, aweary.' This in effect, is the faith of all physicists; the deeper we seek, the more is our wonder excited, the more is the dazzlement for our gaze.
>
> (From Abdus Salam's speech at the Nobel banquet, December 1979)[34]

Fig. 11.17. Abdus Salam. (Courtesy AIP Emilio Segrè Visual Archives.)

1980

Broken mirrors

Let us resume the subject of symmetries in particle physics. As we have seen (p. 260), in 1956 Tsung Dao Lee and Chen Ning Yang discovered that *parity* is violated in the weak interactions of subatomic particles. Twenty-four years later, another symmetry breakdown attracted the Nobel Physics Committee, and they decided to award the 1980 Nobel Prize to the American experimentalists James Cronin and Val Fitch, for having discovered that an even more complex symmetry is violated in the decay of certain particles. It is called the *charge–parity symmetry*. This unexpected discovery emerged from an experiment that Cronin and Fitch carried out in 1963–4 at the Brookhaven National Laboratory.

Magic CP-mirror

Firstly, we are going to talk about two symmetries, which are labelled with the letters C and P. Symmetry C means *charge reflection*, or, in plainer language, 'switch all particles into their corresponding antiparticles'. The letter P stands instead for *parity transformation* or *mirror reflection* (we have already spoken of it, when we met Lee and Yang). Now, let us disport ourselves a little more with these two letters.

As we have seen, P-symmetry is not conserved in weak interactions. Lee and Yang had already predicted, and experiments proved it to be so, that C-symmetry too would be violated in the same kind of interactions; that is, that the laws of the weak interactions do not remain the same when you change all particles with their antiparticles. Then other theorists came along, and invented a more comprehensive symmetry principle. They said: well, even if C and P are not

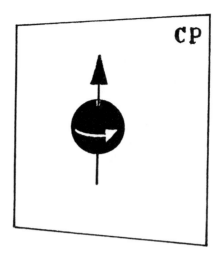

Fig. 11.18. A particle which is rotating clockwise is transformed by a 'magic *CP*-mirror' into its antiparticle which is rotating anticlockwise.

conserved, *charge–parity* (or *CP*) might be conserved. This means that the laws of physics would be unchanged by a mirror reflection of a physical process if, at the same time, all particles in the process were replaced by their antiparticles (the rule is: use a magic *CP-mirror* – like the one illustrated in Fig. 11.18).

Charge–parity symmetry seemed to be respected not only in electromagnetic and strong interactions, but also in weak interactions (though T. D. Lee had noted in his 1957 Nobel lecture that experiments had not 'reached a conclusive stage' at the time). The test came eventually in 1963–4, when Cronin and Fitch investigated the decay of neutral *K*-mesons, members of the same family of subatomic particles which had led to the famous θ–τ *puzzle* in the mid-1950s, which was solved by Lee and Yang (p. 262).

The Brookhaven experiment

In 1963 our two Nobelists were leading a group of experimentalists from Princeton University. They were working at the same accelerator (the AGS) as had been used two years before by Leon Lederman, Melvin Schwartz and Jack Steinberger (Nobel 1988) in an effort to discover the muon neutrino (p. 286). (The same accelerator was to be used ten years later by Samuel Ting (Nobelist in 1976) when discovering the *J*-particle.) In their experiment, the Princeton team bombarded a beryllium target with protons, provided by the AGS at an energy level of 30 billion electron-volts. The protons interacted with the target nuclei and created a large number of secondary particles, among which they selected, with specially designed instruments, one of the two existing types of neutral *K*-mesons (a long-lived variety).

The theory predicted that, if *CP*-symmetry had to be conserved, the long-lived neutral *K*-meson could decay into three pions, but may never do so into only two. To the surprise of the team the experiment proved the opposite. They found that out of about 1000 neutral *K*-mesons, two on average decayed into two pions. Bewildered by this unexpected result, they checked and rechecked their data to be sure that they were not mistaken. ('We kept everything quiet until we were sure', Fitch remembered.) Then in July 1964 they published their results in *Physical Review Letters*. In his Nobel lecture, Cronin wrote:

> Upon learning of the discovery in 1964, the natural reaction of our colleagues was to ask what was wrong with the experiment . . . I remember vividly a special session organised at the 1964 International Conference on High Energy Physics at Dubna in the Soviet Union. There, for an afternoon, I had to defend our experiment before a large group of physicists who wanted to know every detail of the experiment . . .[35]

But Cronin and Fitch were right: 'If you do an experiment that is quite striking, people at other laboratories usually rush to confirm it, and that's what happened . . .', wrote Cronin. 'At the time there were probably about six laboratories in the world where this experiment could have been done . . . Within three or four months the work was confirmed, because people had apparatus that could be adapted to look at this problem. It was nice for someone else to see what we saw. We felt maybe we were right after all.'[36]

Cronin and Fitch's results thus showed for the first time that the left–right asymmetry (P) is not always compensated by transformations from matter into antimatter (C). As a consequence, another symmetry principle, called (in the jargon of physics) *time reversal* (T), must be violated in weak interactions ('time reversal' means 'reverse the motion of all particles involved in a physical process').

Since its discovery, physicists have continued to carry out more and more sophisticated experiments in order to obtain precise measurement of the elusive effect discovered by Cronin and Fitch. They are also trying to reveal ever smaller and smaller effects that operate at the quark level. (These quark transformations under the action of the weak force have been extensively studied by theoreticians like the Italian Nicola Cabibbo, and the Japanese scientists Makoto Kobayashi and Toshihide Maskawa.) Recent experiments (2001) have provided a first indication of a possible *CP*-symmetry violation in decays of particles other than the neutral *K*-mesons used by Cronin and Fitch (they are mesons containing the bottom quark, p. 423).

James Watson Cronin was born in 1931 in Chicago, the son of a university professor of Latin and Greek. He studied physics at the University of Chicago with teachers such as the Nobelists Enrico Fermi, Maria Goeppert Mayer and Murray Gell-Mann; in 1955 he obtained his Ph.D. there. Cronin then joined the Brookhaven National Laboratory, where he worked at the newly completed accelerator – the Cosmotron. In 1958 he moved to Princeton, becoming in 1964 full professor of physics. In 1965, after the experiment that resulted in his being awarded the Nobel Prize, he went to France to work at the *Centre d'Etudes Nucléaires* in

Fig. 11.19. Five Nobelists at the Brookhaven National Laboratory. From left to right (standing): S. Ting, J. Cronin, V. Fitch; seated: I. Rabi (left) and C. N. Yang. (Courtesy Brookhaven National Laboratory.)

Saclay. In 1971 he transferred to Chicago, where he is now a university professor of physics, and where he began to carry out research into high-energy physics at Fermilab. Cronin is the recipient of many honours, among them the Ernest O. Lawrence Award (1977), and membership of the US National Academy of Sciences.

Val Logsdon Fitch was born in 1923 in Cherry County, Nebraska, the son of a ranch owner. During the Second World War he was sent as a soldier to the Los Alamos Laboratory, where he worked for three years as a technician on the Manhattan Project. After the war he went to McGill University in Montreal, Canada, where in 1948 he graduated in electrical engineering. He then went to Columbia University, New York, where in 1954 he obtained his Ph.D. in physics with a thesis under James Rainwater (Nobelist in 1975). The same year Fitch moved to Princeton, where in 1958 he became full professor of physics and head of the physics department. He is now Cyrus Fogg Brackett Professor of Physics at Princeton, and, like Cronin, a member of the US National Academy of Sciences.

Cosmic asymmetry

The physical effect discovered by Cronin and Fitch is important in the study of nature both on the smallest scale of elementary particles, and on the largest scale of the whole universe. It could be one of the key ingredients in understanding an old problem that has troubled astrophysicists for long time, and which has to do with the question: why does our universe appear to be made predominantly of matter? (The role of *CP*-symmetry violation was first pointed out in 1967 by the Russian physicist Andrei Sakharov.) Using particle physics, scientists have managed to work out a model which explains how the universe has expanded and evolved after the Big Bang (p. 353). The model suggests that in the very first instants after the Big Bang the universe was unimaginably hot and dense, and it produced

Fig. 11.20. An aerial view of SLAC. (Courtesy Stanford Linear Accelerator.) The three-kilometre-long linear accelerator (in the upper half of the photo) accelerates electrons and positrons to a speed of 0.999 999 9 times the speed of light. Magnets separate them and bend them around the collider (sited underground – but see the lower half of the photo), where they circulate in opposite directions.

Fig. 11.21. Collider. When a beam of high-energy particles collides with a stationary target, only a small fraction of the energy of the projectile particles is available to be converted into new particle states. But in a collider the beam particles collide head-on, and the available energy derived from these collisions is much larger. Early in 1960 the Austrian physicist Bruno Touschek, then working at the Frascati laboratory near Rome, Italy, first outlined the potentiality of an electron–positron collider. The first prototype, called Anello di *Accumulazione* or *AdA*, was in fact built right there, in 1961, and was followed soon afterwards by a research collider called *ADONE* (meaning 'big AdA').[37] Other electron–positron colliders of increasing energy have since been constructed in France, Russia, Germany, USA, and at CERN, Geneva.

equal amounts of particles and antiparticles, so that the universe was initially symmetrical. As it began to expand and freeze out, *CP*-symmetry violation, which distinguishes matter from antimatter, may have occurred, and so brought about a minute predominance of particles over antiparticles. To that particle–antiparticle asymmetry we and the stars of the universe owe our existence!

Chapter 12
Big physics – small physics

The second half of the twentieth century is the age of *big science*: big accelerators for high-energy particle physics; large-scale facilities for fusion energy research; space probes for astrophysical investigations, and giant telescopes able to explore the very edges of the universe.

From the 1950s on the undisputed leaders in the development of particle physics, thanks to their extraordinary experimental discoveries, have been the Lawrence Berkeley National Laboratory, the Brookhaven National Laboratory and the Stanford Linear Accelerator Center (SLAC) in the USA, and CERN in Europe. The high point, however, comes in 1983 when weak bosons, the decisive proof of the validity of the Glashow–Salam–Weinberg electroweak theory, are produced and detected in the proton–antiproton collider at CERN (see Nobel 1984). All these discoveries will be honoured with eight Nobel Prizes bestowed upon fourteen scientists.

But *small physics*, which involves small research groups, small apparatuses, lower costs, has also grown up during the preceding decades. It includes areas such as atomic and condensed-matter physics, low-temperature physics, and research into semiconductors and superconductors. In the 1980s two achievements were signalled at the IBM Zurich Research Laboratory – and both were soon honoured, with two Nobel Prizes: the *scanning tunnelling microscope*, built in 1981, received the award in 1986. And the discovery, also in 1986, of new superconducting materials with high transition temperatures was similarly honoured, just a year later.

In the field of atomic physics, new techniques, developed since the 1960s, opened the way to a wealth of studies on the structure of atoms and molecules. New types of maser and laser were developed and rapidly found multiple applications in both sciences and technology: the laser soon brought about a new kind of spectroscopy, called *laser spectroscopy*; it initiated the evolution of new studies on the nature of the interaction of light with matter, and the new field of *nonlinear optics*. New techniques were also invented to cool and trap atoms with laser light, so allowing physicists to manipulate them, and perform ultra-precise measurements on their characteristics.

1981

New spectroscopies

Spectroscopy, an old science born around the 1850s thanks to the research work of scientists such as Joseph Fraunhofer, Robert Bunsen and Gustav Kirchhoff (p. 25), is still of the very first importance. The 1981 Nobel Prize for physics was in fact awarded to three experts in spectroscopy: the Americans Nicolaas Bloembergen from Harvard University and Arthur Schawlow from Stanford University, and the Swedish scientist Kai Siegbahn from Uppsala University. Bloembergen and Schawlow developed what is called *laser spectroscopy*, and used it for the analysis of materials, while Siegbahn studied atoms in matter by examining the electrons expelled from them (*electron spectroscopy*).

Laser spectroscopy

The old school of specialists in spectroscopy analysed the optical spectra of atoms; that is, the characteristic colours, or wavelengths, at which atoms emit and absorb light. The science of spectroscopy developed greatly during the twentieth century, eventually covering wider and wider bands of the electromagnetic spectrum. Here are two examples. During the first decades of the twentieth century, X-ray spectroscopy was used to study the innermost structure of atoms (p. 138); whereas after the Second World War radio-frequency spectroscopy evolved enormously and was extended to the study of atomic nuclei (p. 219). We shall now concentrate on laser spectroscopy. The laser produces light with a purity of colour, brightness and directionality superior to all other light sources. It has thus become a powerful tool in the hands of specialists in spectroscopy, making easier the study of the interaction of light with matter. Now let us have a look at Bloembergen and Schawlow's contributions.

Bloembergen

Bloembergen began developing laser spectroscopy in the 1960s, after very productive research work in the maser and laser fields. He then began to exploit lasers in order to study the changes that take place in atoms and molecules within matter when very high-intensity light is shone on them. His studies opened the way to a new field of optics, called in technical language *non-linear optics*. Together with his co-workers at Harvard University, he was able to explain nonlinear optical phenomena which occur in a large number of substances.

Nonlinear optics has been used in several applications. New devices based on these physical processes are used today in optical communication systems, in material science and in metrology. Bloembergen himself commented in his Nobel lecture:

> The field of nonlinear spectroscopy has matured rapidly ... The applications in chemistry, biology, medicine, materials technology, and especially in the field of communications and information processing are numerous. Alfred Nobel would have enjoyed this interaction of physics and technology.[1]

Fig. 12.1. Mrs Deli Bloembergen and Professor Bloembergen. (Courtesy AIP Emilio Segrè Visual Archives, Bloembergen Collection.) 'Bloembergen reminisced about the occasion in 1959 when he and Townes received the Morris Leibmann award for contributions to the maser art. Townes had presented his wife, Frances, with a medallion made from the ruby he used for his maser. When Bloembergen's wife, Deli, admired the medallion and asked for a similar memento of 'his' maser, he said, 'Well, dear, my maser works with cyanide."[2]

Nicolaas Bloembergen was born in Dordrecht, the Netherlands, in 1920, the son of a chemical engineer. He graduated in 1943 from the University of Utrecht. 'The remaining two dark years of the war I spent hiding indoors from the Nazis, eating tulip bulbs to fill the stomach and reading Kramers' book . . . [*Quantum Theory of Electrons and Radiation*] by the light of a storm lamp . . . My parents did an amazing job of securing the safety and survival of the family', he recalled in his Nobel autobiography.[3] In 1946 he went to the United States, to Harvard University, where he began to work with Edward Purcell (Nobelist in 1952) on nuclear magnetic resonance. In 1947 he returned to the Netherlands, and one year later he obtained his Ph.D. from the University of Leiden. He then went back again to Harvard, where he became Associate Professor of Applied Physics in 1951.

Apart from nonlinear optics, Bloembergen's major contributions came in the late 1950s and early 1960s. In those years he proposed a new type of maser. It was a solid-state maser, which worked on three energy levels (that is, atomic electrons could be made to take up two excited energy levels in an atom), and it was much more powerful than all earlier masers. (The first was built at Bell Labs; others were built by Bloembergen himself, and also by Charles Townes.) This new device became widely used in microwave receivers for radar and for radio astronomy (a maser of this type was used as a microwave amplifier by Arno Penzias and Robert Wilson in 1964–5 to detect the cosmic microwave background, p. 295). Bloembergen also devised a method which made it possible to generate laser radiation in both the infrared and ultraviolet regions. Meanwhile, his brilliant academic career continued to develop at Harvard. Here, in 1957, he became Gordon McKay Professor of Applied Physics, in 1980 Gerhard Gade University Professor, and in 1990 emeritus. Since 2000 Bloembergen has been an honorary professor in optical sciences at the University of Arizona in Tucson.

Schawlow

The Swedish Academy of Sciences first pointed out Schawlow's contribution to the invention of the laser. In the Academy's press release of 19 October 1981 we find:

> ... it was primarily the work by Schawlow and Townes [published in 1958, p. 292] which initiated the whole dynamic field which we now associate with the concept of 'laser' ... The 1964 Nobel Prize in Physics was awarded to Townes, Prokhorov and Basov ... Subsequent developments, however – particularly in lasers – have made this field increasingly deserving of additional rewards.[4]

The Academy then emphasised Schawlow's contribution to laser spectroscopy. He and his co-workers developed, at Stanford University, a new spectroscopic method, based on particular physical phenomena which occur when atoms of certain substances absorb laser light of a very high intensity. They utilised this method to improve the sharpness of the observed lines in atomic spectra and thereby the precision with which their wavelengths can be determined. (A limitation to a spectral line sharpness is due to the motion of the atoms; in fact, the familiar *Doppler effect*, p. 116.)

In 1972 Schawlow and his co-workers used a pulsed laser and the new method they had invented to observe infinitesimal shifts in frequency between very narrow lines of the spectrum of the hydrogen atom (the 'Rosetta stone of modern physics', as Schawlow himself called it). They were able to determine one of the most fundamental atomic constants, the so-called Rydberg constant (p. 93), with a precision much higher than had previously been possible. Schawlow was also one of the pioneers in the use of lasers for the detection of trace elements.

Arthur Leonard Schawlow (1921–99) was born in Mount Vernon, New York, USA, the son of an insurance company agent. He studied at the University of Toronto, Canada, obtaining his Ph.D. in 1949. He then went to Columbia University, New York, to work with Charles Townes. ('What a marvellous place Columbia was then, under I. I. Rabi's leadership! There were no less than eight future Nobel laureates in the physics department during my two years there.'[5]) From 1951 to 1961 Schawlow worked at Bell Labs, mostly on superconductivity; here he also collaborated with Townes on the principles of the laser (p. 292). He then went to Stanford University, and began to work on laser spectroscopy. At Stanford, Schawlow was full professor of physics and chairman of the physics

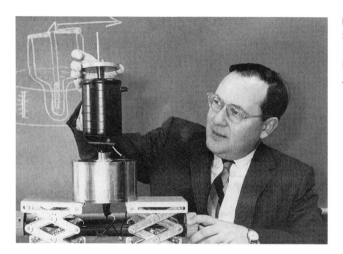

Fig. 12.2. Arthur Schawlow. (Stanford University, courtesy AIP Emilio Segrè Visual Archives.)

department (1966–70). In 1978 he became J. G. Jackson and C. J. Wood Professor of Physics, and Emeritus in 1991.

> Working with Charles Townes was particularly stimulating. Not only was he the leader in research on microwave spectroscopy, but he was extraordinarily effective in getting the best from his students and colleagues . . . Best of all, he introduced me to his youngest sister, Aurelia, who became my wife in 1951.
>
> (Arthur Schawlow)[6]

Electron spectroscopy

Electron spectroscopy uses electrons emitted by atoms instead of electromagnetic radiation. It was developed in the mid-1950s by Kai Siegbahn and his co-workers at Uppsala University. They irradiated samples of matter with X-rays having a specific energy. The rays expelled electrons from atomic shells in the sample. They then measured the energies of the emitted electrons with a high-precision spectrometer, and plotted the distribution of these energies (that is, the number of electrons at different energies, which is technically called the 'energy spectrum'). The spectrum showed well-defined peaks corresponding to the various energy levels of the atoms in the sample. Siegbahn used electron spectroscopy to obtain information on the atomic levels for different elements. The Royal Academy also noted that 'in the development of electron spectroscopy, a practically useful analytical method had been obtained . . . [which] has found several important fields of application, for instance in the study of surface-chemistry processes such as catalysis and corrosion.'[7]

Kai Manne Siegbahn was born in 1918 in Lund, Sweden, the son of Karl Manne Siegbahn (who was awarded the 1924 Nobel Prize in physics for his work on X-ray spectroscopy, p. 138). He studied at Uppsala University, and received his

Fig. 12.3. Kai Siegbahn. (© The Nobel Foundation, Stockholm.) Siegbahn is a member of many academies, amongst them the Swedish Academy of Sciences and the Royal Society of London. Apart from electron spectroscopy, his research activity has covered many other fields such as atomic and molecular physics, nuclear physics, plasma physics and electron optics.

doctorate from the University of Stockholm in 1944. From 1942 to 1951 he was research associate at the Nobel Institute for Physics, and then Professor of Physics at the Royal Institute of Technology in Stockholm. In 1954 Siegbahn became professor and head of the physics department at Uppsala University.

1982

Critical phenomena

The 1982 Nobel Prize for physics was only the ninth, from the whole period of the previous thirty years, to have been awarded to a single individual. The prizewinner was Kenneth Wilson, an American theoretical physicist, from Cornell University, Ithaca, New York. He was awarded the prize in recognition of his work on the study of the behaviour of physical systems when, close to their critical point, they change their state. Most of Wilson's work found applications in condensed-matter physics, but his mathematical techniques showed their worth in many other fields of physics.

Critical points

In order to grasp the essence of Wilson's work, let us consider an example from everyday life. If you heat a mass of water it begins to boil when the temperature of 100 °C is reached (at normal atmospheric pressure), and turns into steam.

Physicists call this change in the state of water a *phase transition*; that is, a transition from the liquid phase to the vapour phase, accompanied by an abrupt change in the water density. This change of state is affected by pressure: if the pressure is higher than that of the atmosphere, the boiling temperature of water is higher than 100 °C (this is the principle that underlies the pressure cooker). There is, however, an upper limit, known as the *critical point* (for water it occurs at a temperature of 374 °C and a pressure of 218 atmospheres), where the two phases merge, and the distinction between liquid and gas vanishes. At a temperature above that of its critical point there is only one single fluid phase, regardless of how great the pressure may be. Another interesting example is given by a ferromagnetic material (p. 316): heating such a material above a certain temperature (770 °C in the case of iron) will cause it to lose its magnetic power. This particular temperature at which the magnetism disappears represents the critical point of the material.

We have picked out these two typical examples from among the many that relate to states of matter which exhibit the same behaviour near their respective critical points (other examples include a superconducting transition, liquid helium when it becomes a superfluid, or metal alloys.) The similarities in the critical behaviour of physically different systems are described as *universality*.

Another feature which distinguishes critical point transitions from most other phenomena is that, in many systems near their critical point, fluctuations over widely different scales of length appear simultaneously. (For example, near its critical point, water develops density fluctuations in the form of drops and bubbles of vapour, whose sizes range from microscopic dimensions up to a dimension equal to the whole volume of the fluid.) Physicists generally use different methods to explain phenomena which take place on different scales of length (as an example, when describing atmospheric currents, physicists completely ignore the motion of each air molecule, on the other hand, when studying molecular dynamics, they do not care if the molecules are in the atmosphere or confined in a vessel). However, in the case of critical phenomena, physicists sought to devise a method of studying them with two key ingredients: to give identical solutions for apparently different critical behaviours; and to cover the entire range of length scales which appear during a critical transition.

Wilson's work

In the late 1930s Lev Landau (Nobel 1962) proposed a fruitful method of describing the phenomena of phase transitions. The accompanying theory was quantitatively accurate in many situations but failed to give quantitatively correct results for transitions occurring very close to the critical point. The same difficulty had been faced much earlier by Johannes van der Waals (Nobel 1910) and, in later years, by other physicists and chemists who tried to explain critical phenomena in a general form.

The problem was then taken up and solved by Wilson in two fundamental papers published in 1971, and was persued in still other papers in subsequent years. Using a method called by theoreticians the *renormalization group* (it had been developed many years earlier in the field of particle physics), together with a sophisticated mathematical technique, Wilson was able to develop an accurate numerical treatment of the critical point phenomena. In simple words, the key to the renormalization group approach consists of breaking down a critical transition with multiple scales into a sequence of more easily studied subunits, each of which is characterized by a single scale of length. (As we have seen (p. 298) the concept of *renormalization* was introduced into physics with the advent of modern QED in order to remove the annoying infinities which cropped up in the calculations.)

Similar ideas had already been suggested many years before by the American physicist Leo Kadanoff. Wilson then took up Kadanoff's ingenious ideas and succeeded in transforming them into a quantitative tool that could be used to predict certain crucial quantities which appear in the formulae describing critical phenomena. Using the renormalization group he obtained identical solutions to the problem of these phenomena for different systems, and at the same time his method was able to deal with multiple scales of length. Later work, done by several people including Wilson himself, used computational and pencil and paper methods in order to calculate, with great precision, the crucial quantities relevant to critical phenomena. These results were fully supported by experiments.

Wilson's method can be applied to different systems which are near their critical points, such as liquids, mixtures of liquids, magnetic materials and alloys. (One notable spin-off of Wilson's method, applied in another, different field, was the definitive explanation for the *Kondo effect* (p. 346), which had remained a mystery in condensed-matter physics for some forty years.) Other fields of science in which Wilson's method can find applications include the turbulent motion of fluids, the physics of polymers and the interactions of quarks confined in subatomic particles.

Kenneth G. Wilson was born in Waltham, Massachusetts, in 1936, the son of a professor of physical chemistry at Harvard University. He studied at Harvard, and then at Caltech, where he obtained his Ph.D. in 1961, under Murray Gell-Mann (Nobel 1969). He became assistant professor at Cornell University in 1963, associate professor in 1965 and full professor in 1970; he was also Director of the Center for Theory and Simulation in Science and Engineering there, one of the most important supercomputer centres in the USA. In 1988 Wilson moved to Ohio State University, where he is now the Hazel C. Youngberg Trustees Distinguished Professor of Physics. Wilson is one of those rare examples of a contemporary theoretical physicist who, not confining himself to a single subject, has mastered vast areas of physics. His present research interests concern the study

Fig. 12.4. Kenneth Wilson. (Courtesy AIP Emilio Segrè Visual Archives.) The day the Swedish Academy of Sciences announced the Nobel Prize to Wilson, he said that he was surprised, and 'especially that I'm getting the prize alone', adding that he 'would have expected to share the prize with Leo Kadanoff (University of Chicago) and Michael Fisher (Cornell University) . . .'[8] In 1980 Wilson had shared with these two colleagues (both pioneers in the field of critical phenomena) the prestigious Israeli Wolf Prize for physics.

of molecules, atoms and elementary particle theory. He is a member both of the US National Academy of Sciences and of the American Academy of Arts and Sciences.

1983

Stellar evolution

For the fourth time in sixteen years the Nobel Physics Committee rewarded studies related to the cosmos: the 1983 Nobel Prize for physics in fact concerned the early life and evolution of stars. The research had been carried out by two American astrophysicists, the Indian-born Subrahmanyan Chandrasekhar, from the University of Chicago, and William Fowler, from Caltech. Chandrasekhar earned the prize for his theories concerning the processes connected with the structure and evolution of stars, and Fowler earned his for his studies on the nuclear reactions which occur when chemical elements are formed in stars during their lifetimes.

The life of stars
Stars are born in the huge clouds of interstellar gas and dust which are diffused throughout galaxies. Briefly, the course of a star's life is the following.

A large gas cloud (mostly hydrogen) begins to contract under the action of gravity, and condenses into a star. During this process the gravitational energy of the particles which form the gas is converted into heat, so raising the temperature of the star core. In time, the temperature becomes high enough to cause hydrogen to burn. Then thermonuclear fusion reactions convert hydrogen into helium (p. 303), so releasing enormous quantities of energy. This outflow of energy creates an internal pressure which supports the immense weight of the overlying layers as they press down, until the star stops contracting. At this point a stable star is born, and it shines for billions of years (as you already know, our own sun will happily go on burning hydrogen for some billions of years!) When all its hydrogen and other elements have been exhausted the star evolves in a dramatic way. Other fusion reactions then take place, and after consuming all its thermonuclear fuels, the star begins to die, following one of two possible different avenues.

Low-mass stars expel their outermost layers, leaving a dead dense, tiny star, which has roughly the same size as our earth. These stars are called *white dwarfs*, because they have both a small size and a high temperature, and they emit a bluish light. The density of matter in a white dwarf is about one million times that of water (a teaspoonful of white-dwarf matter would weigh nearly five tons here on earth!)

Stars with a higher mass die quite differently, because their core is more compressed under the gravitational pull. They become hotter, therefore, so that thermonuclear reactions can continue. These create heavier and heavier elements in the core, the final product being iron. At this point, because iron cannot act as a thermonuclear fuel, fusion reactions stop, and the star reaches its dead end. The star then collapses suddenly with a spectacularly violent explosion that blasts the outer layers away from its core. This is called a *supernova* explosion, which may leave behind a *neutron star* (p. 333) or a *black hole*. (It was this kind of explosion that created the Crab Nebula, p. 330.)

Chandrasekhar's work

In the late 1930s Chandrasekhar, during the period that he was a research student at Cambridge University, England, pioneered theoretical studies precisely on white dwarfs. He used Fermi and Dirac's quantum statistics (p. 176), and the special theory of relativity, to calculate what should be the maximum mass of a white dwarf. This is called the *Chandrasekhar limit*, and it proved to be about 1.4 times the mass of our sun. (This means that to become a white dwarf a star must have a mass less than the Chandrasekhar limit.) As well as his first work on white dwarfs, Chandrasekhar contributed greatly to the study of stellar evolution. Beginning in the 1970s he also developed a theory on black holes using general relativity.

Subrahmanyan Chandrasekhar (1910–95) was born in Lahore, India (now in Pakistan), the nephew of Chandrasekhara Raman (Nobel 1930). He studied at the Presidency College, Madras University. Then he went to Trinity College in

Fig. 12.5. Chandrasekhar (left) talking with the American astronomer H. N. Russell, 1940. (AIP Segrè Visual Archives, Dorothy Davis Locanthi Collection.)

Cambridge, England, on a scholarship, where he studied with Sir Arthur Eddington, and in 1933 he obtained his Ph.D. He remained at Cambridge till 1937, when he joined the University of Chicago; in 1952 he became Professor of Astrophysics there. As Chandrasekhar wrote in his Nobel biographical notes, his research activity can be divided into several periods, which are of his principal study subjects: the structure of stars and the theory of white dwarfs (1929–39); the dynamics of stars (1938–43); the theory of the atmospheres of stars and planets (1943–50); relativistic astrophysics (1962–71); and the theory of black holes (1974–83).

Fowler's work

William Fowler made a great contribution to astronomy by applying nuclear physics to it. His work dealt with the nuclear reactions which take place in stars during their life. In addition to generating energy in the form of light and heat, these nuclear reactions provide the means by which heavy chemical elements are created from the original hydrogen and helium. Fowler carried out careful experimental investigations into the rates of such reactions, as well as conducting extensive theoretical work in the field of nuclear astrophysics.

William Alfred Fowler (1911–95) was born in Pittsburgh, Pennsylvania, USA, the son of an accountant. He graduated in 1933 from Ohio State University, Columbus, and he then went to Caltech, California, where in 1936 he obtained his Ph.D.; in 1939 he was appointed Professor of Physics there. Fowler developed his theory on the formation of heavy elements in stars during the 1950s, together with the famous British astronomer Fred Hoyle and the physicists Margaret and Geoffrey Burbidge. Together they showed how elements are synthesised in stars, starting from light elements and progressively producing heavier elements through nuclear reactions. Fowler and Hoyle also worked in

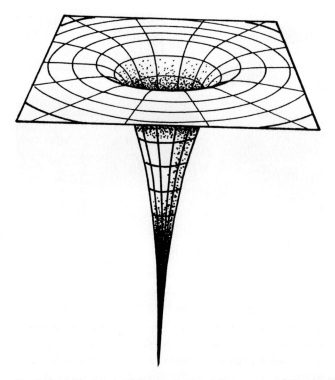

Fig. 12.6. A black hole. If the mass of a stellar remnant following a supernova explosion is greater than about three solar masses, the gravitational force will become so strong that the object collapses to a point (known as a *singularity*). It exerts such a strong gravitational field that no kind of matter, and not even light, can escape its attraction. This object has been named a *black hole*. It is a 'hole' because all matter falls into it and has no hope of getting out. It is 'black' because not even light can escape. In the 1970s, however, the Cambridge University theoretical physicist Stephen W. Hawking, using general relativity and quantum mechanics, showed that, contrary to the commonly accepted view, black holes can indeed emit radiation and particles. (In 1971 a new cosmic object was discovered in a binary X-ray system, located in the constellation Cygnus; it was named Cygnus X-1, and astronomers were unanimous in believing that they had discovered the first black hole.)

the field of radio astronomy, and proposed a theory regarding radio galaxies and quasars.

1984

Electroweak messengers

On 25 January 1983, a press conference was called at CERN, near Geneva. The Italian physicist Carlo Rubbia and the Dutch engineer Simon van der Meer triumphantly announced the discovery of the charged Ws, the long-sought weak bosons predicted by the electroweak theory – the theory which earned Sheldon Glashow, Abdus Salam and Steven Weinberg the 1979 Nobel Prize. On 1 June a second press conference was called: CERN announced the discovery of the neutral Z, the third member of the trio.

After not more than a year had passed, of on 17 October 1984, the Swedish Academy of Sciences announced that they were awarding the year's Nobel Prize for physics jointly to Rubbia, the team leader of the experiment, and van der Meer, the architect of the technique. (This was one of the shortest 'waiting times' in the whole history of the Nobel Prize.)

The proton–antiproton project

In its announcement of the award, the Academy noted that 'the project at CERN ... is the largest that has ever appeared in the context of a Nobel Prize. Two persons in the project are outstanding – Carlo Rubbia, who had developed the idea, and Simon van der Meer, whose invention made it feasible.'[9]

The story of the project started around the mid-1970s, when particle physicists were studying how to provide enough energy to make the heavy W and Z weak bosons. It seemed obvious that it would be necessary to collide particles and antiparticles head-on, so transforming their masses and their kinetic energy into new particles. However, the weak bosons were well beyond the energy range of any existing accelerator machine.

In 1976 Rubbia and two American colleagues of his, David Cline and Peter McIntyre, presented an idea for using the existing large proton accelerators at Fermilab (USA) and at CERN to collide protons and antiprotons that would counter-revolve within the same ring. The CERN management soon picked up Rubbia's idea and decided to convert their Super Proton Synchrotron (SPS) into a proton–antiproton collider. This provided an energy of 270 billion electron-volts for each beam (a total of 540 billions for each head-on collision of protons and antiprotons), enough to produce the W and Z weak bosons – if they existed.

However, this scheme ran into a difficulty. While an intense proton beam was easily achievable, a very intense beam of antiprotons would be required in order to ensure a large number of proton–antiproton collisions, so as to provide a high probability of seeing the predicted new bosons. This was brilliantly solved by van der Meer, who had invented some years earlier an ingenious method for packing and storing antiprotons (the so-called *stochastic cooling technique*), which made intense antiproton beams feasible. (The development of this technique was, as Leon Lederman, Nobelist in 1988, wrote, 'a monumental achievement in accelerator science'.) Thus, Rubbia's idea and van der Meer's invention were successfully combined in a large project, which included two experiments for hunting the weak bosons, one to be led by Rubbia himself, and the other by the French physicist Pierre Darriulat.

Fig. 12.7. A view of the SPS. (Courtesy CERN.)

The UA1 experiment

It was thus that Rubbia picked up the baton for the UA1 experiment. (UA1 stands for 'Underground Area number 1' since its location was a large cavern in the underground proton–antiproton collider.) With endless energy and contagious enthusiasm he gathered around him a group of some 130 physicists and engineers, coming from thirteen universities and research centres in many countries. They assembled high-precision instruments in an immense apparatus of some 2000 tons, called the *multipurpose detector* ('a huge beast, essentially a concentric set of boxes', as Rubbia himself explained); and used the most advanced technology to develop a microprocessor system to handle and analyse the enormous amounts of information produced by the collisions that were being observed through the detector. In the summer of 1981, under Rubbia's leadership, the UA1 team was ready to start the experiment. On 10 July they obtained their first success, when the UA1 detector witnessed the first proton–antiproton collisions.

The discovery of W and Z

The harvest came in the autumn of 1982. The course of events was both exciting and agitated. Let us follow the saga as it was chronicled in the November 1983 issue of the *CERN Courier*.

Between October and December 1982 the UA1 people observed and analysed several thousand million proton–antiproton collisions. From this jungle they were able to select about ten significant events which might correspond to charged *W*

Fig. 12.8. The UA1 Detector. (Courtesy CERN.)

bosons. At a meeting held in Rome in January 1983 Rubbia presented the first tentative evidence for his Ws. ('This is probably an historic meeting which will be discussed in future years . . .' noted Leon Lederman, the chairman.) After the Rome meeting Rubbia held a seminar at CERN (with the main auditorium 'packed to the roof')[10], where he '. . . announced six candidates for W events; UA2 [Darriulat's team] announced four. The presentations were still tentative . . . However, over the weekend of 22–23 January, Rubbia became more and more convinced. As he put it, "They look like Ws, they feel like Ws, they smell like Ws, they must be Ws".'[11] So, on Tuesday 25 January, the discovery of the W bosons was announced at a press conference. ('The UA2 team reserved judgement at this stage, but further analysis also convinced them.')[12] The paper containing the UA1 discovery was immediately sent to the editor of the scientific journal *Physics Letters*, where it was published in the issue of 24 February.[13]

The saga was repeated during the next summer. In June UA1 announced the discovery of the Z, and in July UA2 announced that they, too, had witnessed four good Z-events. It was thus that two independent experiments had at last confirmed Glashow, Salam and Weinberg's electroweak theory.

Fig. 12.9. The computer reconstruction of a proton–antiproton collision where one of the first *W* bosons was produced; the arrowed track is due to an electron, one of the decay products of the *W*. (Courtesy CERN.)

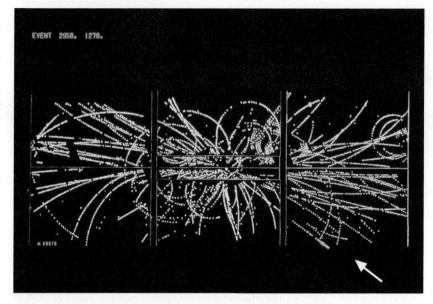

The award-winners

Carlo Rubbia was born in 1934 in Gorizia, Italy. His father was an electrical engineer, and his mother an elementary school teacher. In his Nobel autobiography he describes how his career as a physicist began:

> After completing high school, I applied to the Faculty of Physics at the rather exclusive Scuola Normale in Pisa [the same as that of Enrico Fermi, p. 203] . . . I badly failed the admission tests and my application was turned down. I forgot about physics and I started engineering at the University of Milan (Politecnico). To my great surprise and joy a few months later I was offered the possibility of entering the Scuola Normale. One of the people who had won the admission contest had resigned! I am recollecting this apparently insignificant fact since it has determined and almost completely by accident my career as a physicist.[14]

Rubbia received his doctorate in 1957. He then went to Columbia University as a research fellow, and in 1961 he joined CERN. In 1973 he headed a team working on neutrino interactions at Fermilab near Chicago, and in 1970 was appointed Professor of Physics at Harvard University, USA. Thereafter he commuted between Harvard and CERN, where, during the early 1980s, he directed the UA1 team. From 1989 to 1994 he served as director general of CERN. After retiring in 1999, the University of Pavia, Italy, offered Rubbia a chair of experimental physics (a personal chair); he was also appointed President of ENEA (the Italian agency for alternative energies).

Simon van der Meer was born in 1925, at The Hague, the Netherlands. He studied technical physics at the University of Technology in Delft, where he obtained his engineering degree in 1952. Then he worked at the Philips Research Laboratory, Eindhoven, on high-voltage equipment and electronics for electron microscopes. In 1956 he joined CERN, where he remained until his retirement, 'working ... on many different projects, in an agreeable and stimulating international atmosphere.'[15] His work there was concerned mainly with accelerator design. He developed new methods for accelerating and handling particles, and participated in experiments carried out at CERN's accelerators. The late 1970s marked the starting point of the great adventure which led him to the Nobel Prize, when Rubbia and his colleagues proposed to use the SPS as a proton–antiproton collider, and van der Meer became joint project leader for the construction of the antiproton accumulator.

1985

The quantum Hall effect

From big physics to small physics; from particle physics to condensed-matter physics. The Swedish Academy of Sciences awarded the 1985 Nobel Prize for physics to the German physicist Klaus von Klitzing, for the discovery he made, in the spring of the year 1980, of the so-called *quantum Hall effect*.

Fig. 12.10. S. van der Meer (left) and C. Rubbia while toasting their Nobel Prize. (Courtesy CERN.)

Fig. 12.11. Klaus von Klitzing. (Courtesy AIP Emilio Segrè Visual Archives, *Physics Today* Collection and W. F. Meggers Gallery of Nobel Laureates.) 'Semiconductor research and the Nobel Prize in physics seem to be contradictory since one may come to the conclusion that such a complicated system like a semiconductor is not useful for very fundamental discoveries ... Up to 1980 nobody expected that there exists an effect like the quantized Hall effect, which depends exclusively on fundamental constants and is not affected by irregularities in the semiconductor like impurities or interface effects.' (Klaus von Klitzing)[16]

Klaus von Klitzing was born in 1943 in Schroda, Poland. He first attended the Technical University of Brunswick (1962–9), and then went on to the University of Würzburg, Germany, where in 1972 he obtained his doctorate. In 1980 he was appointed professor of physics at the Technical University of Munich, and in 1985 he became director of the Max Planck Institute for Solid State Research at Stuttgart. In the mid-1970s von Klitzing carried out research in experimental solid-state physics at the Clarendon Laboratory, Oxford, England. He then continued his research work at the High Magnetic Field Laboratory, Grenoble, France (1979–80), where he made his historic discovery, which was honoured with the Nobel Prize.

An old phenomenon

Let us start with the *classical Hall effect*, an old phenomenon discovered by the American physicist Edwin Hall in 1879. If a magnetic field is applied perpendicularly to a metal sheet carrying an electric current, a voltage (called the *Hall voltage*) develops across the sheet, orthogonal to the current flow. If you keep the magnitude of the current fixed, and increase the strength of the magnetic field, then the *Hall voltage* rises smoothly with the field strength.

Exactly a hundred years later von Klitzing was working in co-operation with Gerhard Dorda (from the Siemens Research Laboratory, Munich, Germany) and Michael Pepper (from the Cavendish Laboratory, Cambridge, England), who provided special devices used in microelectronics. He conducted an experiment at the High Magnetic Field Laboratory in Grenoble, a joint facility of the German Max Planck Institute and the French National Research Council (CNRS), in which he observed that under special circumstances the Hall effect did not obey classical rules. Let us briefly describe the new phenomenon.

Klaus von Klitzing used a special silicon transistor of very high quality, and began to investigate the Hall effect of electrons trapped and moving in a very thin layer at the interface between the semiconductor silicon and the insulator silicon dioxide. He placed the transistor in a coil which produced a very strong magnetic field, and cooled it down to temperatures of about 2 kelvins, using liquid helium.

Surprisingly, he found that the *Hall voltage* did not rise smoothly and continuously as the strength of the magnetic field increased, according to the dictates of the usual rules; it increased, on the contrary, in discrete steps. Perhaps more surprising, he found that, at each step, the ratio between the *Hall voltage* and the electric current through the transistor (what is called the *Hall resistance* of the layer) was equal to an integer multiple of a basic unit, which was measured with an astonishingly small uncertainty of about one-millionth. The quantum of the electrical resistance can be represented in terms of two fundamental constants: the Planck constant h, and the elementary electric charge e; the quantized values of this resistance are equal to h/e^2, $h/2e^2$, $h/3e^2$, and so on. This discovery was called the *integral quantum Hall effect*. Because of its extremely high precision, it was adopted as a standard of the electrical resistance, with a fixed value of h/e^2, the so-called *von Klitzing constant* (this constant is also related to the 'fine structure constant', which plays an important role in the fields of atomic and particle physics, p. 111).

In 1982 Daniel Tsui, Horst Störmer and Arthur Gossard, then working at Bell Labs, discovered that the *Hall voltage* had plateaux also at specific fractional values. This phenomenon was called, not surprisingly, the *fractional quantum Hall effect* (see Nobel 1998).

1986

Touching on atoms

Two of the most important inventions of the twentieth century in the field of microscopy were honoured with the 1986 Nobel Prize for physics. The German engineer Ernst Ruska earned one half of the prize for his work in electron optics, and for his invention of the *electron microscope* (the invention dates from the early 1930s, see p. 199). The other half of the prize was awarded jointly to two experimental physicists, the German Gerd Binnig and the Swiss scientist Heinrich Rohrer, for the design of the *scanning tunnelling microscope*, an extremely sensitive instrument that permits individual atoms to be identified, addressed and even 'handled': this invention was made at the IBM Zurich Research Laboratory in Rüschlikon, Switzerland.

Ernst August Ruska (1906–88) was born in Heidelberg, Germany. He first studied at the Technical University of Munich, and then at the Technical University in Berlin. In 1937 he went to work as a research engineer at the Siemens Company. (In 1939 Siemens brought out its first commercial electron microscope.) In 1955 Ruska was appointed director of the Institute for Electron Microscopy of the Fritz Haber Institute, Berlin, where he remained until 1972. He was also a professor at the Technical University of West Berlin.

The electron microscope

The story begins at the Technical University in Berlin, towards the end of the 1920s. Ruska, a research student there, had discovered how a magnetic field could focus a beam of electrons (just as a glass lens focuses a beam of light). Thus, he thought, electrons can be used, as can light, to 'illuminate' a minuscule object and so proceed to form a magnified image of them.

As we have already noted (p. 199), since electrons have the properties of waves, with wavelengths 100 000 times shorter than those of light, they are able to magnify details of the object more distinctly than can a conventional light microscope (they can reach a resolution of some two tenths of a nanometre). By coupling magnetic lenses, Ruska was able to build the first electron microscope around 1933.

Soon after their invention electron microscopes found applications within many areas of science. There are two types of electron microscope: the so-called *transmission electron microscope*, invented by Ruska, and the *scanning electron microscope*. In a transmission electron microscope, electrons pass through the specimen and form a magnified image of it. In a scanning electron microscope, a fine beam of electrons is swept across the specimen, and, scanning its surface bit by bit, this produces a magnified image.

The scanning tunnelling microscope

Binnig and Rohrer in an article that they wrote for the scientific magazine *Reviews of Modern Physics*, defined the *scanning tunnelling microscope* (STM) thus: '[It is a] kind of nanofinger for sensing, addressing and handling individually selected atoms, molecules, and other tiny objects, and for modifying condensed matter on an atomic scale.'[17] Let us try to describe how this technological marvel works.

In an STM a very fine probing stylus, made of a conducting material (tungsten) is used. The stylus scans across the surface of a conducting or semiconducting material, being separated from it by a tiny distance (of the order of one nanometre,

Fig. 12.12. The principle of the STM (Source: G. Binnig and H. Rohrer, *Reviews of Modern Physics*, 71, 1999, p. S324. © 1999 by the American Physical Society.)

or less than ten atom diameters!). An electric voltage is applied across this tiny *vacuum gap* (which constitutes an energy barrier), and causes electrons to 'tunnel' through the gap, and then pass into the stylus tip. A tunnelling electron current thus begins to flow between the surface and the probe. (This is analogous to the electron tunnelling phenomena that we have discussed in the context of the 1973 Nobel Prize.) The stylus probe is so sharp that it is able to follow even the finest details of the surface to be studied, so that individual atoms can be clearly identified. (Modern STMs achieve astonishing resolutions: they can distinguish two points at the minute distance of one hundredth of a nanometre.)

Binnig and Rohrer succeeded in 1981 in building their first STM. They realised, and this is part of their success, that '... a [stylus] tip is never smooth ... unless treated in a special way. This roughness implies the existence of minitips [with one atom at the apex].'[18] A stylus tip like this is sharp on an atomic scale, and gives the 'nanofinger' the required sensitivity to 'touch and play' with atoms. They then placed their apparatus on a superconducting lead bowl, which levitated upon a magnet, so as to eliminate the disturbing vibrations from the environment; by so doing, they succeeded in obtaining the first images of individual atoms on the surface of a material.

Soon after Binnig and Rohrer had built their STM, this began to be used in many sectors of science and technology (the so-called 'nano-technology'). It is used, for example, in surface semiconductor studies, microelectronics, electrochemistry, metallurgy and molecular biology.

Gerd Binnig was born in Frankfurt, Germany, in 1947. A Nobelist of a recent generation, he recalls his youth with these words:

> My brother was responsible for my transition from classics to beat by his perpetually immersing me with the sounds of the Beatles and the Rolling Stones, until I finally really liked that kind of music, and even started composing songs

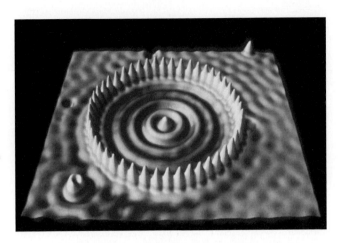

Fig. 12.13. STM image of 48 iron atoms on copper. This 'atom corral' was made in an artificial nanostructure by D. Eigler (IBM Research Center, Almaden, California), using the STM as a 'nanofinger'. (Image courtesy of the IBM Zurich Research Laboratory.)

> and playing in various beat-bands . . . For the time being, nearly all my hobbies, like music (singing, playing the guitar and the violin), and sports (soccer, tennis, skiing, sailing and playing golf) have had to take a back seat. It was in 1978 that Lore – my private psychotherapist – convinced me to accept an offer from the IBM Zurich Research Laboratory to join a physics group. This turned out to be an extremely important decision, as it was here I met Heinrich Rohrer.[19]

Binnig obtained his doctorate from the University of Frankfurt in 1978. He then joined the IBM Zurich Research Laboratory, where he went to work in the group headed by Rohrer which was beginning to design the scanning tunnelling microscope. In 1986 he was appointed IBM Fellow, and established his IBM physics group in Munich (he retired from IBM in 2002). In 1990 Binnig became a member of the Supervisory Board of the Daimler–Benz Holding.

Heinrich Rohrer was born in Buchs, St Gallen, Switzerland, in 1933. He studied at the Swiss Federal Institute of Technology (ETH) in Zurich, with among others, Wolfgang Pauli. Here he received his doctorate in 1960, with a thesis on superconductors. In 1963 he joined the IBM Zurich Research Laboratory, where he remained until his retirement in 1997 (he was appointed IBM Fellow in 1986). Rohrer wrote in his Nobel autobiography:

> In all the years with IBM Research, I have especially appreciated the freedom to pursue the activities I found interesting, and greatly enjoyed the stimulus, collegial cooperation, frankness, and intellectual generosity of two scientific communities, namely, in superconductivity and critical phenomena.[20]

> . . . [The] local study of growth and electrical properties of thin insulating layers appeared to me an interesting problem, and I was given the opportunity to hire a

Fig. 12.14. H. Rohrer (left) and G. Binnig. (Courtesy AIP Segrè Visual Archives, *Physics Today* Collection.)

new research staff member, Gerd Binnig, who found it interesting, too, and accepted the offer. Incidentally, Gerd and I would have missed each other, had it not been for K. Alex Müller [Nobelist in 1987], then head of Physics, who made the first contacts.

(Heinrich Rohrer)[21]

1987

Revolution in superconductivity

Once again the IBM Zurich Research Laboratory occupied the Nobel authorities' attention when they decided to award the 1987 Nobel Prize for physics jointly to Georg Bednorz and Alexander Müller, two IBM researchers. (They worked in the same laboratory as Gerd Binnig and Heinrich Rohrer, who had received the same award one year earlier.) Bednorz and Müller joined the elite of the Nobel laureates for having discovered in 1986 the so-called *high-temperature superconductors*. The award came only one year after their discovery had been reported to the world of science. In December 1987 the American magazine *Physics Today* reported: 'Few doubted that the discovery by Bednorz and Müller merited the Nobel Prize; speculations on what year the prize would be awarded have abounded since last January.'[22] (This is only the third example of a Nobel Prize for physics satisfying the rule of the Nobel Statutes that the discovery recognised should have been made in the preceding year.)

Johannes Georg Bednorz was born in Neuenkirchen, North-Rhine Westphalia, (then West) Germany, in 1950, the son of a primary school teacher. He started his studies in 1968 at the University of Münster. (He later recalled: 'In 1972 . . . my teachers arranged for me to join the IBM Zurich Research Laboratory for three months as a summer student. It was a challenge for me to experience how my scientific education could be applied in reality . . . I soon was impressed by the freedom that even I as a student was given to work on my own, learning from mistakes and thus losing the fear of approaching new problems in my own way.')[23] After graduation in 1976, Bednorz went to the Swiss Federal Institute of Technology (ETH) in Zurich, where he obtained his doctorate in 1982 with Müller as a thesis supervisor. The same year he joined the IBM Zurich Research Laboratory, where he is now an IBM Fellow, working on the development of new materials to be used in microelectronics. As well as the Nobel, Bednorz' researches have been honoured with many other prizes, among them the Hewlett–Packard Europhysics Prize and the APS International Prize for Materials Research.

Karl Alexander Müller was born in Basel, Switzerland, in 1927, the son of a musician. He studied at the ETH in Zurich with Pauli: '[Pauli's] courses and examinations [that] I took, formed and impressed me. He was truly a wise man with a deep understanding of nature and the human being.'[24] He received his master's degree in 1952 and his doctorate in 1958. He then joined the Batelle Memorial Institute in Geneva, where he became the head of the magnetic resonance group. In 1962 he was appointed a lecturer at the University of Zurich, and a professor in 1970. He joined the IBM Zurich Research Laboratory in 1963, where in 1972 he became head of the physics department (here Binnig and Rohrer developed the scanning tunnelling microscope). From 1985 he was relieved of all duties as a manager in order to devote his all work time to studying new superconducting materials, together with Bednorz. From 1982 up to his retirement in 1998, Müller was an IBM Fellow, and from 1989 he has been a member of the US National Academy of Sciences. He has also been rewarded with many honorary degrees, from more than ten universities, and recognition from many industrial and research institutions.

Fig. 12.15. Müller (left) and Bednorz. (Courtesy Professor Alexander Müller.)

High-temperature superconductors

When in 1911 Heike Kamerlingh Onnes, working at the University of Leiden, discovered superconductivity in mercury, he needed to chill the metal to a very low temperature (about 4 degrees above absolute zero) for its superconducting characteristics to be observed. Soon after Kamerlingh Onnes' discovery, scientists began to dream of finding new materials which could become superconducting with zero electrical resistance at higher temperatures, but they obtained little practical success. Superconductors hence remained a mere laboratory curiosity for half a century.

Then in the 1960s experimenters did begin to discover new superconducting materials, and they gradually pushed up the transition temperature at which superconductivity appears. Metal wires made of niobium and tin, and later of niobium and titanium alloy, were developed. These then became widely used in applications of superconductors (in large electromagnets for high-energy particle accelerators, in apparatuses for thermonuclear fusion, and in magnetic resonance imaging devices). By 1973 a transition temperature of 23 kelvins (-250 °C) was reached with an alloy of germanium and niobium. This was the maximum value achieved at that time, and many solid-state theorists were then convinced that the transition temperature would never go above 30 kelvins.

At this point an amazing surprise occurred. In September 1986 Bednorz and Müller reported, in a paper published in *Zeitschrift für Physik*,[25] that they had discovered a truly new material (a mixture of lanthanum, barium, copper and oxygen) which became a superconductor at 35 kelvins (-238 °C), well above the dream of 30 kelvins. And this was not all! The new class of compounds discovered by our Nobelists were ceramics (they are metal oxide compounds, called *perovskites*). This made their discovery even more remarkable, because ceramics are normally insulators, whereas the so-called low-temperature superconductors, discovered before the year 1986, were all metallic or semi-metallic.

The Zurich discovery stimulated a flurry of research activity in hundreds of laboratories around the world. About ten months later, the physicist Paul Chu (from Houston University, USA) announced, in a paper published in *Physical Review Letters*, that his group had discovered a new compound, a mixture of yttrium, barium, copper and oxygen, which became superconducting at 93 kelvins (−180 °C); a year or so later the maximum transition temperature reached 125 kelvins (−148 °C) in a compound of thallium, barium, calcium, copper and oxygen. These discoveries were highly significant because these transition temperatures are well above the boiling point of liquid nitrogen (77 kelvins). This means that superconductivity could become economically viable (cooling to low temperatures with liquid nitrogen is easier and much cheaper than with liquid helium). Since then physicists have discovered many different compounds (more than a hundred), reaching a world record of about 140 kelvins for the transition

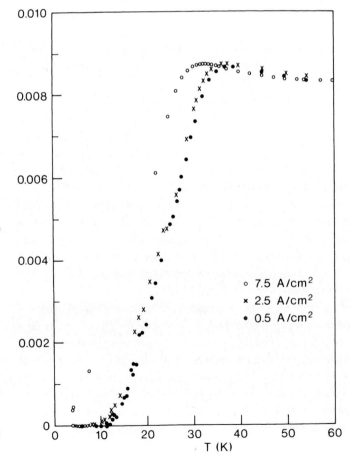

Fig. 12.16. The decrease in the electrical resistance of a high-temperature superconductor sample used by Bednorz and Müller. (Source: J. G. Bednorz and K. A. Müller, *Zeitschrift für Physik B – Condensed Matter*, 64, 1986, p. 190. © Springer-Verlag.) Note the similarity of this temperature descent with that reported in Kamerlingh Onnes' 1911 diagram (see Nobel 1913; Fig. 5.7).

temperature (the current record, about 160 kelvins, has been reached by holding a superconducting compound under pressure).[26]

All these new materials are referred to as high-temperature superconductors. Today, they are moving out of the laboratory and into the worlds of engineering and technology in efforts to explore possible fresh applications. Let us mention just a few of them: very sensitive devices for measuring electric voltage or current; energy storage systems; detectors for extremely feeble magnetic fields; microwave communication systems; power transmission; and bio-magnetism.

A Woodstock for physics

One of the most memorable sessions in physics occurred on March 18, 1987, during a meeting of the American Physical Society. Hastily arranged to accommodate a huge number of postdeadline talks, the gathering in the Hilton Hotel in New York City drew 2000 physicists. Crammed into a ballroom, with many others spilling into the hallway, they struggled to give and to hear five-minute briefings on the latest ideas and investigations. Dubbed the Woodstock of physics, the session started at 7:30 p.m. and continued until

Fig. 12.17. The bright star in the right lower corner is SN1987A. (Courtesy Lawrence Berkeley National Laboratory.)

3 A.M. – and even at that hour the excitement was still evident. The cause of all this academic commotion was the announcement of high-temperature superconductivity. Late in 1986 J. Georg Bednorz and K. Alexander Müller of the IBM Research Laboratory in Zurich had reported that a ceramic called lanthanum barium copper oxide lost all electrical resistance when it was cooled to only – 238 degrees Celsius . . . Although that temperature is still quite frigid, it was nonetheless more than 10 degrees better than the best conventional superconductors, which are made from metals or alloys.[27]

Supernova SN1987A

On 23 February 1987 a supernova was discovered in the Large Magellanic Cloud (one of the companion galaxies of our own Milky Way), about 160 000 light years away from the earth. Its light was seen by astronomers of the European Southern Observatory (located in Cerro La Silla, Chile). The supernova, designated *SN1987A*, was so bright that it could be seen with the naked eye. It was the brightest since the supernova that Johannes Kepler had witnessed in 1604, and the first to be studied with new-technology optical telescopes. (It was the most important supernova event during the twentieth century!)

Apart from the luminous energy, a supernova explosion produces a huge quantity of neutrinos, due to the weak nuclear decays during the collapse of the star core. A typical supernova neutrino pulse corresponds to an energy burst of about ten billion billion billion billion billion joules per second (ten times the luminous energy of all the stars and galaxies in our observable universe). Fortunately, on the same day as the supernova exploded, a neutrino telescope, the Kamiokande detector, located in Japan (see Nobel 2002) was ready and detected no less than a burst of neutrinos, three hours before astronomers saw the supernova light in the sky. (Simultaneously, the Irvine–Michigan–Brookhaven detector, located in a salt mine north-east of Cleveland, Ohio, USA, confirmed the Kamiokande signal.)

1988

Neutrinos at last rewarded!

At long last! Fifty-eight years after its debut in the field of science, and thirty-two years after its actual detection, the neutrino receives its due by gaining a Nobel Prize. The Swedish Academy of Sciences in fact decides to acknowledge, for the first time, the most enigmatic of all the known particles, by awarding the 1988 Nobel Prize for physics jointly to three American physicists: Leon Lederman, Melvin Schwartz and Jack Steinberger. They had demonstrated, in 1962, the existence of two types of neutrinos, the *electron neutrino* and the *muon neutrino* (see p. 286), so opening up an entirely new way to the understanding of elementary particles and their interactions. On hearing the Nobel news, Steinberger said, referring to the delay: 'If you want to get that prize, do your work early!' But in a sense there was nothing unprecedented in this: for example the discovery of the existence of the neutrino itself, made by Clyde Cowan and Frederick Reines as long ago as 1956 (see p. 260) had to wait no less than thirty-nine years, until 1995, before it received the full acknowledgement that it deserved!

History of an experiment

And now back to the year 1962, when the famous *two-neutrino experiment* was carried out at the Brookhaven National Laboratory. Let us start our chronicle by seeing how it was born.

Towards the end of the 1950s, soon after the discovery of parity violation in weak interactions by T. D. Lee and C. N. Yang (see Nobel 1957), particle physicists attempted to transform Fermi's 1934 theory of beta decay into a new theory of weak interactions capable of explaining the many phenomena observed in experiments. They also began thinking of a way of using neutrinos in order to study weak interactions at high energies. At Columbia University this was a hot argument for discussion; Melvin Schwartz thus recalls an afternoon coffee session in the conference room of the Pupin Physics Laboratory in November 1959:

> The Columbia University physics department had a tradition of a coffee hour at which the latest problems in the world of physics came under intense discussions. At one of these Professor T. D. Lee was leading such a discussion of the possibilities for investigating weak interactions at high energies. A number of experiments were considered and rejected as not feasible. As the meeting broke up there was some sense of frustration as to what could ever be done to disentangle the high energy weak interactions from the rest of what takes place when energetic particles are allowed to collide with targets . . . That evening the key notion came to me – perhaps the neutrinos from pion decay could be produced in sufficient numbers to allow us to use them in an experiment. A quick 'back of the envelope' calculation indicated the feasibility of doing this at one or another of the accelerators under construction or being planned at that time. I called T. D. Lee at home with the news and his enthusiasm was overwhelming. The next day planning for the experiment began in earnest. Meanwhile Lee and Yang began a study of what could be learned from the experiment . . . Not long after this point we became aware that Bruno Pontecorvo [one of Fermi's boys in Rome, who at the time was working in the Soviet Union] had also come up with many of the same ideas as we had.[28]

Schwartz wrote down his idea in a short paper which appeared in the March 1960 issue of the *Physical Review Letters*, back to back with a paper by Lee and Yang. In their paper Lee and Yang investigated the physics which should result from an experiment like that described by Schwartz: they proposed, as a first step, to test whether the neutrinos produced with muons in pion decays were the same as those produced with electrons in nuclear beta decays (this also represented a possible explanation of the fact that a certain type of muon decay, as predicted by theory, had never before been observed in experiments).

The two-neutrino hypothesis was thus the conjecture that our three heroes, Lederman, Schwartz and Steinberger (at the time professors at Columbia University), set out to investigate. They worked out a proposal for setting up a neutrino beam and a detector at the Brookhaven AGS accelerator. In the summer of 1961 they were ready to run, and, as we have seen (p. 286), after eight months they were able to prove that neutrinos do indeed come in two forms, electron-like and muon-like. (We now know that there is a third neutrino, the *tau neutrino*, see p. 422.)

The two-neutrino experiment (also known as the 'four-lepton experiment') led to the recognition of two families of leptons, so providing a fundamental piece of information for the development of what was later called the *standard model* of fundamental particles and forces. It was also the first experiment with a high-energy neutrino beam; this transformed the elusive neutrino into becoming one of the main experimental tools of high-energy physics.

The Columbia team

Leon Max Lederman was born in 1922 in New York City, the son of Jewish immigrants. ('My father, Morris, operated a hand laundry and venerated learning. Brother Paul, six years older, was a tinkerer of unusual skill.'[29]) Lederman graduated in chemistry from the City College of New York in 1943, and successively spent three years in the US army, where he rose to the rank of Second Lieutenant in the Signal Corps. After the war he continued his studies at Columbia University, where in 1951 he obtained his Ph.D. He remained working there for twenty-eight years, and became full professor of physics in 1958. From 1961 to 1978 he was director of the cyclotron laboratory at Columbia, and in 1979 he was appointed director of the newly constructed Fermilab in Batavia, near Chicago. In 1989, after retiring from Fermilab, Lederman joined the University of Chicago, concluding there an outstanding career, both as a scientist and as a university professor. He remarked in his Nobel autobiography: 'During my academic career at Columbia (1951–79) I have had 50 Ph.D. students, 14 are professors of physics, one is a university president and the rest with few exceptions, are physicists at national labs, in government or in industry. None, to my knowledge, is in jail.'[30] Lederman is the recipient of many honorary doctorates, and other awards, among them the US National Medal of Science and the 1982 Israeli Wolf Prize for physics.

Melvin Schwartz was born in 1932 in New York City, in a Jewish family. He recalled in his Nobel autobiography:

> Having been born . . . at the peak of the great depression, I grew up in difficult times. My parents worked extraordinarily hard to give us economic stability but at

Fig. 12.18. From left to right: L. Lederman, J. Steinberger and M. Schwartz. (Courtesy CERN.)

the same time they managed to instill in me two qualities which became the foundation of my personal and professional life. One is an unbounded sense of optimism; the other is a strong feeling as to the importance of using one's mind for the betterment of mankind.[31]

In 1949 Schwartz went to Columbia University, where he obtained his Ph.D. in 1958, working under Steinberger. He remained at Columbia for seventeen years, becoming first assistant professor (1958), then associate professor (1960) and finally full professor of physics (1963). Later he moved to Stanford University, where he became involved in new research in the field of particle physics. During the 1970s he founded a company specialising in computer communications. A very enthusiastic and active man, Schwartz thus concluded his Nobel autobiography: 'Although it is difficult to predict the future I still have all the optimism that I had back when I first grew up in New York – life can be a marvellous adventure.'[32]

Jack Steinberger was born in Bad Kissingen, West Germany, in 1921, the son of a Jewish 'cantor and religious teacher'. He emigrated to the United States in 1934:

> In 1933, the Nazis came to power and the more systematic persecution of the Jews followed quickly . . . When, in 1934, the American Jewish charities offered to find homes for 300 German refugee children, my father applied for my older brother and myself. We were on the SS Washington, bound for New York, Christmas 1934. I owe the deepest gratitude to Barnett Faroll, the owner of a grain brokerage house on the Chicago Board of Trade, who took me into his house, parented my high-school education, and made it possible also for my parents and younger

> brother to come in 1938 and so to escape the holocaust . . . I was able to continue my education . . . [studying] chemical engineering. I was a good student, but these were the hard times of the depression, my scholarship came to an end, and it was necessary to work to supplement the family income . . . In the evenings I studied chemistry at the University of Chicago, the weekends I helped in the family store.[33]

During the Second World War Steinberger worked at the MIT Radiation Laboratory; after the war he went to the University of Chicago, where he obtained his Ph.D. in 1948, writing a thesis under the supervision of Enrico Fermi. He was appointed Professor of Physics at Columbia University from 1950 to 1971. He then returned to Europe to work at CERN. In 1986 Steinberger retired from CERN and became a professor at the *Scuola Normale Superiore* in Pisa, Italy.

Discoveries and inventions

***How many neutrinos*?** In 1988 two neutrinos were known (the *electron neutrino* and the *muon neutrino*). Physicists had also inferred the existence of a third kind of neutrino, the τ *neutrino*, associated with the τ *lepton*; it was observed in the year 2000 (p. 422). The number of neutrino species existing in nature plays an important role not only in the field of particle physics but also in cosmology. Their importance in cosmology is due to the fact that, after the Big Bang, the universe expanded and cooled, and chemical elements began to be formed. Astrophysicists then discovered that the amount of elements such as hydrogen, deuterium, or helium, and so on, which would later become part of our present world, depended on how many neutrinos there were. They used the measured abundance of these elements in our stars and calculated that there ought to exist only three kinds of neutrinos. In 1989, experiments carried out at CERN and at SLAC demonstrated that the number of neutrino species in nature is in fact *three*, and only *three*: exactly what astrophysicists had foreseen!

1989

Atomic clocks and particle traps

After eight years, the Nobel Physics Committee once again considered the science of spectroscopy when planning the Nobel awards. There were three prizewinners for the year 1989: the American physicists Norman Ramsey from Harvard University and Hans Dehmelt from the University of Washington, Seattle, and the German scientist Wolfgang Paul from the University of Bonn, West Germany.

Ramsey had invented a method which laid the basis of the *caesium atomic clock* – the standard clock used to define the present-day unit of time. He had also developed the *hydrogen maser*, an unprecedentedly stable source of microwave radiation. Dehmelt and Paul had developed very refined techniques, called *particle traps*, with which physicists could study a single isolated electron or a single charged ion with extreme precision. With these techniques, which received the seal of the Nobel Prize, it became possible to study a number of matters of great scientific and practical importance, ranging from tests on fundamental theories, such as general relativity and quantum electrodynamics, to accurate measurements of physical constants, as well as technical improvements in space communication.

Atomic clocks

In 1947 Norman Ramsey moved from Columbia to Harvard University. At Columbia he had worked with Isidor Rabi (Nobel 1944) in his laboratory, where he had carried out experiments for measuring nuclear magnetic moments. When at Harvard, Ramsey established a new laboratory, and in 1949 he invented a new technique which appreciably modified the magnetic-resonance method developed by Rabi (p. 218), increasing its accuracy. He then applied this new method to studying the internal oscillations of the isotope caesium-133, and succeeded in measuring its transitions between very closely spaced energy levels (the hyperfine structure) with the fantastic accuracy of two parts in a hundred thousand billion. (Caesium-133 has thus become the standard clock, adopted by the General Conference on Weights and Measures, which in 1967 defined the second as the time for 9 192 631 770 oscillations of the hyperfine transition of the caesium atom.) Caesium-133 atomic clocks are also used in many practical applications: for example, they constituted the time standards placed on the satellites of the GPS (Global Positioning System).

The Swedish Academy of Sciences cited Ramsey and his co-workers, among them his former research student Daniel Kleppner, also for their 1961 work in developing a new maser, the so-called hydrogen maser. This device functioned as a source of a very pure beam of microwaves (with a frequency of 1420 million oscillations per second, and a wavelength of 21 centimetres). This was the first atomic maser, and it was used in the study of the hyperfine structure of atomic hydrogen, deuterium and tritium, as well as for time and frequency measurements of extreme precision. Another application of the hydrogen maser has been the verification in 1976 of the gravitational red shift predicted by Einstein's general relativity (see p. 109). The frequency of one hydrogen maser sent aloft on a rocket was compared with that of an analogous maser on the earth. The result was that Einstein's theory was verified with an accuracy of seven parts in a hundred thousand.[34]

Norman Foster Ramsey was born in 1915 in Washington, DC. His father was a military officer and his mother a university mathematics instructor. After graduation at Columbia University in 1935, he went for two years to the Cavendish Laboratory in Cambridge, England, where he obtained his second degree. He then returned to Columbia as a research student, working with Isidor Rabi and Polykarp Kusch (Nobelist in 1955) in the new field of molecular-beam magnetic

Fig. 12.19. Mrs Elinor Ramsey and Professor Ramsey. (AIP Emilio Segrè Visual Archives, Ramsey Collection.) Ramsey thus concluded his Nobel autobiographical notes: 'I have greatly enjoyed my years as a teacher and research physicist and continue do so. The research collaborations and close friendships with my eighty-four graduate students have given me especially great pleasure. I hope they have learned as much from me as I have from them.'[35]

resonance; he obtained his Ph.D. in 1940. During the Second World War, Ramsey worked on radar at the MIT Radiation Laboratory, and at Los Alamos on the Manhattan Project. After the war he returned to Columbia, where he was appointed Professor of Physics. He participated with Rabi in the establishment of the Brookhaven National Laboratory, where in 1946 he became head of the physics department. In 1947 he moved to Harvard University; he taught physics there for forty years, including being appointed in 1966 to the Higgins Chair of Physics.

Particle traps

Physicists have always dreamed of being able to observe a single atom or a single particle for enough time for it to be analysed accurately. In the 1950s Wolfgang Paul, then at the University of Bonn, Germany, using both a specially shaped electric field and radio waves, showed that it was possible to guide and separate ions with different masses. This method was used to develop very sensitive mass spectrometers, which became widely used in research and industrial laboratories.

Later, in the 1970s, he invented a new device, a trap for ions, which used an electric field oscillating at radio frequencies, to confine and hold ions in a small volume of space. The *Paul trap*, as the invention was named, became an important tool for high-precision spectroscopy.

Also in the 1950s Hans Dehmelt, in Seattle, USA, developed another type of trap, which combined both a static, non-uniform electric field and a static, uniform magnetic field (it was called the *Penning trap*). Then, in 1973 Dehmelt used his trap to confine, for the first time, a single electron. He was able to hold the electron in place for several months, and to measure its anomalous magnetic moment, an anomaly discovered by Polykarp Kusch twenty-six years earlier (see p. 252), with an accuracy of four parts in a thousand billion. (Kusch had measured this same anomaly with a precision of four parts in a hundred thousand.) This represented the most precise test of quantum electrodynamics ever performed. Finally, in the late 1970s, Dehmelt succeeded in observing a single ion in a trap, so opening the way to the so-called *single-ion spectroscopy*.

Hans Georg Dehmelt was born in 1922 in Görlitz, Germany. During the Second World War he served in the German army until he was captured in 1945. He spent a year in an American prisoner of war camp in France; then, in 1946, he moved to the University of Göttingen to study physics, where in 1950 he earned his doctorate. He emigrated to the USA in 1952, and went to work at the University of Washington, Seattle, where he began teaching physics in 1955, and became a full professor in 1961.

Wolfgang Paul (1913–93) was born in Lorenzkirch, a small village in Saxony, Germany, the son of a university professor of pharmaceutical chemistry. He studied both at the Technical University of Munich and at the Technical University in Berlin, where in 1939 he received his doctorate in physics. Paul became a lecturer at the University of Göttingen in 1944, and assistant professor in 1950. Two years later he moved to the University of Bonn, where he became a full professor of physics and director of the physics institute. In the period 1964–7 Paul was a research director at CERN (Geneva); and from 1970 to 1973 he served as chairman of the scientific council of DESY, the German national laboratory for particle physics in Hamburg.

1990

Quarks revealed

As you will recall, Murray Gell-Mann (Nobel 1969) introduced, in 1964, the concept of quarks. The two basic constituents of the atomic nuclei, the proton and the neutron, were thought of as being composed of two kinds of quarks, up and down.

In 1967 three American physicists, Jerome Friedman and Henry Kendall, both from MIT, and Richard Taylor, from Stanford University, started a series of experiments at the SLAC linear accelerator. After seven years of careful investigations they obtained clear evidence that 'hard matter grains' resided within the nucleons. Then experiments carried out at other accelerators followed, and all fully confirmed the SLAC results, giving more and more evidence that the 'hard grains' discovered there could be correctly interpreted as Gell-Mann's quarks.

Later investigations carried out between 1970 and 1980 were able to explain how quarks interact; thus the quark model became widely accepted by physicists, and could be fitted into a coherent theoretical scheme, called *quantum chromodynamics* – the modern theory of the strong nuclear forces. The Swedish Academy of Sciences recognised the importance of the pioneering SLAC investigations, and awarded Friedman, Kendall and Taylor the 1990 Nobel Prize for physics.

The SLAC–MIT experiment

The SLAC–MIT team used a beam of high-energy electrons to investigate the internal structure of protons and neutrons. The electrons were supplied by the three-kilometre-long linear accelerator at SLAC, the same one that in 1974 would be used to supply electrons and positrons to SPEAR, for Richter's famous experiment (p. 337). (At the end of the 1960s the Stanford linear accelerator was thus functioning as a gigantic electron microscope.)

The technique used here, which is called *electron scattering*, was the same as that used by Robert Hofstadter (Nobelist in 1961) for studying how the electric charge is distributed in the nucleons. Other experiments were performed after those of Hofstadter, without revealing any more about the inner structure of nucleons. But physicists had up until then used electrons with not enough energy to investigate the innermost part, the (to use Sommerfeld's words) *holy of holies* of the nucleon. Only in the late 1960s were electrons of a high enough energy (up to about 20 billion electron-volts) available, and then only at SLAC.

The SLAC–MIT team fired these electrons at a target consisting of either liquid hydrogen or deuterium, and recorded the scattered electrons at different angles. (To measure the energies of the electrons they used devices called 'magnetic spectrometers', which contained strong magnetic fields.) The first experiment started at the beginning of 1967, and investigated the so-called *elastic scattering*, in which electrons with lower energies bounced up against the nucleons in the target.

The experimenters then decided to study the so-called *deep inelastic scattering*, where the incident electrons possessed a higher energy, and lost a large part of this energy in the collisions with the target nucleons. They found, to their amazement, that the frequency of electrons scattered at large angles was considerably greater than had been expected. It was exactly the same situation as had been

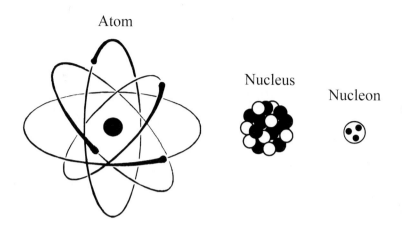

Fig. 12.20. Four layers of matter. In the early 1910s the scattering of alpha particles on atoms revealed the atomic nucleus; in the 1930s physicists discovered that the nucleus is composed of protons and neutrons; in the early 1970s high-energy electrons were fired at protons and neutrons and revealed the quarks within.

found in the Geiger–Marsden experiment (p. 80). At that time, the scattering of alpha particles at large angles was explained by Rutherford in terms of the existence of a 'hard core' within the atom, that is, the atomic nucleus. But now, the large number of high-energy electrons scattered at large angles was explained by the existence of 'hard grains' within nucleons, that is, quarks. These results were confirmed by complementary experiments carried out with neutrino beams at CERN (Geneva), and at Fermilab (Batavia, near Chicago).

More experiments were performed in the following years, which indicated (as the SLAC–MIT experiments had done) that there were also electrically neutral particles within nucleons. These were interpreted as being carriers of the strong forces between quarks, and they were called *gluons*. (The first direct evidence for gluons came in 1979 from experiments performed at the electron–positron collider PETRA, then in operation at the DESY laboratories in Hamberg, Germany.)

The SLAC–MIT leaders

Jerome Isaac Friedman was born in Chicago in 1930, the son of Russian immigrants. In 1950 he went to the University of Chicago to study physics, where in 1956 he obtained his Ph.D. In 1957 he joined Hofstadter at Stanford, where he learned the electron-scattering technique; at the same time he began his collaboration with Henry Kendall. In 1960 he moved to MIT, to a teaching post there, but he returned later to Stanford to work at SLAC. In 1967 Friedman became full professor at MIT, and in the 1980s he was appointed director both of the Laboratory for Nuclear Science and of the physics department there.

Henry Way Kendall (1926–99) was born in Boston, Massachusetts. He earned a Ph.D. in physics at MIT in 1955, and subsequently worked there and at the Brookhaven National Laboratory. He then joined Hofstadter's group at Stanford, and in 1961 he went back to MIT, where he became a full professor in 1967. After the work with Friedman and Taylor, he worked at the proton accelerator at

Fig. 12.21. From left to right: R. Taylor, H. Kendall and J. Friedman. (Courtesy AIP Emilio Segrè Visual Archives, *Physics Today* Collection.)

Fermilab, and later again at SLAC. In addition to his research activity, Kendall was deeply involved with arms control and nuclear-power safety issues; he was a founding member of the Union of Concerned Scientists, and served as its chairman for twenty-five years.

Richard Edward Taylor was born in 1929 in Medicine Hat, a small town in south-western Alberta, Canada. He first studied at the University of Alberta in Edmonton, and then went to the USA, to Stanford University as a graduate student. In 1958 Taylor went to France to the Orsay Laboratory, Paris, to carry out experiments at the new electron accelerator. In 1961 he returned to the USA, and in 1962 he obtained his Ph.D. from Stanford University. Since then he has been a staff member at SLAC. In 1968 he was appointed associate professor at Stanford University, becoming a full professor in 1970.

Quantum chromodynamics

To explain the combinations of quarks which make up hadrons, it was necessary to suppose that they were distinguished by an extra physical property, or quantum number; they had to carry a form of strong charge on which the strong force acts. This strong charge had been given the whimsical name of *colour* (this is a conventional term which has nothing whatsoever to do with chromatic colours.)

By contrast with the electric charge, which can have only two values (positive and negative), quarks can carry three different types of colour charge (this property was first introduced by Yoichiro Nambu), which are called 'red', 'green' and 'blue', by analogy with the primary colours of light. This analogy makes it easier to specify the rules which govern the different combinations within quarks. All the hadrons must be 'white'; that is, 'colourless'. (For example, baryons, which include the proton and the neutron, are made up of a red, a green and a blue quark, since these colours average up to white.)

Fig. 12.22. Big physics. An aerial view of the laboratories at CERN, at its Meyrin (Geneva) site. (Courtesy CERN.) The new collider, called LHC, will be operating towards the end of the first decade of this century. It will be composed of 5000 super-conducting magnets, housed in the 27-kilometre LEP tunnel. LHC will provide proton–proton collisions at energies of (7000 + 7000) billion electron-volts.

Fig. 12.23. Small physics. The IBM Zurich Research Laboratory at Rüschlikon, near Zurich, Switzerland. (Image courtesy of the IBM Zurich Research Laboratory.) Here Gerd Binnig and Heinrich Rohrer (Nobelists in 1986) invented the scanning tunnelling microscope, and Georg Bednorz and Alexander Müller (Nobelists in 1987) discovered high-temperature superconductors.

The *colour charge* plays the same role in strong interactions as the electric charge does in electrodynamics. Because of this analogy the theory of the strong force is called *quantum chromodynamics*, or *QCD*. It is closely modelled on quantum electrodynamics, and like QED it postulates messenger particles, which are exchanged between quarks and carry the strong force. Like the photon of QED, these messengers have zero mass. They are called *gluons* because they provide the 'glue' that holds quarks together in hadrons.

To allow for all the possible interactions between the three colour charges there must be eight such gluons. The absorption or emission of a gluon changes the colour of the quark that it is interacting with. Thus, gluons bear a colour charge, at variance with the photon of QED, which is electrically neutral. Because of this, gluons can interact among themselves, so accounting for the differing behaviours, over distance, of the strong and electromagnetic forces.

In fact, the strong colour force varies with the distance, but in a different way compared with the electric force; instead of weakening with distance the force grows stronger and stronger. This prevents quarks from being separated and becoming free. Quarks are hence permanently *confined* to the interior of hadrons (actually no single *free* quark has ever been observed!). Only at distances of less than one million-billionth of a metre does the colour force diminish enough to permit mutually bound quarks a certain degree of independence (this effect is called in technical jargon *asymptotic freedom*; it was discovered in the early 1970s by David Gross and Franck Wilczek, and independently by David Politzer).

Chapter 13
New trends

The discoveries in 1995 of the 'top quark' and in the year 2000 of the 'tau neutrino', at Fermilab, have to be seen as the achievements of a very long journey which had started off more than forty years previously, in a continuous and widening search for the ultimate building blocks, and the most basic forces, of nature. The end of the story, however, is by no means in view. Seeing how capable their theories are of explaining the subatomic world, particle physicists have begun to set their sights even higher, and we catch them dreaming of one day devising a *final theory*: this would eventually unify all the forces of nature, and describe every one of its basic constituents. At the same time, more practically, they are also dreaming of having at their disposal more and more powerful particle accelerators, so that they can test their theories. (A new collider, named the LHC, is under construction at CERN, and will be the work-horse of the first decades of the twenty-first century.)

During the preceding decades, as we have seen, stars, galaxies and the Big Bang have been explored by astronomers and astrophysicists with ever more powerful instruments. The results obtained have already revolutionised our understanding of the universe. Cosmologists have progressed in their understanding of its birth, and the universe itself is becoming a unique astrophysical laboratory for testing out basic theories in physics (see Nobel 1993 and Nobel 2002).

Finally, advances and breakthroughs in so-called small physics (the physics of matter and light, superconductivity, superfluidity, the transistor, nuclear magnetic resonance, the laser, new materials) have had an enormous impact on science and technology, and on today's society in general. Modern electronics has led us into the information age (see Nobel 2000); new sensitive instruments have been created and used in science, industry and medicine. New, promising lines of research have been opened up (for example *soft-matter physics*, see Nobel 1991). The ability to manipulate super-chilled atoms yielded in 1995 a new state of matter (the 'Bose–Einstein condensate'), and opened the door to the development of new quantum technologies. During the last thirteen years (1991–2003), eight Nobel Prizes, awarded to twenty-one scientists, witness the importance given to the triumphs gained in different areas of small physics.

1991

Soft matter

Soft-matter (or *complex-fluid*) *physics* deals with complex systems such as liquid crystals, polymers, colloids, glues and biological structures. It is a growing and expanding field of research, with many promising perspectives. The Swedish Academy of Sciences underlined the importance of this new trend in physical science research when awarding the 1991 Nobel Prize for physics to the French theorist Pierre-Gilles de Gennes – one of the most prestigious exponents working in the field. The prize was awarded particularly for his studies on liquid crystals and polymers, which he carried out in the 1960s and 1970s.

Liquid crystals

Certain materials have two distinct melting points: by increasing the temperature, they change first into a hazy, and then into a transparent, liquid. They behave both like a liquid and a solid, and are called *liquid crystals*. These materials consist of organic molecules having a rod-like structure, which tend to be oriented in the same direction. In a liquid, molecules have no preferred orientation; on the contrary, in a solid crystal they are highly ordered. The characteristic orientation of molecules in liquid crystals is between that of a solid and that of a liquid. Liquid crystals are employed in many areas of science and technology: their optical properties, for example, have been exploited to construct displays for digital watches, calculators, television sets and computers.

De Gennes made important contributions to knowledge about liquid crystals. He was able to explain their optical properties; he demonstrated, too, the existence of strong analogies which related a certain class of liquid crystals to superconductors. His deductions were largely verified by experiment.

Polymers

Polymers are a large class of materials consisting of many small molecules (about one-billionth of a metre long) that are linked together to form long chains, consisting of tens of thousands of similar links: thus they are known as *macromolecules*. Polymers have been used for centuries in the form of gums, oils and resins. In the 1930s materials such as polystyrene and nylon were developed, and soon an explosion in polymer research took place. Natural and synthetic polymers (cotton, wool, rubber; Teflon, plastics) are used in nearly every industry today.

For years scientists had tried their best to describe the spatial arrangement of the molecules in polymers; for example, by exploiting certain mathematical techniques that had been successful in theoretical particle physics. De Gennes and his colleagues made important contributions in this field. They discovered that the behaviour of polymers displays a deep similarly with that of a system of magnetic moments, when the latter changes from an ordered array to a disordered form. They also explained the dynamics of polymer motions, that is, how polymer chains

Fig. 13.1. Pierre-Gilles de Gennes. (© The Nobel Foundation, Stockholm.) Adopting the classification of physicists used by de Gennes himself ('golf players' or 'tennis players', p. 345), the American journal *Physics Today* (in December 1991) decided to classify de Gennes as a tennis player: 'In his career he has studied condensed matter in many forms: ferromagnets, superconductors, liquid crystals, polymers and, most recently, interfacial phenomena.'[1]

and their single parts move. Experiments performed using neutron-scattering techniques (see p. 418) fully verified de Gennes' predictions.

Pierre-Gilles de Gennes was born in Paris in 1932, the son of a physician. He studied at the *Ecole Normale Supérieure*, and from 1955 to 1959 he worked at the Centre for Atomic Studies in Saclay on neutron scattering and magnetism; in 1958 he obtained his doctorate there. Three years later he became Professor of Physics at the *Université Paris-Sud* in Orsay, where he began to study superconductivity. In the late 1960s he became interested in liquid crystals ('beautiful and mysterious', in his own words); so he gathered together physicists, chemists and material scientists, all of them studying liquid crystals, which were considered useful for industrial applications. In 1971 he became Professor of Condensed-Matter Physics at the *Collège de France*; from then on he has worked on polymers. De Gennes has received several other prizes for his outstanding scientific achievements, among them the prestigious Israeli Wolf Prize for physics (1990). He is a member of important scientific academies, such as the French Academy of Sciences, the Royal Society of London and the US National Academy of Sciences.

1992

New particle detectors

In 1992 CERN, the international laboratory for particle physics (situated near Geneva) was honoured with a second Nobel Prize (the first had gone to Carlo Rubbia and Simon van der Meer in 1984), and France won its eleventh Nobel laureate in physics, and its second one in a row. One of CERN's most distinguished experimentalist, the French physicist Georges Charpak, earned the prize for having invented new particle detectors for high-energy physics – in particular the *multiwire proportional chamber*.

A virtuoso instrument maker

In 1959 Charpak joined CERN as a research staff member. He had acquired experience in laboratory instrumentation during the mid-1950s, when he had worked as a graduate student in Frédéric Joliot's laboratory in Paris. ('I was, by nature, inclined to understand what was inside the instruments we used . . . That was not too common among particle physicists', he remarked later.)[2] In the 1960s the most-used detector in experiments at high-energy accelerators was the bubble chamber (p. 273), but now the spark chamber began to appear. In his Nobel lecture Charpak describes how particle detectors evolved in those years:

> Our study of multiwire proportional chambers, which began in 1967, was triggered by the problems with spark chambers which then faced us. The [spark chamber], introduced in 1959 . . . beautifully supplemented the bubble chamber. Whereas the latter was still peerless in the quality of the information which it provided and from which one single exposure could on its own lead to an interesting discovery, the spark chamber gave a repetition rate more than 100 times higher . . . Nevertheless, the need to store the information on photographic films led to a bottleneck: beyond a few million photographs per year or per experiment, the exposure analysis equipment was saturated.[3]

Physicists, including Charpak himself, then began (using electronic means) to invent new methods of 'reading' the sparks produced in a chamber by the passing particles. But with these new methods not more than some hundreds of events could be registered per second. This could not keep up with the big accelerators of the time, which began to produce billions of collisions and particles each second. So Charpak invented a new detector, called the *multiwire proportional chamber*. The first prototype was inaugurated in 1968.

A multiwire proportional chamber can record up to one million particles per second, and track their paths in three dimensions, with an accuracy of a tenth of a millimetre; it has the additional advantage of sending all the data directly to a computer for analysis. Its speed and precision revolutionised particle physics: it was used in the 1974 discovery of the charm particle J/ψ (see Nobel 1976); and in the 1983 discovery of the weak bosons W and Z (see Nobel 1984). Charpak's chamber and its descendants have also found useful applications in various fields of medicine, biology and industry.

Georges Charpak was born in 1924 in Dąbrowica, Poland, into a Jewish family, which fled Poland for France in 1931. He studied at high school, first in Paris and then in the South of France: 'At the same time, I was active in resistance groups', he recalled, 'and in 1943 I was arrested. After a year in jail I went to Dachau. I had to work with a shovel there, and I think since that time I haven't done any gardening. I didn't like it very much. When I came back after the war, I instinctively decided to go forward. I practically devoted my life to learning.'[4] After the war he went to study at the *Ecole des Mines* in Paris; here in 1948 he obtained the degree of mining engineer, and in 1954 he earned his doctorate in nuclear physics from the *Collège de France*. From 1959 to 1991 he was a staff member at CERN. Charpak was made a member of the French Academy of Sciences in 1985.

Discoveries and inventions

In April 1992 scientists from the University of California at Berkeley, and from the Lawrence Berkeley National Laboratory, presented the results of observations made with a NASA satellite named *Cosmic Background Explorer* (COBE), which had been placed in orbit about the earth in 1989. COBE carried very sensitive microwave receivers to map, with extreme accuracy, the *cosmic microwave background radiation*. (This is the frozen radiation relic left over from the Big Bang thirteen or so billion years ago; it had been discovered in 1964 by Arno Penzias and Robert Wilson; see Nobel 1978.)

Fig. 13.2. Georges Charpak. (© The Nobel Foundation, Stockholm.) 'When I was seven, I came as an immigrant to France from Poland. Although there were very difficult conditions, my parents organised themselves so their two children could have a good education . . . Their great pleasure was to be with their children . . . I had my three children and also had a relationship with a lot of other children who were educated in my house. I always had some pleasure in doing that . . . [And] when I look back at the way we lived when they were young, when we went on holiday in Corsica, we often had a dozen of their friends camping with us. So we had a kind of tribal life. I think that comes from my parents too.' (Georges Charpak)[5]

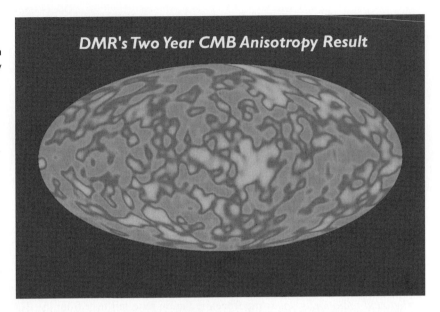

Fig. 13.3. The cosmic microwave background. (Courtesy NASA.) This is a map of the microwave sky produced from data taken by COBE. It shows the variations in the intensity of the cosmic microwave background which correspond to variations in the temperature of the universe. Light regions are warmer regions, just about 0.0003 of a degree warmer than the average temperature of 2.73 kelvins; dark grey regions are cooler regions, 0.0003 of a degree cooler than the average.

The spectacular photographs it produced showed very slight differences in the intensity of the radiation coming from different directions across the sky. These minute variations represent the ripples in the energy of the universe, which were formed when the universe itself was some 300 000 years old. These ripples correspond to a slightly non-uniform distribution of matter at that time – the 'seeds' of the giant galaxies we are seeing in the sky today. (The English theoretical physicist Stephen W. Hawking, from Cambridge University, said in the paper the *Daily Mail*: 'It is a discovery of equal importance to the discovery that the universe is expanding, or the original discovery of the background radiation.')

1993

Binary pulsars

The 1993 Nobel Prize went for the fifth time to astronomy and astrophysics. It was shared between two American scientists: Russell Hulse and Joseph Taylor. They were awarded the prize for their discovery in 1974 of the first *binary pulsar*. This discovery, as the Swedish Academy of Sciences noted, opened up new possibilities for the study of relativistic gravitation, and in particular for testing Einstein's general theory of relativity with extreme accuracy.

An amazing discovery

By 1973, six years after the discovery of the first pulsar (see Nobel 1974), about a hundred pulsars had already been discovered. At this point astronomers and astrophysicists knew fairly well what a pulsar was: a rapidly spinning neutron star having a very strong magnetic field, and emitting radio-wave pulses (p. 233).

So it was that Hulse and Taylor decided to carry out a systematic search for pulsars, and in December 1973 they went off to the Arecibo observatory in Puerto Rico. They started to sweep continuously across the Milky Way with the Arecibo radio telescope, the largest in the world. This detected the radio pulses coming from pulsars and sent them to a computer, which examined them. When a pulsar candidate was found, they searched ceaselessly in that direction for days and weeks on end, aiming to measure precisely all the parameters that characterised its radio signal; in particular they wanted to carry out refined measurements of the pulse period.

On 25 August 1974 Hulse was alone at Arecibo. Looking again at a pulsar that he had first detected in July, he noted something unusual. The pulse period (the time between two radio flashes – about fifty-nine thousandths of a second) at the beginning of an observation time of two hours was different from the same period measured at the end of the two hours themselves: this last was shorter by about twenty-seven millionths of a second. A further attempt two days later to carry out the same observation resulted in an even greater disagreement. Hulse soon began to improve the ability of his instruments to analyse the radio signals, and he also wrote out a new computer program. But on 1 September, when he looked at his pulsar again, the strange pulse period was still there! ('It looked as if I was seeing the Doppler variation of a pulsar in a *binary* system . . .', he recalled later.)

Hulse was very excited: normal binary stars are quite common, but no one had ever seen a binary pulsar. Then he started to believe that he must have made an important discovery. He waited to see if the period started to increase, as it would have to do if the object really was a binary pulsar. Finally on 16 September the conclusive evidence, the 'smoking gun', actually appeared, so Hulse was able to make a long-distance call to Taylor to give him the astonishing news; as he recalled, Taylor 'was on a plane to Arecibo in very short order'. The two continued to analyse their data throughout the following months, until it was time to write up their paper to announce the discovery (the binary pulsar was dubbed PSR 1913+16, PSR denoting a pulsar and the numbers standing for its position in the sky). The paper was published in the June 1975 issue of the *Astrophysical Journal*.[6]

Russell Alan Hulse was born in 1950 in New York City, where in 1970 he obtained his bachelor's degree in physics from Cooper Union College. He then went to the University of Massachusetts in Amherst, where in 1975 he obtained his Ph.D. Then, for three years, he worked at the National Radio Astronomy Observatory in Charlottesville, Virginia; and in 1977 he joined the Plasma Physics Laboratory at Princeton University, where he began research into plasma physics.

Joseph Hooton Taylor, Jr, was born in 1941 in Philadelphia, Pennsylvania, the son of Quaker farmers. He studied at Haverford College, Philadelphia, and

416 New trends

Fig. 13.4. PSR 1913 + 16. The pulsar, emitting radio pulses, orbits around an invisible companion, which has almost the same mass. The pulsar approaches the earth while the companion recedes; then it recedes while the companion approaches. Thus the pulsar undergoes alternate Doppler shifts: when it approaches, the pulse period diminishes (with a shift towards higher frequencies); when it recedes, the period increases (with a shift towards lower frequencies).

Fig. 13.5. R. Hulse (left) and J. Taylor (right). (© The Nobel Foundation, Stockholm.)

at Harvard University, where in 1968 he received a Ph.D. in astronomy. He then taught at the University of Massachusetts, Amherst, from 1969 to 1981, when he was appointed full professor at Princeton University; here in 1986 he became the James S. McDonnell Professor of Physics. (The year before he was honoured with the Nobel Prize he received the Israeli Wolf Prize for physics.)

General relativity again vindicated!

The orbital velocity of PSR 1913+16 is about a thousandth of the speed of light; its orbital size is of the order of the radius of our sun, and its orbital period is about eight hours. Each companion star has a mass equal to about one solar mass; and each one experiences a gravitational field 10 000 times stronger than, for example, the sun's field as experienced by Mercury. So this astronomical system has favoured interesting studies of relativistic gravitational phenomena. Let us consider two of these phenomena, cited by our Nobelists in their Nobel lectures.

Orbit precession

One of the earliest triumphs of Einstein's general theory of relativity was its ability to explain the advance of the perihelion of Mercury's orbit (an observed excess of forty-three seconds of arc per century, p. 107). In the PSR 1913+16 system, the stronger gravitational field, the higher orbital velocity and the short orbital period (about eight hours, as compared with Mercury's three months) give a precession rate of its orbit of about four degrees per year. (This means that in 100 years Mercury's orbit rotates by about 1/100 degrees, whereas the binary pulsar orbit rotates more than 360 degrees.) The measured precession for PS 1913 + 16 is known to be 4.2266 degrees per year. From this value, and assuming general relativity to be correct, astrophysicists have calculated the total mass of the two companion pulsars, obtaining '2.8286 solar masses' – within an accuracy of one part in 100 000. (Notice that this total mass corresponds roughly to twice as much as the 'Chandrasekhar mass' for neutron stars!)

Gravitational waves

Einstein's general theory of relativity also predicted gravitational waves. What are these?

Let us consider this simple analogy. When an electric charge oscillates, electromagnetic waves are emitted; this is what Maxwell's theory predicts. In the same way, according to general relativity, when an object with a mass oscillates, gravitational waves must be emitted. This is because massive accelerated objects have to emit these waves. However, they carry unimaginably small quantities of energy, and are therefore very difficult to detect. Only objects with a huge mass, such as immense cosmic objects, might hopefully produce detectable bursts of gravitational radiation.

During the last thirty years physicists have devoted great efforts to the pursuit of a 'Holy Grail' – the actual detection of gravitational waves (coming, for example, from the gravitational collapse of a massive star during a supernova explosion). But a real burst of gravitational waves has not yet been detected. (A number of sensitive laser detectors, functioning like interferometers, and set up with the purpose of picking up signals of extraterrestrial gravitational waves, are now in operation in many locations all over the globe; they depend upon the collaboration of scientists from many different countries around the world.)

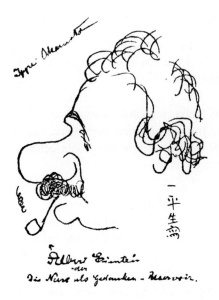

Fig. 13.6. Albert Einstein. (Drawing by Ippei Okamoto, from the Japanese newspaper *Asahi Shimbun*, courtesy AIP Emilio Segrè Visual Archives.)

Nevertheless, scientists have blazed new trails in their efforts to test general relativity. They consider neutron-star binary systems (like PSR 1913 + 16) as sources of gravitational waves. In fact, the two neutron stars composing the system have strong gravitational fields, and they are orbiting very rapidly. As a consequence the system should emit gravitational waves which carry away part of its orbital kinetic energy. As a result its orbital period should become shorter. In fact, in the late 1970s Taylor and his co-workers found that the orbital period of PSR 1913+16 gets 76 millionths of a second shorter every year, so confirming Einstein's prediction of gravitational waves to about three parts in a thousand.

1994
Mapping materials

After more than forty years, the 1994 Nobel Prize for physics honoured pioneering contributions made in the development of techniques that used neutrons to study condensed matter. The prize was shared between two physicists: the Canadian Bertram Brockhouse from McMaster University in Hamilton, Ontario, and the American Clifford Shull from MIT. They had carried out their prizewinning research at nuclear reactors respectively in Canada and in the USA, in periods ranging from the late 1940s to the early 1960s.

Brockhouse and Shull
Brockhouse and Shull used neutrons for 'mapping' materials; that is, for studying their atomic structure. As we have seen (p. 189), in the mid-1930s Enrico Fermi

and his young co-workers had discovered that slow neutrons could be used to probe deeply within matter. This is because neutrons are uncharged particles; as a result, they are scattered not by the external electrons, but only by the nuclei of the atoms. On the contrary, X-rays are bounced off by the electronic clouds, so that they can better 'see' heavy atoms which have more electrons (for example, hydrogen, with only one electron, is practically invisible to X-rays, whereas it can be easily identified by neutrons).

Neutron diffraction

In 1946 Shull went to the Oak Ridge National Laboratory in Tennessee, USA, to work with Ernest Wollan, who was developing a technique, called *neutron diffraction*, at the nuclear reactor there. The principle of this techniques is derived from the Braggs' X-ray diffraction method. Let us remind you of it.

Quantum mechanics has taught us that neutrons, like all other subatomic particles, can behave either as particles or as waves. When a beam of single-energy neutrons (or single-wavelength neutron waves) collides 'elastically' with the atoms of a sample material, they change their directions of motion, without losing energy. As a consequence, part of the neutrons traversing the material are scattered, and move out from the sample. These scattered waves, as a result of interference between themselves, form a diffraction pattern, which yields detailed information about the relative positions of the atoms in the sample.

Shull and Wollan also exploited the fact that neutrons have a magnetic moment, and used them to probe the structure of many magnetic materials. (They provided the first experimental evidence of the existence of antiferromagnetism, which had been predicted a long time before by Louis Néel, Nobelist in 1970.)

Neutron spectroscopy

Brockhouse developed a variant technique, based on 'inelastic' scattering; that is, based on the fact that in inelastic collisions between neutrons and the atoms of a sample, the neutrons change both their directions and their energies. The energies of the scattered neutrons then give information on how atoms oscillate in the material. Brockhouse's technique, called *neutron spectroscopy*, was used for studying the dynamics of atoms in solids and liquids. It was also used for investigating how crystal lattices vibrate, by measuring how much the scattered neutrons have changed their energies, due to their interactions with the lattice itself (from these measurements one can infer much about the properties of the material).

Brockhouse and Shull's *neutron scattering* techniques made crucial contributions to many fields of research, such as condensed matter, material science, chemistry and biology. Their methods are also used for studying several kinds of materials, including metals, superconducting ceramics, polymers, plastics and proteins.

Bertram Neville Brockhouse (1918–2003) was born in Lethbridge, Alberta, Canada. During the Second World War he served in the Royal Canadian Navy Volunteer Reserve; and after the war he went to study physics and mathematics at the University of British Columbia, where in 1947 he obtained his master's degree. He then moved to the University of Toronto, where he earned a Ph.D. in 1950, becoming soon afterwards a lecturer there. He then joined the Chalk River Nuclear Laboratories, where he performed most of his research work on neutron spectroscopy. From 1962 on he was Professor of Physics at McMaster University, and became emeritus in 1984. As well as the Nobel Prize, Brockhouse received many other honours, including Fellowships of the Royal Societies of Canada and of London.

Clifford Glenwood Shull (1915–2001) was born in Pittsburgh, Pennsylvania, USA. He studied at the Carnegie Institute of Technology, Pittsburgh, where he graduated in 1937. He received his Ph.D. in nuclear physics in 1941 from New York University. During the Second World War he worked at the Texas Company (later to become Texaco) in the field of solid-state physics. In 1946 he moved to the Oak Ridge National Laboratory, where he carried out his prizewinning research with Ernest Wollan. ('I regret very much that Wollan's death in 1984 precluded his sharing in the Nobel honour that has been given to Brockhouse and me since his contributions were certainly deserving of recognition', he noted in his Nobel autobiography.)[7] From 1955 until 1986 Shull was a professor of physics at MIT. He was elected to the US National Academy of Sciences in 1975.

1995

Particle hunters

Wolfgang Pauli introduced the idea of the neutrino in 1930 (p. 169), and in 1956 Clyde Cowan and Frederick Reines succeeded in the actual detection of the elusive particle (p. 260). In December 1995 Reines was seventy-seven years old, and was professor emeritus at the University of California at Irvine. He was alone when he flew to Stockholm to receive the Nobel Prize for physics from the hands of the King of Sweden, for Cowan had died in 1974. His companion on the journey was Martin Perl, from Stanford University, who shared with him the coveted award for his discovery of one of the two members of the third lepton family – the *tau lepton*.

Frederick Reines (1918–98) was born in Paterson, New Jersey, the son of Jewish emigrants from Russia. After obtaining his Ph.D. in 1944 from New York University, he went to Los Alamos to participate in the Manhattan Project, and until 1959 he worked there principally in the field of nuclear weaponry. Then he became professor and head of the department of physics at the Case Institute of Technology in Cleveland, Ohio. Here he turned to the fields of particle physics and

Fig. 13.7. Frederick Reines. (Courtesy AIP Emilio Segrè Visual Archives; W. F. Meggers Gallery of Nobel Laureates.)

cosmic rays. He later carried out an experiment in a gold mine in South Africa, aiming at observing, for the first time, the neutrinos produced in the atmosphere by cosmic rays. From 1966 until 1988 Reines was a professor at the University of California at Irvine.

Three lepton families

Leptons are the most fundamental of particles, and can interact only through the gravitational electromagnetic and weak forces – they do not feel the strong force. Leptons are considered 'elementary'; that is, they are among the smallest existing subatomic particles, with dimensions of less than one hundred-million-billionth of a metre, and with no detectable size, or internal structure.

Physicists speak of three *lepton families*, each family being composed of two particles (plus their antiparticles): one particle has a mass, an electric charge and a weak charge; the other particle is a neutrino which has an unimaginably minute mass, a weak charge but a zero electric charge. All members of the three families are fermions with spin equal to $\frac{1}{2}$. (To distinguish the three families from each other, physicists have invented a characteristic which they call *flavour*: each family has its own flavour.) Let us briefly examine their stories.

> *Electron*. It is the first particle to have been discovered. It was identified in the late 1890s thanks to the work of many researchers, which culminated in the famous experiments carried out by J. J. Thomson (see Nobel 1906). It is the lightest among all free electrically charged particles. It is stable (it can exist by itself for ever), and

it carries a negative elementary charge. Its antiparticle is the *positron*, which was discovered in 1932 (see Nobel 1936). Electrons, along with protons and neutrons, are the building blocks of all the atoms that compose matter. Positrons, on the contrary, are produced only in high-energy particle collisions (in cosmic rays and in accelerators); they live only for brief time spans before they annihilate with electrons in matter.

Electron neutrino. It is associated with the positron in the first family, whereas the *electron antineutrino* is associated with the electron.

Muon. It was unexpectedly discovered in 1936 by Carl Anderson and Seth Neddermeyer in cosmic rays (p. 195). It is the heavier cousin of the electron (about two hundred times heavier), and just like the electron it carries a unit of negative charge. It is unstable: a negative muon decays, with a lifetime of some millionths of a second, into three leptons: an electron, an electron antineutrino and a muon neutrino. Its antiparticle is its positive replica.

Muon neutrino. This is another type of neutrino which belongs to the second family: it is produced, for instance, when a negative pion breaks down along with a negative muon. (The *muon antineutrinos* instead are produced in decays of positive pions.) The muon neutrino was discovered in 1962 by the Columbia team (see Nobel 1988).

τ lepton. The pattern of two lepton families encouraged physicists to search for an additional (third) family. They were wondering how heavier leptons could possibly be produced in high-energy collisions. At Stanford, SPEAR (the same collider which in 1974 had produced the first charm particle J/ψ, see Nobel 1976) provided electron–positron collisions at energies of up to eight billion electron-volts. Here, during the same year as the J/ψ discovery, Martin Perl captained a team of experimenters coming from SLAC and from Berkeley University. They began an experiment with instrumentation that they had designed to select only those special events which would show signs of being heavy leptons. By the summer of 1975 Perl's team had examined some 40 000 events, and had come up with a total of about 90 special events, a fraction of which bore a close resemblance to heavy leptons.

During the summer of 1976 Perl presented his results at a conference in Europe where his European colleagues were still quite sceptical about his hypothetical lepton. But about one year later the new lepton was seen at DORIS (the new electron–positron collider at DESY, Hamburg, Germany). So Perl was at last able to announce the birth of the *tau lepton* in a paper published in *Physical Review Letters*. (It emerged that the third heavy lepton has a mass some 3500 times greater than that of the electron, and some years later experiments showed that it is unstable: it breaks down with a lifetime of a few ten-thousand-billionths of a second – ten million times shorter than that of its lighter cousin, the muon.)

τ neutrino. For the τ lepton to fit into the picture of the lepton families, it had to be accompanied by its own neutrino. Experiments performed at the existing colliders during the late 1980s and the 1990s did not succeed in observing any tau neutrinos. At last, in the autumn of the year 2000 it was announced that an experiment at the Tevatron (Fermilab) had gathered direct evidence for the elusive particle.

Martin Lewis Perl was born in 1927 in New York City, the son of Russian Jews, who had immigrated to the USA. After graduating in chemical engineering in 1948, he worked for two years at General Electric. He then went to Columbia University to study physics, and in 1955 he obtained a Ph.D. there, under the supervision of Isidor Rabi (Nobel 1944). He then joined the University of Michigan, Ann Arbor, and in 1963 moved to Stanford University, where he became a full professor of physics in 1967.

Three quark families

Quarks are the fundamental particles which feel all the three forces; that is, the electromagnetic, weak and strong forces. Just as for leptons, so for quarks, too, there exist *three families*; they, too, are considered 'elementary' – not larger than one hundred-million-billionth of a metre in size. The six quarks (plus their antiquarks) have a mass, an electric charge, a colour charge and a weak charge. Each of the three families has two quarks, one with a positive electric charge (equal to $+\frac{2}{3}$, in units of the elementary charge), and the other with a negative charge (equal to $-\frac{1}{3}$). All the six (3×2) quarks are fermions (with a spin equal to $\frac{1}{2}$). The strong colour force binds the quarks together to form hadrons (baryons and mesons). Here are their stories.

> *First family – quarks up and down*. These quarks were postulated in 1964 by Gell-Mann (see Nobel 1969). They are the fundamental units that make up protons and neutrons, as was dramatically confirmed by a series of experiments carried out by Jerome Friedman, Henry Kendall and Richard Taylor in the late 1960s and early 1970s at SLAC (see Nobel 1990).
>
> *Second family – strange and charm quarks*. In addition to the quarks up and down, Gell-Mann proposed a third quark, the *strange quark*, to explain the properties of the strange hadrons. Subsequently, Sheldon Glashow and colleagues came out with the idea of a fourth quark, the *charm quark*; the experimental evidence for it came in 1974, when the J/ψ particle was discovered (see Nobel 1976).
>
> *Third family – bottom quark*. With the discovery of the charm quark the pattern of four leptons and four quarks was completed. So Perl's tau lepton found itself an intruder when it peeped out in 1976. But one year later the first sign of a fifth quark, called a *bottom quark*, emerged at Fermilab, near Chicago. Here a team of physicists led by Leon Lederman (Nobelist in 1988) had bombarded the atomic nuclei of a metal target with high-energy protons, which were produced by a 400 billion electron-volt accelerator. They had found two hadron resonances which

were interpreted as bound states of the new bottom quark and its antiquark. The same resonances were observed two years later at the DORIS electron–position collider. So the existence of a fifth quark was confirmed.

Third family – *top quark*. While the third lepton family had to wait until the year 2000 to be completed with the tau neutrino, the sixth member of the quark families was discovered in 1995. It was christened the *top quark*. Briefly, this is its story.

Particle physicists had known that the top quark must exist ever since 1977, when its partner, the bottom quark, was discovered. Soon 'top-hunting' began at the most powerful accelerators of the time, at SLAC (Stanford) and at DESY (Hamburg), but neither of them turned up anything. After the discovery of the weak bosons at CERN in 1983 (see Nobel 1984), physicists believed that the top quark would soon be discovered. Experimenters continued research until the close of the 1980s, but still no top quark was observed. They concluded that its mass must be greater than 77 billion electron-volts, the maximum particle mass that CERN's energy could create.

Meanwhile, in the mid 1980s, a new accelerator, the Tevatron, was activated at Fermilab; it produced protons with an energy of 900 billion electron-volts.

Fig. 13.8. The Fermilab (Fermi National Accelerator Laboratory), at Batavia near Chicago. (Fermilab Photo.) Its accelerator, the Tevatron, at present is the most powerful accelerator in the world, with 1000 super-conducting magnets, cooled by liquid helium.

Later it was transformed into a proton–antiproton collider with a total energy of 1800 (900 + 900) billion electron-volts (more than three times CERN's energy), so making possible a world record for high-energy collisions between protons and antiprotons. Two complementary experiments competed in the search for the top quark; they were called CDF (Collider Detector at Fermilab) and D-zero. These experiments involved an international collaboration of about a thousand physicists, and countless engineers and technicians. After three years of gathering data the leaders of the two teams finally announced, in March 1995, the discovery of the top quark; its mass turned out to be 175 times the mass of the proton, or 350 000 times the mass of the electron (it is the heaviest particle ever discovered).

1996

Superfluid helium-3

Three American experimentalists, David Lee, Douglas Osheroff and Robert Richardson, discovered, at the beginning of the 1970s, that the isotope helium-3 (as well as helium-4) can be made a superfluid. As we will see (p. 426), this was a phenomenon that had always been regarded as extremely difficult to demonstrate, until new sophisticated techniques had been developed in the field of ultra-low temperatures. The breakthrough earned the three of them the 1996 Nobel Prize for physics.

Discovering the phenomenon
In the late 1960s David Lee and Robert Richardson began to develop a research programme at the low-temperature physics laboratory of Cornell University, Ithaca, USA. The goal of their research was to cool helium-3 to temperatures within a few thousandths of a degree above absolute zero, so as to explore certain magnetic properties of that isotope in its solid state. They had decided to exploit a new cooling method, suggested by the Russian theorist Isaac Pomeranchuk, which was based on a peculiar property of helium-3. By compressing this isotope in its liquid state with no heat entering or leaving the system, part of the helium transforms into a solid, and this, drawing thermal energy from the surrounding liquid, is able to cool it. Douglas Osheroff was then a graduate student in the team, and he had built a refrigerator to be used in the experiment.

On 24 November 1971, the day before the American holiday of Thanksgiving, Osheroff was alone in the laboratory. He had begun a test which consisted of continually measuring the pressure within the helium-3 sample as a function of time, while compressing the liquid helium at a constant rate. While checking the curve that traced variations in the internal pressure over time, he noted, with surprise, minute jumps. ('At a temperature that we estimated later to be 2.6 mK [2.6 thousandths of a kelvin, about a thousand times lower than the critical temperature of superfluid helium-4], there was this sudden decrease in the rate of increase of pressure with time – a very sharp kink in the curve. Even a blind man could probably have seen this', he remembered later.[8]) Lee, Osheroff and Richardson interpreted the sudden jumps as corresponding to superfluid phase transitions in helium-3. They presented their amazing results in a paper published in 1972 in *Physical Review Letters*.

Subsequent experiments revealed that helium-3 can in fact condense into three different superfluid phases: two occurring at 2.7 thousandths of a kelvin (one of these appearing only when a magnetic field is applied to the sample); the third phase occurring at 1.8 thousandths of a kelvin. Later research also showed that superfluid helium-3 has different properties in different directions in space, and that when rotated, it produces microscopic complex vortices. All this demonstrated how far superfluid helium-3 differs from helium-4, which has only one superfluid phase, has the same properties in all directions and produces simpler vortices.

Explaining the phenomenon

When theorists learned about the Cornell findings, they soon began to wonder how to explain the phenomenon. Here are their basic ideas.

Let us start by considering superfluid helium-4. An atom of helium-4 consists of an 'even' number of particles with a spin equal to $\frac{1}{2}$ (two protons and two neutrons in its nucleus, and two electrons around it). This results in a total spin of the atom equal to zero; thus helium-4 is classified as a boson (remember: bosons are particles either with zero or with an integer spin, p. 175). Now bosons obey the Bose–Einstein statistics, the rules of which permit all the helium-4 atoms in a sample to condense into a single quantum state with the least possible energy; that is, all the atoms lose their individuality, and act as a single entity which is described by a single wave function (a manifestation of what is called a *Bose–Einstein condensate*, p. 448).

And now let us turn to helium-3. An atom of helium-3 instead consists of an 'odd' number of particles (two protons and one neutron in its nucleus, and two electrons around it); thus, its total spin being equal to $\frac{1}{2}$, it is classified as a fermion (particles with a half-odd spin, p. 176). As we know, fermions obey different rules from bosons; they follow the Fermi–Dirac statistics, which forbid them sharing the same quantum state. This is the reason why theorists were convinced that a collection of helium-3 atoms could not all condense into a single state. So it was that even for years after the discovery of superfluidity in helium-4 (by Peter Kapitza, Nobelist in 1978, and by Jack Allen and Donald Misener), physicists believed that superfluidity would be very difficult to achieve in helium-3.

In 1957, however, John Bardeen, Leon Cooper and Robert Schrieffer were able to explain superconductivity in metals, by simply arguing that electrons, which are in fact fermions, can pair up when they are in a superconducting metal, and when this is reduced below its transition temperature: each pair turns into a system having in many respects the characteristics of a boson (in the pair, the two electrons have opposite spins, so that the total spin of the pair comes to equal zero.) Thus, these *Cooper pairs* can actually undergo condensation – such as do bosons when forming a Bose-Enstein condensate – and they can then roam freely through the material (see Nobel 1972). In the same way atoms of helium-3 can pair up and form ordered pairs with an integer total spin, which resemble bosons in many ways: thus, at extremely low temperatures, these 'boson-like' pairs can undergo condensation, collapse into a single macroscopic quantum state, and finally become a superfluid (see also Nobel 2003). This was exactly what our three Cornell heroes managed to demonstrate with their experiments!

The three prizewinners
David Morris Lee was born in 1931 in Rye, New York, USA. He graduated in 1952 from Harvard University; and in 1955 he went to Yale University, Connecticut, as a graduate student ('. . . I learned a great deal about experimental low-temperature physics and the life of an experimental physicist. As time went on my growing fascination with low-temperature physics led me to the decision that this would be my area of specialisation . . .', he wrote in his Nobel autobiography.)[9] It was at Yale that he earned his Ph.D. in physics in 1959, with a thesis which involved precisely research into liquid helium-3. He then moved to Cornell University, where in 1968 he was appointed full professor of physics, and in 1999 became the White Distinguished Professor of the Physical Sciences.

Douglas Dean Osheroff was born in 1945 in Aberdeen, Washington, USA. He obtained a master's degree at Cornell University in 1969, and a Ph.D. in physics there in 1973. From 1972 to 1987 he worked as a researcher at Bell Labs, where he continued his studies on helium-3. He then went on to study first solid helium-3, and then certain electrical properties in disordered two-dimensional conductors. In 1982 Osheroff became head of the Bell Labs solid state and low-temperature physics laboratory. In 1987 he left Bell Labs and moved to Stanford University, where he is now the J. G. Jackson and C. J. Wood Professor of Physics.

Robert Coleman Richardson was born in 1937 in Washington, DC. He obtained a master's degree in 1960 from Virginia Polytechnic Institute, and in 1966 a Ph.D. in physics at Duke University, Durham, North Carolina. Richardson then joined Cornell University, where he began to work with David Lee in the field of low-temperature physics. ('The research environment at Cornell has been superb with an unbroken string of talented graduate students, close colleagues in both theory and experiment, and a team of technical support specialists who helped make everything work.')[10] At Cornell, Richardson became a full professor in 1975. He was later appointed to the Floyd R. Newman chair of physics and director of the laboratory of atomic and solid-state physics.

1997

Super-chilled atoms

Three physicists, extremely skilled in manipulating atoms by means of laser light, received the 1997 Nobel Prize for physics: the Americans Steven Chu and William Phillips, and the French Claude Cohen-Tannoudji. They were awarded the prize for having developed sophisticated techniques which use laser light to 'chill' atoms to extremely low temperatures (near absolute zero), and then trap them. At such low temperatures the atoms move slowly enough to be examined in detail and with a very high precision. These techniques were developed around the mid-1980s: Chu worked at Bell Labs, and Phillips at the National Institute of Standards and Technology (NIST, formerly the National Bureau of Standards), both in the USA. Cohen-Tannoudji's prizewinning researches were carried out in Paris – at the *Ecole Normale Supérieure*.

Chilling and trapping

Air atoms and molecules in your room move around in a frenetic motion at a speed of a few hundred metres per second. Our three prizewinners were able to slow atoms down to speeds below a few centimetres per second, and then hold them in place. Their methods led to a deeper understanding of the interaction between light and matter, and to a wide variety of applications: high-precision atomic spectroscopy, new atomic clocks, ultra-precise atomic interferometers and atom lasers that might one day be used in the manufacture of very small electronic components. ('It's remarkable how simple curiosity can lead to a lot of practical things', commented Chu.) Let us now follow the chronology of the discoveries of our three Nobelists.

At Bell Labs

In the early 1980s Steven Chu was working at Bell Labs in Holmdel, New Jersey. Here he studied how atoms could be first slowed down, and then trapped. Chu and his co-workers used light photons to slow down sodium atoms placed in a vacuum chamber. The principle is this: when a photon with the right energy (or right frequency) collides with an atom, the atom absorbs the photon and receives a minuscule kick in the direction that the photon is moving; it then recoils in the opposite direction, thus reducing its kinetic energy and speed. To slow atoms down, the Bell Labs team used a method suggested in the mid-1970s by Theodor Hänsch and Arthur Schawlow (Nobelist in 1981); and, independently, by David Wineland and Hans Dehmelt (Nobelist in 1989); the former used this method for neutral atoms, whereas the latter exploited it for trapped ions.

By varying the frequency of the laser photons, Chu and his team were able to slow down sodium atoms with different speeds, and hence to create a cloud of chilled atoms. They then surrounded these atoms with six laser beams which shone photons through the cloud, exactly at the point where the beams intersected (see Fig. 13.9). They were thus able to obtain a cloud of chilled atoms at a temperature of about 240 millionths of a kelvin; this corresponds, for the atoms, to an average speed of about thirty centimetres per second. The cloud was as small as a pea, and it contained about a million atoms. Chu coined the name 'optical molasses' for it, because it behaved like a viscous fluid (this was in fact what had suggested the term 'molasses').

But the chilled atoms tended to move out of the molasses (this in fact did not provide any force capable of trapping them). To keep the atoms in place, Chu decided to use a technique suggested by one of his co-workers, Arthur Ashkin: together they developed what was called a *laser trap*. With this they shot a powerful, tightly focused laser beam, called *optical tweezers*, through the optical molasses, and so held the atoms stationary. (Optical tweezers later became a tool widely used for manipulating living cells, polymers and biological molecules like DNA.) All this happened between 1985 and 1987.

At NIST

During the same years, William Phillips was also working on the slowing down of atoms at NIST, in Gaithersburg, Maryland. But here he was able to devise another method: rather than vary the frequency of the laser photons, Phillips thought of applying a varying magnetic field along the path of the atom beam, in order to alter the energy levels of the atoms themselves (he exploited the widely known 'old Zeeman effect', p. 40). By altering the energy levels of the travelling atoms, he changed the frequency at which they could absorb photons, so tuning this frequency to the frequency of the laser photons. As a consequence the photons were able progressively to slow the atoms down. In 1985 Phillips and his

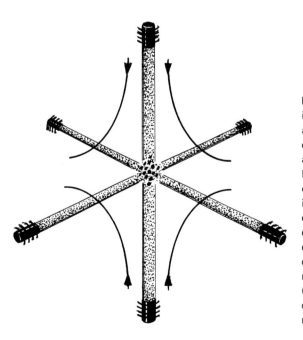

Fig. 13.9. Schematic illustration of the apparatus used for chilling and trapping atoms: light from six lasers comes from six directions to form a trap in the middle of the chamber. Electric currents, circulating in coils surrounding the chamber, create the magnetic field (represented by the curved lines) for the magnetic trap.

co-workers, by using this method, succeeded first in stopping all the sodium atoms in a beam, and then in keeping them in a trap made from a magnetic field.

Next steps

In 1987 Chu at Bell Labs improved his trapping of atoms. He just used laser light, and added a magnetic field to his array of six laser beams, so as to be able to trap the chilled atoms (see Fig. 13.9). When the atoms, slowed-down by the laser photons, started straying from the centre of the magnetic trap, they met a force that pushed them back toward the centre. Thus the chilled atoms were firmly held in place, and could be carefully studied. This stronger trap was named the *magneto-optical trap* (MOT) (Phillips called it the 'work-horse of atomic cooling').

Finally, in 1988 Phillips and his co-workers made a very careful measurement of the temperature of atoms in optical molasses, and obtained a figure of about 40 millionths of a kelvin; this was much lower than the theoretically calculated limit (which is called the *Doppler limit*).

At the ENS

At the ENS (*Ecole Normale Supérieure*) in Paris, another team, headed by Claude Cohen-Tannoudji, was also working on atom chilling and trapping. They came up with a theoretical explanation of Phillips' deep cooling, and devised new methods for reaching even lower temperatures. In 1995 the ENS team succeeded with six laser beams in cooling helium atoms to about two-tenths of a millionth of a kelvin, corresponding to a speed for the atoms of only about two centimetres per second.

Over the years, researchers progressively refined the techniques of chilling and trapping neutral atoms. These new techniques, in the mid-1990s, led to the first demonstration of the *Bose–Einstein condensate*, a fascinating novel macroscopic quantum system, in which thousands, or sometimes even millions, of identical atoms act in an unusual, collective way, just as if they were one single 'super-atom' (see Nobel 2001).

The prize-winning trio

Steven Chu was born in 1948 in St Louis, Missouri, USA. His father was a university professor, who had emigrated from China to the USA during the Second World War. In his Nobel autobiography he recalls: 'Education in my family was not merely emphasised, it was our raison d'être. Virtually all of our aunts and uncles had PhD's in science or engineering, and it was taken for granted that the next generation of Chu's were to follow the family tradition . . . In this family of accomplished scholars, I was to become the academic black sheep.'[11] Chu graduated from the University of Rochester, New York, and in 1976 received his Ph.D. in physics from the University of California, Berkeley. He joined Bell Labs in Murray Hill, New Jersey, in 1978, and five years later he became the head of the quantum electronics research department at Bell Labs in Holmdel; here he remained until 1987. ('Life at Bell Labs, like Mary Poppins, was "practically perfect in every way". However, in 1987, I decided to leave my cosy ivory tower... My management at Bell Labs was successful in keeping me at Bell Labs for nine years, but... the urge to spawn scientific progeny was growing stronger.'[12]) Chu then moved to Stanford University, where he is now the Theodore and Francis Geballe Professor of Physics and Applied Physics.

William Daniel Phillips was born in 1948 in Wilkes-Barre, near Kingston, Pennsylvania, USA. His father and mother were both professional social workers. He wrote in his autobiographical notes:

> I clearly remember the value my parents placed on reading and education... Dinner table conversations included discussions of politics, history, sociology, and current events. We children were heard and respected, but we had to compete for the privilege of expressing our opinions. In these discussions our parents transmitted important values about respect for other people, for their cultures, their ethnic backgrounds, their faith and beliefs, even when very different from our own. We learned concern for others who were less fortunate than we were. These values were supported and strengthened by a maturing religious faith.[13]

Phillips received his Ph.D in physics from the MIT in 1976. Two years later he went to the NIST laboratory in Gaithersburg, where he conducted his Nobel-winning research.

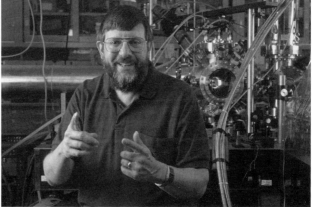

Fig. 13.10. S. Chu (left) (courtesy Lucent Technologies/Bell Labs), and W. Phillips (courtesy NIST – © Robert Rathe).

Claude Cohen-Tannoudji was born in 1933 in Constantine, Algeria (then part of metropolitan France). He introduced his Nobel self-presentation with these words:

> My family, originally from Tangiers, settled in Tunisia, and then in Algeria in the 16th century, after having fled Spain during the Inquisition. In fact, our name, Cohen-Tannouidji, means simply the Cohen family from Tangiers . . . My father was a self-taught man but had a great intellectual curiosity, not only for biblical and talmudic texts, but also for philosophy, psychoanalysis and history. He passed on to me his taste for studies, for discussion, for debate, and he taught me what I regard as being the fundamental features of the Jewish tradition – studying, learning and sharing knowledge with others.[14]

Cohen-Tannoudji went to Paris in 1953 to study at the *Ecole Normale Supérieure*. ('This French "grande école", founded during the French Revolution, . . . selects the top high-school students . . . I attended a series of fascinating lectures in mathematics given by Henri Cartan and Laurent Schwartz, and in physics by Alfred Kastler [Nobel 1966]. Initially, I was more interested in mathematics but Kastler's lectures were so stimulating, and his personality so attractive, that I ended up changing to physics.'[15]) In 1962 he received his doctorate, with a thesis written under the supervision of Kastler himself. He then continued to work in the department of physics at the *Ecole Normale*. He taught at the University of Paris VI from 1964 to 1973, when he was appointed Professor of Physics at the *Collège*

Fig. 13.11. Claude Cohen-Tannoudji. (Courtesy Professor Claude Cohen-Tannoudji.) 'We were a small group, but the enthusiasm for research was exceptional and we worked hard. Brossel and Kastler were in the lab nearly day and night, even at weekends. We had endless discussions on how to interpret our experimental results . . . I think that what I learned during that period was essential for my subsequent research work . . .'[16]

de France (an institution for higher studies created in 1530 by the reigning king, *François I*).

1998

Fractional quantum Hall effect

For the second year in a row, Bell Labs had been honoured with a Nobel Prize in physics. It was their sixth! The 1998 prize was actually awarded to Horst Störmer and Daniel Tsui, two Bell Labs experimenters, and to the theorist Robert Laughlin (he had worked at Bell Labs until 1981). In 1982 Störmer and Tsui discovered a new state of matter with unexpected properties. The phenomenon seemed analogous to that discovered in 1980 by Klaus von Klitzing (Nobel 1985). In 1983 Laughlin provided the theoretical interpretation of this new 'Bell Labs effect', which was called the *fractional quantum Hall effect*.

From experiment to theory
Störmer and Tsui, the experimenters
Klaus von Klitzing had discovered the *integral quantum Hall effect* by investigating the flow of electrons in a flat layer, which took place in the interface of a special silicon transistor that he had placed in a strong magnetic field, at very low temperatures (p. 387). At the same time physicists at Bell Labs had developed high-quality slabs of a new semiconductor material in which electrons could wander even more freely than in devices such as those used by von Klitzing (the electrons moved in the flat interface at the junction of a *heterostructure* semiconductor, p. 445). So Tsui and Störmer, using this new device, set out to see whether some hitherto unknown phenomenon occurred at still lower temperatures and higher magnetic fields than those used by von Klitzing.

They performed an experiment at the MIT Francis Bitter National Magnet Laboratory, and discovered fractional plateaux; that is, they found that the Hall electrical resistance not only obeyed the formula h/ne^2 (p. 387) for integer values of n (such as 1, 2, 3), as discovered by von Klitzing – but it did so also for fractional values such as $\frac{1}{3}$, $\frac{2}{3}$, with a precision better than one part in a million. From then on this effect was in fact called the 'fractional quantum Hall effect'.[17] In 1981 Robert Laughlin had already succeeded in explaining the integral quantum Hall effect, but the new phenomenon discovered by Tsui and Störmer called for new ideas.

Laughlin, the theorist
Just a few months on from Bell Labs' discovery Robert Laughlin enters the picture again. While working at the Livermore National Laboratory, California, he invented a new wave function (the ψ of Schrödinger), in order to describe the behaviour of a two-dimensional electron gas under such extreme conditions

(very strong magnetic field and very low temperatures). His ingenious theoretical model included the concepts of gauge invariance (the same concept which plays a crucial role in the electroweak theory, p. 359), and that of localisation of electrons by disorder (the concept put forward by Philip Anderson, p. 347). With these ingredients, which he had already used to explain the integral quantum Hall effect, and especially with the new ψ (the real focus of the theory), he was able to explain the puzzling effect that his former Bell Labs colleagues had found.[18] According to Laughlin, the two-dimensional electron gas (some hundred billion particles per square centimetre) condenses and becomes a new incompressible quantum liquid. In this liquid the interacting electrons combine with vortices of the magnetic field and form *excitations* (or *quasiparticles*). These are entities which carry a fractional electric charge, and obey a new set of statistical rules – a generalisation of Fermi–Dirac and Bose–Einstein statistics, called *fractional quantum statistics*. (For example, for a fractional value $n = \frac{1}{3}$, the excitations of the quantum liquid are quasiparticles with a charge $\frac{1}{3}$ that of the electron charge.)

Three men from Bell Labs
Daniel Chee Tsui was born in 1939 in the province of Honan in central China:

> My childhood memories are filled with the years of drought, flood and war which were constantly in the consciousness of the inhabitants of my over-populated village . . . Like most other villagers, my parents never had the opportunity to learn how to read and write. They suffered from their illiteracy and their suffering made them determined not to have their children follow the same path at any and whatever cost to them.[19]

In 1958 Tsui went to the USA, where in 1967 he obtained his Ph.D. from the University of Chicago. He then joined Bell Labs, and in 1982 moved to Princeton University, where he became a professor of electrical engineering ('Many of my friends and esteemed colleagues had asked me: "Why did you choose to leave Bell Laboratories and go to Princeton University?" Even today, I do not know the answer. Was it to do with the schooling I missed in my childhood? Maybe. Perhaps it was the Confucius in me, the faint voice I often heard when I was alone, that the only meaningful life is a life of learning. What better way is there to learn than through teaching!').[20]

Horst Ludwig Störmer was born in 1949 in Sprendlingen, in the region of Frankfurt, Germany, the son of a shop-keeper. He graduated from Goethe University in Frankfurt, and in 1977 he obtained his doctorate in physics from the University of Stuttgart. He then moved to the USA to carry out solid-state research at Bell Labs, where in 1991 he became head of the physical research laboratory. In 1997 Störmer moved to Columbia University, where he was appointed Professor of Physics and Applied Physics.

Fig. 13.12. D. Tsui (left) and H. Störmer (right). (Courtesy Lucent Technologies/Bell Labs.)

Robert B. Laughlin was born in Visalia, California, in 1950. He studied at the University of California, Berkeley, and obtained his Ph.D in physics from the MIT in 1979. He carried out research at Bell Labs, from 1979 to 1981, and at the Livermore National Laboratory, from 1981 to 1982. In 1985 he became an associate professor of physics at Stanford University, and a full professor in 1989. Thus he recalled his experience at Bell Labs:

> Bell Labs had been a kind of holy place of solid-state physics . . . I did notice during my job talk that everybody understood what I was saying immediately – this had never happened before – and that the audience had an irresistible urge to interrupt, heckle, and argue about the subject matter loudly among themselves during the talk so as to lob hand grenades into it, just like back-benchers do in the House of Commons . . . I later came to understand that this heckling was a sign of respect from these people, that the ability to handle it was a test of a person's worth, and that polite silence from them was an extremely bad sign . . .[21]

1999

Fighting against infinities

Starting in 1965, a string of twelve Nobel Prizes have marked the amazing achievements that have been effected in the field of elementary particles, and the most fundamental forces of nature. The last landmark in this series was the 1999 Nobel Prize for physics, awarded jointly to the Dutch theorists Gerardus 't Hooft and Martinus Veltman, 'for having placed particle physics theory on a

Fig. 13.13. Robert Laughlin. (© The Nobel Foundation, Stockholm.) 'The data immediately suggested to me objects of charge 1/3 *e*', recalls Laughlin, 'but I took a wrong first step . . . Thank God the Phys. Rev. Letters referee rejected this first effort, pointing out why it couldn't be right. Only after the Nobel Prize was announced did I learn that this providential referee had been Steve Kivelson . . . I could kiss him.'[22]

firmer mathematical foundation' – as the Swedish Academy of Sciences pointed out. Veltman and 't Hooft had invented a powerful mathematical technique, which served, among other things, to consolidate the electroweak theory, which had won the 1979 Nobel Prize for Sheldon Glashow, Abdus Salam and Steven Weinberg; it also made possible detailed predictions concerning particle physics, all of which were successively confirmed by actual experiments.

Utrecht, late 1960s – early 1970s

We must first go back to the late 1960s, to a time when Steven Weinberg and Abdus Salam proposed their electroweak model by building on the Yang–Mills theory (see Nobel 1979). As you recall, they employed the concept of *spontaneous symmetry breaking* (or the 'Brout–Englert–Higgs mechanism') to show how the three messengers of the weak force (Ws and Z) could acquire 'spontaneously' a mass; thus leaving the photon, the messenger of the electromagnetic force, without a mass. However, it was not clear how their theory could resolve the *problem of the infinities*, which was rooted in it. It was at that period of time that 't Hooft and Veltman entered the story.

The professor

Martinus Veltman in 1966 had become Professor of Theoretical Physics at the University of Utrecht, Holland, where he had once been a student. During the preceding years he had devised a powerful computer program, which he had called *Schoonschip* (in Dutch it '. . . means cleaning a ship's deck; but colloquially it refers to clearing up an issue')[23], in order to compute complicated Feynman diagrams. The program had been completed at the end of 1963, and Veltman had begun to use it for obtaining the values of certain physical quantities contained in the weak interaction theory in vogue at that time.

In 1968 Veltman entered the field of Yang–Mills type theories, because he was convinced that some form of these theories might better describe the weak interactions he was at that moment studying. He recalled in his Nobel autobiography: 'A turning point in my scientific life occurred during a one-month visit (April 1968) to Rockefeller University. In the quietness of that institution I embarked on the scientific venture that has now been honoured with the Nobel Prize. I am still indebted to [Abraham] Pais who got me there and counseled me in that period.'[24]

Veltman, on starting his new venture, however, soon realised that he was facing a seemingly insurmountable barrier: it was the above-mentioned problem of the infinities. In fact one of the main defects of the Yang–Mills theories (such as the Salam–Weinberg electroweak theory) was that it was impossible to make precise predictions, because calculations of very refined corrections ended up always in infinities. And, still worse, it seemed that these infinities, rooted in the theory, were different from those hidden in quantum electrodynamics. Here, the techniques invented more than twenty years before by Richard Feynman, Julian Schwinger and Sin-Itiro Tomonaga (see Nobel 1965) had been able to cancel all the infinities, but the same technique could not now be used to cancel the new infinities. Moreover, the charged Yang–Mills force messengers had zero masses, whereas, in fact, they must have a mass. It was thus that a host of theoreticians, among them Veltman, concentrated on developing new mathematical techniques in order to solve these vexing problems.

Soon after returning from the USA, Veltman went to Orsay, Paris, for a sabbatical year. Here he started to analyse what happened to a Yang–Mills theory in which massive force messengers had been introduced *ad hoc*; at the same time, he invented a new technique for cancelling out the unwanted infinities. He worked hard (in spite of the 'French May '68 student revolution'!), and in a few months he brilliantly succeeded in cleaning his model out of almost all the infinities – for the moment only at a first level of approximation. These results were soon published in the September 1968 issue of *Nuclear Physics*. (This paper is among those cited by the Nobel Physics Committee as their motivating for awarding the Nobel Prize to our two heroes.) As Veltman recollected:

> To me, and some others, that [success] stimulated further work and it just made me feel very sure that I was on the right track. You see, if you make some sophisticated reasoning and then things fall into place, there is after that nothing that will stop you. This is a very personal feeling; it is the moment of discovery. The technical virtues of [that] paper . . . were . . . essential for the subsequent development of the theory. A barrier had been overcome.[25]

The graduate student

After his 1968 seminal paper, Veltman continued to make progress in looking for a final solution to the problem of infinities in Yang–Mills type theories. Then the year 1969 arrived, and another person entered the picture. He was a twenty-two-year-old student, of the name of Gerardus 't Hooft, who had asked Veltman for a doctoral thesis. From then on things started to change.

As suggested by his supervisor, young 't Hooft began to study Yang–Mills theories. In the summer of 1970 he went to Cargèse, Corsica, to attend a summer school on particle physics. He remembered later: 'It was one of those caprices of fate that brought me, as a young student of Veltman's, to the 1970 Cargèse Summer Institute. (I had first applied to Les Houches, where my application was turned down.)'[26] At Cargèse there were outstanding theoreticians who were discussing a certain theory concerning elementary particles. Here 't Hooft learned about the *spontaneous symmetry breaking* method, which could give masses to messenger particles included in the theory, without ruining the likelihood of cancelling the infinities hidden in the theory itself. Inspired by this, he at once thought that he could use a very similar procedure for the Yang–Mills theories he was fighting with.

He returned to Utrecht, and soon went to discuss his idea with his superior. He then found out how to reshape the 1968 Veltman model by using spontaneous symmetry breaking; in less than one year, he was able to demonstrate that the Yang–Mills theories incorporating that mechanism could actually be corrected (or *renormalised*), that is, they could be cleaned out of the worrisome infinities. The decisive test came when 't Hooft's model was checked at a computer with the *Schoonschip* program. This was performed at CERN, and the results fully confirmed that all the infinities had been cancelled. Student 't Hooft could thus present his spectacular results at a conference on elementary particles which was held in Amsterdam in July 1971 (Veltman was the chairman), and soon afterwards he was able to publish them in *Nuclear Physics*. It was then that the war against infinites, frequently characterized by mistakes, misconceptions, missed chances and other obstacles, was happily concluded.

After his discovery, 't Hooft, who in the meantime had obtained his doctorate *cum laude*, continued to collaborate with Veltman on the subject. Many other theoreticians in other universities participated in an effort to perfect 't Hooft's results. His method was immediately applied to eliminating the infinities in the Salam–Weinberg electroweak theory. And so particle physicists started to become tremendously excited. (Weinberg's 1967 paper on the electroweak theory grew into the most cited paper in the whole history of particle physics! As it was said at the time, 't Hooft's work had 'turned the Salam–Weinberg frog into an enchanted prince!') Meanwhile, in 1970, Sheldon Glashow and his co-workers had involved leptons as well as hadrons in the theory (see Nobel 1979). Finally, the discovery of neutral currents in 1973, and of the W and Z weak bosons in 1983 (see Nobel 1984), both at CERN, sealed the triumph of the electroweak theory.

Fig. 13.14. G. 't Hooft, left (© The Nobel Foundation, Stockholm) and M. Veltman, right (courtesy CERN).

In 1989 LEP, the large electron–positron collider at CERN, started to produce precise data on the electroweak theory, and on the properties of the Ws and the Z. Veltman and 't Hooft's mathematical technique was extensively used to predict the physical quantities relevant to these properties, and it resulted in a perfect agreement between calculations and measurements. The same technique was also used in order to predict the mass of the sixth quark – the top quark. When it was discovered in 1995 at Fermilab (p. 424), its mass exactly agreed with what had previously been calculated theoretically!

Gerardus 't Hooft was born in 1946 in Den Helder, the Netherlands, the son of a naval engineer. He wrote in his Nobel autobiography:

> 'A man who knows everything'. This, reportedly, was my reply to a school teacher asking me what I'd like to become when I grew up. I was eight years old . . . [What] I really meant was a 'scientist', someone who unravels the secrets of the fundamental Laws of Nature. This perhaps was not such a strange wish. Science, after all, was in my family. Just about at that time, 1953, my grand-uncle, Frits Zernike had earned his Nobel Prize [see p. 242] . . . My grandmother, Zernike's sister . . . had married her professor, a well-known zoologist, Pieter Nicolaas van Kampen at the University of Leyden . . . My uncle, Nicolaas Godfried van Kampen was appointed Professor of Theoretical Physics at the State University of Utrecht . . . Was it the environment or was it in my genes to choose to become a physicist? My grandmother adored scientists and by that she may have further determined my choice, but I think that my mind was made up long before I could talk.[27]

In 1964, 't Hooft went to study at the University of Utrecht, where in 1972 he obtained his doctorate, and five years later became a full professor. He was also a visiting professor at numerous other research institutions, including Stanford, Harvard, and Boston in the USA. (In 1982 't Hooft shared, with Victor Weisskopf and Freeman Dyson, the Israeli Wolf Prize for physics.)

Martinus Veltman was born in 1931 in Waalwijk, the Netherlands, the son of the headmaster of the local primary school. In 1948 he went to the University of Utrecht to study physics:

> ... my high school physics teacher came to my home and suggested to my parents to send me to the university ... As the money situation was very tight the main point was to find a university where I could go to by train. This was possible with the University of Utrecht. For three years I commuted back and forth from Waalwijk to Utrecht, a 90-minute trip each way. I am still grateful to this high school teacher, Mr Beunes, as he did the extra thing, going to my parents' house. Since then I have found out that many physicists owe their career to a good high-school teacher.[28]

In 1963 Veltman received his doctor's degree in physics, and three years later he became professor of theoretical physics at the University of Utrecht. In 1981 he moved to the USA, where he was appointed full professor at the University of Michigan in Ann Arbor, and where he became emeritus in 1997.

The standard model

The *standard model* of fundamental particles and forces, one of the crowning achievements of twentieth-century physics, is a theoretical construct which gives an accurate mathematical description of all particles and the interactions between them. It describes the properties of matter down to one billion-billionth of a metre, which is the smallest distance that has been probed at today's most powerful particle accelerators.

The standard model is characterised by twelve fundamental particles (and their antiparticles): six 'quarks' and six 'leptons'. Quarks and leptons appear in three families, each family consisting of two quarks (one with charge $\frac{2}{3}$, and the other with charge $-\frac{1}{3}$), and two leptons – one carrying a unit of electric charge, and a neutrino (with zero electric charge). The standard model also explains how the forces between these particles work.

> *Electroweak theory* (p. 359). This is a *quantum field theory* which describes all of the known phenomena of electromagnetism and the weak interactions. It represents the combination of *quantum electrodynamics* (QED, p. 297) and *quantum weak dynamics* (the theory of the weak nuclear force). All the twelve fundamental particles feel the electroweak force. This is produced by the exchange

between the interacting particles of the massless 'photon', carrying the electromagnetic force, and of three varieties of heavy 'weak bosons', carrying the weak force.

Quantum chromodynamics (QCD, p. 406). This is the theory of the *strong force* acting on quarks and hadrons. Like the electroweak theory, QCD, too, is a quantum field theory, which describes how the strong force is transmitted between quarks by eight species of massless 'gluons'.

The standard model also includes electrically neutral massive particles, called 'Higgs particles', which pervade the whole of space, and which give the other particles of the model their masses (p. 356).

Despite the astonishing success of the standard model, particle physicists believe that it does not represent the ultimate theory of fundamental particles and forces. They want to go beyond it; many talented physicists work all out in their search for a theory capable of unifying the electromagnetic and weak forces with the strong force (the unification of these three fundamental forces is often called the *Grand Unification*, and it should occur at energies of more than a million billion billion electron-volts and distances less than ten thousand-billion-billion-billionths of a metre). The next step, the 'Holy Grail', would be a theory which would also include the fourth force – gravity. This would be very likely to involve more revolutionary concepts, such as those used in *superstring theories* – modern theories which describe the elementary particles as one-dimensional space-time strings, having a length of about one hundred-million-billion-billion-billionth of a metre!

Dreams of a final theory

'The [twentieth century] . . . has seen in physics a dazzling expansion of the frontiers of scientific knowledge . . . [We] have built a successful theory of electromagnetism and the weak and strong nuclear interactions of elementary

Fig. 13.15. Plato's dodecahedron – symbolising the quintessence of the heavens.

Table 13.1. *Twelve fundamental particles (fermions)*

Leptons	Quarks
τ neutrino	top
τ lepton	bottom
muon neutrino	charm
muon	strange
electron	down
electron neutrino	up

Table 13.2. *Twelve force messengers (bosons)*

8 gluons	colour (strong) force	QCD
photon	electromagnetic force	
3 weak bosons	weak force	electroweak theory

particles. Often we have felt as did Siegfried after he tasted the dragon's blood, when he found to his surprise that he could understand the language of birds . . . Our present theories are of only limited validity, still tentative and incomplete. But behind them now and then we catch glimpses of a final theory, one that would be of unlimited validity and entirely satisfying in its completeness and consistency. We search for universal truths about nature, and, when we find them, we attempt to explain them by showing how they can be deduced from deeper truths . . . [A] final theory would not end scientific research, not even pure scientific research, nor even pure research in physics . . . A final theory will be final in only one sense – it will bring to an end a certain sort of science, the ancient search for those principles that cannot be explained in terms of deeper principles.'

(Steven Weinberg)[29]

2000

The information age

As the Swedish Academy of Sciences entered the twenty-first century it decided to acknowledge Information Technology (IT). The Nobel Prize for physics of the year 2000 in fact honoured inventions which have been the driving force behind the development of those twin symbols of today's information age: *computers* and *communication systems*. The prize went to the American engineer Jack Kilby 'for his part in the invention of the integrated circuit', and to two physicists – the Russian Zhores Alferov and the German-born American Herbert Kroemer – for having devised new kinds of semiconductor structures which have led to the development of semiconductor lasers and fast transistors.

The integrated circuit

The invention of the transistor in 1947 by John Bardeen, Walter Brattain and William Shockley (see Nobel 1956) opened the way to modern electronics. A decade later Jack Kilby invented the *integrated circuit*, also named the *chip*, the root of almost every electronic device used today (p. 269). Briefly, here is the chronicle of Kilby's invention; it has almost become a legend!

Jack Kilby, an engineer who had previously worked in the field of electronics, in 1958 joined Texas Instruments in Dallas, Texas. As a new employee he was unable to take a holiday that summer, so, during the month of July, he was alone in his laboratory. Here he conceived a new electronic circuit, in which all of the circuit elements, such as diodes, transistors, resistors, capacitors, were combined in a single piece of semiconducting material. Kilby himself recalls:

> Further thought led me to the conclusion that semiconductors were all that were really required – that resistors and capacitors, in particular, could be made from the same material as the active devices. I also realised that, since all the components could be made of a single material, they could be made *in situ*, interconnected to form a complete circuit. I then quickly sketched a proposed design for a [circuit] using these components.[30]

In September Kilby was able to present, to a group of executives from Texas Instruments, three prototypes of the first integrated circuit ever conceived. In March 1959 the invention was announced at a press conference in New York City.

Meantime, Robert Noyce, one of the founders of Silicon Valley, then at Fairchild Semiconductor International, invented a revolutionary method for fabricating integrated circuits (it was based on the so-called 'planar technology', which had been developed by a Fairchild physicist of the name of Jean Hoerni in order to develop a new type of transistor). The Swedish Academy of Sciences recognised the dual parentage: 'Jack S. Kilby and Robert Noyce [who died in 1990] are both considered as the inventors of the integrated circuit. Kilby was the one who built the first circuit. Noyce developed the circuit as it was later to be manufactured in practice . . . '.[31]

It was now possible to build up the equivalent of many circuit components on a single chip, and this opportunity completely revolutionised the electronic industry. In 1965 Gordon Moore, who with Noyce had founded Fairchild Semiconductor, predicted that the number of transistors and other circuit elements that could be built on a chip would double every eighteen to twenty-four months with the same cost for each chip. Moore's law held up for more than forty years: in 1965 a chip could contain about fifty individual elements; twenty years later, some hundreds of thousands; and today over a hundred million! Chips are the heart of all pocket calculators and present-day computers, large and small; they are used

Fig. 13.16. Jack Kilby holding chips. (Courtesy Texas Instruments.)

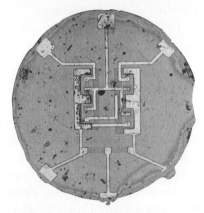

Fig. 13.17. One of the earliest integrated circuits printed on a silicon wafer: it contains two transistors. (Photo courtesy Fairchild Semiconductor International.) The first microprocessor, the engine driving the ongoing IT revolution, was introduced by the US company Intel in 1971 (it was a single silicon chip, no larger than a fingernail, which contained 2300 transistors). The most recent microprocessors used in PCs contain some ten million transistors.

in communications, control systems, medical instrumentation, and millions and millions of other areas.

Jack St Clair Kilby was born in 1923 at Jefferson City, Missouri, USA. He received his master's degree in electrical engineering from the University of

Wisconsin in 1950. He then worked for Centralab in Milwaukee until 1958, when he joined Texas Instruments. As well as the integrated circuit, he co-invented the world's first electronic handheld calculator and the thermal printer (he holds more than sixty US patents). From 1978 to 1984 Kilby was Professor of Electrical Engineering at Texas A&M University in College Station. For his inventions he has received many prestigious honours in science and engineering, amongst them the US National Medal of Science (1969).

Two pioneers of heterostructures

A *heterostructure* semiconductor consists of several thin layers (with a thickness varying from a few millionths to some thousandths of a millimetre) of semiconducting materials of varying composition and doping. All of them are jointed together, and have their atomic structures fitting perfectly into a single crystalline pattern. Each layer has different electronic properties, which have profound effects on the behaviour of the electrons and holes at the interfaces of the structure. Here are the principal steps that, in less than fifteen years, marked the development of heterostructures, which were pioneered by Zhores Alferov and Herbert Kroemer.

- In 1957 Kroemer was working at RCA in Princeton, New Jersey. Here he first conceived the idea of joining together two semiconductor layers with different electronic properties, so as to control better the flow of charge carriers (electrons and holes) moving at their junction; in this way he laid down the basis for designing and fabricating heterostructures. Kroemer noted that such an innovative structure should be capable of powering electronic devices, like high-speed transistors, which it would be impossible to produce with a single-structure semiconductor. (In the 1970s Kroemer himself greatly contributed to the development of high-speed transistors made with heterostructures.)
- In the early 1960s, solid-state scientists in the USA, and also in the Soviet Union (among them Nikolai Basov, Nobelist in 1964), focused their attention on developing a laser, by using a semiconducting material. Success was achieved in 1962, when groups of American researchers, working at the General Electric Company, at the IBM Thomas J. Watson Research Center, and at MIT, were able to produce laser light in a diode: this was made of two regions of a single semiconducting compound, called *gallium arsenide*, having different concentrations of positive and negative charge carriers. Across the junction between these two regions, laser light was produced by the stimulated emission process, which took place when a photon, from the recombination of a hole–electron pair, induced the emission of another photon from a pair that had not recombined spontaneously.

  The new laser, unlike its predecessors, could be made very small, and was cheap and simple to operate. It suffered, however, from a major disadvantage: it required too high an electric current to operate, so that, in order to prevent overheating, it was forced to function at very low temperatures. This limited its usefulness in many important applications, such as fibre-optic communications.

- At this point our two pioneers entered the scene. Alferov was then working at the Ioffe Institute in Leningrad (now St Petersburg), and Kroemer at Varian Associates in Silicon Valley, California. In 1963 they independently suggested the use of a heterostructure semiconductor to construct a laser, which would be superior in efficiency to the semiconductor lasers first demonstrated. Their proposal was to utilise crystals with a region of a semiconducting material, sandwiched between two regions of another type of semiconductor, possessing different electronic properties; this was called a *double heterostructure* semiconductor.
- Four years later, in 1967, Alferov and his Ioffe co-workers developed a sophisticated technique, and succeeded in obtaining for the first time, and to a high degree of perfection, a junction between two very thin layers (less than a thousandth of a millimetre), one of gallium arsenide and the other one of an alloy called *aluminium gallium arsenide*. (The same result was achieved in the same year at the IBM Research Center, in the USA.) This success opened the door to the construction of double heterostructure lasers. At the same time Alferov proceeded to study how electrons and holes, injected by the two sandwiching layers, are trapped in the very thin region between the two junctions of a double heterostructure – the so-called active layer (here electrons and holes recombine, giving off laser light by stimulated emission). The effectiveness of this he succeeded in proving by experiments in 1968. And in the following year he was able to announce the creation of the first *double-heterostructure laser* (an active layer of gallium arsenide sandwiched between two doped layers of aluminium gallium arsenide), operating at room temperature, and capable of emitting short pulses of coherent light.
- Finally, in 1970, Alferov and his co-workers – and soon afterwards, researchers at Bell Labs in the USA – reported the invention of a double-heterostructure laser, emitting coherent, monochromatic, infrared light at room temperature, and in a continuous way. (At last this made possible its use in optical-fibre communications.)

'Heterostructures for everything'

This is the title of a talk that Kroemer gave in 1980, and it well expresses how heterostructures have become the basic building blocks of semiconductor devices. Here are some examples:

- Heterostructures are used in the field of high-speed microelectronics. Heterostructure transistors are some hundred times faster than the best ordinary transistors, and they are used in satellite systems for wireless communications, in cellular phones and in many other applications.
- Heterostructures are also applied in many areas of optics and electronics: in devices like solar cells; and in light-emitting diodes (LEDs), which are used in displays of all kinds. Heterostructure lasers are used in optical-fibre networks on the Internet, and in compact disc and digital video disc players, supermarket scanners, laser printers, fax machines, laser pointers, and numerous scientific instruments.
- Heterostructures have also become of fundamental importance for physics research: as an example, the fractional quantum Hall effect (see Nobel 1998) was observed in a two-dimensional sheet of electrons, moving in a heterostructure interface.

Zhores Ivanovich Alferov was born in Vitebsk, Belorussia, in 1930. He graduated in 1952 at the Lenin Electro-Technical Institute in Leningrad (now St Petersburg). He then began to carry out research work at the Ioffe Physico-Technical Institute in the same city, where in 1987 he became its director. In 1973 he was appointed Professor of Optical Electronics at the Leningrad (now St Petersburg) Technical University, where in 1988 he became Dean of the Faculty of Physics and Technology. Alferov received many honours, among them the Franklin Institute gold medal (1971) from the USA, and the Lenin Prize (1972) – the highest Soviet scientific award of that time. He has also served as vice-president of the Russian Academy of Sciences, and as a member of the Russian Parliament.

Herbert Kroemer was born in 1928 in Weimar, Germany, the son of a civil servant. He obtained his doctorate in 1952 at the University of Göttingen. He then went to work in a laboratory of the German postal service until 1954, when he moved to the USA. Here he was employed at the RCA Laboratories in Princeton, New Jersey (1954–7); at the Varian Associates in Palo Alto (1959–66); and at the University of Colorado in Boulder, where, from 1968 to 1976, he was a professor of physics. He then moved to the University of California at Santa Barbara, where he is a full professor in the Department of Electrical and Computer Engineering, and also in the materials department.

Discoveries and inventions

Nanotechnology is the science of super-tiny things – things on the scale of atoms and molecules, which have superior physical, chemical and biological properties because of their 'nanosize'.

Fig. 13.18. Zhores Alferov (left) and Herbert Kroemer (right). (© The Nobel Foundation, Stockholm.)

Semiconductor nanotechnology continues to make impressive advances, opening new opportunities for investigating condensed matter down to two, or even one dimension. One typical example is the so-called *quantum wells*. These are ultra-thin semiconductor layers (with a thickness of some nanometres) contained in heterostructures. In these layers charge carriers, like electrons, are trapped, and quantum effects occur there, which restrict what the mobile electrons themselves can do. These two-dimensional quantum wells have many useful properties because one can exactly design how to confine the electrons. As a result, they are widely used to make semiconductor lasers and other useful devices.

Physicists and engineers have taken this kind of quantum confinement further, and they have made other kinds of semiconductor *nanostructures*, which are called *quantum wires*; here, mobile electrons are confined in a sort of one-dimensional electron gas. They also speak of *quantum dots* (crystals containing only a few hundred atoms), or *nanotubes* (miniature straws made up of pure carbon, praised for their excellent mechanical, electrical and chemical properties). All these marvels of nanotechnology are awaiting new and unexpected applications in fields such as electronics, telecommunications and material manufacture.

Biomedical nanotechnology, too, is a rapidly advancing field of science (for instance, scientists are using quantum dots to develop molecular 'nanoprobes'). It is believed that this science will soon produce major advances in molecular diagnostics, molecular biology and bio-engineering.

2001

An extreme state of matter

In 1995, after years of unsuccessful attempts carried out by experimentalists in different laboratories, three American physicists, Eric Cornell and Carl Wieman, in Boulder, Colorado, and the German-American Wolfgang Ketterle at MIT, succeeded in creating a new form of matter. It is known as *Bose–Einstein condensate*, from the names of two scientists: the Indian theoretical physicist Satyendra Nath Bose, who had had the underlying idea some seventy years earlier; and Albert Einstein, who had proposed the new state of matter. After this discovery, a flood of experiments began to engage the community of atomic physicists. In more than twenty laboratories around the world Bose–Einstein condensates began to be created: theorists and experimentalists all rushed to conquer the new quantum territory. So, six years later, the amazing discovery was honoured by the award of the 2001 Nobel Prize for physics jointly to Cornell, Ketterle and Wieman. (All three belong to the post-war generation of physicists who have been honoured with Nobel Prizes.)

Bose–Einstein condensate

In 1924 Satyendra Nath Bose was a young theoretical physicist at the University of Dacca (then in India). He had carried out calculations, and had shown that

Planck's old black-body radiation law (p. 37) could be deduced in a new, original way, entirely by applying statistical rules to the light photons. Bose sent the paper containing his theory to Einstein in Berlin, who soon realised its importance, and sent it to the *Zeitschrift für Physik*, where it was published.

Einstein did not stop there, but extended Bose's idea to all particles or particle systems with an integer spin (the so-called *bosons*, p. 175), and in 1924–5 he published two papers concerning the subject. In these papers, Einstein predicted that if a large assembly of identical massive bosons, which do not interact between themselves, were cooled down to an extremely low temperature, all of them would drop rapidly into the lowest possible energy state. In other words, they would undergo a phase transition, 'condensing' into a new state – like a gas when it condenses into a drop of liquid. This is what is called a Bose–Einstein condensate: a new, extreme state of matter, which displays the effects of quantum mechanics on a macroscopic scale. Let us try to describe this with an analogy used in some textbooks.

Imagine that you are looking at a crowd in a distant square on market day: the crowd is wandering around at random, and each individual is doing something different. Suppose now that the crowd is replaced by a battalion of soldiers performing a parade: here every soldier is doing the same thing at the same time. And now the physics analogy: a normal gas is like the market-day crowd, every atom flitting around at random, whereas in a Bose–Einstein condensate the atoms are all forced to behave in a perfectly synchronised fashion, 'to do the same thing at the same time'. (Physicists like to say that the atoms are all forced to stay in a common 'quantum state'.) Using a more sophisticated quantum language, we can say that in a normal gas, every atom has its own wave function (ψ) describing what it is doing at each instant, whereas in a Bose–Einstein condensate it is the assembly of atoms that is described by a single wave function, just as if it were a single 'super-atom' in a 'macroscopic' quantum state.

We have already met systems somewhat similar to a Bose–Einstein condensate, when we discussed the BCS theory for superconductivity (p. 323), the superfluid helium-4 and the superfluid helium-3 (p. 426). In fact, the electron pairs, which are created in a superconductor below its transition temperature, can be considered as bosonic systems with spin zero behaving collectively as bosons do in a Bose–Einstein condensate; the atoms of helium-4 are also bosons, again with spin zero; and the pairs of helium-3 atoms, in their superfluid state, are considered particles each with an integer spin, which can undergo condensation as well. But all these three phenomena are much more complicated than the Bose–Einstein condensate in a gas of genuine bosons, where direct comparison with first principle theories is possible. Physicists thus thought of using a gas of bosonic atoms in the hope of seeing a persuasive manifestation of such a condensate.

A challenging competition

At this point a question arises: how do you go about forming a Bose–Einstein condensate with a gas of bosonic atoms? Well, firstly you have to reduce the separations existing between the atoms composing the gas, and you have to make those separations as small as possible (this is in order to achieve a high density). Secondly, you have to chill the gas to extremely low temperatures, so that the motion of the atoms is suddenly halted, and they collapse into the lowest possible energy state. However, all this was easier said than done. It had to await the development of suitable cooling and trapping techniques, essential to confining and grooming the atomic states. (The 1997 Nobel Prize for physics, you will remember, was awarded to Steven Chu, William Phillips and Claude Cohen-Tannoudji for having developed just these very techniques.)

Starting the race

The race started out in the mid-1970s at MIT with two pioneers in the field, Thomas Greytak and Daniel Kleppner; also in the race were Isaac Silvera and Jook Walraven, both working at the University of Amsterdam, the Netherlands. They were all attempting to condense a gas of hydrogen atoms. To chill the gas down to temperatures where the atoms form a condensate, Harald Hess, working with Greytak and Kleppner, developed a new technique, called *evaporative cooling*. In this technique the hottest atoms (the most energetic) were kicked out from the magnetic trap where they had been held, so that only the cooler ones were left behind (this is like the cooling process in a cup of hot tea, where the most energetic molecules are given off as steam, leaving the less energetic ones in the cup).

The scene moves to Boulder

In the late 1980s the scene moved to Boulder, Colorado, to the Joint Institute for Laboratory Astrophysics (JILA) – a joint research centre between the National Institute of Standards and Technology (NIST) and the University of Colorado. Here, Carl Wieman, an *alumnus* of MIT, had started a research programme capable of producing a Bose–Einstein condensate. He had thought of using alkali metal atoms instead of hydrogen, and he had devised a way of combining laser cooling and evaporative cooling. His idea was this: first, pre-cool the atoms with laser light in a magneto-optic trap (the so-called MOT, see Nobel 1997); second, transfer the pre-cooled atoms into a purely magnetic trap, where by the evaporative-cooling method they would reach the desired low temperature needed for them to collapse into a condensate.

In 1990 a second person arrived at Boulder: Eric Cornell. He had earned a Ph.D. at MIT, and had accepted a postdoctoral position in Wieman's group. Two years later he was appointed as a scientist at NIST in Boulder. The two experimenters continued to work together, applying Wieman's basic ideas. Cornell developed a method of obtaining a more efficient evaporative cooling, and devised an original technique to retain the atoms better in the magnetic trap – which was crucial for the success of the experiment.

At last, on 5 June 1995, the JILA team succeeded for the first time in creating and photographing the long-sought Bose–Einstein condensate – a 'super-atom' with a diameter of 2/1000 of a millimetre, containing a condensate of about 2000 atoms of rubidium, which had been super-chilled to a temperature of 170 billionths of a degree above absolute zero. These results were promptly published in the 14 July issue of the scientific magazine *Science*.

Success at MIT

At MIT, Wolfgang Ketterle and his co-workers followed close on the heels of Cornell and Wieman. On 29 September they were able to obtain a condensate of about 500 000 sodium atoms, all held in a special magnetic trap, inside an ultra-high-vacuum chamber. (Since that time Ketterle and his team have succeeded in creating a condensate with no less than ten million atoms.)

The higher number of atoms in Ketterle's condensate enabled him to investigate the phenomenon still further. Making two separate clouds of Bose–Einstein condensate merge into one another, Ketterle obtained very clear interference patterns, showing that the condensate contained *coherent matter waves* (just like the interference fringes produced by two beams of coherent light, as first seen by Thomas Young about two hundred years earlier, p. 21). And, in 1997, he was also able to produce a stream of small drops of the condensate – a first rudimentary *atom laser* which used atoms rather than light photons.

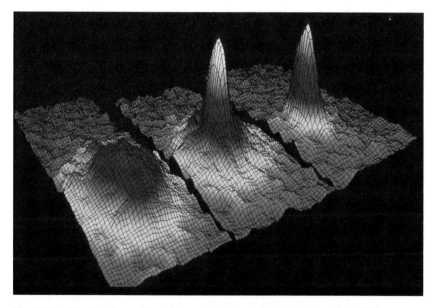

Fig. 13.19. The first Bose–Einstein condensate produced by Cornell and Wieman at JILA. (Image courtesy of Mike Matthews, JILA research team.) The three drawings show how rubidium atoms come together into the same quantum state (represented by the height of the central spike) as the temperature is decreased (reading from left to right) to 400, 200 and 50 billionths of a degree above absolute zero.

Fig. 13.20. Interference pattern obtained by Ketterle at MIT, by merging the coherent atom waves of two Bose–Einstein condensates. (Courtesy Professor Wolfgang Ketterle.)

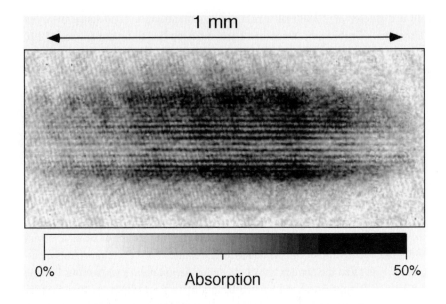

Following on the pioneer experiments at JILA and MIT, Bose–Einstein condensates have been created in dozens of laboratories worldwide. Possible applications that physicists often speak about include atomic clocks, quantum computing and nanotechnology. Moreover, several other groups have tried to create an even more exotic state of matter – a condensate of 'fermionic' atoms (these, having half-integer spins, are ordinarily forbidden, by Pauli's exclusion principle, to share the same quantum state). Amazingly enough, on December 2003, researchers at the same JILA as Cornell and Wieman announced that the long-standing goal had been achieved![32]

Three ingenious experimenters

Eric Allin Cornell was born in 1961 in Palo Alto, California. He graduated at Stanford University in 1985, and received his Ph.D. in physics from MIT in 1990, writing his thesis under David Pritchard. He then joined JILA in Boulder to work with Wieman on laser cooling and trapping. In 1992 he was appointed senior scientist at NIST and assistant professor at the physics department of the University of Colorado, becoming professor in 1995. He is a member of the US National Academy of Sciences.

Carl E. Wieman was born in 1951 in Corvallis, Oregon, USA. He graduated at MIT in 1973 and received his Ph.D. from Stanford University in 1977. He was assistant research scientist (1977–9), and later Assistant Professor of Physics (1979–84) at the University of Michigan. He then moved to the University of Colorado, where he was associate professor of physics, becoming professor in 1987, and distinguished professor in 1997. He is a Fellow of JILA, where he

Fig. 13.21. C. Wieman (left) and E. Cornell atop Macky Auditorium on the campus of the University of Colorado at Boulder. (Photograph by Ken Abbott/University of Colorado at Boulder.)

served as Chairman from 1993 to 1995. Like Cornell, Wieman too is a member of the US National Academy of Sciences.

Wolfgang Ketterle was born in 1957. He received his master's degree in 1982 from the Technical University of Munich, and a Ph.D. from the Ludwig Maximilians University of Munich in 1986. He then worked as a postdoctoral fellow at the Max-Planck Institute for Quantum Optics in Garching, and at the University of Heidelberg. In 1990 he left Germany, and went on to the USA, to MIT, where he started to work on ultra-cold atomic matter in a group headed by David Pritchard. In 1993 Ketterle joined the physics faculty at MIT, where he was appointed assistant professor, and in 1998 he was promoted John D. MacArthur Professor of physics. He is a member of the European Academy of Sciences and Arts, the Academy of Sciences in Heidelberg and the American Academy of Arts and Sciences.

A beautiful summer night

At long last! After more than twenty years of striving, Greytak and Kleppner were rewarded for their perseverance. Three years after the 1995 breakthrough of our Nobel prizewinners, they succeeded in achieving Bose–Einstein condensation

Fig. 13.22. Wolfgang Ketterle (right) and his co-workers gathered around the apparatus they had used in order to demonstrate the first atom laser. (Courtesy Donna Coveney/MIT.) 'Today, if you have a demanding job for light, you use an optical laser. In the future, if there is a demanding job for atoms, you may be able to use an atom laser . . . ' (Wolfgang Ketterle.)[33]

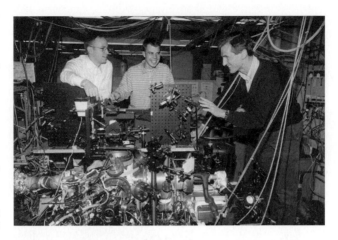

in a gas of hydrogen atoms – the most difficult substance for creating such a condensate. With these words Kleppner describes a wonderful night of June 1998:

> Late one night last June, a phone call from the lab implored me to come quickly. I had a pretty good idea of what was up because [a Bose–Einstein condensate] in hydrogen had seemed imminent for more than a week. As I drove in the deep night down Belmont Hill toward Cambridge [Massachusetts], still dopey with sleep, the blackness of the sky suddenly gave way to a golden glow. I was not surprised because I had a premonition that the heavens would glow when [a Bose–Einstein condensate] first occurred in hydrogen. Abruptly, streamers of Bose–Einstein condensates shot across the sky, shining with the deep red of rubidium and the brilliant yellow of sodium. Small balls of lithium condensates flared and imploded with a red pop . . . The spectacle was exhilarating but totally baffling until I realised what was happening: The first Bose–Einstein condensates were welcoming hydrogen into the family![34]

2002

Watching the heavens

At a press conference held on 8 October 2002 the Swedish Academy of Sciences announced that the Nobel Prize for physics would be awarded – it was only for the sixth time in a century – to astronomy and astrophysics. The prize was shared between three scientists, two Americans, Raymond Davis and Riccardo Giacconi, and a Japanese physicist, Masatoshi Koshiba. Davis and Koshiba received the prestigious award for having detected and studied cosmic neutrinos, whereas Giacconi was nominated for his discovery of the first X-ray cosmic source and for his contribution to X-ray astronomy.

Cosmic neutrinos

The neutrinos which earned the Nobel Prizes for the years 1988 and 1995 were man-made neutrinos, produced in laboratories. Clyde Cowan and Frederick Reines had first detected the neutrino in their experiment at a nuclear reactor (p. 260), while the Columbia team (Leon Lederman, Melvin Schwartz and Jack Steinberger) had discovered the muon neutrino at an accelerator (p. 286). Today, people talk not only about neutrino physics, but also about *neutrino astrophysics* and *neutrino astronomy*. These new sciences involve many kinds of extraterrestrial neutrino sources, which include our sun and exploding stars (supernovae); there exist also neutrinos that are a relic from the *Big Bang*, high-energy neutrinos from galaxies, and cosmic-ray neutrinos produced in the earth's atmosphere. Let us start our story by talking about *solar neutrinos*, because these heralded the beginning of a fascinating adventure which, starting more than forty years ago, ended up by taking Davis to Stockholm in December 2002.

Davis' experiment

We have already mentioned Davis' pioneer neutrino experiment in the context of the 1967 Nobel Prize awarded to Hans Bethe (p. 306). Here we want to tell you something about its story, which began in the late 1950s, soon after the discovery of the neutrino by Cowan and Reines.

In those years the astrophysicist William Fowler (Nobelist in 1983) and his colleague Al Cameron had pointed out that, among the nuclear fusion reactions which power the sun, there might exist one particular reaction that generated neutrinos with energies high enough to be detected with the techniques then available (the majority of the reactions in the sun produce neutrinos with an energy that was too low to be detected at that time). Later on, during the period 1962–4, the American theorist John Bahcall performed careful theoretical calculations, and showed how the Fowler and Cameron solar neutrinos could be detected if they were interacting with a huge target, made up of material containing chlorine atoms (the idea of using chlorine with the function of a target had been put forward in the 1940s by the Italian-born physicist Bruno Pontecorvo, and independently by Luis Alvarez, Nobel 1968).

At that period Raymond Davis was a member of the chemistry department of the Brookhaven National Laboratory – the same laboratory where the muon neutrino had been discovered. He promptly took up the idea, and gathered round him a small group of researchers to embark on building an experimental apparatus, whose heart was a huge tank filled with 615 tons of a common cleaning liquid containing chlorine. By 1967 Davis' team was ready to install the tank in the Homestake Gold Mine, Lead, South Dakota, at an underground depth of 1500 metres, in order to shield the apparatus from cosmic rays. In fact within a few months they were able to begin the experiment.

This is the method they used. If one of the billion billion solar neutrinos that daily traversed the tank happened to interact with a chlorine atom, it would convert this into a radioactive argon atom. Theorists had calculated that about one or two argon atoms a day should have been created in the tank. (This truly represented a formidable experimental challenge: to single out very few argon atoms from more than one thousand billion billion billion chlorine atoms, and to count each of them, one by one!) For this purpose our experimenters developed a very sophisticated technique with which they succeeded in removing and collecting, every month or two, the argon atoms from their tank; they then detected their individual radioactive decays.

The first results came a few years later; they proved that a handful of neutrinos had definitely been detected. This was the first observation of solar neutrinos, and it confirmed that nuclear fusion reactions are actually the basis for the energy produced by the sun, as had been predicted by Hans Bethe thirty years earlier.

But the story does not end here. To his surprise, Davis noticed that his results revealed a difficulty: the number of neutrinos detected was only about one third of the value that the theorists had told him he ought to be detecting! For more than twenty-five years Davis continued, with few interruptions, to monitor with great care the flux of solar neutrinos reaching his detector; and always the flux he measured was below the one that had been calculated. (Over such a long time, Davis was able to extract from his tank, and to count one by one, about 2000 argon atoms!) This low neutrino flux became a vexing mistery which troubled physicists for a long time (it was called the *solar neutrino problem*).

What was to be done about this matter? Physicists thought of various possibilities: it could mean that Davis' experiment was wrong; or that the solar theory (called the *standard solar model*) was wrong; or that some strange thing was happening to the neutrinos on their way from the sun to the detector. The solution had to wait many years before emerging. Meanwhile, the puzzle gave birth to many experiments, with scientists working all around the world (in the USA, in Russia, in Italy, in Japan and in Canada).

The scene shifts to Japan

In the early 1980s Masatoshi Koshiba was a professor of physics at Tokyo University. He was then leading a team of physicists who had built a huge detector aimed at investigating the instability of the proton, one of the most fundamental issues of modern particle physics. They placed their detector in a mine 1000 metres underground, located in Kamioka-cho, Gifu, in the Japanese Alps. Kamiokande – so the detector was named, consisted of a tank containing 3000 tons of pure water, surrounded by about 1000 light detectors, called 'phototubes', which were placed on the inner wall of the tank. These phototubes were a sort of super-sensitive artificial 'eyes' which collected the bluish Cherenkov light that was emitted by charged particles travelling through the water at very high speeds.

In 1985 Koshiba and his co-workers decided to upgrade his apparatus, so as to be able to detect neutrinos of astrophysical origin. They used the following technique: each time a neutrino hit an atom in the tank, an electron was pushed away. This electron, while traversing the water, produced a brief Cherenkov light flash, which was captured by the watching phototubes. This technique enabled the Kamioka lab. team not only to measure the energy of the incoming neutrinos, but also to trace their direction, so that they were able to ascertain the region in the sky from which the neutrinos were coming. In the 1990s the same team built another detector, a super-detector called Super-Kamiokande. It contained 50 000 tons of water, and it was surrounded by over 10 000 phototubes. From 1996 on, a new series of experiments began, and they are actually still running today. Let us briefly describe the most important results so far obtained with Kamiokande and Super-Kamiokande, results which have earned Koshiba the Nobel Prize.

Solar neutrinos
The first result from the upgraded Kamiokande came towards the late 1980s and confirmed the correctness of Davis' observations: the detected neutrinos did actually came from the sun, and their flux was lower than that calculated by means of the theoretical solar models. This solar-neutrino paradox focused the attention of physicists on a hypothesis which had been put forward in 1968 by Bruno Pontecorvo and Vladimir Gribov in the Soviet Union. They had guessed that if neutrinos had even only a tiny mass, they would change type, or flavour, as they were travelling – or, as theorists like to say, they would 'oscillate' from one variety to another. It would be a kind of metamorphosis, so that, on their way from the sun to a detector on earth, an electron neutrino would transform itself into a muon neutrino, which would transform itself into a tau neutrino, and so on. This could eventually explain why the Davis and Koshiba experiments, which were sensitive only to electron neutrinos, had measured a deficit in the neutrino flux.

Supernova SN1987A
A really staggering breakthrough came from Kamiokande in 1987. On 23 February, a sudden burst of neutrinos lasting for about thirteen seconds was detected; the neutrinos were coming from the supernova SN1987A – a star twenty times more massive than our sun, which had collapsed and died about 170 000 years ago in the Large Magellanic Cloud (p. 396). Kamiokande detected twelve out of the ten million billion neutrinos composing the burst which had hit the detector. We can say that this exceptional event represented the dawn of a new era in neutrino astronomy, and provided a strong argument in favour of the theoretical models concerning the evolution of stars that had been previously worked out by astrophysicists.

Fig. 13.23. This picture shows the interior wall and the top of Super-Kamiokande with about 11 000 phototubes watching the water where cosmic neutrinos interact. (Courtesy Kamioka Observatory, ICRR (Institute for Cosmic Ray Research), The University of Tokyo.)

Neutrino oscillations

In 1998 the Super-Kamiokande group reported the first evidence for neutrino oscillations relative to the so-called *atmospheric neutrinos* – neutrinos produced in the collisions of high-energy cosmic rays (principally protons) with the atoms of the earth's atmosphere. And recently (2001–3), scientists working at the Sudbury Neutrino Observatory in Ontario, Canada, and using a large detector containing heavy water, have reported results which in turn provide clear evidence of oscillations for *solar neutrinos*.

The significance

All these discoveries are of the greatest significance for astrophysics and cosmology, and for particle physics as well. They give scientists the key to a better understanding of how the sun, and the other stars, shine, and how matter is distributed throughout the universe. Secondly, if neutrinos do possess a mass, then

Fig. 13.24. Raymond Davis (left, courtesy Brookhaven National Laboratory), and Masatoshi Koshiba (right, courtesy AIP Emilio Segrè Visual Archives.)

particle physicists will be obliged to modify their theories regarding these elusive particles, over and beyond the current standard model of particle physics (for this model, in the present form, assumes that neutrinos have zero mass!)

Raymond Davis, Jr was born in 1914 in Washington, DC. In 1940 he graduated at the University of Maryland, and two years later he received a Ph.D. in physical chemistry from Yale University, Connecticut. During the Second World War he was enrolled in the US Army Air Force (1942–6). He later went to work at the Monsanto Chemical Company, until 1948 when he joined the chemistry department at the Brookhaven National Laboratory. After retiring from the Laboratory in 1984, he moved to the University of Pennsylvania, Philadelphia, where he is now professor emeritus. A member of the US National Academy of Sciences and the American Academy of Arts and Sciences, Davis has been honoured with many awards, including the US National Medal of Science and the Israeli Wolf Prize for physics (shared with Masatoshi Koshiba).

Masatoshi Koshiba was born in 1926 in Toyohashi, Japan. After graduating from the University of Tokyo in 1951, he went to the USA, where he received a Ph.D. in physics from the University of Rochester. In 1970 Koshiba was appointed Professor of Physics at the University of Tokyo; he remained there until his retirement in 1987, when he became emeritus. Between 1987 and 1997 he was a professor at Tokai University. Koshiba has been the recipient of numerous honours and prizes, among them the Order of Cultural Merit conferred in 1997 by the Emperor of Japan in person, as well as the 2000 Israeli Wolf Prize for physics.

X-rays from the cosmos

The first person to see a heavenly object, not through the light that it emitted, but through X-rays, was an American physicist of the name of Herbert Friedman. In 1949 he and his co-workers placed a Geiger–Müller counter on a rocket, which then flew at a very high altitude above the ground. It was thus that they were able to observe, for the first time, X-rays coming from the sun. (X-rays from the cosmos are almost entirely absorbed by the earth's atmosphere; so if you want to detect them you must take your detectors to sufficiently high altitudes, aboard rockets or satellites.)

At that time Riccardo Giacconi was a teenage student, at the end of his high-school studies in Milan, Italy. After taking his diploma, he went to the University of Milan to study physics. Here, in 1954, he obtained his doctor's degree, writing a thesis under the supervision of Giuseppe Occhialini, the well-known physicist who had participated in the cosmic-ray discoveries made by Patrick Blackett and Cecil Powell during the 1930s and 1940s (see Nobel 1948 and Nobel 1950). (For these discoveries Occhialini was awarded the 1979 Wolf Prize for physics.) After two years of teaching at the university, Giacconi asked his professor to advise him about a future research task; it is said that Occhialini answered: 'Go west!' Giacconi followed this advice and went to the USA, first to Indiana and then to Princeton University.

The turning point of his career came towards the end of the 1950s, when he met Bruno Rossi, who had fled Italy before the war ('another one of those marvellous gifts of the Fascist regimes in Europe to the United States')[33]. In those years Rossi was a professor at MIT, and at the same time was serving as chairman of the board of American Science and Engineering (ASE), a private research company near Boston. The president of ASE (Martin Annis) recruited Giacconi, and appointed him to be responsible for a space research programme. Soon thereafter, Rossi suggested that he search for cosmic X-ray sources. Three years had to pass before Giacconi and his group were to succeed in discovering the first X-ray source from beyond the solar system. Let us briefly describe the experiment.

Giacconi's group loaded three Geiger–Müller counters aboard a US Air Force rocket. The aim was to see whether the lunar soil could emit X-rays under the influence of the sun. After two failures, on 18 June 1962 the rocket finally soared to 224 kilometres above the earth, and the payload instruments began sending back amazing data. These indicated that the counters were recording a strong flux of X-rays (100 per second), coming from a source different from the moon, and located outside the solar system. At first Giacconi, Rossi and all their colleagues were sceptical, but they began to check and recheck the data, until they became convinced that the results were not mistaken. Two months later, Giacconi was able to present their amazing results at a symposium at Stanford University in California, and to publish them in the *Physical Review Letters*. (They named this, the first discovered X-ray source, *Scorpio X-1*, after the Scorpio constellation in which it was found.) This discovery opened doors onto the new the field of X-ray astronomy, which soon witnessed an astonishing expansion (in a few years, about fifty new X-ray sources were identified; among these was the Crab pulsar, which was found to emit not only optical and radio pulses, but also X-rays, with a flux ten billion times greater than that from the sun).

Giacconi and his group then began to develop new types of X-ray telescopes to be placed aboard satellites. The first X-ray satellite, by the name of UHURU, was launched on 12 October 1970 from Kenya and it composed the first maps of the X-ray sky. The group discovered 339 X-ray sources, among them binary stars, and clusters of galaxies all emitting strong fluxes of X-rays; they were also able to furnish evidence of the existence of X-ray sources due to matter falling on to either a neutron star or a black hole (an important black hole candidate was Cygnus X-1, discovered in the constellation of Cygnus). In the 1970s Giacconi worked on the construction of an X-ray observatory, known as the *Einstein Observatory*, which was launched on November 1978, and remained in operation until April 1981. He was also the director of the Einstein mission, which made more than five thousand X-ray observations. This produced detailed maps of extended X-ray sources, and discovered X-ray jets in radio galaxies, and also X-ray quasars.

Following on his pioneering researches, Giacconi has been involved in most of the major projects of X-ray astronomy, which have given astronomers crucial information about star formation, galactic nuclei, black holes and other cosmic objects, and so succeeded in drastically modifying our view of the universe. In 1976 he and Harvey Tananbaum proposed the construction of the *Chandra X-ray Observatory* (named for the 1983 Nobelist, the astrophysicist Subrahmanyan Chandrasekhar). It was launched in 1999, and it has already provided remarkable images of the X-ray universe.

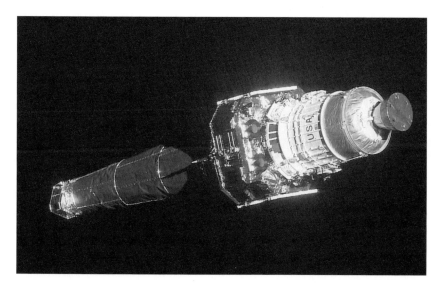

Fig. 13.25. The Chandra X-ray Observatory. (Courtesy NASA/CXC/SAO.)

Fig. 13.26. Riccardo Giacconi. (Courtesy AIP Emilio Segrè Visual Archives, *Physics Today* Collection.) Giacconi has been honoured with numerous awards. In addition to the 2002 Nobel Prize, he has received the Dannie Heinemann Prize in Astrophysics, and the 1987 Israeli Wolf Prize for physics (this last shared with Herbert Friedman and Bruno Rossi). He also holds memberships in numerous scientific societies.

Ricardo Giacconi was born in 1931 in Genoa, Italy. He studied physics at the University of Milan, where he obtained his doctor's degree in 1954. Two years later he went to the USA, first to the University of Indiana and then to Princeton. In 1959 he joined American Science and Engineering, where he began work on X-ray astronomy. In 1973 he moved to the Harvard–Smithsonian Center for Astrophysics, and in 1981 he became director of the Space Telescope Science Institute. In 1993 Giacconi was appointed director of the European Southern Observatory, and from 1999 he has been president of Associated Universities, Inc., holding simultaneously the positions of Professor of Physics and Astronomy (1982–97), and research professor (from 1998 on) at Johns Hopkins University.

2003

Super-chilled matter

Our chronicle makes its debut in 1901, at the old Royal Academy of Music in Stockholm, when Wilhelm Röngten was presented with the first Nobel Prize for physics; and it reaches its finale on 7 October 2003 at noon, when the announcement of the awards for the year 2003 was published on the website of the Nobel Foundation. It is on this day that the Swedish Academy of Sciences made public its decision to award the 2003 Nobel Prize for physics to three theoreticians: the Russian Vitaly Ginzburg and the Russian-American Alexei Abrikosov (for their theories on superconductivity), and the British-American Anthony Leggett (for his theories on superfluidity).

A hundred-year old story

Let us first have a look at the principal events that we have witnessed during our journey within the realm of Nobel Physics, at least in so far as they concern the two typical phenomena of low-temperature physics – *superconductivity* and *superfluidity*. These phenomena are protagonists in a hundred-year drama, which can trace its origin back to the Netherlands at the beginning of the twentieth century.

Superconductivity

In 1908 Heike Kamerlingh Onnes (see Nobel 1913), after years of trying, succeeded for the first time in obtaining liquid helium in his laboratory at the University of Leiden, the Netherlands. Three years later he discovered that mercury, when chilled to the temperature of liquid helium (about 4 kelvins), becomes a superconductor (that is, it acquires the ability to conduct an electric current with no resistance or loss of energy).

The next breakthrough occurred in 1933, when the German physicists Walter Meissner (then director of the PTR at Berlin–Charlottenburg) and Robert Ochsenfeld discovered, as a result of experiments, this rather strange fact: that when a metallic superconductor is placed in a magnetic field and is chilled below the temperature at which it acquireds its superconductivity (the so-called 'transition temperature'), the field is abruptly repelled (that is, it is not allowed to penetrate into the interior of the material): this phenomenon is known as the *Meissner effect* (see Fig. 11.1).

Superconducting materials which exhibit the Meissner effect are called *type-I* (or *soft*) *superconductors*; they are mostly metals. Moreover, they are characterised by the fact that they lose their superconductivity when an applied magnetic field exceeds a certain critical value; in this situation the field itself suddenly penetrates into their interior.

In the decades following on the discovery of superconductivity many other superconducting materials were found. And among these, there were many alloys and chemical compounds which exhibited a different behaviour compared with type-I superconductors. For instance, these materials allow a partial penetration by a magnetic field; they achieve higher transition temperatures and can conduct the most powerful currents; moreover, they retain their superconductivity even when placed in intense magnetic fields. They have huge potential for practical applications, and are called *type-II* (or *hard*) *superconductors*. This class of materials also includes the so-called high-temperature superconductors – the first of which was discovered by Georg Bednorz and Alexander Müller in 1986 (see Nobel 1987).

You will likewise remember that our understanding of the origin of superconductivity on a microscopic scale evolved in 1957, thanks to the work of John

Bardeen, Leon Cooper and Robert Schrieffer, when they conceived a novel quantum theory – the so-called BCS theory (see Nobel 1972).

Superfluidity

In 1938, thirty years after the liquefaction of helium, the Russian physicist Peter Kapitza in Moscow, and independently Jack Allen and Donald Misener in Cambridge, England, discovered that ordinary helium, namely the isotope helium-4, when chilled below 2.17 kelvins, exhibits superfluidity (that is, the ability to flow through thin capillaries without apparent friction and loss of energy, see Nobel 1978).

More than forty years had to pass before superfluidity was also discovered in a rare isotope of helium, helium-3, when chilled at temperatures below 0.003 kelvin. In fact this dramatic discovery took place in 1971 at Cornell Univertsity, USA, in experiments carried out by the American physicists David Lee, Douglas Osheroff and Robert Richardson (see Nobel 1996).

Moreover, you will recall how superfluidity in helium-4 was explained in the 1940s by the Russian theoretician Lev Landau, and how this theory was subsequently developed by other scientists (see Nobel 1962).

The prizewinners' contributions

Let us now take a closer look at the contributions that our three prizewinners have made to the related phenomena of superconductivity and superfluidity.

Ginzburg

In 1950 Vitally Ginzburg was working at the Lebedev Institute of Physics in Moscow. He was collaborating with Lev Landau, then at Kapitza's Institute for Physical Problems in the same city. Together they worked out a macroscopic quantum theory which gave an excellent description of the behaviour of a superconducting material near its transition temperature, including also when it is in the presence of a magnetic field. (The paper presenting the theory was the only one that Ginzburg ever wrote with Landau.) Their equations also predicted that there could exist two kinds of superconducting states: one state in which the penetration of a magnetic field into the interior of the material is not allowed, and another one in which such a penetration can take place.

As we know, seven years later (1957) the BCS theory burst on to the scene, affording us a full explanation of the microscopic origin of superconductivity. It also permitted us to understand the meaning of several physical quantities which formed part of the Ginzburg–Landau theory.

Abrikosov

In those same years, the 1950s, Alexei Abrikosov, who had been a student of Landau's, was a research scientist at the Institute for Physical Problems. He was studying the Ginzburg–Landau theory, and was trying to understand how this theory could explain the physical condition which allowed a magnetic field to partially penetrate into the interior of a superconducting material (a possibility that had not been fully investigated by Ginzburg and Landau themselves).

Starting from the Ginzburg–Landau equations, Abrikosov developed his own calculations and also extended their theory, demonstrating that, apart from ordinary type-I superconductors, there exist in nature superconducting materials of another type, which he proposed to call 'superconductors of the second group': these are the above mentioned type-II superconductors.

Abrikosov continued to investigate deeper and deeper into the matter, asking himself how a magnetic field could penetrate into the interior of these new kinds of materials and coexist with their superconducting state. He thought that when the strength of the field is in a certain range of values, it can penetrate as tiny *vortices* of circulating superconducting currents, which arrange themselves in the interior of the material in the form of an ordered lattice. Within each vortex the magnetic field is high, and the material is not superconducting, whereas outside the vortices the material remains superconducting. Using these concepts, he was also able to explain how ever-stronger magnetic fields can change the vortex structure and destroy superconductivity even in type-II superconductors.

Abrikosov published his vortex theory in 1957, the same year as the BCS theory; the vortices he had predicted and their ordered structure were first actually observed in experiments in the late 1960s. (Two years later, in 1959, another student of Landau's, Lev Gor'kov, succeeded in deriving the Ginzburg–Landau equations from the BCS theory, so demonstrating that the ideas of Landau, Ginzburg and Abrikosov were firmly rooted in the microscopic nature of superconductivity.)

The vortex theory grew in importance when high-temperature superconductors were discovered, because many of their physical properties are dominated by the behaviour of vortices as those discovered by Abrikosov.

Leggett

Unlike helium-4, superfluid helium-3 has three superfluid phases (see p. 426); moreover, all three phases exhibit magnetic properties and are anisotropic (this means that superfluid helium-3 displays different physical properties when measured in different directions in space).

Anthony Leggett, a recognised world leader in the field of low-temperature physics, has greatly contributed to our understanding of superfluid helium-3 and its properties. The Swedish Academy of Sciences cited in particular a theory that he developed in the early 1970s, when he was working at the University of Sussex in England. This theory helped to provide a quantitative understanding of the physical properties of the three phases of superfluid helium-3, and also to explain how it is anisotropic. Let us try to describe the unusual behaviour of this superfluid in a few plain words.

Helium-3, when it exhibits one of its three superfluid phases, can be thought of as being made up of pairs of atoms which act in some ways like bosonic particles, each pair having an integer spin (see p. 427). Moreover, the two atoms composing a pair rotate, each one relative to the other at a distance. As a consequence, the relations between spin and orbital motion result in the pairs of atoms not having the same appearance when viewed from different directions. Thus superfluid helium-3 can be said to have directionality, and this is the origin of the anisotropy observed in it.

While studying the interactions among the 'boson-like' pairs of helium-3 in their co-ordinated motion through the superfluid, Leggett succeeded in explaining how such interactions should leave a magnetic signature; that is, how the superfluid phases should present different magnetic properties. This effect had been actually observed in the experiments carried out by the Cornell team, the Nobelists for the year 1996. (They had used a nuclear magnetic resonance technique, and, by measuring the magnetic properties of superfluid helium-3, they had identified its three superfluid phases).

Leggett's theories were also successfully applied to high-temperature superconductors, and to other superfluid systems. In addition, they have had a significant impact on our understanding of different phenomena in several other fields of science, including liquid crystals, particle physics and cosmology.

The prizewinners' curricula vitarum

Alexei A. Abrikosov was born in 1928 in Moscow. He studied at Moscow State University and at the Institute for Physical Problems. Here in 1951 he obtained his doctor's degree in physics, and was a senior research scientist there until the year 1965, when he became the Head of the Department of the Theory of Solids at the Landau Institute of Theoretical Physics. In 1988 he moved to Troitsk, near Moscow, where he served for three years as director of the Institute for High Pressure Physics. In 1991 Abrikosov left Russia and emigrated to the USA. He went to work at the Argonne National Laboratory, where he is now a distinguished Argonne scientist. Abrikosov has received many honours for his research work in the field of condensed-matter theory, including memberships of the Russian Academy of Sciences, the American Academy of Arts and Sciences and the US National Academy of Sciences.

Vitaly L. Ginzburg was born in 1916 in Moscow. He graduated from Moscow State University in 1940, and two years later obtained his doctor's degree there. In the same year (1940) he joined the Lebedev Institute of Physics, where he started his long scientific career. He has also been Professor of Physics at Gorky State University (1945–68), and from 1968 on at the Moscow Institute of Physics and Technology. His research interests have covered many areas of the physical sciences, ranging from superconductivity (the field giving rise to his Nobel Prize), up to radio astronomy, plasma physics and also theories on the origin of cosmic rays. Ginzburg's prestigious career has been acknowledged by numerous awards, among them the Israeli Wolf Prize for physics (1994–5). He is a member of the Russian Academy of Sciences and of nine other foreign academies, including the Royal Society of London and the US National Academy of Sciences.

Anthony J. Leggett was born in 1938 in London. He studied at the University of Oxford, where he obtained his doctor's degree in 1964. He remained at Oxford until 1967, when he moved to the University of Sussex, Brighton, where he became Professor of Physics in 1978. Five years later Leggett transferred to the USA and joined the University of Illinois at Urbana-Champaign; here he is now the John D. and Catherine T. MacArthur Professor, and the Center for Advanced Study Professor of Physics. At Urbana-Champaign Leggett has continued his research work in the field of theoretical condensed-matter physics, low-temperature phenomena, and also his studies on the foundations of quantum mechanics. He is a member of many scientific institutions, including the US National Academy of Sciences, the Russian Academy of Sciences and the Royal Society of London. The year 2003 was an astonishingly fortunate one for him, for as well as the Nobel Prize, he also received the Israeli Wolf Prize for physics.

Appendix
Nobel Prizes for physics

1901. **Wilhelm Conrad Röntgen.** 'In recognition of the extraordinary services he has rendered by the discovery of the remarkable rays subsequently named after him.'
1902. **Hendrik Antoon Lorentz**, **Pieter Zeeman.** 'In recognition of the extraordinary service they rendered by their researches into the influence of magnetism upon radiation phenomena.'
1903. **Antoine Henri Becquerel.** 'In recognition of the extraordinary services he has rendered by his discovery of spontaneous radioactivity.' **Pierre Curie**, **Marie Skłodowska Curie.** 'In recognition of the extraordinary services they have rendered by their joint researches on the radiation phenomena discovered by Professor Henri Becquerel.'
1904. **John William Strutt, Third Baron Rayleigh.** 'For his investigations of the densities of the most important gases and for his discovery of argon in connection with these studies.'
1905. **Philipp Lenard.** 'For his work on cathode rays.'
1906. **Joseph John Thomson.** 'In recognition of the great merits of his theoretical and experimental investigations on the conduction of electricity by gases.'
1907. **Albert Abraham Michelson.** 'For his optical precision instruments and the spectroscopic and metrological investigations carried out with their aid.'
1908. **Gabriel Lippmann.** 'For his method of reproducing colours photographically based on the phenomenon of interference.'
1909. **Guglielmo Marconi, Karl Ferdinand Braun.** 'In recognition of their contributions to the development of wireless telegraphy.'
1910. **Johannes Diderik van der Waals.** 'For his work on the equation of state for gases and liquids.'
1911. **Wilhelm Wien.** 'For his discoveries regarding the laws governing the radiation of heat.'
1912. **Nils Gustaf Dalén.** 'For his invention of automatic regulators for use in conjunction with gas accumulators for illuminating lighthouses and buoys.'
1913. **Heike Kamerlingh Onnes.** 'For his investigations on the properties of matter at low temperatures which led, inter alia, to the production of liquid helium.'
1914. **Max von Laue.** 'For his discovery of the diffraction of X-rays by crystals.'
1915. **William Henry Bragg, William Lawrence Bragg.** 'For their services in the analysis of crystal structure by means of X-rays.'
1916. Not awarded.

1917. **Charles Glover Barkla.** 'For his discovery of the characteristic Röntgen radiation of the elements.'
1918. **Max Karl Planck.** 'In recognition of the services he rendered to the advancement of physics by his discovery of energy quanta.'
1919. **Johannes Stark.** 'For his discovery of the Doppler effect in canal rays and the splitting of spectral lines in electric fields.'
1920. **Charles Edouard Guillaume.** 'In recognition of the service he has rendered to precision measurements in physics by his discovery of anomalies in nickel steel alloys.'
1921. **Albert Einstein.** 'For his services to theoretical physics, and especially for his discovery of the law of the photoelectric effect.'
1922. **Niels Henrik Bohr.** 'For his services in the investigation of the structure of atoms and of the radiation emanating from them.'
1923. **Robert Andrews Millikan.** 'For his work on the elementary charge of electricity and on the photoelectric effect.'
1924. **Karl Manne Siegbahn.** 'For his discoveries and research in the field of X-ray spectroscopy.'
1925. **James Franck, Gustav Ludwig Hertz.** 'For their discovery of the laws governing the impact of an electron upon an atom.'
1926. **Jean Baptiste Perrin.** 'For his work on the discontinuous structure of matter, and especially for his discovery of sedimentation equilibrium.'
1927. **Arthur Holly Compton.** 'For his discovery of the effect named after him'. **Charles Thomson Rees Wilson.** 'For his method of making the paths of electrically charged particles visible by condensation of vapour.'
1928. **Owen Willans Richardson.** 'For his work on the thermionic phenomenon and especially for the discovery of the law named after him.'
1929. **Louis Victor de Broglie.** 'For his discovery of the wave nature of electrons.'
1930. **Chandrasekhara Venkata Raman.** 'For his work on the scattering of light and for the discovery of the effect named after him.'
1931. Not awarded.
1932. **Werner Karl Heisenberg.** 'For the creation of quantum mechanics, the application of which has, inter alia, led to the discovery of the allotropic forms of hydrogen.'
1933. **Erwin Schrödinger, Paul Adrien Maurice Dirac.** 'For the discovery of new productive forms of atomic theory.'
1934. Not awarded.
1935. **James Chadwick.** 'For the discovery of the neutron.'
1936. **Victor Francis Hess.** 'For his discovery of cosmic radiation.' **Carl David Anderson.** 'For his discovery of the positron.'
1937. **Clinton Joseph Davisson, George Paget Thomson.** 'For their experimental discovery of the diffraction of electrons by crystals.'
1938. **Enrico Fermi.** 'For his demonstrations of the existence of new radioactive elements produced by neutron irradiation, and for his related discovery of nuclear reactions brought about by slow neutrons.'
1939. **Ernest Orlando Lawrence.** 'For the invention and development of the cyclotron and for results obtained with it, especially with regard to artificial radioactive elements.'
1940. Not awarded.
1941. Not awarded.

1942. Not awarded.
1943. **Otto Stern.** 'For his contribution to the development of the molecular ray method and his discovery of the magnetic moment of the proton.'
1944. **Isidor Isaac Rabi**. 'For his resonance method for recording the magnetic properties of atomic nuclei.'
1945. **Wolfgang Pauli.** 'For the discovery of the exclusion principle, also called the Pauli principle.'
1946. **Percy Williams Bridgman.** 'For the invention of an apparatus to produce extremely high pressures, and for the discoveries he made therewith in the field of high pressure physics.'
1947. **Edward Victor Appleton.** 'For his investigations of the physics of the upper atmosphere especially for the discovery of the so-called Appleton layer.'
1948. **Patrick Maynard Blackett.** 'For his development of the Wilson cloud chamber method, and his discoveries therewith in the fields of nuclear physics and cosmic radiation.'
1949. **Hideki Yukawa.** 'For his prediction of the existence of mesons on the basis of theoretical work on nuclear forces.'
1950. **Cecil Frank Powell.** 'For his development of the photographic method of studying nuclear processes and his discoveries regarding mesons made with this method.'
1951. **John Douglas Cockcroft, Ernest Thomas Walton.** 'For their pioneer work on the transmutation of atomic nuclei by artificially accelerated atomic particles.'
1952. **Felix Bloch, Edward Mills Purcell.** 'For their development of new methods for nuclear magnetic precision measurements and discoveries in connection therewith.'
1953. **Frits Zernike.** 'For his demonstration of the phase contrast method, especially for his invention of the phase contrast microscope.'
1954. **Max Born.** 'For his fundamental research in quantum mechanics, especially for his statistical interpretation of the wavefunction.' **Walther Bothe.** 'For the coincidence method and his discoveries made therewith.'
1955. **Willis Eugene Lamb, Jr.** 'For his discoveries concerning the fine structure of the hydrogen spectrum.' **Polykarp Kusch.** 'For his precision determination of the magnetic moment of the electron.'
1956. **William Bradford Shockley, John Bardeen, Walter Houser Brattain.** 'For their researches on semiconductors and their discovery of the transistor effect.'
1957. **Chen Ning Yang, Tsung Dao Lee.** 'For their penetrating investigation of the so-called parity laws which has led to important discoveries regarding the elementary particles.'
1958. **Pavel Aleksejevic Cherenkov, Ilja Michailovich Frank, Igor Eugenevich Tamm.** 'For the discovery and the interpretation of the Cherenkov effect.'
1959. **Emilio Gino Segrè, Owen Chamberlain.** 'For their discovery of the antiproton.'
1960. **Donald Arthur Glaser.** 'For the invention of the bubble chamber.'
1961. **Robert Hofstadter.** 'For his pioneering studies of electron scattering in atomic nuclei and for his thereby achieved discoveries concerning the structure of the nucleons.' **Rudolf Ludwig Mössbauer.** 'For his researches concerning the resonance absorption of gamma radiation and his discovery in this connection of the effect which bears his name.'
1962. **Lev Davidovich Landau.** 'For his pioneering theories for condensed matter, especially liquid helium.'
1963. **Eugene Paul Wigner.** 'For his contributions to the theory of the atomic nucleus and the elementary particles, particularly through the discovery and application of

fundamental symmetry principles.' **Maria Goeppert Mayer, Johannes Hans Jensen.** 'For their discoveries concerning nuclear shell structure.'

1964. **Charles Hard Townes, Nikolai Gennadievich Basov, Alexander Mikhailovich Prokhorov.** 'For fundamental work in the field of quantum electronics, which has led to the construction of oscillators and amplifiers based on the maser–laser principle.'

1965. **Sin-Itiro Tomonaga, Julian Seymour Schwinger, Richard P. Feynman.** 'For their fundamental work in quantum electrodynamics, with deep-ploughing consequences for the physics of elementary particles.'

1966. **Alfred Kastler.** 'For the discovery and development of optical methods for studying Hertzian resonances in atoms.'

1967. **Hans Albrecht Bethe.** 'For his contributions to the theory of nuclear reactions, especially his discoveries concerning the energy production in stars.'

1968. **Luis Walter Alvarez.** 'For his decisive contributions to elementary particle physics, in particular the discovery of a large number of resonance states, made possible through his development of the technique of using hydrogen bubble chamber and data analysis.'

1969. **Murray Gell-Mann.** 'For his contributions and discoveries concerning the classification of elementary particles and their interactions.'

1970. **Hannes Olof Alfvén.** 'For fundamental work and discoveries in magneto-hydrodynamics with fruitful applications in different parts of plasma physics.' **Louis Eugène Néel.** 'For fundamental work and discoveries concerning antiferromagnetism and ferrimagnetism which have led to important applications in solid state physics.'

1971. **Dennis Gabor.** 'For his invention and development of the holographic method.'

1972. **John Bardeen, Leon N Cooper, John Robert Schrieffer.** 'For their jointly developed theory of superconductivity, usually called the BCS theory.'

1973. **Leo Esaki, Ivar Giaever.** 'For their experimental discoveries regarding tunnelling phenomena in semiconductors and superconductors, respectively.' **Brian David Josephson.** 'For his theoretical predictions of the properties of a supercurrent through a tunnel barrier, in particular those phenomena which are generally known as the Josephson effects.'

1974. **Martin Ryle, Antony Hewish.** 'For their pioneering research in radio astrophysics: Ryle for his observations and inventions, in particular of the aperture synthesis technique, and Hewish for his decisive role in the discovery of pulsars.'

1975. **Aage Niels Bohr, Ben Roy Mottelson, Leo James Rainwater.** 'For the discovery of the connection between collective motion and particle motion in atomic nuclei and the development of the theory of the structure of the atomic nucleus based on this connection.'

1976. **Burton Richter, Samuel Ting.** 'For their pioneering work in the discovery of a heavy elementary particle of a new kind.'

1977. **Philip Warren Anderson, Nevill Francis Mott, John Hasbrouck Van Vleck.** 'For their fundamental theoretical investigations of the electronic structure of magnetic and disordered systems.'

1978. **Peter Leonidovich Kapitza.** 'For his basic inventions and discoveries in the area of low-temperature physics.' **Arno Allan Penzias, Robert Woodrow Wilson.** 'For their discovery of cosmic microwave background radiation.'

1979. **Sheldon Lee Glashow, Abdus Salam, Steven Weinberg.** 'For their contributions to the theory of the unified weak and electromagnetic interactions between elementary particles, including, inter alia, the prediction of the weak neutral current.'
1980. **James Watson Cronin, Val Logsdon Fitch.** 'For the discovery of violations of fundamental symmetry principles in the decay of neutral K-mesons.'
1981. **Nicolaas Bloembergen, Arthur Leonard Schawlow.** 'For their contribution to the development of laser spectroscopy.' **Kai Manne Siegbahn.** 'For his contribution to the development of high-resolution electron spectroscopy.'
1982. **Kenneth G. Wilson.** 'For his theory for critical phenomena in connection with phase transitions.'
1983. **Subrahmanyan Chandrasekhar.** 'For his theoretical studies of the physical processes of importance to the structure and evolution of the stars.' **William Alfred Fowler.** 'For his theoretical and experimental studies of the nuclear reactions of importance in the formation of the chemical elements in the universe.'
1984. **Carlo Rubbia, Simon van der Meer.** 'For their decisive contributions to the large project, which led to the discovery of the field particles W and Z, communicators of weak interaction.'
1985. **Klaus von Klitzing.** 'For the discovery of the quantized Hall effect.'
1986. **Ernst August Ruska.** 'For his fundamental work in electron optics, and for the design of the first electron microscope.' **Gerd Binnig, Heinrich Rohrer.** 'For their design of the scanning tunnelling microscope.'
1987. **Johannes Georg Bednorz, Karl Alexander Müller.** 'For their important breakthrough in the discovery of superconductivity in ceramic materials.'
1988. **Leon Max Lederman, Melvin Schwartz, Jack Steinberger.** 'For the neutrino beam method and the demonstration of the doublet structure of the leptons through the discovery of the muon neutrino.'
1989. **Norman Foster Ramsey.** 'For the invention of the separated oscillatory fields method and its use in the hydrogen maser and other atomic clocks.' **Hans Georg Dehmelt, Wolfgang Paul.** 'For the development of the ion trap technique.'
1990. **Jerome Isaac Friedman, Henry Way Kendall, Richard Edward Taylor.** 'For their pioneering investigations concerning deep inelastic scattering of electrons on protons and bound neutrons, which have been of essential importance for the development of the quark model in particle physics.'
1991. **Pierre-Gilles de Gennes.** 'For discovering that methods developed for studying order phenomena in simple systems can be generalized to more complex forms of matter, in particular to liquid crystals and polymers.'
1992. **Georges Charpak.** 'For his invention and development of particle detectors, in particular the multiwire proportional chamber.'
1993. **Russell Alan Hulse, Joseph Hooton Taylor, Jr.** 'For the discovery of a new type of pulsar, a discovery that has opened up new possibilities for the study of gravitation.'
1994. **Bertram Neville Brockhouse, Clifford Glenwood Shull.** 'For pioneering contributions to the development of neutron scattering techniques for studies of condensed matter.' 'For the development of neutron spectroscopy.' (**Brockhouse**). 'For the development of the neutron diffraction technique.' (**Shull**).

1995. **Martin Lewis Perl, Frederick Reines.** 'For pioneering experimental contributions to lepton physics.' 'For the discovery of the tau lepton.' (**Perl**) 'For the detection of the neutrino.' [**Reines**]

1996. **David Morris Lee, Douglas Dean Osheroff, Robert Coleman Richardson.** 'For their discovery of superfluidity in helium-3.'

1997. **Steven Chu, Claude Cohen-Tannoudji, William Daniel Phillips.** 'For development of methods to cool and trap atoms with laser light.'

1998. **Robert B. Laughlin, Horst Ludwig Störmer, Daniel Chee Tsui.** 'For their discovery of a new form of quantum fluid with fractionally charged excitations.'

1999. **Gerardus 't Hooft, Martinus Veltman.** 'For elucidating the quantum structure of electroweak interactions in physics.'

2000. **Zhores Ivanovich Alferov, Herbert Kroemer, Jack St. Clair Kilby.** 'For basic work on information and communication technology.' 'For developing semiconductor heterostructures used in high-speed- and opto-electronics.' (**Alferov, Kroemer**) 'For his part in the invention of the integrated circuit.' (**Kilby**)

2001. **Eric A. Cornell, Wolfgang Ketterle, Carl E. Wieman.** 'For the achievement of Bose–Einstein condensation in dilute gases of alkali atoms, and for early fundamental studies of the properties of the condensates.'

2002. **Raymond Davis, Jr, Masatoshi Koshiba.** 'For pioneering contributions to astrophysics, in particular for the detection of cosmic neutrinos.' **Riccardo Giacconi.** 'For pioneering contributions to astrophysics, which have led to the discovery of cosmic X-ray sources.'

2003. **Alexei A. Abrikosov, Vitaly L. Ginzburg, Anthony J. Leggett.** 'For pioneering contributions to the theory of superconductors and superfluids.'

Glossary of terms

(Particle) Accelerator. A device that increases the speed (and the energy) of a beam of charged particles such as electrons and protons. Physicists use an accelerated beam to study collisions between particles.

Accelerator, linear. In a linear accelerator the particles move in a straight line; they are accelerated by receiving small boosts of energy from electric fields placed along the accelerator.

Accelerator – collider. A collider is a cyclic accelerator where two beams of particles circulate in opposite directions. The beams intersect each other at specific points where head-on collisions between two particles can then occur.

Accelerator, cyclic. In a cyclic accelerator magnetic fields hold the particles in a circular path, while electric fields, placed along the accelerator, give the particles a boost of energy and speed each time the beam goes round.

Anode. The positive electrode in a vacuum tube.

Antiparticle. To every subatomic particle there corresponds an antiparticle which has the same mass and spin as the corresponding particle, while possessing opposite charges (such as the electric charge or the colour charge), and other quantum numbers (such as strangeness). (See Nobel 1936 and Nobel 1959.)

Astrophysics. The science which applies the laws of physics to the understanding of astronomical objects and phenomena.

Atom. Every single object, such as this book, your body, the sea, or a flower, is made of atoms – the smallest possible unit of any chemical element. Atoms are very tiny objects: in a grain of sand there are some thousands of billion billions of atoms. An atom is made up of a very small and dense, positively charged *nucleus*, surrounded by a tenuous cloud of negatively charged *electrons*; these are held bound to the nucleus by the *electric force*. (See Nobel 1926.)

Atom – angular momentum. When electrons are combined together to form an atom there results an *angular momentum*: this is the combination of all the angular momenta due to the electron motions around the nucleus and to their *spins*: it is called the *total angular momentum* of the atom.

Atom – atomic number (symbol: Z). The number of positive protons in the nucleus, which is the same as the number of external negative electrons. Examples: the simplest, lightest atom in nature is *hydrogen*, with one proton in its nucleus, and one electron around it ($Z = 1$); one of the heaviest atoms is *uranium*, with 92 protons in its nucleus and 92 electrons around it ($Z = 92$).

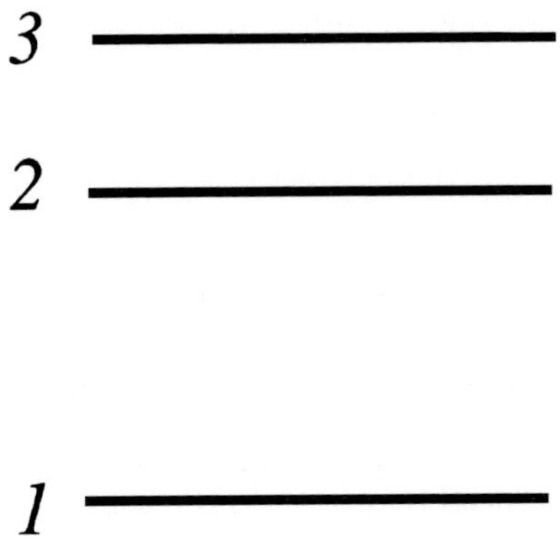

Fig. G.1. Energy levels.

3 ⎯⎯⎯⎯⎯⎯

2 ⎯⎯⎯⎯⎯⎯

1 ⎯⎯⎯⎯⎯⎯

Atom – Bohr magneton. A fundamental unit of the *magnetic moment* of an atom. It represents the smallest unit for the magnitude of a magnetic moment due to the orbital motion of an electron in the atom. (After Niels Bohr, Nobel 1922.)

Atom – energy levels. Set of specific values of the internal energy possessed by an atom, corresponding to different quantum states of the atom itself. An atom can have internal energy equal to any one of these values, but it cannot have an energy intermediate between any two of them. They are graphically represented in Fig. G.1: each line corresponds to an energy level. (See Nobel 1925 and Nobel 1966.)

Atom – magnetic moment. Electrons orbiting around a nucleus give rise to a magnetic field. Their intrinsic spins too give rise to a magnetic field. A resultant magnetic field of the atom is hence generated whose strength is measured by its *magnetic moment*.

Atom – mass number (symbol: A). The total number of particles in the nucleus (protons + neutrons). Example: in a nucleus of the isotope uranium-238, $A = 238$ particles (92 protons, and $238 - 92 = 146$ neutrons).

Atomic clock. An extremely accurate clock based on the natural vibrations of an atomic system such as the isotope caesium-133. (See Nobel 1989.)

Aurora. Light radiated by atoms and ions in the earth's upper atmosphere, mostly in the polar regions.

Balmer lines. Emission or absorption lines in the spectrum of hydrogen caused by transitions of the electron between the second and higher energy levels. The entire pattern of Balmer lines is called a *Balmer series*. (After Johann Balmer.)

Baryons. A subclass of *hadrons* made up of three quarks (examples: protons or neutrons). All baryons are fermions.

BCS (Bardeen–Cooper–Schrieffer) theory. A microscopic theory, developed in the late 1950s, which explains the phenomenon of superconductivity in metals. (After John Bardeen, Leon Cooper and Robert Schrieffer, Nobel 1972.)

Beta decay. Decay of a radioactive atom, emitting a fast electron (*beta particle*).

Big Bang. The titanic explosion involving all space, in a state of enormous density and temperature, which is believed to have taken place about 13 billion or so years ago – and from which subsequently our universe emerged and expanded.

Binary pulsar. A pulsar in a binary system consisting of two neutron stars. (See Nobel 1993.)

Binary stars. Two stars revolving about each other and held together by their mutual gravitational attraction.

Black body. An ideal body which totally absorbs the electromagnetic radiation which falls upon it.

Black-body radiation. Electromagnetic radiation emitted from a black body, also called *thermal radiation*. (See Nobel 1911 and Nobel 1918.)

Black-body spectrum. The distribution of the intensity of black-body radiation at the different wavelengths (or frequencies). (See Nobel 1911 and 1918.)

Black hole. A region of space where gravity is so strong that even electromagnetic radiation (light) cannot escape from it.

Bose–Einstein condensate. An extreme state of matter that occurs when a large assembly of massive bosons attract each other and in so doing reach a single quantum state of lowest energy. (After Satyendra Bose and Albert Einstein; see Nobel 2001.)

Bose–Einstein statistics. A set of quantum rules governing the behaviour of a collection of bosons. It allows many bosons to occupy the same quantum state. (After Satyendra Bose and Albert Einstein.)

Bosons. Particles whose spin is equal either to zero or to a whole number (1, 2, 3, and so on), measured in natural units $h/2\pi$. Bosons do not obey the Pauli exclusion principle. Examples of bosons: the photon, the weak bosons Ws and Z, the gluons and the helium-4 nucleus. (After Satyendra Nath Bose.)

Bottom. The fifth flavour of quarks, belonging to the third quark family.

Brout–Englert–Higgs mechanism. (Also called *spontaneous symmetry breaking*). A mechanism which enables the massive messenger particles of the electroweak theory to 'spontaneously' acquire a mass. (After Robert Brout, François Englert and Peter Higgs.)

Brownian motion. The erratic motion of particles in a fluid due to their impacts with the molecules of the fluid itself. (After John Brown; see Nobel 1926.)

Bubble chamber. A device for detecting high-energy subatomic particles by observing their tracks through a chamber containing superheated liquid. (See Nobel 1960.)

Cathode. The negative electrode in a vacuum tube.

Cathode rays. Streams of electrons through a vacuum tube (like a television tube) from cathode to anode when a high electric voltage is applied between the two electrodes. (See Nobel 1905 and Nobel 1906.)

Centripetal acceleration. The acceleration of an object moving in a circular path. It is directed, like the centripetal force, towards the centre of the circle.

Centripetal force. The force acting on an object moving in a circular path; it is directed towards the centre of the circle. Examples: our earth orbiting around the sun because of the attractive gravitational force; an electron orbiting around a proton because of the electric attractive force.

Chandrasekhar limit. The maximum mass of a white dwarf; its value is about 1.4 solar masses. (After Subrahmanyan Chandrasekhar, Nobelist in 1983.)

Charge reflection (or charge conjugation). The replacement, in a particle system, of all particles with their antiparticles.

Charm. The fourth flavour of quarks, belonging to the second quark family. (See Nobel 1976.)

Cherenkov counter. A device for detecting high-energy subatomic particles by observation of the Cherenkov light produced by the particles themselves in a transparent medium.

Fig. G.2. Left: incoherent light. Right: coherent light.

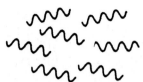

Cherenkov radiation. Electromagnetic radiation emitted in a transparent medium by particles travelling with a speed which is greater than the speed of light in the medium itself. (After Pavel Cherenkov, Nobelist in 1958.)

(Wilson) Cloud chamber. A device for detecting subatomic particles by observing their tracks through a chamber containing a supersaturated vapour. (After C.T.R. Wilson, Nobelist in 1927.)

Coherent light. In a coherent light beam all the waves that make up it move in the same direction, with all the troughs and crests of the waves aligned (see Fig. G.2).

Colour (strong) charge. The property possessed by quarks and gluons which represents the source of the colour (or strong) force between quarks. A small residue of the colour force acts to hold protons and neutrons together as nuclei.

Compton effect. The scattering of photons by atomic electrons. (After Arthur Compton, Nobelist in 1927.)

Conductor. Any material that easily allows the flow through it of an electric charge (electric current). Metals are common conductors.

Conservation law. A law which says that the total value of some physical quantity, such as energy, electric charge, linear or angular momentum, does not change in any kind of physical process.

Cooper pair. A pair of bound fermions (such as two electrons) which behaves as a boson. A collection of Cooper pairs can condensate into a single quantum state. (After Leon Cooper, Nobelist in 1972.)

Copenhagen interpretation (C. I.). Interpretation of the quantum theory built around Bohr's correspondence principle, Heisenberg's uncertainty principle, Bohr's complementary principle and Born's statistical interpretation of the wave function.

C. I. – Bohr's complementarity principle. The uncertainty principle shows that both light and matter have a dual, wave–particle, nature. The two aspects are like coins that can display either face, but not both simultaneously. Bohr argued that the ideas of wave and of particle complemented each other. Example: if we devise an experiment to reveal the wave character of an electron, its particle character will be fuzzy; if we modify the experiment to reveal its particle character, its wave character will become fuzzy. (After Niels Bohr, Nobel 1922.)

C. I. – Bohr's correspondence principle. The quantum description of a quantum system must merge into the classical description when the energy involved is large compared with the energy corresponding to individual quantum jumps. (After Niels Bohr, Nobel 1922.)

C. I. – Born's interpretation of the wave function. The wave-function ψ can be used in order to calculate the probability of finding a particle in a certain location in space. (After Max Born, Nobelist in 1954.)

C. I. – Heisenberg's uncertainty principle. Certain pairs of physical quantities such as position and momentum, or energy and time, are intrinsically linked, so that any attempts

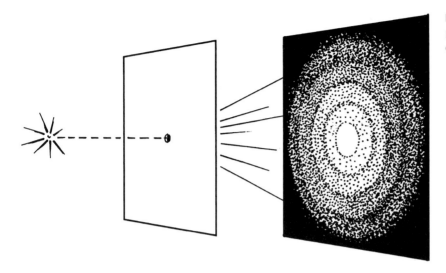

Fig. G.3. Light diffraction pattern formed by a circular aperture.

to make a precise measurement of one quantity will increase the uncertainty of measuring the other one. (After Werner Heisenberg, Nobel 1932.)

CNO cycle. A series of nuclear reactions, involving carbon as a catalyst, by which hydrogen is transformed into helium. (See Nobel 1967.)

Cosmic microwave background radiation (CMBR). Isotropic microwave radiation, with a black-body temperature of about 2.73 degrees above absolute zero (2.73 kelvins); it permeates the entire universe. (See Nobel 1978.)

Cosmic rays. High-energy particles (predominantly protons) that enter the earth's atmosphere from outer space. (See Nobel 1936.)

Critical phenomena. The phenomena which occur in the neighbourhood of a continuous phase transition. The temperature at which a critical phenomenon occurs is called a *critical temperature*. (See Nobel 1982.)

Crystal. A solid with a regular geometric shape in which plane faces meet at definite angles, and with a characteristic symmetry.

Crystal, liquid. A substance having properties intermediate between those of liquids and crystals. (See Nobel 1991.)

Cyclotron. A circular accelerator in which charged particles spiral under the effect of a magnetic field, and are accelerated by an oscillating electric field. (See Nobel 1939.)

Diffraction. The spreading out of waves. Examples: the spreading out of light waves passing the edge of an obstacle or through a tiny opening (light diffraction; see Fig. G.3); the spreading out of X-rays by crystals (X-ray diffraction; see Nobel 1914 and Nobel 1915), or of electrons by a thin foil of metals (electron diffraction; see Nobel 1937); or neutron diffraction (see Nobel 1994.)

Diode. Electronic device that allows the passage of an electric current only in one direction. The first diodes were thermionic vacuum-tube diodes; today's diodes used in electronic circuits are semiconductor diodes.

Diode, light emitting (LED). A semiconductor device that emits light when an electric current is applied to it.

Fig. G.4. Wave diagram.

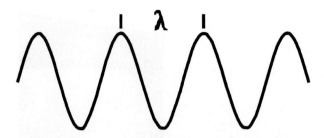

Diffraction grating. A piece of glass containing a very large number of grooves ruled closely together. It is used to disperse light into a spectrum and to measure the wavelength of its different spectral components.

Dirac equation. An equation in relativistic quantum mechanics that describes how the wave function of an electron or other spin $\frac{1}{2}$ particle evolves in space and time. Dirac's equation combines quantum mechanics and special relativity. (After Paul Dirac, Nobelist in 1933.)

Disordered system. A solid in which atoms are not in their regular lattice position. (See Nobel 1977.)

Doppler effect. The apparent change in wavelength (or frequency) of radiation caused by a relative motion of source and receiver. (After Christian Doppler.)

Down. The second flavour of quarks, belonging to the first quark family. (See Nobel 1969 and Nobel 1990.)

Eightfold way. A mathematical scheme to classify hadrons in groups of eight (octets) and ten particles (decuplets). (See Nobel 1969.)

Einstein's equivalence principle. The principle that states the equivalence of acceleration and gravity. The principle is demonstrated in its most familiar form by the observation that all bodies fall equally fast under gravity; it also asserts the identity of the inertial mass and the gravitational mass of an object. (After Albert Einstein, Nobel 1921.)

Electric charge. The property possessed by particles which represents the source of the *electromagnetic force*.

Electric charge, elementary. The basic unit of the electric charge, which is the magnitude of the charge of an electron or a proton. No free charge in nature can have a value less than that of the elementary charge. (See Nobel 1923.)

Electric current. A flow of electric charge from one place to another. Example: when a lamp lights, electric charges (electrons) flow through the copper wires connecting the lamp with a source of electric energy (say a battery) and then pass on through the light-bulb filament.

Electromagnetic wave. This consists of oscillating electric and magnetic fields, which are mutually linked with each other at right angles and are perpendicular to the direction of the wave motion.

Electromagnetic wave – wavelength. The distance between two successive crests of a wave (symbol: λ, see Fig. G.4).

Electromagnetic wave – frequency. The number of wave crests passing a given point every second, measured in cycles per second (symbol: ν).

Electromagnetic wave – speed. All electromagnetic waves have the same speed (symbol: c) in empty space, equal to 299 729 (roughly 300 000) kilometres per second. They can differ in frequency, and thus in wavelength, but their speed is always the same: *speed = wavelength × frequency* ($c = \lambda \times \nu$).

Electron. The lightest charged subatomic particle. It carries one unit of negative electric charge, and its mass is about 10^{-30} (one-thousand-billion-billion-billionth) of a kilogram. Unlike protons and neutrons, electrons are elementary particles (they belong to the first lepton family). (See Nobel 1906.)

Electroweak theory. A mathematical framework which describes both the weak nuclear and the electromagnetic force. (See Nobel 1979 and Nobel 1999.)

Energy. A physical quantity which can exist in different forms: *chemical energy* (as a result of chemical reactions); *kinetic energy* (possessed by an object when it is moving); *heat energy* (due to the motion of the atoms or molecules in a body); *electrical energy* (due to electric charges moving through a wire); *radiant energy* (like that carried by light or radio waves); *gravitational energy* (like that stored by an apple on a tree); *nuclear energy* (when released as the result of nuclear reactions).

Energy – unit. The usual unit in physics is called a *joule*. Example: a 100-watt lamp converts, each second, 100 joules of electrical energy into light and heat.

Energy, kinetic. This is the 'movement energy' of an object; it is equal to the 'product of half its mass and the square of its velocity'. Example: an athlete running in the Olympic 100-metre sprint has a kinetic energy of about 4000 joules.

Energy – electron-volt. The usual energy unit in the fields of atomic, nuclear and particle physics. It is extremely small: about a ten-billion-billionths of a joule. Examples: the energy per atom involved in chemical reactions is typically some electron-volts; protons in the future LHC at CERN, will reach some hundred billions of electron-volts.

Energy band. A continuous range of energies in a solid which correspond to possible quantum states of the atomic electrons.

Esaki (or tunnel) diode. A semiconductor diode based on the tunnelling effect of electrons through the energy barrier at the junction between two differently doped regions of a semiconductor. (After Leo Esaki, Nobelist in 1973.)

Ether (also aether). A hypothetical invisible medium which was believed to fill all space and support the propagation of light and other electromagnetic waves. This concept was made impossible by Einstein's special theory of relativity.

Evaporative cooling. A technique used to cool a collection of bosonic atoms and so obtain a *Bose–Einstein condensate*. (See Nobel 2001.)

(Pauli) Exclusion principle – The principle stating that no two fermions can exist in the same quantum state. It is not obeyed by bosons. (After Wolfgang Pauli, Nobel 1945.)

Fermi–Dirac statistics. A set of quantum rules governing the behaviour of a collection of fermions. It does not allow two fermions to occupy the same quantum state. (After Enrico Fermi, Nobel 1938; and Paul Dirac, Nobelist in 1933.)

Fermions. Particles whose spin is an odd-half integer – that is, equal to $\frac{1}{2}, \frac{3}{2}, \frac{5}{2}$, and so on (in units of $h/2\pi$) – and which obey the *Pauli exclusion principle*. Examples of fermions: all fundamental particles of matter (leptons and quarks), the nucleons (proton and neutron).

Feynman diagrams. Pictorial representations in the *Minkowski space-time* for studying subatomic particle interactions. (After Richard Feynman, Nobelist in 1965.)

Field (of force). In physics, a force can be described as an invisible *field* which fills the whole space around a body, within which it can exert a force on other similar bodies not in contact with it.

Field, electric. If a body carries an electric charge, physicists say that it is surrounded by an invisible *electric field*. Any charged particle, moving or not, in an electric field experiences an electric force.

Glossary of terms

Table G.1. *The four fundamental forces of nature*

Force	Strength
Gravity	1
Weak	10^{25}
Electromagnetic	10^{36}
Strong	10^{38}

(10^{38} = one hundred billion billion billion billions)

Field, gravitational. Physicists say that the earth is surrounded by an invisible *gravitational field* which exerts the weight force on any heavy body which is in the field. Any heavy body is surrounded by its own gravitational field.

Field, magnetic. Every magnet, or every wire in which a current is flowing, generates an invisible *magnetic field* in the space around it. In general, a charged particle moving in a magnetic field experiences a magnetic force.

Fine structure constant. Fundamental numerical constant of atomic physics and quantum electrodynamics. Its value is equal to about 1/137.

Flavour. The name given to the characteristic that distinguishes the six known varieties of quarks and leptons.

Fluorescence (Phosphorescence). The emission of light or other radiation from certain substances that are excited by radiation; these substances re-emit the radiation soon after being excited. A fluorescence that persists after the excitation has stopped is called a 'phosphorescence'.

Force. Whenever you are pushing or pulling an object you are exerting a force on it. Forces either can change the speed of an object or the direction of its movement. A familiar force is *weight*: it is the force of *gravity*, due to the earth's gravitational attraction on heavy objects. In contemporary language the term *interaction* is also used instead of force.

Force, gravitational **(or *gravity*).** This is the force of attraction that arises between objects carrying a mass: it has an infinite range, and it decreases with the square of the distance between two objects carrying a mass. The property of an object that renders it sensitive to this force is called the *gravitational mass*.

Force, electromagnetic. This arises between particles carrying an electric charge. When particles are at rest it is expressed by the Coulomb law (after Charles Augustin de Coulomb). Like the gravitational force, the range of the electric force is infinity, and (in empty space) it decreases with the square of the distance between two point electric charges.

Force, strong **(*nuclear*).** The force that binds protons and neutrons together in an atomic nucleus. It acts between hadrons, and originates from the strong *colour force* between quarks. Its range is about one-million-billionth of a metre.

Force, weak **(*nuclear*).** The force which governs certain radioactive decays in the atomic nuclei. It acts on all particles of matter. It is a short-range force, acting within a distance less than about one-hundred-million-billionth of a metre.

Frame of reference. A conceptual framework used in determining the position of bodies in space. Imagine you want (as the philosopher René Descartes did) to describe the position, at any moment of time, of a fly buzzing around your room. You simply use three numbers; that is, the distances (commonly labelled with the letters x, y, z) of the fly from two walls and the floor where it meets them in a corner of your room. Thus the two walls and the floor will become three orthogonal planes, and will constitute a *frame of reference*, or a *coordinate system* (the three meet each other in three orthogonal straight lines called *coordinate axes*).

Frame of reference, inertial. A frame of reference in which force-free objects move uniformly along straight lines obeying Newton's First Law. (They are also named *Galilean frames of reference*). Examples: Galileo's ship travelling with a constant velocity in a calm sea; a spaceship moving at a constant velocity. Accelerated reference frames are *non-inertial*. Examples: your car when travelling on a curved road; an aeroplane during take-off or landing.

Galaxy. An immense assembly of stars (some hundreds of billions) held together by the attractive gravitational force. Example: our own Milky Way, which contains the solar system.

Galilean transformations (also called **Newtonian transformations**). A set of mathematical equations, in classical physics, which allow one to relate the spatial coordinates (x, y, z) and the time (t) from one inertial frame of reference to another frame.

Gauge symmetries. Mathematical symmetries possessed by the fields describing matter particles and force messengers, which allow certain quantities in the theory to change (*gauge transformations*), at each position, and at each time, without affecting the values of the observable physical quantities calculated by the theory itself (*gauge invariance*).

Gauge (field) theories. Theories in which the weak, electromagnetic and strong forces are described in terms of fields which concern matter particles (leptons and quarks) and the force messengers. These fields possess certain symmetry properties called *gauge symmetries*.

Geiger–Müller counter. A device for detecting radiation such as X-rays, gamma rays or beta rays, by measuring the ionisation produced by such particles in the gas contained in the device itself. (After Hans Geiger and Walther Müller.)

Ginzburg–Landau theory. A macroscopic theory which explains many aspects of the behaviour of superconductors when they are near their transition temperature, and also when they are placed in a magnetic field. (After Vitaly Ginzburg, Nobelist in 2003, and Lev Landau, Nobel 1962.)

Gluons. Messenger particles that convey the strong *colour force* between quarks. There are *eight* types of gluons with different colour charges; all of them are massless bosons with zero electric charge. (See Nobel 1990.)

Grand unified theories (GUTs). Theories which attempt to provide a unified description of the electroweak and the strong forces.

Gravitational radiation. Radiation carried by gravitational waves.

Gravitational waves. Waves in the gravitational field, analogous to the electromagnetic waves in the electromagnetic field. They travel at the speed of light in empty space (c). Their existence is required by Einstein's general theory of relativity. (See Nobel 1993.)

Graviton. The hypothetical quantum of the gravitational field; it conveys the gravitational force between heavy bodies.

Hadrons. Subatomic particles that experience the strong force. They are made up of quarks: three quarks form the *baryons*; a quark and an antiquark may combine together to form *mesons*.

Hall effect. The creation of an electric field (and its resulting electric voltage, called the *Hall voltage*) by a magnetic field acting in a direction perpendicular to an electric current flowing in the material. The ratio of the *Hall voltage* to the current is called the *Hall resistance*. (After Edwin Hall.)

Hall effect, quantum. In a two-dimensional electron system at temperatures near absolute zero and in a high magnetic field the *Hall resistance* is a integer multiple of the *von Klitzing resistance*. (After Klaus von Klitzing, Nobel 1985.)

Hall effect, fractional quantum. A type of quantum Hall effect where the *Hall resistance* is a fraction with odd denominator of the *von Klitzing resistance*. (See Nobel 1998.)

Heat. A form of internal energy, also called *thermal energy*, which is transferred from one body to another because of a temperature difference between them.

Heat, specific. The quantity of heat needed per unit mass (for example a kilogram) to raise its temperature by one degree Celsius.

Heterostructure. Very thin layers of semiconducting materials of varying composition and doping, all of them joined together, with their atomic structures fitting into a single crystalline pattern. Each layer has different electronic properties, which have effects on the behaviour of the electrons and holes at the interfaces (*heterojunctions*) of the structure. (See Nobel 2000.)

Higgs particles. The quanta of a quantum field pervading all space, which interacts with the fields of the electroweak theory and produces the Brout–Englert–Higgs mechanism. (See Nobel 1979.)

Holography. The technique of producing a photographic record (called a *hologram*) by illuminating the object with coherent light, and exposing the film to light reflected by the object itself and by a reference beam of coherent light. When the film is illuminated by coherent light a three-dimensional image is produced. (See Nobel 1971.)

Insulator. Any material that does not allow a flow of electric charge. Glass and rubber are common insulators.

Integrated circuit. A collection of transistors and electrical circuits all built onto a single semiconducting crystal. Today's integrated circuits are no more than a centimetre long, and they can carry millions of microscopic transistors. (See Nobel 2000.)

Ion. An atom or a group of atoms that carries a net electric charge.

Ionisation. The formation of ions as the result of a chemical reaction, heat, electrical discharge (as in a cathode-ray tube) or radiation (as the ionisation of the air molecules by X-rays).

Ionosphere. A region of the earth's atmosphere extending about from 60 to 100 kilometres above the earth's surface, in which there is a high concentration of free electrons and ions formed as a result of ionising radiation entering the atmosphere from space. (See Nobel 1947.)

Isotopes. Atoms made of nuclei with the same atomic number (Z) but different mass number (A). For example, uranium-235 ($Z = 92$, $A = 235$) is an isotope of uranium-238 ($Z = 92$, $A = 238$).

Josephson effect. Tunnelling of Cooper pairs (electron pairs) through a thin insulating layer sandwiched between two superconductors. (After Brian Josephson, Nobelist in 1973.)

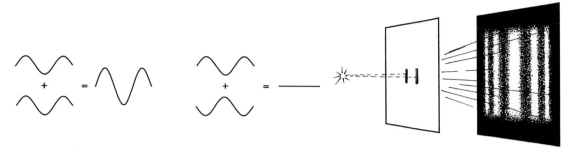

Fig. G.5. Left: interference of two waves. Right: interference fringes produced by light waves passing through two slits.

J/ψ particle. A meson composed of a charm quark and a charm antiquark. (See Nobel 1976.)

K-mesons (Kaons). A variety of strange mesons containing a strange quark or a strange antiquark.

Lamb shift. Minute shift of the first excited energy level of the hydrogen atom predicted by modern QED. (After Willis Lamb, Nobelist in 1955.)

Laser. A device which produces a very narrow, nearly monochromatic and coherent beam of light. (See Nobel 1964.)

Laser, atom. A device which produce a nearly mono-energetic and coherent beam of atoms. (See Nobel 2001.)

Lattice, crystal. An array of atoms (or ions) in a periodic pattern in three dimensions.

Leptons. A class of six fundamental elementary particles which can experience electromagnetic and weak forces, but not the strong force. All leptons are fermions.

Light. Electromagnetic waves which affect the human eye, and produce the sensation of colours (red, orange, yellow, green, blue, indigo and violet). Colours are actually light waves of different wavelengths. Light of a single wavelength or colour is called *monochromatic* light.

Light interference. If coherent light waves of the same wavelength coming from two narrow slits (see Fig. G.5) arrive in the same place at the same time, they may add together and produce an effect known as *interference*. In some places they produce reinforcement (constructive interference); in others they cancel each other out (destructive interference): alternate bright and dark bands appear on a screen; they are called *interference fringes*.

(Bohr's) Liquid drop model. A model which explains how the nucleons in an atomic nucleus attract neighbouring nucleons, just as molecules in a drop of liquid are attracted by neighbouring molecules to form a drop. (After Niels Bohr, Nobel 1922.)

Lorentz transformations. A set of four mathematical equations in the special theory of relativity which allow one to relate the spatial coordinates (x, y, z) and the time (t) of occurrence of an event from one inertial frame of reference to another frame. (After Hendrik Lorentz, Nobel 1902.)

Magic numbers. Characteristic numbers of protons or neutrons corresponding to particularly stable atomic nuclei. (See Nobel 1963.)

Magnetic dipole. A magnet, equivalent to a flow of electric charge around a loop. The strength of its magnetic field is represented by a physical quantity called a *magnetic*

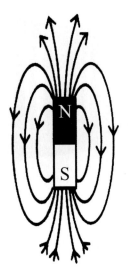

Fig. G.6. Magnetic field lines of a bar magnet.

moment. Examples of macroscopic magnetic dipoles: a bar magnet (see Fig. G.6); a compass needle. Examples of microscopic magnetic dipoles: an electron circling around an atomic nucleus; a spinning electron.

Magnetosphere. The region surrounding the earth in which the behaviour of charged particles is dominated by the earth's magnetic field.

Maser. A device which amplifies microwaves and produces a nearly monochromatic and coherent microwave beam. It works on the same principle as a laser. (See Nobel 1964.)

Matrix mechanics. A version of quantum mechanics developed by Werner Heisenberg, Max Born and Pascual Jordan.

Matter wave. A moving object behaves as though it has a wave character. The wave representing it is called a *matter wave* or a *De Broglie wave*. (After Louis de Broglie, Nobel 1929.)

Matter wave – De Broglie wavelength. The wavelength of a matter wave. For macroscopic objects this wavelength is exceedingly small, so that quantum effects are irrelevant. On the other hand, the wavelength for an electron in an atom is comparable to the size of the atom, so that here quantum effects are paramount.

Maxwell equations. A set of four mathematical equations that describe how the electromagnetic field evolves in space and time. (After James Clerk Maxwell.)

Meissner effect. The phenomenon in which a superconducting metal, cooled through its transition temperature, in the presence of a magnetic field completely repels the field. (After Walter Meissner.)

Mesons. A class of hadrons, made up of quark–antiquark pairs. All mesons are bosons.

Microscope. An instrument that produces a magnified image of a small object. Examples: *optical microscope* (it uses light to illuminate the object); *electron microscope* (it uses electrons, see Nobel 1986); *scanning tunnelling microscope* (it uses a tunnelling current, see Nobel 1986).

Molecule. The simplest unit of a chemical compound, consisting of two or more atoms held together by electric forces.

Mössbauer effect. This concerns the fact that, under certain conditions, nuclei of radioactive atoms emitting gamma rays when embodied in a crystal can produce a beam of gamma rays of sharply defined frequency (or energy). (After Rudolf Mössbauer, Nobel 1961.)

Motion. The change in the position of an object with time.

Motion – acceleration. The rate of change of the velocity of an object.

Motion – angular momentum. A physical quantity of the rotational motion, analogous to the momentum in linear motion, which reflects the rate of rotation of a body in its orbital motion. For a body moving in a circular orbit, the greater the number of revolutions per seconds, the greater is its angular momentum. A spinning body too has an *intrinsic angular momentum* (the greater its speed of spin, the greater is its intrinsic angular momentum).

Motion – free fall. Four hundred years ago, Galileo found – as the result of careful measurements made on balls rolling down inclined planes – that all freely falling bodies have the same acceleration at the same place near the earth's surface (it is about 10 metres per second-squared).

Motion – (linear) momentum. A physical quantity which depends on both the mass and the velocity of an object: *(linear) momentum = mass × velocity*.

Motion – Newton's first law. When an object is at rest, it needs a force to make it move; in a similar way, a moving object needs a force to stop it moving. Newton's first law states that the object, once started in motion, persists in its state of linear motion with constant velocity in the absence of applied forces.

Motion – Newton's second law. It is expressed as *force = mass × acceleration*. Physicists call the ratio of the force to the acceleration in Newton's second law the *inertial mass* of the object. The greater the inertial mass of an object, the more the object 'resists' being accelerated.

Motion – velocity (also ***speed***). The rate at which a moving object covers distance. (Examples: an athlete running in an Olympic 100-metre sprint travels at a speed of about 12 metres per second; an electron in a modern accelerator travels at a speed of about 299 000 000 metres per second).

Mott transition. The change of the electrical conductivity of a material from metal to insulator due to parameters such as composition or pressure. (After Nevill Mott, Nobelist in 1977.)

Multiwire proportional chamber (MPC). A device for detecting high-energy subatomic particles. It consists of a large number of thin parallel wires arranged in a plane, placed between two cathode plates. The wires detect the ionisation produced by a charged particle in the gas between the cathode plates themselves. All of the information concerning particle tracks is recorded electronically. (See Nobel 1992.)

Muon. A charged lepton, analogue of the electron, belonging to the second lepton family.

Nanometre. One nanometre is equal to one billionth (10^{-9}) of a metre.

Nanostructure. Structure in size range of a few nanometres.

Neutral current. Weak interaction where no change takes place in the charges of the participating particles. (See Nobel 1979.)

Neutrino. Electrically neutral lepton. There are three known varieties (flavours), one in each of the three lepton families, associated respectively with the electron (first family), the muon (second family) and the tau lepton (third family). The neutrinos have an exceedingly small mass, no electric charge, and they only take part in weak and gravitational interactions. (See Nobel 1988 and Nobel 1995.)

Neutrino oscillation. The switching of a neutrino from one variety (flavour) to another. (See Nobel 2002.)

Neutron. Electrically neutral particle with roughly the same mass as a proton, found along with protons in atomic nuclei. (See Nobel 1935.)

Neutron star. A very dense star composed solely of neutrons, which has collapsed under its own gravity. (See Nobel 1974.)

Noble gases (also inert gases). A group of six chemical elements with the outermost electron shell closed. It includes helium, neon, argon, krypton, xenon and radon. (See Nobel 1904.)

Non-linear optics. A branch of optics which is concerned with the optical properties of matter that do not depend linearly on the radiation strength. (See Nobel 1981.)

Nuclear collective model. This concerns the structure of atomic nuclei. In this model there is an interplay between the motion of individual nucleons and the 'collective' motion of the nucleus as a whole. It incorporates aspects of both the liquid-drop model and the nuclear shell model. (See Nobel 1975.)

Nuclear fission. Some heavy nuclei (such as uranium-235 or plutonium-239) can break into two with the release of a large amount of energy. Commercial nuclear reactors use nuclear fission, and atom bombs are fission weapons.

Nuclear fusion. The joining together of light nuclei (for example deuterium and tritium) to form heavier nuclei and release large amounts of energy. Fusion reactions are responsible for the creation of heavy elements in the cores of stars, and they provide the energy source which causes them to shine. (See Nobel 1967.)

Nuclear magnetic resonance (NMR). A technique based on the fact that many atomic nuclei possess magnetic properties. Molecules exposed to a magnetic field will absorb certain frequencies of radio waves; from such absorption information concerning the structure of these molecules can be obtained. (See Nobel 1952.)

Nuclear reactions. Reactions changing the identity of the participating atomic nuclei. (Examples: radioactive nuclei produced by the bombardment of neutrons; nuclear fission reactions.)

Nuclear shell model. This concerns the structure of atomic nuclei. In this model protons and neutrons in a nucleus move in spherical shells just as electrons do in atoms. (See Nobel 1963.)

Nucleons. The particles, protons and neutrons, found in atomic nuclei.

Nucleosynthesis. The nuclear processes by which the chemical elements are synthesised. It takes place in the central regions of stars where, for example, helium is synthesised from protons and neutrons.

Nucleus. In the nucleus is concentrated over 99% of the mass of the atom. In size it is about ten thousand times smaller than the atom itself. The nucleus is composed of two kinds of subatomic particles, protons and neutrons, which are held together by the strong nuclear force.

Parity. The operation of studying a physical phenomenon in an inverted coordinate system – a system where all the three spatial coordinates (x, y, z) are inverted, becoming $(-x, -y, -z)$. In certain cases this inversion is like the reflection of the phenomenon itself in a mirror. Example: a particle spinning clockwise around a vertical direction (z) in the system (x, y, z) is viewed as spinning anticlockwise in the inverted system $(-x, -y, -z)$: it is as if the same spinning particle were reflected in a vertical mirror. (See Nobel 1957 and Nobel 1980.)

Particle trap. A device that confines particles or atoms away from containing walls. Physicists use particle traps to store particles, to cool them, and to hold them in place to be studied for long periods of time. (See Nobel 1989.)

Perihelion. The point in its orbit where a planet is nearest the sun.

Phase transition. A change of state such as that which occurs when a liquid is freezing or boiling, or when a magnetic solid changes from a ferromagnetic state to a non-ferromagnetic one. (See Nobel 1982.)

Photoelectric effect. A flow of electrons into the space surrounding a metal surface caused by light (photons) of sufficiently high frequency, which supply energy to the electrons in the metal. (See Nobel 1921 and Nobel 1923.)

Photon. The quantum of light (or of the electromagnetic field) characterised by the energy $E = h\nu$ and momentum $p = h/\lambda$ (ν and λ are respectively the frequency and the wavelength of the radiation). It is a boson (spin equal to one) with zero mass, and zero electric charge. It is the messenger of the electromagnetic force.

Physics. The branch of science concerned with the properties of matter and energy and the relationships between them.

Pion (π-meson). The lightest of the mesons. It explains the strong nuclear force that binds protons and neutrons in an atomic nucleus. (see Nobel 1950.)

Planck's constant. A fundamental constant of nature which characterises quantum physics. It assumes great importance in microscopic systems, but for the macroscopic ones it is negligibly small. The quantity $h/2\pi$, the value of which is about 6×10^{-34} (ten-million-billion-billion-billionths) joule $\times$ second, represents a fundamental unit of angular momentum in atomic and subatomic systems. (See Nobel 1918.)

Plasma. A kind of very hot ionised gas made up of electrons and positive ions. (See Nobel 1970.)

Polymer. A long molecule consisting of many smaller molecules joined together. (See Nobel 1991.)

Power-of-ten notation. This is a useful shorthand system of writing numbers. In this book the following notation is used: 10^3 = one thousand, 10^6 = one million, 10^9 = one billion (one thousand million); 10^{-3} = one thousandth, 10^{-6} = one millionth, 10^{-9} = one billionth.

Precision. Physical measurements always have uncertainties. Physicists often indicate the precision of the value of a physical quantity in terms of its fractional or percentage *uncertainty*. Examples: if we measure a distance of 1000 metres (1 kilometre) with an uncertainty of 1 metre, we say that the precision (or fractional uncertainty) of our measure is 10^{-3} (1/1000, or 1 part in 1000). If we measure the same distance with a smaller uncertainty (higher precision) – for example, of 1 millimetre – the precision increases to 10^{-6} (1/1 000 000, or 1 part in a million).

Principle of relativity (Einsteinian). This principle, first stated by Einstein in 1905, states that *all the laws of physics* must be exactly the same in any inertial frame of reference.

Principle of relativity (Newtonian). This principle, first described by Galileo and subsequently formulated by Isaac Newton in his *Principia*, states that *all the laws of mechanics* must be exactly the same in any frame of reference moving rectilinearly and uniformly (*inertial or Galilean frame of reference*).

Proton. Positively charged particle found along with neutrons in atomic nuclei. Its mass is about 10^{-27} (one-billion-billion-billionth) of a kilogram.

Proton–proton chain. A sequence of thermonuclear fusion reactions by which hydrogen nuclei are built up into helium nuclei. (See Nobel 1967.)

Pulsar. A rapidly spinning neutron star which emits pulses of radio waves (and in some cases light and X-rays). (See Nobel 1974.)

Quantum. The smallest amount by which some physical property of a quantum-mechanical system can change.

Quantum chromodynamics (QCD). The modern theory which describes how the 'coloured' quarks are held together by the *strong colour force* to form hadrons; and how nucleons are held together to form the atomic nuclei. (See Nobel 1990.)

Quantum electrodynamics (QED). Theory which describes the electromagnetic interactions between electrically charged particles (such as electrons) and photons. (See Nobel 1965.)

Quantum electronics. Parts of quantum optics which have practical device applications.

Quantum field theory. A modern theory which is used to describe the microworld. It is a combination of quantum mechanics and special relativity, its main physical ingredient being the *quantum field*, which brings together particles and fields. Example: the quantum electromagnetic field can be reduced to photons, or to waves, this last described by the classical electromagnetic field.

Quantum (mechanical) system. A physical system, like a molecule, an atom, a nucleus, a composite subatomic particle, or a collection of atoms (as those in a Bose–Einstein condensate) described by the laws of quantum mechanics.

Quantum mechanics. The fundamental physical theory developed in the 1920s as a replacement of classical (or Newtonian) mechanics, in order to describe the behaviour of quantum mechanical systems. (In quantum mechanics waves and particles are two aspects of the same underlying entity.)

Quantum state. The state of a quantum mechanical system described by a wave-function and characterised by a definite energy.

Quarks. Fundamental elementary particles, of which all hadrons are supposed to be composed. They are fermions. (See Nobel 1969.)

Quasar (quasi-stellar object). A star-like astronomical object with a very small size, but with a very large red shift. The true nature of quasars is as yet still unknown.

Radar. A device for detecting the position and velocity of a distant object, by means of extremely high-frequency radio pulses which are transmitted and then reflected back to the transmitter by the object itself.

Radio astronomy. The astronomy carried out by radio telescopes in a wide interval of the electromagnetic spectrum, which extends from low-frequency radio waves (some tens of metres wavelength) to centimetre and millimetre wavelengths. (See Nobel 1974.)

Radioactivity. When an unstable atomic nucleus disintegrates to acquire a more stable state it emits radiation: the phenomenon is called *radioactivity*, or *radioactive decay*. (See Nobel 1903.)

Radioactivity – alpha particles. Alpha particles (or alpha rays) consist of helium nuclei (two protons + two neutrons). They are emitted from certain radioactive nuclei in a process called *alpha decay*. They are double-charged and so they are deflected by electric or magnetic fields.

Radioactivity – beta particles. Beta particles (or beta rays) are fast electrons emitted along with antineutrinos when, in certain radioactive nuclei, a neutron becomes a proton. The emitted beta particles have a continuum range of energies (*continuum spectrum*). They are deflected by electric or magnetic fields.

Radioactivity – gamma rays. Electromagnetic radiation of very short wavelength emitted by radioactive nuclei. They are not deflected by electric or magnetic fields.

Radioactivity – half-life. The time taken for half the nuclei of a radioactive substance to disintegrate. The half-lives of different substances vary widely – from fractions of a second up to millions of years.

Raman effect. The change in the wavelength of a light beam when scattered by molecules. (After Chandrasekhara Raman, Nobel 1930.)

Red shift. The shifting toward longer wavelengths, caused by the Doppler effect for a receding light source.

Red shift, gravitational. The wavelength of the light emitted in a strong gravitational field is larger compared with the wavelength of the same light emitted in a weaker gravitational field.

Relativity, general. The general theory of relativity is a theory of gravitation in which the gravitational force is explained as a distortion of space-time. It was introduced by Albert Einstein (Nobel 1921) in 1915–16.

Relativity, special. The special theory of relativity is a theory of space-time in which the speed of light is the same for everybody, no matter how they are moving. It was introduced by Albert Einstein (Nobel 1921) in 1905.

Renormalization. Mathematical procedure for circumventing infinite quantities (such as mass or electric charge) in a quantum field theory – like, for example, quantum electrodynamics (QED). (See Nobel 1965.)

Renormalization group. A mathematical method to study the behaviour of physical quantities under changes of scale in the fields of condensed-matter and particle physics. (See Nobel 1982.)

Resistance, electrical. A metal wire tends to resist the movement of electrons in it: it has a certain *electrical resistance* to the current. This means that energy has to be supplied to force electrons through the wire.

Resolution. The degree to which fine details in an optical image can be distinguished.

Resonance particles (resonances). Extremely short-lived systems of subatomic particles which decay via the strong nuclear force. (See Nobel 1968.)

Rest (or proper) mass. The mass of an object measured at rest.

Scattering. The deflection of photons or particles (for example, alpha particles, electrons) by other particles.

Schrödinger's equation. A mathematical equation of the same sort as the equations used to study waves of sound or light; it describes how a particle wave function (ψ) evolves in space and time. (After Erwin Schrödinger, Nobelist in 1933.)

Scintillator. Transparent material in which an energetic charged particle causes emission of ultra-short flashes of light. These materials are used in the so-called scintillating detectors.

Semiconductor. A material that allows a flow of charge (electric current) more than an insulator but less than a conductor. Some common semiconductors are silicon, germanium and gallium arsenide.

Semiconductor laser. Laser in which the lasing medium is a semiconductor *heterostructure*. (See Nobel 2000.)

Solid-state detector. Radiation detectors in which a semiconducting material constitutes the detecting medium.

(Minkowski) Space-time. The four-dimensional continuum having the three spatial dimensions (x, y, z) fused with the time dimension (t). (After Hermann Minkowski.)

Spark chamber. A device for detecting high-energy subatomic particles consisting of parallel conducting plates between which a high electric voltage is applied. An electrical discharge occurs whenever a charged particle traverses the gap between the plates; the particle spark tracks can thus be either photographed or recorded electronically.

Spectral lines. Because of the slit-shaped aperture generally used in spectroscopes, an optical spectrum can consist either of a pattern of narrow coloured bands of light on a dark background (emission spectrum), or of narrow black bands on a coloured background (absorption spectrum). These bands are the images of the slit and are called *spectral lines* (see Fig. G.7).

Spectral lines – fine structure. The splitting of atomic spectral lines into two or more components resulting from the magnetic interactions between the orbital magnetic moments and the spin magnetic moments of the electrons in an atom.

Spectral lines – hyperfine structure. A splitting of the atomic spectral lines much smaller than that corresponding to the fine structure. It results from the interactions between the magnetic moment of the nucleus and the orbital and spin magnetic moments of the electrons.

Fig. G.7. A line spectrum produced by a glass prism.

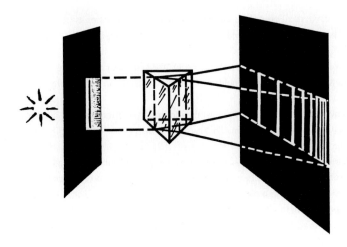

Spectroscope. An instrument that splits a beam of light into components of different frequencies (or wavelengths) on a screen or on a detector.
Spectroscopy. The science of analysing and studying spectra.
Spectrum, electromagnetic (*e. m.*). The interval of wavelengths (or frequencies) of electromagnetic waves. It extends from low values over the range of radio waves, television waves, microwaves, and infrared rays to visible light and so beyond to higher values of ultraviolet radiation, X-rays and gamma rays.
E. m. spectrum, infrared. Electromagnetic waves with a wavelength between about 0.001 and 0.1 of a millimetre.
E. m. spectrum, microwave. Electromagnetic waves with a wavelength between about 0.01 and 10 centimetres.
E. m. spectrum, radio wave. Electromagnetic waves with a wavelength between about 10 centimetres and more than 1000 metres.
E. m. spectrum, visible. Part of the electromagnetic spectrum corresponding to visible light. Its limits are about 400 and 700 nanometres. When white light is dispersed by a glass prism it forms a continuous pattern of colours, which is called a *continuum spectrum*.
Spin. A physical property representing the intrinsic angular momentum of a particle. According to the rules of quantum mechanics, the spin can take on only certain special values, equal to a whole number or half a whole number multiplied by the Planck constant h.
Spontaneous emission. In a spontaneous emission an excited atom radiates a photon in a random way; the process is independent of external radiation.
Stark effect. The splitting of the lines of an atomic spectrum when the atoms are placed in an electric field. (After Johannes Stark, Nobel 1919.)
Stimulated emission. When an incident photon of the right frequency is not absorbed by an excited atom, but induces the atom to emit a second photon of the same frequency. Stimulated emission is the driving force behind the laser.
Strange. The third flavour of quarks, belonging to the second quark family.

Strangeness. A property (quantum number) possessed by all subatomic particles containing a strange quark. These particles are called *strange particles*. (See Nobel 1969.)

Superconductivity. The property of certain materials that have almost no electrical resistance at temperatures close to absolute zero. (See Nobel 1913, Nobel 1972, Nobel 1987 and Nobel 2003.)

Superconductivity – critical magnetic field. Value of an externally applied magnetic field at which a superconductor becomes non-superconducting.

Superconductivity – transition (or critical) temperature. The highest temperature at which superconductivity occurs in a superconducting material.

Superfluidity. A phenomenon occurring in liquid helium-4 below about 2.17 kelvin, in which the liquid flows through thin capillaries without apparent friction and displays many other anomalous properties (see Nobel 1962 and Nobel 1978). Liquid helium-3 instead becomes super-fluid below about 3×10^{-3} kelvin (see Nobel 1962, Nobel 1978, Nobel 1996 and Nobel 2003).

Supernova. A titanic explosion in which all but the inner core of a star is blown off into space. The matter ejected is called a *supernova remnant*.

Superstring theory. The latest attempted unified field theory of particles and forces in which the fundamental particles are not point-like, but one-dimensional strings only about 10^{-32} (one hundred-thousand-billion-billion-billionth) of a millimetre long.

Symmetry. A system possesses a symmetry if it looks the same when changes are performed on it, for example, a sphere looks the same when it is rotated. A physical law possesses a symmetry if it remains the same when changes are performed on the phenomenon it describes.

Symmetry, charge (or charge conjugation invariance). When the same laws of physics describe a phenomenon involving either particles or their antiparticles. (See Nobel 1980.)

Symmetry, mirror (or parity invariance). When the same physical laws describe a phenomenon both in a coordinate system (x, y, z) and in a system in which all the three spatial coordinates are inverted $(-x, -y, -z)$. In certain cases the phenomenon in the inverted system is the mirror image of the original phenomenon (hence *parity invariance* is also called *mirror symmetry*). (See Nobel 1957.)

τ **Lepton.** The electrically charged lepton, belonging to the third lepton family. It is a heavier analogue of the electron and the muon. (See Nobel 1995.)

τ **Neutrino.** The electrically neutral lepton, belonging to the third lepton family; it is associated with the tau lepton.

Temperature. This is a measure of hotness or coldness of a body or substance on some scale. Heating a certain quantity of water, you transfer a certain amount of thermal energy to the water molecules, so that they move faster. As a result you have raised the water temperature.

Temperature – absolute zero. As the temperature diminishes, gases condense into liquids and liquids freeze into solids. If the temperature continues to get lower and lower, molecules and atoms vibrate less and less until eventually they have their lowest possible energy; this happens at about $-273\ °C$ (or zero kelvin).

Temperature, Celsius. On this temperature scale the numbers are called degrees Celsius (symbol: $°C$). Pure ice melts at $0\ °C$; pure water boils at $100\ °C$ (provided always that atmospheric pressure is exactly one atmosphere).

Temperature, kelvin. On this temperature scale the degree is called a 'kelvin'. This means that 0 °C = 273 kelvins, and 100 °C = 373 kelvins.

Thermionic emission. The flow of electrons into the space surrounding a heated, negatively charged, metal surface. (See Nobel 1928.)

Top. The sixth flavour of quarks, belonging to the third quark family.

Transistor. A tiny electrical device, made of semiconducting material, that can either amplify an electrical signal, or switch it on and off, letting current through or blocking it as necessary. (See Nobel 1956.)

Tunnelling. A quantum-mechanical phenomenon in which, if a particle reaches an energy barrier without sufficient energy to pass over it, it has a non-zero probability of passing through the barrier itself.

Up. The first flavour of quarks, belonging to the first quark family. (See Nobel 1969 and Nobel 1990.)

Virtual particle. A particle that exists only for an extremely short time, without being observed, for example, while being exchanged during an interaction between other particles. Example: the electromagnetic force between two electrons is caused by the exchange of virtual photons between the electrons themselves.

Voltage (potential difference). When a battery is connected to a lamp there is a *voltage* or a *potential difference* across the lamp: the higher the voltage, the more energy is given to the electrons flowing in the wires. The unit used to measure a voltage is called a *volt*.

Wave-function. A mathematical entity, generally denoted with the Greek letter ψ, used in quantum mechanics to describe the states of a quantum-mechanical system. (Example: using the wave function of the hydrogen atom one can calculate the probability of finding the electron at a given point around its nucleus).

Wave mechanics. A version of quantum mechanics developed by Erwin Schrödinger (Nobelist in 1933).

Wave–particle dualism. The quantum-mechanical concept that radiation and particles have both wave-like and particle-like properties.

Weak bosons. Massive messenger particles that convey the weak force. There are three types of weak bosons: two are electrically charged (W^+ and W^-), whereas the third is neutral (Z). (See Nobel 1979 and Nobel 1984.)

White dwarfs. Stars that have consumed all their nuclear fuel and have contracted to a size roughly the same as that of the earth; they have a mass less than the *Chandrasekhar limit*. (See Nobel 1983.)

X-ray astronomy. Astronomy carried out in the interval of the electromagnetic spectrum roughly between 0.01 and 10 nanometres. (See Nobel 2002.)

X-rays. Short-wavelength electromagnetic waves (typical wavelengths between 0.001 and 1 nanometre) emitted by fast electrons when they impinge on matter. (See Nobel 1901.)

Zeeman effect. The splitting of the atomic spectral lines when atoms are placed in a magnetic field. A more complicated Zeeman splitting is called the *anomalous Zeeman effect*. (After Pieter Zeeman, Nobelist in 1902.)

Notes

Chapter 1 Introduction
1. The country cited is the one where the discoveries or inventions rewarded with a Nobel Prize were carried out. (CERN is excluded from this. In this case the country corresponds to the nationality of the prizewinners.)
2. The university cited is the one where the Nobel prizewinners were working at the time when their discoveries or inventions that were rewarded with a Nobel Prize were actually carried out.

Chapter 2 Founding fathers
1. Quoted by I. Bernard Cohen, *Revolution in Science*, Cambridge, MA; London: Harvard University Press, 1985, p. 140.
2. Galileo Galilei, *Dialogue Concerning the Two Chief World Systems, Ptolemaic and Copernican* (translated by Stillman Drake), Berkeley; Los Angeles: University of California Press, 1967, pp. 187–8.
3. S. W. Hawking, 'Newton's *Principia*', in S. W. Hawking and W. Israel, eds., *Three Hundred Years of Gravitation*, Cambridge: Cambridge University Press, 1987, p. 3.
4. Isaac Newton, *The Principia*: *Mathematical Principles of Natural Philosophy* (translated by I. Bernard Cohen and Anne Whitman), Berkeley; Los Angeles: University of California Press, 1999, pp. 408–9.
5. Ibid., p. 416.
6. Ibid., p. 793
7. Ibid., p. 810.
8. Ibid., p. 811.
9. Sir Isaac Newton, *Opticks*, New York: Dover Publications, 1979, p. 370.

Chapter 3 Highlights of classical physics
1. W. D. Niven, ed. *The Scientific Papers of James Clerk Maxwell*, New York: Dover Publications, 1965, vol. 1, pp. 527–34.
2. Ibid., p. 535.
3. William Francis Magie, ed. *A Source Book in Physics*, New York; London: McGraw-Hill, 1935, pp. 310–11.
4. Quoted by Steven Weinberg, *Dreams of a Final Theory*: *The Scientist's Search for the Ultimate Laws of Nature*, New York: Vintage Book, 1992, p. 13.

Chapter 4 New foundations
1. The Nobel Foundation, *Nobel Lectures: Physics 1901–1921*, Amsterdam: Elsevier, 1967, p. 3.

2. Robert L. Weber, *Pioneers of Science: Nobel Prize Winners in Physics*, Bristol; London: The Institute of Physics, 1980, p. 9.
3. Morris H. Shamos, *Great Experiments in Physics*, New York: Dover Publications, 1987, p. 201.
4. Ibid., p. 205.
5. Max Planck, 'Über das Gesetz der Energieverteilung im Normalspektrum', *Annalen der Physik*, 4, 1901, pp. 553–63.
6. Quoted by Jagdish Mehra and Helmut Rechenberg, *The Historical Development of Quantum Theory*, New York: Springer–Verlag, 1962, vol. 1, Part 1, pp. 49–50.
7. The Nobel Foundation, *Nobel Lectures*: *Physics 1901–1921*, pp. 31–2.
8. Owen W. Richardson, 'Hendrik Antoon Lorentz', *Journal of the London Mathematical Society*, 4, 1929, pp. 183–92.
9. Elisabeth Crawford, *The Beginnings of the Nobel Institution: The Science Prices, 1901–1915*, Cambridge: Cambridge University Press, 1984, p. 140.
10. Ibid.
11. In the rationale for the 1903 Nobel Prize for physics no mention was made of the new radioactive elements, polonium and radium, discovered by the Curies. This, in 1911, permitted the Swedish Academy of Sciences to award the Nobel Prize for chemistry to Marie Curie.
12. Antoine H. Becquerel, 'On radioactivity, a new property of matter', *Nobel Lectures*: *Physics 1901–1921*, p. 52.
13. Quoted by Per F Dahl, *Flash of the Cathode Rays: A History of J. J. Thomson's Electron*, Bristol, PA: Institute of Physics Publishing, 1997, p. 146.
14. Nanny Fröman, 'Marie and Pierre Curie and the discovery of polonium and radium', p. 4. (See: www.nobel.se/physics/articles/index.html.)
15. Ibid., p. 5.
16. Elisabeth Crawford, *The Beginnings of the Nobel Institution*, p. 141.
17. Nanny Fröman, 'Marie and Pierre Curie and the discovery of polonium and radium', p. 3. (See: www.nobel.se/physics/articles/index.html.)
18. Ibid., p. 2.
19. Albert Einstein, *Ideas and Opinions*, London: Souvenir Press (Educational & Academic), 1973, p. 77.
20. The Nobel Foundation, *Nobel Lectures*: *Physics 1901–1921*, pp. 137–8.
21. Albert Einstein, *The Collected Papers of Albert Einstein* (Translation), Princeton: Princeton University Press, 1992, vol. 5, p. 20. © 1989 Hebrew University of Jerusalem. Reprinted by permission of Princeton University Press.
22. Quoted by Banesh Hoffmann, *Albert Einstein: Creator and Rebel*, London: Hart-Davis, MacGibbon, 1972, p. 88.
23. Albert Einstein, *The Collected Papers of Albert Einstein* (Translation), Princeton: Princeton University Press, 1992, vol. 2, p. 20. © Hebrew University of Jerusalem. Reprinted by permission of Princeton University Press.
24. Ibid., p. 99. Albert Einstein, 'Über einen die Erzeugung und Verwandlung des Lichtes betreffenden heuristischen Gesichtspunkt', *Annalen der Physik*, 17, 1905, pp. 132–48. (Einstein's paper concerning the photoelectric effect was published on 9 June 1905.)
25. Albert Einstein, 'Zur Elektrodynamik bewegter Körper', *Annalen der Physik*, 17, 1905, pp. 891–921. (Einstein's paper on the special theory of relativity was published on 26 September 1905.) Albert Einstein, 'Ist die Trägheit eines Körpers von seinem

Energieinhalt abhängig?', *Annalen der Physik*, 18, 1905, pp. 639–41. (Einstein's paper concerning the equivalence of mass and energy was published on 21 November 1905.)
26. Paul Arthur Schilpp, ed., *Albert Einstein: Philosopher-Scientist*, La Salle, IL: Open Court, 1970, p. 53.
27. Albert Einstein, *The Collected Papers of Albert Einstein*, vol. 2, p. 140.
28. Quoted by Abraham Pais, *Subtle is the Lord . . . ,: The Science and the Life of Albert Einstein*, New York: Oxford University Press, 1982, p. 153.
29. Wolfgang Pauli, *Theory of Relativity*, London; New York; Paris; Los Angles: Pergamon Press, 1958, p. 123.
30. Irwin I. Shapiro, 'A century of relativity', *Reviews of Modern Physics*, 71, 1999, p. S42.
31. Quoted by Abraham Pais, *Subtle Is the Lord . . .*, p. 139.
32. Quoted by Per F Dahl, *Flash of the Cathode Rays*, p. 160.
33. Per F. Dahl, *Flash of the Cathode Rays*, p. 162, quoting J. J. Thompson, *Recollections and Reflections*, London: Bell, 1936, p. 335.
34. J. J. Thomson, 'Cathode Rays', *Philosophical Magazine*, 44, 1897, p. 310
35. Ibid., pp. 293–316.
36. J. J. Thomson, *Recollections and Reflections*, pp. 338–9.
37. Quoted by Sir G. P. Thomson, 'J. J. and the Cavendish', in *A Hundred Years and More of Cambridge Physics* (see: www.phy.cam.ac.uk/cavendish/history/years/jjandcav.asp).
38. Abraham Pais, *Niels Bohr's Times*, In *Physics, Philosophy, and Polity*, Oxford: Clarendon Press, 1991, pp. 117–18.
39. Quoted by George Shiers, 'The First Electron Tube', *Scientific American*, March 1969, p. 110.
40. The Nobel Foundation, *Nobel Lectures: Physics 1901–1921*, pp. 164–5.
41. Albert Einstein, *The Collected Papers of Albert Einstein*, vol. 2, p. 301.
42. Quoted by R. S. Shankland, 'The Michelson–Morley Experiment' *Scientific American*, November 1964, p. 109.
43. Albert Einstein, *The Collected Papers of Albert Einstein*, vol. 2, p. 302. Einstein's 1907 paper concluded with a paragraph entitled 'Principle of relativity and gravitation', which contains the principle of equivalence.
44. Irwin I. Shapiro, 'A century of relativity', *Reviews of Modern Physics*, 71, 1999, p. S44.
45. Isaac Newton, *The Principia*: *Mathematical Principles of Natural Philosophy*, pp. 403–4.
46. Quoted by Abraham Pais, *Subtle Is the Lord . . .*, p. 178.
47. The Nobel Foundation and W. Odelberg, eds., *Nobel: The Man and His Prizes*, New York: American Elsevier, 1972, p. 309.
48. Quoted by Elisabeth Crawford, *The Beginnings of the Nobel Institution*, p. 134.
49. Quoted by Per F Dahl, *Flash of the Cathode Rays*, p. 334.
50. Ibid., p. 334.
51. Ibid., p. 133.
52. Ibid., p. 134.
53. Dennis Gabor, 'Holography, 1948–1971', *Les Prix Nobel en 1971*, Stockholm: Imprimerie Royale P. A. Norstedt & Söner, 1972, p. 190.
54. Richard Rhodes, ed., *Visions of Technology*, New York: Touchstone, 2000, p. 32.
55. Abraham Pais, *Subtle Is the Lord . . .*, p. 505.
56. Ibid., p. 505.

57. Lord Rayleigh, *The Life of Sir J. J. Thomson, OM*, Cambridge: Cambridge University Press, 1943, p. 50.

Chapter 5 The quantum atom

1. Quoted by Abraham Pais, *Subtle Is the Lord* . . . , p. 509.
2. Ibid., p. 502.
3. Wilhelm Wien, 'On the laws of thermal radiation', *Nobel Lectures: Physics 1901–1921*, p. 281.
4. Ibid., p. 283.
5. Per F Dahl, *Flash of the Cathode Rays*, p. 335, quoting E. Marsden, *Rutherford at Manchester*, New York: Benjamin, 1963, p. 8.
6. Ibid., p. 336, quoting E. N. da C. Andrade, *Rutherford and the Nature of the Atom*, New York: Anchor, 1964, p. 111.
7. Quoted by David Wilson, *Rutherford: Simple Genius*, London: Hodder and Stoughton, 1983, p. 295.
8. Ibid., p. 296.
9. Rutherford's paper was published in the *Philosophical Magazine*, 21, 1911, pp. 669–88.
10. Quoted by E. N. da C. Andrade, 'The birth of the nuclear atom', *Scientific American*, November 1956, pp. 96–8.
11. Hans Geiger and Ernest Marsden, 'The laws of deflection of α-particles through large angles', *Philosophical Magazine*, 25, 1913, pp. 604–23.
12. James Chadwick, *Nature*, 140, 1937, p. 750.
13. Albert Einstein, *The Collected Papers of Albert Einstein*, vol. 5, pp. 190–2.
14. Charles T. R. Wilson, 'On the cloud method of making visible ions and the tracks of ionising particles', The Nobel Foundation *Nobel Lectures: Physics 1922–1941*, Amsterdam: Elsevier, 1965, p. 194.
15. Ibid., p. 196.
16. Bruno Rossi, *Cosmic Rays*, New York; London: McGraw-Hill, 1964, pp. 1–2.
17. Heike Kamerling Onnes, 'Investigations into the properties of substances at low temperatures, which have led, amongst other things, to the preparation of liquid helium', *Nobel Lectures: Physics 1901–1921*, p. 333.
18. Ibid., p. 325.
19. Quoted by L. Rosenfeld, ed., *Niels Bohr: Collected Works, Volume 1: Early Work (1905–1911)*, Amsterdam: North-Holland, 1972, p. 559.
20. Bohr's celebrated paper consisted of three parts which appeared in *Philosophical Magazine*, 26, 1913: in July, 'On the constitution of atoms and molecules (Part I), pp. 1–25; in September, 'Part II. Systems containing only a single nucleus', pp. 476–502; and in November, 'Part III. Systems containing several nuclei', pp. 857–75.
21. L. Rosenfeld, ed., *Niels Bohr: Collected Works, Volume 2: Work on Atomic Physics (1912–1917)*, Amsterdam: North-Holland, 1981, p. 585.
22. N. Bohr, 'On the constitution of atoms and molecules', p. 2.
23. Ibid., p. 5.
24. Ibid., p. 7.
25. Quoted by Niels Bohr, 'The Rutherford Memorial Lecture 1958', in Finn Aaserud, ed., *Niels Bohr: Colleted Works, Volume 10: Complementary Beyond Physics (1928–1962)*, Amsterdam: Elsevier, 1999, p. 393.

26. L. Rosenfeld, ed., *Niels Bohr: Collected Works*, vol. 1, p. 567.
27. Ibid., vol. 2, p. 532.
28. S. Rozental, ed., *Niels Bohr: His Life and Work As Seen by His Friends and Colleagues*, Amsterdam: North-Holland, 1967, pp. 23–4.
29. Max von Laue, 'Concerning the detection of X-ray interferences', *Nobel Lectures: Physics 1901–1921*, pp. 350–1.
30. Ibid., p. 351.
31. Ibid., p. 352.
32. Ibid., p. 353.
33. H. G. J. Moseley, 'The high-frequency spectra of the elements', *Philosophical Magazine*, 26, 1913, p. 1031.
34. James Franck, 'Transformations of kinetic energy of free electrons into excitation energy of atoms by impacts', *Nobel Lectures: Physics 1922–1941*, Amsterdam: Elsevier, 1965, p. 104.
35. William Lawrence Bragg, 'The diffraction of X-rays by crystals', *Nobel Lectures: Physics 1901–1921*, p. 371.
36. *Proceedings of the Cambridge Philosophical Society*, 17, 1913, pp. 43–57.
37. William Lawrence Bragg, 'The diffraction of X-rays by crystals', *Nobel Lectures: Physics 1901–1921*, p. 373.
38. W. H. Bragg and W. L. Bragg, 'The reflection of X-rays by crystals', *Proceedings of the Royal Society*, A88, 1913, pp. 428–38.
39. Quoted by Sheldon L. Glashow, *From Alchemy to Quarks*, Pacific Grove, CA: Books/Cole Publishing, 1994, p. 460.
40. Robert A. Millikan, 'The electron and the light-quant from the experimental point of view', *Nobel Lectures: Physics 1922–1941*, p. 63.
41. Abraham Pais, *Subtle Is the Lord . . .* , pp. 211–12.
42. Ibid., p. 212.
43. Quoted by C. P. Snow, 'Albert Einstein 1879–1955', in A. P. French, ed., *Einstein: A Centenary Volume*, London: Heinemann Educational Books, 1979, p. 4.
44. Albert Einstein, 'Die Grundlage der allgemeinen Relativitätstheorie' *Annalen der Physik*, 49, 1916, pp. 769–822. English translation in Albert Einstein, *The Principle of Relativity*, New York: Dover Publications, 1952, pp. 111–64.
45. Quoted by Abraham Pais, *Subtle Is the Lord . . .* , p. 257.
46. Sir Isaac Newton, *Opticks*, p. 369.
47. See B. Bertotti L, Iess and P. Tortora, *Nature*, 425, 2003, pp. 374–6.
48. See Clifford M. Will, 'Einstein, relative and everyday life', in (http://www.physicscentral.com/writers/writers-00-2.html).
49. The Nobel Foundation, *Nobel Lectures: Physics 1922–1941*, p. 75.
50. Quoted by Jagdish Mehra and Helmut Rechenberg, *The Historical Development of Quantum Theory*, vol. 1, Part 1, p. 240.
51. Abraham Pais, *Subtle Is the Lord . . .* , p. 502.
52. The Nobel Foundation, *Nobel Lectures: Physics 1901–1921*, pp. 405–6.
53. Max Planck, 'The genesis and present state of development of the quantum theory', *Nobel Lectures: Physics 1901–1921*, p. 407.
54. Quoted by Jagdish Mehra and Helmut Rechenberg, *The Historical Development of Quantum Theory*, vol. 1, Part 1, p. 44.

55. Max Planck, 'The genesis and present state of development of the quantum theory', *Nobel Lectures: Physics 1901–1921*, p. 411.
56. Max Planck, *Where Is Science Going?*, Woodbridge, CT: Ox Bow Press, 1981, p. 19.
57. Ibid., pp. 19–21.
58. Emilio Segrè, *From X-rays to Quarks: Modern Physicists and Their Discoveries*, Berkeley: Emiliosegrè, University of California, 1980, p. 76.
59. Ibid., p. 76.
60. *Philosophical Magazine*, 37, 1919, pp. 581–7.
61. Quoted by David Wilson, *Rutherford: Simple Genius*, pp. 408–9.
62. Ibid., p. 411.
63. Quoted by Banesh Hoffmann, *Albert Einstein: Creator and Rebel*, 1972, p. 131.
64. Robert Marc Friedman, 'Nobel Physics Prize in perspective', *Nature*, 292, 1981, p. 795.
65. Sir E. Rutherford, 'Nuclear constitution of atoms – Bakerian Lecture', in Sir James Chadwick, ed., *The Collected Papers of Lord Rutherford*, London: George Allen & Unvoin, 1965, vol. 1, p. 34.
66. Abraham Pais, 'Introducing atoms and their nuclei', in Laurie M Brown *et al.*, eds., *Twentieth Century Physics*, Bristol, PA: Institute of Physics Publishing; New York: American Institute of Physics Press, 1995, vol. I, p. 114.
67. Abraham Pais, *Subtle Is the Lord . . .* , pp. 239–40.

Chapter 6 The golden years
1. Robert Marc Friedman, 'Nobel Physics Prize in perspective', *Nature*, 292, 1981, p. 795.
2. Ibid., p. 794.
3. Ibid., p. 795.
4. The Nobel Foundation, *Nobel Lectures: Physics 1901–1921*, p. 479.
5. Ibid., pp. 480–1.
6. Quoted by Abraham Pais, *Subtle Is the Lord . . .* , p. 503.
7. Ibid., pp. 504–5.
8. I. I. Rabi, 'Albert Einstein: 1879–1955', *Scientific American*, June 1955, p. 32.
9. Walther Gerlach and Otto Stern, 'Der experimentelle Nachweis der Richtungsquantelung', *Zeitschrift für Physik*, 9, 1922, pp. 349–52. Walther Gerlach recalled the discovery with these words: '[The experiment] started in the evening and went on during the whole night. And the next morning . . . I immediately developed [the plate] and this figure came out: absolutely nothing in the middle, only something on the right and left . . .' He immediately sent a telegram to Stern, reading: 'Bohr is right after all!' (quoted by Jagdish Mehra and Helmut Rechenberg, *The Historical Development of Quantum Theory*, vol. 1, Part 2, p. 442). See also *Physics Today*, December 2003, pp. 53–9.
10. Abraham Pais, *Niels Bohr's Times. In Physics, Philosophy, and Polity*, p. 18.
11. The Nobel Foundation, *Nobel Lectures: Physics 1922–1941*, p. 6.
12. Niels Bohr, 'The structure of the atom', *Nobel Lectures: Physics 1922–1941*, p. 43.
13. Abraham Pais, *Niels Bohr's Times, In Physics, Philosophy, and Polity*, p. 210.
14. Ibid., p. 171.
15. Niels Bohr, 'The Rutherford Memorial Lecture 1958', in Finn Aaserud, ed., *Niels Bohr: Colleted Works*, vol. 10, p. 400.
16. L. Rosenfeld, ed., *Niels Bohr: Collected Works*, vol. 1, p. XLVIII.

17. The Nobel Foundation, *Nobel Lectures: Physics 1922–1941*, p. 53.
18. Quoted by Sheldon L. Glashow, *From Alchemy to Quarks*, p. 431.
19. R. A. Millikan, 'The isolation of an ion, a precision measurement of its charge, and the correction of Stokes' Law', *Physical Review*, 32, 1911, pp. 349–97.
20. Arthur H. Compton, 'X-rays as a branch of optics', *Nobel Lectures: 1922–1941*, p. 184.
21. Quoted in *Quantum Theory Centenary*, Berlin: Deutsche Physikalische Gesellschaft, 2000, p. 102.
22. Quoted by Abraham Pais, 'George Uhlenbech and the discovery of electron spin', *Physics Today*, December 1989, p. 39.
23. W. Pauli, Über den Zusammenhang des Abschlusses der Elektronengruppen im Atom mit der Komplexstruktur der Spektren', *Zeitschrift für Physik*, 31, 1925, p. 776.
24. Werner Heisenberg, 'Über die quantentheoretische Umdeutung kinematischer und mechanischer Beziehungen', *Zeitschrift für Physik*, 33, 1925, pp. 879–93.
25. Max Born, Werner Heisenberg and Pascual Jordan, 'Zur Quantenmechanik II', *Zeitschrift für Physik*, 35, 1926, pp. 557–615.
26. Quoted by Jan Duck and E. C. G. Sudarshan, *100 years of Planck's Quantum*, Singapore: World Scientific, 2000, p. 169.
27. Jagdish Mehra and Helmut Rechenberg, *The Historical Development of Quantum Theory*, vol. 2, pp. 291, 295.
28. Robert Marc Friedman, *The Politics of Excellence: Behind the Nobel Prize in Science*, New York: Times Books, 2001, p. 159.
29. Ibid., p. 161
30. Jean Baptiste Perrin, 'Discontinuous structure of matter', *Nobel Lectures: Physics 1922–1941*, p. 156.
31. Schrödinger's first three papers were published in *Annalen der Physik*, 79, 1926: 'Quantisierung als Eigenwertproblem' (I. Mitteilung), pp. 361–76; 'Quantisierung als Eigenwertproblem' (Zweite Mitteilung), pp. 489–527; 'Über das Verhältnis der Heisenberg–Born–Jordanschen Quantenmechanik zu der meinen', pp. 734–56. Fourth paper: 'Quantisierung als Eigenwertproblem. (Dritte Mitteilung: Störungstheorie, mit Anwendung auf den Starkeffekt der Balmerlinien', 80, 1926, pp. 437–90.
32. Walter Moore, *Schrödinger: Life and Thought*, Cambridge: Cambridge University Press, 1989, p. 200.
33. Quoted by Abraham Pais, *Niels Bohr's Times. In Physics, Philosophy, and Polity*, p. 281.
34. Ibid., p. 281.
35. Quoted by Helge Kragh, *Dirac: A Scientific Biography*, Cambridge: Cambridge University Press, 1990, p. 17.
36. P. A. M. Dirac, 'The fundamental equations of quantum mechanics', *Proceedings of the Royal Society*, A109, 1925, pp. 642–53.
37. Abraham Pais, *The Genius of Science: A portrait Gallery*, p. 54.
38. Quoted by Walter Moore, *Schrödinger: Life and Thought*, p. 209.
39. Ibid., p. 209.
40. Ibid., p. 222.
41. Werner Heisenberg, 'Quantum theory and its interpretation', in S. Rozental, ed., *Niels Bohr: His Life and Work As Seen by His Friends and Colleagues*, p. 103.
42. Arthur Holly Compton and Alfred W. Simon, 'Direct quanta of scattered X-rays', *Physical Review*, 26, 1925, pp. 289–99. Compton and Simon took cloud chamber

photographs showing the recoil electron which had been produced in a Compton scattering (see Fig. 6.21). In the same year Walther Bothe and Hans Geiger performed a counter experiment in which they demonstrated that the scattered X-ray and the electron were emitted 'simultaneously'. This was a clear proof in favour of Compton's explanation of his effect, which included the photon concept; this definitely closed the door on the opponents of the quantum nature of radiation.

43. Gösta Ekspong, 'The dual nature of light as reflected in the Nobel archive' (see: www.nobel.se/physics/articles/ekspong/index.html), 1999, p. 6.
44. Arthur H. Compton, 'X-rays as a branch of optics', *Nobel Lectures: Physics 1922–1941*, p. 189. Compton published his Nobel-Prize-winning work in *Physical Review*, 21, 1923, pp. 483–502; and, 22, 1923, pp. 409–13.
45. Nobel Foundation, *Nobel Lectures: Physics 1922–1941*, p. 242.
46. Roger Penrose, 'Newton, quantum theory and reality', in S. W. Hawking and W. Israel, eds., *Three Hundred Years of Gravitation*, p. 25.
47. Clinton J. Davisson, 'The discovery of electron waves', *Nobel Lectures: 1922–1941*, p. 391.
48. Werner Heisenberg, 'Über den anschaulichen Inhalt der quantentheoretischen Kinematik und Mechanik', *Zeitschrift für Physik*, 43, 1927, pp. 172–98.
49. Werner Heisenberg, 'Quantum theory and its interpretation', in S. Rozental, ed., *Niels Bohr*, pp. 105–6.
50. C. T. R. Wilson, 'On a method of making visible the paths of ionising particles through a gas', *Proceedings of the Royal Society*, 85, 1911, p. 286. Reprinted by permission of the Royal Society, London.
51. G. P. Thomson, 'Experiments on the diffraction of cathode rays', *Proceedings of the Royal Society*, 117, 1928, p. 604. Reprinted by permission of the Royal Society, London.
52. Werner Heisenberg, *The Physical Principles of the Quantum Theory*, New York: Dover Publications, 1949, p. 8.Reprinted by permission of Dover Publications.
53. Arthur H. Compton and Alfred W. Simon, 'Directed quanta of scattered X-rays', *Physical Review*, 26, 1925, p. 292 (Fig. 3). © 1925 by the American Physical Society.
54. Werner Heisenberg, 'Quantum theory and its interpretation', in S. Rozental, ed., *Niels Bohr*, pp. 107–8.
55. Owen W. Richardson, 'Thermionic phenomena and the laws which govern them', *Nobel Lectures: Physics 1922–1941*, p. 225.
56. Ibid., p. 227.
57. Louis de Broglie, *Physics and Microphysics*, New York: Pantheon Books, 1955, p. 70.
58. Quoted by Abraham Pais, *Niels Bohr's Times. In Physics, Philosophy, and Polity*, p. 240.
59. Louis de Broglie, *Physics and Microphysics*, p. 148.
60. Sir Chandrasekhara V. Raman, 'The molecular scattering of light', *Nobel Lectures: Physics 1922–1941*, p. 267.
61. Quoted by Edoardo Amaldi, 'From the discovery of the neutron to the discovery of nuclear fission', *Physics Reports*, vol. 111, 1984, p. 76.
62. Ibid., p. 306. In the reference [277] Amaldi remarked: 'The word [neutrino] came out in a humorous conversation at the Istituto di Via Panisperna. Fermi, Amaldi and a few others were present and Fermi was explaining Pauli's hypothesis about his "little neutron". For distinguishing this particle from the Chadwick neutron Amaldi jokingly used this funny name . . .'.

63. Victor F. Weisskopf, *Physics in the Twentieth Century: Selected Essays*, Cambridge, MA; London: The MIT Press, 1972, p. 55.

Chapter 7 The thirties

1. P. A. M. Dirac, 'The quantum theory of the emission and absorption of radiation' *Proceedings of the Royal Society*, A114, 1927, pp. 243–65.
2. P. A. M. Dirac, 'The quantum theory of the electron', *Proceedings of the Royal Society*, A117, 1928, pp. 610–24.
3. Walter Moore, *Schrödinger: Life and Thought*, p. 289.
4. Quoted by Jagdish Mehra and Helmut Rechenberg, *The Historical Development of Quantum Theory*, vol. 2, p. 12.
5. Ibid., p. 64.
6. Ibid., p. 72.
7. Abraham Pais, *Niels Bohr's Times. In Physics, Philosophy, and Polity*, p. 276.
8. Ibid., p. 275.
9. Quoted by Jagdish Mehra and Helmut Rechenberg, *The Historical Development of Quantum Theory*, vol. 3, pp. 7–9.
10. Felix Bloch, 'Heisenberg and the early days of quantum mechanics', *Physics Today*, December 1976, p. 27.
11. Chadwick published his discovery of the neutron first in *Nature*, 219, 1932, p. 312; and then in J. Chadwick, 'The existence of a neutron', *Proceedings of the Royal Society*, A136, 1932, pp. 692–708.
12. Cockcroft and Walton's experiment was described in two papers published in *Proceedings of the Royal Society*, A136, 1932, pp. 619–30; A137, 1932, pp. 229–42.
13. The first announcement of the existence of the positron was given by Carl Anderson in an article published in *Science*, 76, 1932, p. 238. He then published a more extensive paper: Carl D. Anderson, 'The positive electron', *Physical Review*, 43, 1933, pp. 491–4.
14. P. M. S. Blackett and G. P. S. Occhialini, 'Some photographs of the tracks of penetrating radiation', *Proceedings of the Royal Society*, A139, 1933, pp. 699–726.
15. Walter Moore, *Schrödinger: Life and Thought*, pp. 194–6.
16. Ibid., p. 3.
17. Robert L. Weber, *Pioneers of Science: Nobel Prize Winners in Physics*, p. 100.
18. Quoted by Walter Moore, *Schrödinger: Life and Thought*, p. 6.
19. Ibid., pp. 290–1.
20. Fritz Stern, *Einstein's German World*, London: Penguin Books, 2001, pp. 152–3.
21. F. Joliot and I. Curie, 'Artificial production of a new kind of radio-element', *Nature*, 133, 1934, pp. 201–2.
22. The Nobel Foundation, presentation speech, Nobel for Chemistry 1935. (See www.nobel.se/laureates/1935/press.html.)
23. Emilio Segrè, *From X-rays to Quarks*, p. 204.
24. Ibid., pp. 204–5. Edoardo Amaldi, in his paper entitled 'From the discovery of the neutron to the discovery of nuclear fission' (*Physics Reports*, vol. III, 1984, p. 152), recalled the discovery of slow neutrons with these words: 'On the morning of October 22 most of us were busy doing examinations and Fermi decided to proceed in making the measurements. Bruno Rossi from the University of Padua and Enrico Persico [a bosom friend of Fermi's] from the University of Turin were around in the

[institute] and Persico was, I believe, the only eyewitness of what happened. At the moment of using lead, Fermi suddenly decided to try with a wedge of some light element, and paraffin was used first. The results of the measurements are recorded on pages 8 and 9 of the same note-book B1 . . . They are written at the beginning by Fermi and toward the end by Persico. Toward noon we were all summoned to watch the extraordinary effect of the filtration by paraffin: the activity was increased by an appreciable factor.'

25. Quoted by Walter Gratzer, *Eurekas and Euphorias: The Oxford Book of Scientific Anecdotes*, New York: Oxford University Press, 2002, p. 47.
26. The letter from Rutherford to Fermi is reproduced in Edoardo Amaldi, 'From the discovery of the neutron to the discovery of nuclear fission', *Physics Reports*, vol. III, 1984, p. 131.
27. George Gamow, *Thirty Years That Shook Physics: The Story of Quantum Physics*, New York: Dover Publications, 1966, p. 213.
28. James Chadwick, 'The neutron and its properties', *Nobel Lectures: Physics 1922–1941*, pp. 339–40.
29. Quoted by Mark Oliphant, *Rutherford: Recollections of the Cambridge Days*, Amsterdam: Elsevier, 1972, pp. 75–6.
30. Ibid., pp. 76–7.
31. Fermi sent his paper entitled 'Attempt at a theory of emission of beta rays' to *La Ricerca Scientifica*, the journal of the Italian National Research Council, where it appeared in the issue of December 1933. He also sent a short note to *Nature*, but the editor refused it because he thought it 'contained speculations that were too remote from physical reality'. The German translation of the Italian paper was then published in the *Zeitschrift für Physik*, 88, 1934, pp. 161–71.
32. Carl David Anderson, *The Discovery of Anti-matter*, Singapore; New Jersey; London; Hong Kong: World Scientific, 1999, p. 34.
33. Ibid., p. 35.
34. Clinton J. Davisson, 'The discovery of electron waves', *Nobel Lectures: Physics 1922–1941*, p. 390. Davisson and Germer presented their first results in a short paper published in the April issue of *Nature*, 119, 1927, pp. 558–60. Thomson and Reid, however published their first note on electron diffraction in the June issue of *Nature*, 119, 1927, p. 890.
35. Emilio Segrè, *From X-rays to Quarks*, p. 118.
36. Samuel Devons, 'Recollections of Rutherford and the Cavendish', *Physics Today*, December 1971, p. 39.
37. Emilio Segrè, *From X-rays to Quarks*, p. 205–6.
38. Ruth Moore, 'Niels Bohr as a political figure', in A. P. French and P. J. Kennedy, eds., *Niels Bohr: A Centenary Volume*, Cambridge, MA; London: Harvard University Press, 1985, p. 255.
39. Valentine L. Telegdi, 'Enrico Fermi in America', *Physics Today*, June 2002, p. 38.
40. Emilio Segrè, *Enrico Fermi: Physicist*, Chicago: University of Chicago Press, 1970, p. 56.
41. Quoted by Emilio Segrè, *Enrico Fermi: Physicist*, pp. 59–60.
42. Jack Steinberger, 'Enrico Fermi, my master and teacher', in *Il nuovo saggiatore*, Bologna, Italy: Editrice Compositori, 17, 2001, pp. 38–9.

43. *Science*, 72, 1930, p. 376.
44. The description of Lawrence and Livingston's cyclotron appeared in Ernest O. Lawrence and M. Stanley Livingston, 'The production of high speed light ions without the use of high voltages', *Physical Review*, 40, 1932, pp. 19–35.
45. Ernest O. Lawrence, 'The evolution of the cyclotron', The Nobel Foundation, *Nobel Lectures: Physics 1942–1962*, Amsterdam: Elsevier, 1964, pp. 430–1.
46. Otto R. Frisch, 'How it all began', in Otto R. Frisch and John A. Wheeler, 'The discovery of fission', *Physics Today*, November 1967, p. 47.
47. Ibid., p. 47.
48. Emilio Segrè, *From X-rays to Quarks*, p. 208.
49. Patricia Rife, *Lise Meitner and the Dawn of the Nuclear Age*, Boston: Birkhäuser, 1999, pp. 225–6.
50. C. P. Snow, *The Physicists*, London; Basingstoke: Macmillan, 1981, pp. 79–80.
51. Emilio Segrè, *Enrico Fermi: Physicist*, p. 53.

Chapter 8 The nuclear age

1. Quoted by Sheldon L. Glashow, *From Alchemy to Quarks*, pp. 566–7.
2. Emilio Segrè, *From X-rays to Quarks*, p 195.
3. Victor F. Weisskopf, *Physics in the Twentieth Century: Selected Essays*, pp. 5–6. (Note that one *nuclear magneton* is equal to 1/1840 of a *Bohr magneton*.)
4. Abraham Pais, *The Genius of Science: A portrait Gallery*, New York: Oxford University Press, 1992, p. 217.
5. Ibid., pp. 216–17.
6. Wolfgang Pauli, 'Exclusion principle and quantum mechanics' *Nobel Lectures: Physics 1942–1962*, pp. 27–8.
7. Ibid., p. 29.
8. Quoted by Jagdish Mehra and Helmut Rechenberg, *The Historical Development of Quantum Theory*, vol. 1, Part 2, p. 377.
9. Abraham Pais, *The Genius of Science: A portrait Gallery*, p. 215.
10. Victor F. Weisskopf, 'Personal memories of Pauli', *Physics Today*, December 1985, p. 41.
11. Victor F. Weisskopf, *Physics in the Twentieth Century: Selected Essays*, pp. 10–2.
12. Patrick M. S. Blackett, 'Cloud chamber researches in nuclear physics and cosmic radiation', *Nobel Lectures: Physics 1942–1962*, pp. 97–8.
13. Ibid., pp. 104–5.
14. Val L. Fitch, 'Elementary particle physics: The origins', *Reviews of Modern Physics*, 71, 1999, p. S26.
15. Richard Rhodes, ed., *Visions of Technology*, pp. 185–6.
16. Hideki Yukawa, 'Meson theory in its developments', *Nobel Lectures: Physics 1942–1962*, p. 129.
17. Hideki Yukawa, *Tabibito* [*The Traveller*], Singapore: World Scientific, 1982, p. 202.
18. Ibid., p. 203.
19. C. M. G. Lattes, H. Muirhead, G. P. S. Occhialini, and C. F. Lattes, 'Processes involving charged mesons', *Nature*, 159, 1947, pp. 694–7.

Chapter 9 Wave of inventions

1. John D. Cockcroft, 'Experiments on the interaction of high-speed nucleons with atomic nuclei', *Nobel Lectures: Physics 1942–1962*, p. 167.
2. Mark Oliphant, *Rutherford: Recollections of the Cambridge Days*, pp. 85–6.
3. C. P. Snow, *The Physicists*, p. 89.
4. Edward M. Purcell, 'Research in nuclear magnetism', *Nobel Lectures: Physics 1942–1962*, p. 219.
5. Felix Bloch, 'The principle of nuclear induction', *Nobel Lectures: Physics 1942–1962*, pp. 209–10.
6. The Nobel Foundation, *Nobel Lectures, Physics 1942–1962*, p. 237.
7. Gösta Ekspong, ed., *Nobel Lectures: Physics 1996–2000*, Singapore: World Scientific, 2002, p. 347.
8. Frits Zernike, 'How I discovered phase contrast', *Nobel Lectures: Physics 1942–1962*, p. 240.
9. Gösta Ekspong, ed., *Nobel Lectures: Physics 1996–2000*, p. 347.
10. James D. Watson, *The Double Helix*, London: Penguin Books, 1999, pp. 19–20.
11. Ibid., pp. 171–2.
12. Abraham Pais, *The Genius of Science*: *A Portrait Gallery*, p. 37,
13. Ibid., p. 37.
14. Ibid., p. 42.
15. Quoted by Jagdish Mehra and Helmut Rechenberg, *The Historical Development of Quantum Theory*, vol. 3, p. 12.
16. Walther Bothe, 'The coincidence method', *Nobel Lectures: Physics 1942–1962*, pp. 272–3.
17. Samuel K. Allison, 'Enrico Fermi: 1901–1954', *Physics Today*, January 1955, pp. 9–10.
18. Willis E. Lamb, Jr., 'Fine structure of the hydrogen atom', *Nobel Lectures: Physics 1942–1962*, p. 286.
19. Ibid., pp. 287–8.
20. Polykarp Kusch, 'The magnetic moment of the electron', *Nobel Lectures: Physics 1942–1962*, p. 302.
21. Abraham Pais, *Subtle is the Lord . . .* , pp. 476–7.
22. Quoted from William F. Brinkman, 'The Transistor' (see: www.lucent.com/minds/transistor/brinkman/content/brink1.html).
23. Michael Riordan, Lillian Hoddeson and Conyers Herring, 'The invention of the transistor', *Reviews of Modern Physics*, 71, 1999, p. S340.
24. Sir Brian Pippard, 'Electrons in solids', in Laurie M Brown *et al.*, eds., *Twentieth Century Physics*, vol. III, p. 1324.
25. Gösta Ekspong, ed., The Nobel Foundation *Nobel Lectures*: *Physics 1991–1995*, Singapore: World Scientific, 1997, p. 199.
26. Quoted in *Physics Today*, December 1995, p. 19.
27. Ibid., p. 19.
28. T. D. Lee, 'Elementary particles', *Physics Today*, October 1960, p. 34.
29. Chen Ning Yang, 'The law of parity conservation and other symmetry laws of physics', *Nobel Lectures: Physics 1942–1962*, p. 398.
30. Ibid., p. 399.

31. Ibid., p. 400.
32. Richard P. Feynman *et al.*, *The Feynman Lectures on Physics*, Reading, MA: Addison-Wesley, 1996, Vol.1, p. 52/9.
33. Tsung Dao Lee, 'Weak interactions and nonconservation of parity' *Nobel Lectures: Physics 1942–1962*, p. 417.
34. The Nobel Foundation, *Nobel Lectures: Physics 1942–1962*, p. 419.
35. Chen Ning Yang, *Selected Papers 1945–1980 With Commentary*, San Francisco: W. H. Freeman and Company, 1983, p. 4.
36. Ibid., p. 30.
37. Abraham Pais, *The Genius of Science*: *A Portrait Gallery*, p. 174.
38. George C. Sponsler, 'Sputniks over Britain', *Physics Today*, July 1958, p. 16.
39. Owen Chamberlain, 'The early antiproton work', *Nobel Lectures: Physics 1942–1962*, p. 492. The discovery of the antiproton was announced in O. Chamberlain, E. Segrè, C. Wiegand and T. Ypsilantis, *Physical Review*, 100, 1955, pp. 947–50.
40. The Nobel Foundation, *Nobel Lectures: Physics 1942–1962*, p. 488. Soon after the first antiproton experiment, other experiments were carried out using the Berkeley Bevatron. These experiments used photographic emulsion techniques and confirmed *ad oculos* the existence of the antiproton. Thus Segrè recalled these experiments in his Nobel lecture: 'In the early investigations with photographic emulsions carried out in my group especially by Gerson Goldhaber and by a group in Rome led by [Edoardo] Amaldi, we soon found stars [of particles] . . . giving conclusive evidence of the annihilation in pairs of proton and antiproton.' (See: Emilio G. Segrè, 'Properties of antinucleons', *Nobel Lectures: Physics 1942–1962*, pp. 515–16.)
41. Emilio G. Segrè, 'Properties of antinucleons', *Nobel Lectures: Physics 1942–1962*, p. 508.
42. The antineutron was observed soon after the antiproton in an experiment carried out again at the Berkeley Bevatron. Its discovery was announced in a paper published in *Physical Review*, 104, 1956, pp. 1193–7.
43. Donald A. Glaser, 'Elementary particles and bubble chambers', *Nobel Lectures: Physics 1942–1962*, p. 538.
44. Source: Theodore Maiman, *The Laser Odyssey*, Laser Press, 2001.
45. Lillian Hartmann Hoddeson, 'The roots of solid-state research at Bell Labs', *Physics Today*, March 1977, p. 26.
46. Ibid., p. 30.

Chapter 10 New vistas on the cosmos

1. Robert Hofstadter, 'The electron-scattering method and its application to the structure of nuclei and nucleons', *Nobel Lectures: Physics 1942–1962*, p. 579.
2. Gennady Gorelik, 'The top-secret life of Lev Landau', *Scientific American*, August 1997, p. 54.
3. Ibid., p. 52.
4. The discovery of the muon neutrino was announced in G. Danby, J.-M. Gaillard, K. Goulianos, L. M. Lederman, N. Mistry, M. Schwartz, and J. Steinberger, 'Observation of high-energy neutrino reactions and the existence of two kinds of neutrinos', *Physical Review Letters*, 9, 1962, pp. 36–44.
5. Abraham Pais, *Niels Bohr's Times*. In *Physics, Philosophy, and Polity*, p. 536.

6. Niels Bohr, *Nature*, 140, 1937, p. 753.
7. Erich Vogt, 'Eugene Paul Wigner: a towering figure of modern physics', *Physics Today*, December 1995, p. 42.
8. The Nobel Foundation, *Nobel Lectures: Physics 1963–1970*, Singapore: World Scientific, 1998; p. 4.
9. Erich Vogt, 'Eugene Paul Wigner: a towering figure of modern physics', *Physics Today*, December 1995, p. 40.
10. Quoted by Mario Bertolotti, *Masers and Laser: An Historical Approach*, Bristol; Philadelphia, PA: Adam Hilger, 1983, pp. 76–7.
11. Quoted in W. E. Lamb, W. P. Schleich, M. O. Scully and C. H. Townes, 'Laser physics: quantum controversy in action', *Reviews of Modern Physics*, 71, 1999, p. S266.
12. Ibid., p. S266.
13. Quoted from 'Laser – the invention of the LASER at Bell Labs: 1958–1988' (see: www.bell-labs.com/history/laser/invention/).
14. *Physical Review*, 112, 1958, pp. 1940–9.
15. Orazio Svelto, one of the foremost experts in the field of optical electronics, is full professor of quantum electronics at the Polytechnic of Milan, Italy. In 1998 he has received the prestigious Quantum Electronics Prize of the Europhysics Society for his invention of the hollow fibre shown in Fig. 10.7.
16. Yuval Ne'eman and Yoram Kirsh, *The Particle Hunters*, Cambridge: Cambridge University Press, 1996, p. 59.
17. Feynman's paper containing the description of quantum electrodynamics in space-time (the so-called 'Feynman diagrams') was published in *Physical Review*, 76, 1949, pp. 769–89.
18. Val L. Fitch and Jonathan L. Rosner, 'Elementary particle physics in the second half of the twentieth century', in Laurie M Brown *et al.*, eds., *Twentieth Century Physics*, vol. II, p. 647.
19. Ibid., p. 647.
20. See F. J. M. Farley and E. Picasso, in T. Kinoshita, ed., *Quantum Electrodynamics*, Singapore: World Scientific, 1990, Ch. 11.
21. See *Physics Today*, March 2004, p. 9.
22. Paul Davies, 'Introduction', in Richard P. Feynman, *The Character of Physical Law*, Harmondsworth, Middlesex: Penguin Books, 1992, p. 9.
23. Silvan S. Schweber, *QED and the Men Who Made It: Dyson, Feynman, Schwinger, and Tomonaga*, Princeton, NJ: Princeton University Press, 1994, pp. 572–3.
24. Robert Marc Friedman, *The Politics of Excellence: Behind the Nobel Prize in Science*, New York: Times Books, 2001, p. 148.
25. The Nobel Foundation, *Nobel Lectures: Physics 1963–1970*, Singapore: World Scientific, 1998, p. 214.
26. Emilio Segrè, *Enrico Fermi: Physicist*, p. 59.
27. Luis W. Alvarez, 'Recent developments in particle physics', *Nobel Lectures: Physics 1963–1970*, p. 241.
28. Ibid., pp. 248–9.
29. *Physics Today*, August 2000, p. 43.
30. George Johnson, *Strange Beauty: Murray Gell-Mann and the Revolution in Twentieth-Century Physics*, New York: Alfred A. Knopf, 2000, pp. 222–3.

31. The concept of 'strangeness' was presented by Gell-Mann in *Physical Review*, 92, 1953, pp. 833–4.
32. The eightfold-way scheme was presented by Gell-Mann in *Physical Review*, 125, 1962, pp. 1067–84.
33. Geoffrey F. Chew, Murray Gell-Mann, and Arthur H. Rosenfeld, 'Strongly interacting particles', *Scientific American*, February 1964, p. 89.
34. The concept of quark was presented by Gell-Mann in *Physics Letters*, 8, 1964, pp. 214–15.
35. Murray Gell-Mann, *The Quark and the Jaguar, Adventures in the Simple and the Complex*, London: Abacus, 1995, pp. 180–1.

Chapter 11 The small, the large – the complex

1. Quoted from 'Transistorized', Part 3, p. 1. (See www.pbs.org/transistor/album1/bardeen/bardeen3.html.)
2. Ibid., Part 2, p. 2.
3. John Bardeen, 'Electron–phonon interactions and superconductivity', *Nobel Lectures: Physics 1971–1980*, pp. 54–5.
4. J. R. Schrieffer, 'Macroscopic quantum phenomena from pairing in superconductors', The Nobel Foundation *Nobel Lectures: Physics 1971–1980*, Stig Lundqvist, ed., Singapore: World Scientific, pp. 100–1.
5. *Physical Review*, 108, 1957, pp. 1175–204.
6. *Physics Letters*, 1, 1962, pp. 251–3.
7. Donald G. McDonald, 'The Nobel laureate versus the graduate student', *Physics Today*, July 2001, pp. 50–1.
8. Martin Ryle, 'Radio telescopes of large resolving power', *Nobel Lectures: Physics 1971–1980*, p. 187.
9. Antony Hewish, 'Pulsars and high density physics', *Nobel Lectures: Physics 1971–1980*, p. 175.
10. Ibid., pp. 175–7.
11. Ibid., pp. 177–8.
12. Quoted by Paul Gorenstein and Wallace Tucker, 'Supernova remnants', *Scientific American*, July 1971, p. 77
13. Aage Bohr, 'Rotational motion in nuclei', *Nobel Lectures: Physics 1971–1980*, p. 215.
14. Ibid., pp. 215–16.
15. Stig Lundqvist, ed., *Nobel Lectures: Physics 1971–1980*, p. 211.
16. Ibid., p. 211.
17. Ibid. p. 235.
18. Richter's discovery was published in J.-E. Augustin *et al.*, 'Discovery of a narrow resonance in $e^+ e^-$ annihilation', *Physical Review Letters*, 33, 1974, pp. 1406–8. Ting's discovery was published in J. J. Aubert *et al.*, 'Experimental observation of a heavy particle J', *Physical Review Letters*, 33, 1974, pp. 1404–6.
19. The electron–positron collider ADONE in Frascati, Italy, had been designed for a maximum energy value of three billion electron-volts (only three per cent less than the energy needed for 'seeing' the J/ψ). During the night of 14 November 1974 the Frascati experimenters pushed the ADONE energy up to 3.1 billion electron-volts ('by running all their magnets hot'), and so they were able to 'see' the resonance peak of the J/ψ too!

20. Stig Lundqvist, ed., *Nobel Lectures: Physics 1971–1980*, Singapore: World Scientific, 1992, pp. 351–2.
21. Lillian Hoddeson *et al.*, eds., *Out of the Crystal Maze: Chapters from the History of Solid State Physics*, New York: Oxford University Press, 1992, p. 195.
22. Pierre-Gilles de Gennes, 'Les surprises du désordre', *La Recherche*, December 1977, p. 1082.
23. Silvan S. Schweber, 'Physics, community and the crisis in physical theory', *Physics Today*, November 1993, pp. 34–5.
24. P. W. Anderson, 'More is different', *Science*, 177, 1972, p. 393.
25. Grace Marmor Spruch, 'Pyotr Kapitza, octogenarian dissident', *Physics Today*, September 1979, pp. 34, 36.
26. A. A. Penzias and R. W. Wilson, 'A measurement of excess antenna temperature at 4080 Mc/s', *Astrophysical Journal*, 142, 1965, pp. 419–20.
27. R. H. Dicke, P. J. E. Peebles, P. G. Roll and D. T. Wilkinson, 'Cosmic black-body radiation', *Astrophysical Journal*, 142, 1965, p. 416.
28. *Nuclear Physics*, 22, 1961, pp. 579–88.
29. *Physical Review Letters*, 19, 1967, pp. 1264–6 (Weinberg); *Proceedings of the Eighth Nobel Symposium*, 1968, pp. 367–77 (Salam).
30. *Physical Review*, D2, 1970, pp. 1285–92.
31. Stig Lundqvist, ed., *Nobel Lectures: Physics 1971–1980*, p. 491.
32. Ibid., p. 491.
33. Ibid., p. 507.
34. Quoted by Z. Hassan and C. H. Lai, eds., *Ideals and Realities: Selected Essays of Abdus Salam*, Singapore: World Scientific, 1984, pp. 116–17.
35. James W. Cronin, 'CP symmetry violation: the search for its origin', *Nobel Lectures: Physics 1971–1980*, p. 570.
36. James W. Cronin and Margaret Stautberg Greenwood, 'CP symmetry violation', *Physics Today*, July 1982, p. 42.
37. The history of electron and proton colliders (which began early in the 1960s) is described by Burton Richter, 'The rise of colliding beams', in Hoddeson, Lillian *et al.*, eds., *The Rise of the Standard Model,: Particle Physics in the 1960s and 1970s*, Cambridge: Cambridge University Press, 1999, pp. 251–84. See also: Fernando Amman, *The Early Times of Electron Colliders*, Singapore: World Scientific, 1989, pp. 449–76.

Chapter 12 Big physics – small physics
1. Nicolaas Bloembergen, 'Nonlinear optics and spectroscopy', *Nobel Lectures: Physics 1981–1990*, Singapore: World Scientific, 1993, p. 29.
2. Quoted in *Physics Today*, December 1981, p. 18.
3. Gösta Ekspong, ed., The Nobel Foundation *Nobel Lectures: Physics 1981–1990*, Singapore: World Scientific, 1993, p. 8.
4. Quoted from the Nobel Foundation 1981 press release (see: www.nobel.se/physics/laureates/1981/press.html).
5. Gösta Ekspong, ed., *Nobel Lectures: Physics 1981–1990*, p. 34.
6. Ibid., p. 34.
7. Quoted from the Nobel Foundation 1981 press release (see: www.nobel.se/physics/laureates/1981/press.html).

8. Quoted in *Physics Today*, December 1982, p. 17.
9. Quoted from the Nobel Foundation 1984 press release (see: www.nobel.se/physics/laureates/1984/press.html).
10. See *CERN Courier*, November 1983, p. 361.
11. Ibid., p. 361.
12. Ibid., p. 361.
13. *Physics Letters*, 122B, 1983, pp. 103–16.
14. Gösta Ekspong, ed., *Nobel Lectures: Physics 1981–1990*, p. 237.
15. Ibid., pp. 289.
16. Klaus von Klitzing, 'The quantized Hall effect', *Nobel Lectures: Physics 1981–1990*, p. 316.
17. G. Binnig and H. Rohrer, 'In touch with atoms', *Reviews of Modern Physics*, 71, 1999, pp. S324–30.
18. Gerd Binnig and Heinrich Rohrer, 'Scanning tunneling microscopy: from birth to adolescence', *Nobel Lectures: Physics 1981–1990*, p. 393.
19. Gösta Ekspong, ed., *Nobel Lectures: Physics 1981–1990*, pp. 383–4.
20. Ibid., p. 388.
21. Gerd Binnig and Heinrich Rohrer, 'Scanning tunneling microscopy: from birth to adolescence', *Nobel Lectures: Physics 1981–1990*, p. 389.
22. *Physics Today*, December 1987, p. 17.
23. Gösta Ekspong, ed., *Nobel Lectures: Physics 1981–1990*, p. 417.
24. Ibid., p. 421.
25. J. G. Bednorz and K. A. Müller, 'Possible high T superconductivity in the Ba–La–Cu–O system', *Zeitschrift für Physik B, Condensed Matter*, vol. 64, 1986, pp. 189–93.
26. A. J. Leggett, 'Superfluids and superconductors', in Laurie M Brown *et al.*, eds., *Twentieth Century Physics*, vol. II, p. 959.
27. John R. Kirtley and Chang C. Tsuei, 'Probing high-temperature superconductivity', *Scientific American*, August 1996, p. 50.
28. Melvin Schwartz, 'The first high energy neutrino experiment', *Nobel Lectures: Physics 1981–1990*, pp. 469–70.
29. Gösta Ekspong, ed., *Nobel Lectures: Physics 1981–1990*, p. 509.
30. Ibid., pp. 509–10.
31. Ibid., p. 465.
32. Ibid., p. 466.
33. Ibid., pp. 481–2.
34. Wayne M. Itano and Norman F. Ramsey, 'Accurate measurement of time', *Scientific American*, July 1993, p. 53.
35. Gösta Ekspong, ed., *Nobel Lectures: Physics 1981–1990*, p. 550.

Chapter 13 New trends
1. *Physics Today*, December 1991, p. 17.
2. Quoted in *The Nobel Prize Annual: 1992*, New York: International Merchandising Corporation, 1993, p. 36.
3. Georges Charpak, 'Electronic imaging of ionising radiation with limited avalanches in gases', *Nobel Lectures: Physics 1991–1995*, Singapore: World Scientific, 1993, pp. 23–4.
4. Quoted in *The Nobel Prize Annual: 1992*, p. 40.

5. Ibid., p. 40.
6. *Astrophysical Journal*, 195, 1975, p. L51–3.
7. Gösta Ekspong, ed., *Nobel Lectures: Physics 1991–1995*, p. 144.
8. Quoted in *Physics Today*, December 1996, p. 17.
9. Gösta Ekspong, ed., The Nobel Foundation, *Nobel Lectures: Physics 1996–2000*, p. 9.
10. Ibid., p. 89
11. Gösta Ekspong, ed., *Nobel Lectures: Physics 1996–2000*, p. 115.
12. Ibid., p. 120.
13. Ibid., pp. 193–4.
14. Ibid., p. 161.
15. Ibid., p. 161.
16. Ibid., pp. 161–2.
17. D. C. Tsui, H. L. Störmer and A. C. Gossard were the authors of the prizewinning paper published in *Physical Review Letters*, 48, 1982, pp. 1559–62.
18. Laughlin's paper was published in *Physical Review Letters*, 50, 1983, pp. 1395–8.
19. Gösta Ekspong, ed., *Nobel Lectures: Physics 1996–2000*, p. 327.
20. Ibid., pp. 328–9.
21. Ibid., p. 255
22. Quoted in *Physics Today*, December 1998, p. 18.
23. *Physics Today*, December 1999, p. 17.
24. Gösta Ekspong, ed., *Nobel Lectures: Physics 1996–2000*, p. 378.
25. Martinus Veltman, 'The path to renormalizability', in Lillian Hoddeson *et al.*, eds., *The Rise of the Standard Model: Particle Physics in the 1960s and 1970s*, p. 156.
26. Gerard 't Hooft, 'Renormalization of gauge theories', in Lillian Hoddeson *et al.*, eds., *The Rise of the Standard Model: Particle Physics in the 1960s and 1970s*, p. 185.
27. Gösta Ekspong, ed., *Nobel Lectures: Physics 1996–2000*, p. 347.
28. Ibid., p. 374.
29. Steven Weinberg, *Dreams of a Final Theory*, pp. 3, 4, 6, 18.
30. Jack S. Kilby, 'Invention of the integrated circuit', *IEEE Transactions on Electronic Devices*, ED-23, 1976, p. 650.
31. Quoted from www.nobel.se/physics/laureates/2000/public.html.
32. See *New Scientist*, 29 January 2004.
33. Quoted from web.mit.edu/newsoffice/nr/1997/43242.html.
34. Daniel Kleppner, 'The yin and the yan of hydrogen', *Physics Today*, April 1999, pp. 11–13.

Select bibliography

Aaserud, Finn, ed., *Niels Bohr: Collected Works, Volume 10: Complementary Beyond Physics (1928–1962)*, Amsterdam: Elsevier, 1999.

Brown, Laurie M, Pais, Abraham and Pippard, Sir Brian, eds., *Twentieth Century Physics*, vols. I, II, III, Bristol; Philadelphia, PA: Institute of Physics Publishing; New York: American Institute of Physics Press, 1995.

Chadwick, Sir James, ed., *The Collected Papers of Lord Rutherford of Nelson*, vols. 1, 2, 3, London: George Allen & Unwin Ltd, 1965.

Crawford, Elisabeth, *The Beginnings of the Nobel Institution: The Science Prizes, 1901–1915*, Cambridge: Cambridge University Press, 1984.

Dahl, Per F, *Flash of the Cathode Rays: A History of J. J. Thomson's Electron*, Bristol; Philadelpia, PA: Institute of Physics Publishing, 1997.

Davies, Paul, ed., *The New Physics*, Cambridge: Cambridge University Press, 1989.

Einstein, Albert, *The Collected Papers of Albert Einstein* (Translation), vols. 2, 5, Princeton, NJ: Princeton University Press, 1993.

Feynman, Richard P., Leighton, Robert B. and Sands, Matthew, *The Feynman Lectures on Physics*, vols.1, 2, 3, Reading, MA: Addison-Wesley, 1966.

Gell-Mann, Murray, *The Quark and the Jaguar: Adventures in the Simple and the complex*, London: Abacus, 1995.

Glashow, Sheldon L., *From Alchemy to Quarks*, Pacific Grove, CA: Brooks/Cole Publishing, 1994.

Hoddeson, Lillian, Brown, Laurie, Riordan, Michael and Dresden, Max, eds., *The Rise of the Standard Model: Particle Physics in the 1960s and 1970s*, Cambridge: Cambridge University Press, 1999.

Hoddeson, Lillian, Braun, Ernest, Teichmann, Jürgen and Weart, Spencer, eds., *Out of the Crystal Maze: Chapters from the History of Solid State Physics*, New York: Oxford University Press, 1992.

Mehra, Jagdish and Helmut, Rechenberg, *The Historical Development of Quantum Theory*, vols. 1 (Part 1), 1 (Part 2), 2, 3, 4, 5, New York: Springer-Verlag, 1962.

Moore, Walter, *Schrödinger: Life and Thought*, Cambridge: Cambridge University Press, 1989.

Ne'eman, Yuval and Kirsh, Yoram, *The Particle Hunters*, Cambridge: Cambridge University Press, 1996.

Oliphant, Mark, *Rutherford: Recollections of the Cambridge Days*, Amsterdam: Elsevier, 1972.

Pais, Abraham, *Niels Bohr's Times. In Physics, Philosophy, and Polity*, Oxford: Clarendon Press, 1991.

Pais, Abraham, *'Subtle Is the Lord . . .': The Science and the Life of Albert Einstein*, New York: Oxford University Press, 1982.

Pais, Abraham, *The Genius of Science: A Portrait Gallery*, New York: Oxford University Press, 1992.

Rosenfeld, L., ed., *Niels Bohr: Collected Works, Volume 1. Early Work (1905–1911)*, Amsterdam: North-Holland, 1972.

Rosenfeld, L., ed., *Niels Bohr: Collected Works, Volume 2: Work on Atomic Physics (1912–1917)*, Amsterdam: North-Holland, 1981.

Rozental, S., ed., *Niels Bohr: His Life and Work As Seen by His Friends and Colleagues*, Amsterdam: North-Holland, 1967.

Schilpp, P. A. ed., *Albert Einstein: Philosopher–Scientist*, New York: Tudor, 1949.

Schweber, Sylvan S., *QED and the Men Who Made It: Dyson, Feynman, Schwinger, and Tomonaga*, Princeton: Princeton University Press, 1994.

Scurlock, G. Ralph, *History and Origins of Cryogenics*, Oxford: Clarendon Press, 1992.

Segrè, Emilio, *From X-rays to Quarks: Modern Physicists and their Discoveries*, New York: W. H. Freeman, 1980.

Segrè, Emilio, *Enrico Fermi: Physicist*, Chicago: University of Chicago Press, 1970.

Snow, C. P., *The Physicists*, London; Basingstoke: Macmillan, 1981.

The Nobel Foundation, *Nobel Lectures: Physics 1901–1921*, Amsterdam: Elsevier, 1967.

The Nobel Foundation, *Nobel Lectures: Physics 1922–1941*, Amsterdam: Elsevier, 1965.

The Nobel Foundation, *Nobel Lectures: Physics 1942–1962*, Amsterdam: Elsevier, 1964.

The Nobel Foundation, *Nobel Lectures: Physics 1963–1970*, Singapore: World Scientific, 1998.

The Nobel Foundation, *Nobel Lectures: Physics 1971–1980*, Stig Lundqvist, ed., Singapore: World Scientific, 1992.

The Nobel Foundation, *Nobel Lectures: Physics 1981–1990*, Gösta Ekspong, ed., Singapore: World Scientific, 1993.

The Nobel Foundation, *Nobel Lectures: Physics 1991–1995*, Gösta Ekspong, ed., Singapore: World Scientific, 1997.

The Nobel Foundation, *Nobel Lectures: Physics 1996–2000*, Gösta Ekspong, ed., Singapore: World Scientific, 2002.

The Nobel Foundation, *Nobel: The Man and His Prizes*, W. Odelberg, ed., New York: American Elsevier, 1972.

Weinberg, Steven, *Dreams of a Final Theory: The Scientist's Search for the Ultimate Laws of Nature*, New York: Vintage Books, 1992.

Weisskopf, Victor F., *Physics in the Twentieth Century: Selected Essays*, Cambridge, MA; London: The MIT Press, 1972.

Further reading

Aczél, Amir D. *God's Equation: Einstein, Relativity, and the Expanding Universe*, New York: Delta Publishing, 1999.

Crease, Robert P. and Mann, Charles C., *The Second Creation: Makers of the Revolution in Twentieth-Century Physics*, New York: Macmillan, 1986.

Davies, P. C. W. and Brown, Julian, eds., *Superstrings: A Theory of Everything?*, Cambridge: Cambridge University Press, 2000.

Hawking, Stephen, *A Brief History of Time*, New York: Bantam, 1988.

Lederman, Leon and Schramm, David N., *From Quarks to the Cosmos: Tools of Discovery*, New York: The Scientific American Library, 1989.

Lederman, Leon and Teresi, Dick, *The God Particle*, New York: Dell, 1993.

Pais, Abraham, *Inward Bound of Matter and Forces in the Physical World*, Oxford: Clarendon Press, New York: Oxford University Press, 1986.

Riordan, Michael, and Hoddeson, Lillian, *Crystal Fire: The Invention of the Transistor and the Birth of the Information Age (Sloan Technology Series)*, W. W. Norton & Company, 1998.

't Hooft, Gerard, *In Search of the Ultimate Building Blocks*, Cambridge: Cambridge University Press, 1997.

Townes, Charles H., *How the Laser Happened: Adventures of a Scientist*, New York: Oxford University Press, 1999.

Weinberg, Steven, *The First Three Minutes*, New York: Basic Books, 1988.

Weinberg, Steven, *The Discovery of Subatomic Particles*, Cambridge: Cambridge University Press, 2003.

Weisskopf, Victor, F., *The Privilege of Being a Scientist*, New York: W. H. Freeman, 1989.

Will, C M., *Was Einstein Right? Putting General Relativity to the Test*, New York: Basic Books, 1986.

Index

Abrikosov, Alexei A., 474
   biographical notes on, 466
   Nobel Prize (2003) to, 462
   superconductivity, 464–465
Academy of Sciences, French, 13, 43, 44, 138, 139, 165, 316, 411, 413
Academy of Sciences, Prussian, 104–105, 115, 122, 123, 128, 187
Academy of Sciences, Russian, 268, 284, 349, 447, 466–467
Academy of Sciences, Soviet Union, *see* Academy of Sciences, Russian
Accademia dei Lincei, 348
accelerators, *see* particle accelerators
AGS (BNL), 272, 286, 313, 337, 363, 398
Alembert Le Rond Jean d', 14
Alferov, Zhores Ivanovich, 447, 474
   biographical notes on, 447
   heterostructures, 445–446
   Nobel Prize (2000) to, 442
Alfvén, Hannes Olof, 472
   biographical notes on, 315
   Nobel Prize (1970) to, 315
   plasma physics, 315
Alighieri, Dante, 271
Allen, Jack, 284, 285, 350–351, 426, 464
Allison, Samuel K., 248
alpha decay, 176
alpha particles, 118, 160, 490
   as helium nuclei, 69
   discovery of, 28
alpha rays, *see* alpha particles
Alpher, Ralph, 353
Alvarez, Luis Walter, 214, 240, 309, 455, 472
   biographical notes on, 308
   Nobel Prize (1968) to, 307–308
   resonance particles, 308–309
Alvarez, Walter, 309
Amaldi, Edoardo, 170, 188, 189, 201

American Physical Society, 395
Ampère, André-Marie, 19, 28
Anderson, Carl David, 58, 183, 184, 187, 194, 196, 226, 231–234, 275, 422, 470
   biographical notes on, 195–197
   muon, 195
   Nobel Prize (1936) to, 193
   positron, 183, 197
Anderson, Philip Warren, 328, 344–345, 348, 356, 434, 472
   biographical notes on, 345–346
   challenge of reductionism by, 347–349
   condensed-matter physics, 346–347
   Nobel Prize (1977) to, 340–341
Anderson localisation, 344–347, 434
Anderson model, 346
Ångström, Anders, 25, 73
Ångström, Knut, 73
angular momentum, atomic, 110, 111, 475
Annalen der Physik, 37, 49, 51, 105, 147
annihilation, electron–positron, 195
Annis, Martin, 460
anode, 475
antielectron, *see* positron
antiferromagnetism, 316, 346, 419
antineutron, 271, 310
antiparticle, 176, 475
antiproton, 269–270, 310, 381
   discovery of, 270
Appleton, Edward Victor, 82, 302, 471
   biographical notes on, 224
   ionosphere, 224
   Nobel Prize (1947) to, 223
Appleton layer, 223–224
Arecibo observatory, 415
argon, 47
Argonne National Laboratory, 271, 466
Aristotle, 7

Armstrong, Neil, 279
Arnold, Harold, 163
Arrhenius, Svante, 34, 69–70, 120, 126–139, 302
Aston, Francis William, 82, 200
   mass spectrograph, 121–122
astronomy, 329, 414
Astrophysical Journal, 351, 415
astrophysics, 279, 302, 319, 329, 414, 458, 475
asymptotic freedom, 406
AT&T, 6, 163, 278
Atkinson, Robert, 303
atom, 17, 25, 26–27, 475
   Bohr's model of, 91
   nuclear, 77, 80, 81, 90–91
   plum pudding model of, 81
atom smasher, 174
atomic clock, 400–401, 452, 476
atomic number, 98, 192, 475
atomic physics, 369
Aurivillius, Christopher, 127
aurora, 315, 476
Avogadro, Amedeo, 147
Avogadro number, 147

Bahcall, John, 455
Baker, Jim, 133
Baker, Nicholas, 133
Balmer, Johann, 25
Balmer-α line, 65, 250
Balmer series, 25, 476
Bank of Sweden, 3
Bardeen, John, 226, 257, 325, 328–329, 426, 443, 463, 471–472
   BCS theory, 321–323
   biographical notes on, 257, 324
   Nobel Prize (1956) to, 255
   Nobel Prize (1972) to, 321
   transistor, 255–256

517

Barkla, Charles Glover, 138–139, 470
    biographical notes on, 110
    characteristic X-rays, 98, 109–110
    Nobel Prize (1917) to, 109–110
baryons, 310, 314, 404, 476
Basov, Nikolai Gennadievich, 292, 372, 445, 472
    biographical notes on, 294
    maser, 291
    Nobel Prize (1964) to, 290
BCS theory, 322–323, 327–328, 449, 476
Becker, Herbert, 181
Becquerel, Alexandre Edmond, 43
Becquerel, Antoine César, 43
Becquerel, Antoine Henri, 27–28, 34, 41, 43, 44, 469
    biographical notes on, 43
    Nobel Prize (1903) to, 42
    radioactivity, 42–43
Becquerel rays, 42, 43
Bednorz, Johannes Georg, 391–396, 463, 473
    biographical notes on, 392
    high-temperature superconductors, 393–395
    Nobel Prize (1987) to, 391–392
Bell, Alexander, 29
Bell, Jocelyn, 331
Bell Labs (Bell Telephone Laboratories), 5, 156, 163, 197–198, 226–227, 253, 255–259, 276–278, 291–295, 321, 328–329, 340–347, 349–353, 371–373, 387, 428, 430–431, 433–435, 446
Benedicks, Carl, 144–146
Bergson, Henri Louis, 126
Berkeley Radiation Laboratory, see Lawrence Berkeley National Laboratory (LBNL)
Bernoulli, Daniel, 14
Bernoulli, Johann, 14
beryllium, radiation of, 181–182
Besso, Michele, 50, 59, 111, 254
beta decay, 169, 193, 258, 262–264, 476
    Fermi's theory of, 193
beta particles, 28, 490
beta radioactivity, see beta decay
beta rays, see beta particles

Bethe, Hans Albrecht, 110, 210, 214, 222, 253, 297, 303, 305, 329, 343, 353, 455–456, 472
    biographical notes on, 304–305
    CNO cycle, 304
    Nobel Prize (1967) to, 301–302
    on Enrico Fermi, 202
    proton–proton chain, 303–304
Bevatron (LBNL), 269–272, 308
Bieler, Etienne, 122
Big Bang, 296, 351–354, 366, 400, 409, 413, 455, 476
big physics, 407
big Science, 369
binary pulsar PSR 1913+16, 414–418, 476
binary stars, 477
    gravitational waves from, 417–418
    orbit precession, 417
Binnig, Gerd, 388–390, 407, 473
    biographical notes on, 389–390
    Nobel Prize (1986) to, 388
    scanning tunnelling microscope, 388–389
black body, 26, 37, 477
black-body radiation, 25–26, 353, 477
black-body radiation law, 26, 37, 477
black-body spectrum, see black-body radiation law
black hole, 319, 378–380, 461, 477
Blackett, Patrick Maynard, 15, 58, 82, 155, 171, 187, 194–195, 197, 200, 226, 231, 275, 460, 471
    biographical notes on, 224–225
    Nobel Prize (1948) to, 224
    nuclear and particle physics, 225–226
    positron, 183
Bloch, Felix, 170, 177, 180, 241, 308, 471
    biographical notes on, 240
    Nobel Prize (1952) to, 240
    nuclear magnetic resonance, 240, 241–242
    on Werner Heisenberg, 180
Bloembergen, Deli, 371
Bloembergen, Nicolaas, 370, 371, 473
    biographical notes on, 371–372
    Nobel Prize (1981) to, 370
    non linear optics, 370–371
Blumenfeld, Kurt, 254
Bohr, Aage Niels, 133, 336, 472

biographical notes on, 336
Nobel Prize (1975) to, 333–334
nuclear collective model, 334–336
Bohr, Harald, 90, 93–94
Bohr, Margrethe, 336
Bohr, Niels Henrik, 37, 77, 82, 91–94, 98, 99, 110–113, 114, 120, 125, 130, 134, 140, 148–152, 154, 158, 159, 161, 169, 170–171, 179–180, 185, 190, 199, 201, 207–208, 210, 214, 217, 218, 219, 220, 238, 248, 283, 293, 296, 297, 333–336, 343, 470
    atomic model, 89–91, 133
    biographical notes on, 133
    complementarity principle, 158–160
    correspondence principle, 131–132
    hydrogen atom, 92–93
    liquid-drop model, 208, 334
    Nobel Prize (1922) to, 130–131
    November, 1962, 287
    nuclear fission, 209–210
    on Max Planck, 114
Bohr magneton, 216, 251–252, 476
Bohr–Sommerfeld theory, 130, 141, 145, 148–150
Boltzmann, Ludwig, 18, 26, 113
bomb, nuclear, 213, 214, 215
Bonaparte, Napoleon, 19, 203
Born, Max, 83, 125, 152, 159–161, 171, 177, 178–180, 187, 198, 210, 247, 287, 289, 290, 297, 343, 471
    biographical notes on, 246–247
    matrix mechanics, 143–145, 246
    Nobel Prize (1954) to, 245–246
    on Erwin Schrödinger, 185
    wave function, 149, 246
Bose, Satyendra Nath, 175, 448–449
Bose–Einstein condensate, 322, 409, 426–427, 430, 448–454, 477
Bose–Einstein statistics, 128, 168–169, 175, 426, 434, 477
bosons, 175, 284, 426–427, 449, 477
Bothe, Walther, 153, 194, 225, 471, 502
    biographical notes on, 248
    coincidence method, 247–248
    Nobel Prize (1954) to, 245–246
    radiation of beryllium, 181
Boyle, Robert, 26
Bragg, William Henry, 15, 98, 101, 109, 138, 156, 419, 469

biographical notes on, 101
Nobel Prize (1915) to, 100–101
X-ray spectrometer, 100
X-rays and Crystal Structure, 101
Bragg, William Lawrence, 97, 98, 101, 102, 109, 138, 156, 161, 220, 225, 245, 265, 343–344, 419, 469
biographical notes on, 101–102
Nobel Prize (1915) to, 100–101
reflection method, 100–101
X-rays and Crystal Structure, 101
Brahe, Tycho, 7
Branly, Édouard, 71
Brattain, Walter Houser, 226, 257, 321, 443, 471
biographical notes on, 257
Nobel Prize (1956) to, 255
transistor, 255–256
Braun, Karl Ferdinand, 72–73, 469
biographical notes on, 72
Nobel Prize (1909) to, 71
Brazinsky, Vladimir, 67
Breit, G., 251
Bridgman, Percy Williams, 223, 257, 471
biographical notes on, 223
Nobel Prize (1946) to, 223
Brockhouse, Bertram Neville, 420, 473
biographical notes on, 420
neutron spectroscopy, 419
Nobel Prize (1994) to, 418–419
Broek, Antonius van der, 98
Broglie, Louis Victor de, 125, 139, 150, 153–161, 165, 167, 184, 197, 297, 470
biographical notes on, 164–165
Nobel Prize (1929) to, 164
Broglie, Maurice de, 84, 138, 164
Brookhaven National Laboratory (BNL), 5, 218, 267, 272, 286, 298, 313, 338, 355, 362–364, 369, 397–398, 400, 405, 455–459
Alternating Gradient Synchrotron of, *see* AGS (BNL)
Cosmotron of, *see* Cosmotron (BNL)
Brossel, Jean, 301, 432
Brout, Robert, 356
Brout-Englert-Higgs mechanism, 357–359, 436, 484
Brown, Robert, 146
Brownian motion, 146–147, 477

bubble chamber, 237, 273–274, 275, 276, 412, 477
Buckley, Oliver, 278
Buddha, 313
building-up principle (*Aufbauprinzip*), 132
Bumstead, H. A., 62
Bunsen, Robert, 24–25, 88, 370
Burbidge, Geoffrey, 379
Burbidge, Margaret, 379
Butler, Clifford, 231

Cabibbo, Nicola, 364
Caltech (California Institute of Technology), 5, 134, 173, 183, 195–197, 245, 258, 274, 282, 292, 298, 310–314, 317, 337, 353, 376, 377–379
Cameron, Al, 455
canal rays, 116–117
carbon–nitrogen–oxygen cycle (CNO cycle), 304, 479
Carnot, Sadi, 18
Cartan, Henri, 432
Caruso, Enrico, 64
Casimir, Hendrik, 283
cathode, 23, 477
cathode-ray oscilloscope, 72
cathode rays, 17, 23–24, 48, 477
Cavendish, Henry, 14, 46, 47
Cavendish Laboratory, 28, 39, 41, 59, 62, 74, 82, 85–86, 90, 101, 109–110, 150, 155, 162–163, 174, 181–183, 190–192, 194–197, 199–200, 224, 226, 230, 238–239, 244–245, 325–328, 330, 346, 349, 387, 401
CERN, 5, 58, 133, 180, 272, 273, 275, 286, 319, 337, 340, 355–359, 369, 380–385, 400, 403, 405, 407, 409, 411–413, 423–425, 438–439
Large Electron–Positron collider of, *see* LEP (CERN)
Large Hadron Collider of, *see* LHC (CERN)
proton–antiproton collider of, *see* proton–antiproton collider (CERN)
Proton Synchrotron of, *see* PS (CERN)
Super Proton Synchrotron of, *see* SPS (CERN)

caesium-133, 400–401
Chadwick, James, 82, 122, 169, 182, 191, 192, 200, 229, 232, 350, 470
biographical notes on, 192
neutron, 181–182, 190–192
Nobel Prize (1935) to, 190–192
on Ernest Rutherford, 82
chain reaction, nuclear, 209, 210, 214–215
Chalk River Nuclear Laboratories, 420
Chamberlain, Owen, 203, 272, 471
antiproton, 270
biographical notes on, 271
Nobel Prize (1959) to, 269
Chandra X-ray Observatory, 461
Chandrasekhar, Subrahmanyan, 190, 379, 461, 473
astrophysics, 378
biographical notes on, 378–379
Nobel Prize (1983) to, 377
Chandrasekhar limit, 378, 477
charge
electric, 134–135, 404–406, 421–423, 480
elementary, 135, 480
strong (colour), 404–406, 423, 478
weak, 421–423
charge reflection, *see* symmetry
Charles II, King, 15
Charpak, Georges, 413, 473
biographical notes on, 412–413
multiwire proportional chamber, 412
Cherenkov, Pavel Aleksoyevich, 268, 456–457, 471
biographical notes on, 268
Nobel Prize (1958) to, 267
Nobel Prize (1992) to, 412
Cherenkov counter, 268, 276, 477
Cherenkov effect, 267–268
Cherenkov radiation, 478
Christianson, Gale, 10
Christoffel, Elwin Bruno, 105
Chu, Paul, 394
Chu, Steven, 431, 450, 474
atom chilling, trapping, 428–430
biographical notes on, 430–431
Nobel Prize (1997) to, 428
classical physics, 17, 22, 33
Clausius, Rudolf, 18, 34
Cline, David, 381

cloud chamber, *see* Wilson cloud chamber
CNO cycle, *see* carbon–nitrogen–oxygen cycle (CNO cycle)
COBE, *see* Cosmic Background Explorer (COBE)
Cocconi, Giuseppe, 340
Cockcroft, John Douglas, 58, 82, 174, 175, 200, 204, 239, 271, 471
   artificial nuclear disintegration, 182–183, 238–239
   biographical notes on, 239
   Nobel Prize (1951) to, 238
   voltage multiplier, 174
Cohen, Morrel, 328
Cohen-Tannoudji, Claude, 301, 432, 450, 474
   atom chilling, trapping, 430
   biographical notes on, 432–433
   Nobel Prize (1997) to, 428
collective model, nuclear, *see* nucleus, atomic
Collège de France, 411, 413, 432
collider, *see* particle collider
Columbia University, 5, 134, 173, 201–203, 209, 214, 217–218, 228, 233, 249–253, 260–264, 266, 276, 286, 289, 291–292, 299, 324, 335–337, 340, 352, 360, 365, 370, 384, 397–400, 401–402, 423, 434
communication system, 442
complementarity principle, 159, 478
Comptes Rendus de l'Académie des Sciences, 181
Compton, Arthur Holly, 58, 125, 135–137, 155, 159, 161, 171, 214, 215, 248–249, 470, 501–502
   biographical notes on, 154
   Nobel Prize (1927) to, 153–154
Compton effect, 125, 135–137, 153, 167, 181, 247, 282, 478
computer, 213, 410, 446
condensed-matter physics, 319, 340, 369, 374–376
Condon, Edward, 177
conductor, 478
Confucius, 434
conservation laws, 478
Conversi, Marcello, 231
Cooksey, D., 206

Cooper, Leon N, 257, 325, 426, 464, 472
   BCS theory, 321–323
   biographical notes on, 324
   Nobel Prize (1972) to, 321
Cooper pairs, 322–323, 328, 427, 478
Copenhagen interpretation, 160, 161–162, 478
Copenhagen spirit, 170
Copernicus, Nicolaus, 7, 51
Cornell, Eric A., 451, 453, 474
   biographical notes on, 452
   Bose–Einstein condensate, 450–451
   Nobel Prize (2001) to, 448
Cornell University, 5, 218, 298, 302–305, 360, 372–377, 425–427, 464
correspondence principle, 478
Cosmic Background Explorer (COBE), 413, 414
cosmic microwave background radiation, 279, 295–296, 319, 349–353, 371, 413, 414, 479
cosmic rays, 193–194, 479
   discovery of, 86–87
cosmology, 458, 466
Cosmotron (BNL), 272, 273, 364
Coster, Dirk, 132
Coulomb, Charles Augustin de, 14
Count Rumford, 18
Cowan, Clyde, 267, 285, 395, 420, 455
   neutrino detection, 260
Cowan, Eugene, 231
Crab Nebula, 330, 334, 378, 461
Crick, Francis, 102, 244–245
Critchfield, Charles, 303
critical phenomena, 374–376, 479
critical point, 372–374
Cronin, James Watson, 203, 365, 366, 473
   biographical notes on, 364–365
   charge–parity violation, 363–364
   Nobel Prize (1980) to, 362
Crookes, William, 24, 29, 35, 42
Crookes tubes, 24
crystal, 479
   liquid, 410, 466, 479
   structure analysis, 97
Curie, Eve, 44
Curie, Irène, 44, 188, 191
   and radiation of beryllium, 181
   artificial radioactivity, 187–188

   Nobel Prize for chemistry (1935) to, 188
Curie Skłodowska, Marie, 5, 28, 41, 44, 45, 84, 161, 181, 187–188, 287, 469
   biographical notes on, 46
   Nobel Prize for chemistry (1911) to, 46
   Nobel Prize for physics (1903) to, 42, 44
   polonium and radium, 43–44
Curie, Pierre, 28, 41, 45, 46, 187–188, 469
   biographical notes on, 44
   Nobel Prize (1903) to, 42, 44
   polonium and radium, 43–44
current, electric, 19, 480
cyclotron, 174, 175, 204–205, 271, 479
Cygnus X-1, 378, 461

D'Agostino, Oscar, 189
Dalén, Nils Gustaf, 469
   Nobel Prize (1912) to, 85
Dalton, John, 27
Dam, H. J. W., 35
Darriulat, Pierre, 381–383
Darwin, Charles Galton, 80
Davies, Paul, 299
Davis, Raymond, 306, 457, 459, 474
   biographical notes on, 459
   Nobel Prize (2002) to, 454
   solar neutrinos, 455–456
Davisson, Clinton Joseph, 156–157, 187, 198, 216, 255, 470
   biographical notes on, 198
   electron diffraction, 156, 197–198
   Nobel Prize (1937) to, 197
Davisson–Germer experiment, 156–157
Daily Mail, 414
de Broglie wavelength, 139, 199, 280
de Broglie wave, 125, 139, 156–157, 486
Debye, Peter, 37, 83, 110–111, 137, 157, 161, 171, 240
De Forest, Lee, 63–64, 163
Dehmelt, Hans Georg, 428, 473
   biographical notes on, 403
   Nobel Prize (1989) to, 400–401
   particle trap, 403
Delbrück, Max, 170
Democritus, 26
Denisyuk, Yu. N., 70

deoxyribose nucleic acid (DNA), 102, 244–245, 429
Descartes, René, 483
Deslandres, Henri, 302
DESY, *see* Deutsches Electronen Synchrotron (DESY)
detector, *see* particle detector
deuterium, 173, 179, 303–307, 399
Deutsches Electronen Synchrotron (DESY), 339–340, 401, 422–424
    DORIS of, *see* DORIS (DESY)
    PETRA of, *see* PETRA (DESY)
Dewar, James, 74, 88
Dicke, Robert, 67, 351–353
diffraction grating, 25, 480
diode, 479
    junction, 326
    light-emitting (LED), 446, 479
    thermionic, 63
    tunnel, *see* Esaki diode
dipole, magnetic, 129, 485, 486
Dirac, Paul Adrien Maurice, 125, 132, 147, 152, 161, 165, 171, 173, 176, 178, 183, 186, 190, 195, 220, 250–253, 269, 283, 297, 470
    biographical notes on, 186
    Nobel Prize (1933) to, 184
    on Erwin Schrödinger, 185
    The Principles of Quantum Mechanics, 186
    quantum electrodynamics, 176
    quantum mechanics, 150–151
    relativistic electron theory, 176
Dirac equation, 176, 480
disordered materials, 347, 480
displacement law 26, 78
DNA, *see* deoxyribose nucleic acid (DNA)
dodecahedron, 441
Doppler, Christian, 116, 430
Doppler effect, 116–117, 166, 372, 480
Dorda, Gerhard, 385
DORIS (DESY), 422–424
Dulong–Petit rule, 83
Dyson, Frank, 119–120
Dyson, Freeman, 297, 440

eclipse, solar (1919), 108, 119–120
Ecole Normale Supérieure, 70, 146, 203, 301, 316, 411, 428–432

Ecole Polytechnique, 43
Eddington, Arthur, 108, 119, 302–303
Edison, Thomas Alva, 29, 63
Ehrenfest, Paul, 170
eightfold way, 311, 312–313, 480
Eigler, D., 390
Einstein, Albert, 10–11, 12, 23, 33, 37, 39, 49, 50, 52, 77, 84, 85, 93, 97, 107, 112–115, 116, 120, 122–123, 125, 127, 130, 134, 137, 139, 143, 147–150, 151–152, 153–154, 160–162, 164–166, 169, 175–176, 183, 185, 195, 208, 209, 210, 216, 220–221, 239, 246, 272, 282, 290, 303, 401, 414–418, 461, 470
    and the Nobel Prize, 57, 120
    April, 1955, 254–255
    biographical notes on, 128
    Bose–Einstein condensate, 448–449
    Brownian motion, 147
    Einstein, 1933, 187
    general theory of relativity, 104–106, 128
    light quanta, 49–50, 125, 126–128, 153, 176
    Nobel Prize (1921) to, 126–128
    on Madame Curie, 46
    photoelectric effect, 50, 51, 102–103, 126
    principle of equivalence, 65–68, 104–106
    relativity of time, 59
    special theory of relativity, 51–59, 128
    specific heats, 83
    stimulated emission, 111–112, 128
Einstein, Hans Albert, 254
Einstein Observatory, 461
Ekspong, Gösta, 153
Ekstrand, A. G., 112
electricity, 17, 19–20
electromagnetic waves, 20, 417, 480
electron, 17, 27, 28, 232, 400–402, 421–425, 481
    charge–mass ratio (e/m) of, 40, 59–61
    discovery of, 28, 59
    g-factor, 252, 254, 298
electron diffraction, 156–157, 160, 197–199

electron localisation, *see* Anderson localisation
electron microscope, 199, 385–386
electron pairs, *see* Cooper pairs
electron scattering, 280, 404
    deep inelastic, 404
    elastic, 404
electron theory, *see* Lorentz, Hendrik
electronics, 29, 64
electroweak theory, 319, 355–359, 361, 369, 381–383, 436, 440–441, 481
elements, chemical, 17, 27
    periodic system of, 27
    radioactive, 17, 28
Elsasser, Walter, 156
emission, stimulated, 111–112, 290–293, 492
emulsion, photographic, 230, 275
energy, 15, 17, 18, 481
    conservation of, 18
    kinetic, 18, 481
    mechanical (work), 19
    thermal, *see* heat
energy band, 177, 481
energy levels, 93, 476
Englert, François, 356
Eötvös, Roland von, 67
Epstein, Paul, 111
Esaki, Leo, 472
    biographical notes on, 326
    Nobel Prize (1973) to, 325
Esaki diode, 326, 481
Estermann, Immanuel, 207
ETH, *see* Swiss Federal Institute of Technology (ETH)
ether, 20, 22–23, 24, 54, 57, 481
Euler, Leonhard, 14
European Center for Particle Physics, *see* CERN
evaporative cooling, 450, 481
Everett, E., 76, 86
Ewald, Paul Peter, 95
exclusion principle, 142–143, 203, 481
    see also Wolfgang Pauli

Fairchild Semiconductor, 269, 443
Faraday, Michael, 19, 23, 27, 28, 46, 82
    lines of force, 19
Faroll, Barnett, 399
Feinberg, Gary, 359

Fermi, Enrico, 154, 169, 171, 175, 189, 193, 202, 207–208, 209, 210, 211, 214, 233, 258, 265–267, 269–271, 289, 303, 305, 337, 364, 384, 397–400, 418, 470
  and the Rome School, 188, 201–202
  biographical notes on, 203
  first nuclear reactor, 214–215
  neutron irradiation and reactions, 188–190, 503–504
  Nobel Prize (1938) to, 200–201
  November, 1954, 248–249
  theory of beta decay, 193, 504
Fermi, Laura, 201
Fermi–Dirac statistics, 169, 176, 378, 426, 434, 481
Fermilab, 358, 365, 381–384, 398, 405, 409, 423–425
  Tevatron collider of, *see* Tevatron collider (Fermilab)
fermions, 176, 421–423, 426, 481
ferrimagnetism, 316
ferromagnetism, 316
Feshbach, H., 299
Feynman, Richard P., 8, 214, 253, 264–267, 285, 297, 299, 311–312, 317, 437, 472
  biographical notes on, 298–299
  Nobel Prize (1965) to, 296
  QED, 297–298
Feynman diagrams, 297, 437
field, 481
  electric, 19–20
  electromagnetic, 19–20
  gravitational, 106, 417–418
  magnetic, 19–20, 482
fine structure, *see* spectral lines; hydrogen spectrum
fine structure constant, 111, 387, 482
Finnegans Wake, 314
Fisher, Michael, 375
fission, nuclear, 173, 487
  discovery of, 207–208
Fitch, Val Logsdon, 365, 366, 473
  biographical notes on, 365–366
  charge–parity violation, 363–364
  Nobel Prize (1980) to, 362
Fitzgerald, George, 23
flavour, 421, 482

Fleming, John Ambrose, 63, 163
  thermionic diode, 63, 64
fluorescence, 35–36
Foley, Henry, 252
force, 482
  electromagnetic, 354, 355–359, 363, 408, 421–423, 436, 441, 482
  electroweak, 354, 359
  gravitational, 354, 441, 482
  strong (colour), 288, 354, 363, 406–408, 423–425, 441, 482
  weak, 354, 355–359, 362–363, 423–425, 438, 440–441, 482
fountain effect, 285
Fowler, Ralph, 150, 161, 197
Fowler, William Alfred, 455, 473
  biographical notes on, 379–380
  Nobel Prize (1983) to, 377
  nuclear astrophysics, 379
fractional quantum statistics, 434
frame of reference, 11, 483
  inertial, 11, 483
Frampton, P., 360
François I, King, 433
Franck, Ingrid, 141
  Nobel Prize (1925) to, 139–140
Franck–Hertz experiment, 99, 131, 139
Franck, James, 98–99, 132, 141, 171, 178, 214, 247, 470
  biographical notes on, 140
Frank, Ilja Mikhailovich, 282, 471
  biographical notes on, 268
  Nobel Prize (1958) to, 267
Frankfurter Zeitung, 36
Franklin, Benjamin, 14
Franklin, Rosalind, 245
Frascati National Laboratory, 338, 367
  AdA of, 367
  ADONE of, 338, 367, 509
Fraunhofer, Joseph, 24–25, 370
Fresnel, Augustin-Jean, 22
Friedman, Jerome Isaac, 203, 403, 404, 423, 473
  biographical notes on, 405
  Nobel Prize (1990) to, 403–404
  SLAC–MIT experiment, 404–405
Friedman, Herbert, 460
Friedmann, Alexandre, 166
Friedrich, Walther, 96

Frisch, Otto, 206–208, 209
fusion, nuclear, 303–307, 315, 488

Gabor, Dennis, 70, 472
  biographical notes on, 320
  Nobel Prize (1971) to, 320
galaxy, 483
Galilean transformations, 55, 483
Galilei, Galileo, 7, 8–9, 10, 11, 51, 54, 67
  Dialogue, 8
  falling bodies, 8
  principle of inertia, 9
  principle of relativity, 8
  Two New Sciences, 8, 9
gallium aluminium arsenide, 446
gallium arsenide, 445–446
gamma rays, 41, 490
Gamow, George, 132, 170, 176, 238, 303, 326, 353
Gargamelle (bubble chamber), 358
gases
  inert (noble), 47, 487
  kinetic theory of, 18
gauge field theories, 355, 483
gauge invariance, 355–359, 434
Gauss, Carl Friedrich, 31, 104–105, 171
Geiger, Hans, 69, 80–81, 114, 117, 122, 153, 192, 247–248, 275, 280, 502
Geiger–Marsden experiment, 80, 404
Geiger–Müller counter, 183, 194, 225, 268, 275, 483
Geissler, Johann Heinrich, 23–24
Geissler tubes, 23
Gell-Mann, Murray, 267, 299, 309–310, 311–312, 337–339, 364, 374, 403–404, 423, 472
  biographical notes on, 310–311
  eightfold way, 311, 312–313
  Nobel Prize (1969) to, 310, 311
  quarks, 311, 314–315
  strangeness, 311, 312
General Conference on Weights and Measures, 401
General Electric Research and Development Center, 326
generator, electric, 28
Gennes, Pierre-Gilles de, 345, 411, 473
  biographical notes on, 411
  Nobel Prize (1991) to, 410
  soft-matter physics, 410–411

George II, King, 31
Gerlach, Walther, 129, 216, 217, 500
Germer, Lester, 156, 197–198, 216
Giacconi, Riccardo, 462, 474
   biographical notes on, 462
   Nobel Prize (2002) to, 454
   X-ray astronomy, 460–461
Giaever, Ivar, 327, 472
   biographical notes on, 326
   energy gap, 327
   Nobel Prize (1973) to, 325
Gibbs, Josiah Willard, 18
Ginzburg, Vitaly L., 474
   biographical notes on, 466–467
   Nobel Prize (2003) to, 462
   superconductivity, 464
Ginzburg–Landau theory, 464–465, 483
Glaser, Donald Arthur, 274, 308, 471
   biographical notes on, 274
   bubble chamber, 273–274
   Nobel Prize (1960) to, 273
Glashow, Sheldon Lee, 339, 360, 379–381, 423, 436–438, 473
   biographical notes on, 359–360
   electroweak theory, 355–358
   Nobel Prize (1979) to, 355
Glashow–Salam–Weinberg model, *see* electroweak theory
Global Positioning Satellite System (GPS), 58, 399
gluons, 403–406, 441, 483
Goeppert Mayer, Maria, 5, 171, 290, 333–334, 364, 472
   biographical notes on, 289
   Nobel Prize (1963) to, 287
   nuclear shell model, 289
Goethe, Johann Wolfgang von, 112, 190
Goldhaber, Maurice, 267
Goldschmidt, O., 84
Goldstein, Eugen, 23, 29, 116
Gordon, James, 291
Gor'kov, Lev, 328, 465
Gossard, Arthur, 385
Göttingen theory, *see* matrix mechanics
Goudsmit, Samuel, 141–142
Grand Unification, 441
gravitational radiation, 483
gravitational waves, *see* relativity, general theory of

Greytak, Thomas, 450–453
Gribov, Vladimir, 457
Gross, David, 408
Grossmann, Marcel, 104–105, 254
Groth, Paul von, 95
Guillaume, Charles Edouard, 470
   biographical notes on, 121
   Nobel Prize (1920) to, 120–121
Gullstrand, Allvar, 78, 126
Gurney, Ronald, 177
Gustav V, King, 128
Gustav VI, King, 321

Habicht, Conrad, 49–51
hadrons, 310, 356–357, 406, 441, 484
Hahn, Otto, 58, 68, 82, 206–208
   nuclear fission, 207–208
Hale, George Ellery, 302
Hall, Edwin, 386
Hall effect, 386–387, 484
Hänsch, Theodor, 428
Harvard University, 5, 173, 223, 240, 282, 300, 339, 340–342, 355–361, 370–372, 376, 384, 400–402, 415, 427
Hasselberg, Bernhard, 121
Hawking, Stephen W., 10, 380, 414
heat, 15, 17, 18, 484
Heaviside, Oliver, 224
Heisenberg, Werner Karl, 110, 125, 132, 144, 147–152, 153, 158–160, 161, 170–171, 173, 176–177, 179, 185, 186, 192–193, 218, 240, 245–246, 287, 297, 300, 336, 470
   biographical notes on, 180
   Copenhagen interpretation, 161–162
   matrix mechanics, 143–145
   Nobel Prize (1932) to, 177–179
   uncertainty principle, 157–159
Heitler, Walter, 177
helium, 47
helium-3, superfluid, 427, 464–466
helium-4, superfluid, 284–285, 349, 464–465
   liquid, 33, 77, 88, 463
Helmholtz, Hermann von, 18, 79, 115, 303
Henry, Joseph, 19
Herman, Robert, 353

Hertz, Ellen, 141
Hertz, Gustav Ludwig, 98–99, 141, 470
   biographical notes on, 140
   Nobel Prize (1925) to, 139–140
Hertz, Heinrich, 20, 24, 29, 48–49, 61, 71, 140, 296
Herzberg, Gerhard, 251
Hess, Harald, 450
Hess, Victor Francis, 87, 183, 470
   biographical notes on, 194
   cosmic rays, 86–87, 194
   Nobel Prize (1936) to, 193
heterostructure, 433, 445–446, 484
Hevesy, George de, 82, 90, 93, 132, 170
Hewish, Antony, 332, 472
   biographical notes on, 332
   first pulsar, 331–332
   Nobel Prize (1974) to, 329
Higgs, Peter, 356
Higgs particle, 356, 441, 484
Hilbert, David, 171, 247
Hitler, Adolf, 49, 115, 247
Hittorf, Johann, 23, 35
Hoerni, Jean, 443
Hofstadter, Robert, 402–403, 471
   biographical notes on, 282
   electron scattering, 280–281
   Nobel Prize (1961) to, 280
holography, 320, 484
House of Commons, 435
Houtermans, Fritz, 303
Hoyle, Fred, 353, 379
Hubble, Edwin, 166, 302
Hubble law, 166
Hulse, Russell Alan, 416, 473
   binary pulsar PS 1913+16, 417–418, 417
   biographical notes on, 415
   Nobel Prize (1993) to, 414
Hulthén, E., 270
Huygens, Christiaan, 13–14, 21
   Traité de la Lumière, 14
   wave theory of light, 13–14
hydrogen atom, 91, 249
   Bohr's model of, 92–93
hydrogen spectrum, 25
   fine structure of, 110, 250
   hyperfine structure of, 251–252, 401

IBM Thomas J. Watson Research Center, 326, 445
IBM Zurich Research Laboratory, 6, 388–389, 390–396, 407
Iliopoulos, John, 339, 357
induction, electromagnetic, 19
Industrial Revolution, 28
Information Technology (IT), 442, 444
Institut du Radium, 46, 181, 187
Institute for Advanced Study (Princeton), 128, 186–187, 220–221, 228, 260–265, 310, 322–324, 336
Institute for Physical Problems, 5, 283, 349, 464–466
insulator, 484
integrated circuit, 237, 259, 269, 442–443, 444, 484
Intel, 444
interaction, *see* force
interference fringes, 21, 485
interferometer, *see* Michelson interferometer
Internet, 446
International Bureau of Weights and Measures, 121
Ioffe Physico-Technical Institute, 446–447
ion, 27, 401–403, 484
ionisation, 484
ionosphere, 223–224, 315, 484
Irvine–Michigan–Brookhaven detector, 396
isotopes, 121, 484

Jansky, Karl, 329–330
Janssen, Jules, 47
Jeans, James, 84, 89
Jensen, Johannes Hans, 333–334, 472
  biographical notes on, 289
  Nobel Prize (1963) to, 287
  nuclear shell model, 289
Jodrell Bank, 225, 330
Joint Institute for Laboratory Astrophysics (JILA), 450–452
Joliot, Frédéric, 44, 188, 191, 201, 209, 412
  and the radiation of beryllium, 181
  artificial radioactivity, 187–188
  Nobel Prize for chemistry (1935) to, 188

Jordan, Pascual, 132, 143, 170, 171, 246, 287
Josephson, Brian David, 328, 472
  biographical notes on, 327
  Josephson effect, 327–329
  Nobel Prize (1973) to, 325
Josephson constant, 329
Josephson effect, 328, 347, 484
Joule, James Prescott, 18
J/ψ particle, 337–339, 412, 422–423, 485
Joyce, James, 314
Jung, Carl Gustav, 221

Kadanoff, Leo, 376, 377
Kamerlingh Onnes, Heike, 39, 40, 41, 74, 84, 85, 90, 120, 242, 322, 393, 394, 463, 469
  biographical notes on, 88
  liquid helium, 88
  Nobel Prize (1913) to, 87–88
  superconductivity, 88–89
Kamerilngh Onnes Laboratory, 88, 349
Kamiokande, 394, 456–457
Kampen, Nicolaas Godfried van, 439
Kampen, Pieter Nicolaas van, 439
Kapitza, Peter Leonidovich, 82, 192, 200, 239, 283–284, 350, 426, 464, 472
  biographical notes on, 349
  Nobel Prize (1978) to, 349
  superfluid helium-4, 349
Kastler, Alfred, 252, 432, 472
  biographical notes on, 301
  Nobel Prize (1966) to, 301
Kaufmann, Walther, 59
Keesom, Annie, 349
Keesom, Willem, 349
Kelly, Mervin, 255, 321
Kendall, Henry Way, 405, 406, 423, 473
  biographical notes on, 405–406
  Nobel Prize (1990) to, 403–404
  SLAC–MIT experiment, 404–405
Kendrew, John, 102
Kennelly, Arthur, 224
Kepler, Johannes, 7, 51, 221, 394
Ketterle, Wolfgang, 452, 454, 474
  biographical notes on, 453
  Bose–Einstein condensate, 451–452
  Nobel Prize (2001) to, 448

Kilby, Jack St Clair, 269, 444, 474
  biographical notes on, 444–445
  integrated circuit, 443
  Nobel Prize (2000) to, 442
Kirchhoff, Gustav, 24–25, 88, 115, 368
Kivelson, Steve, 436
Klein, Felix, 171, 247
Klein, Oskar, 133, 170, 302, 336
Kleppner, Daniel, 401, 450–454
Klitzing, Klaus von, 386, 433, 473
  biographical notes on, 386
  Nobel Prize (1985) to, 385
  quantum Hall effect, 385
K-mesons, 231, 261–262, 363–364, 485
Knipping, Paul, 96
Kobayashi, Makoto, 364
Kohlhörster, Werner, 194
Kohlrausch, Friedrich, 34
Kondo effect, 346, 376
Koshiba, Masatoshi, 306, 459, 474
  biographical notes on, 459
  cosmic neutrinos, 456–458
  Nobel Prize (2002) to, 454
Kramers, Hendrik, 132, 161, 170, 248, 336
Kroemer, Herbert, 446, 447, 474
  biographical notes on, 447
  heterostructures, 445–446
  Nobel Prize (2000) to, 442
Kundt, August, 34
Kurlbaum, Ferdinand, 113
Kusch, Polykarp, 214, 253–254, 291–292, 399–401, 471
  biographical notes on, 253
  electron magnetic moment, 251–252
  Nobel Prize (1955) to, 249

Labouisse, Henry R., 44
Lagrange, Joseph Louis, 14
Lamb, Willis Eugene, 214, 252, 253, 298, 299, 471
  biographical notes on, 252–253
  Nobel Prize (1955) to, 249
Lamb shift, 250–251, 253, 298, 485
Landau, Lev Davidovich, 132, 283, 351, 375, 464–465, 471
  biographical notes on, 283–284
  Nobel Prize (1962) to, 282
  superfluidity, 284–285

Landé, Alfred, 246
Langevin, P., 84
Laplace, Pierre Simon de, 14
Large Magellanic Cloud, 396, 457
Larmor, Joseph, 118–119
laser, 112, 277, 290, 294, 295, 370, 485
   atom, 451, 485
   invention of, 276–277
   semiconductor, 443–446, 491
Laue, Max Theodor von, 56, 97, 100, 114, 115, 138, 151, 156, 217, 469
   biographical notes on, 97
   Nobel Prize (1914) to, 95
   X-ray diffraction, 95–97
Laughlin, Robert, 436, 474
   biographical notes on, 435
   fractional quantum Hall effect, 433–434
   Nobel Prize (1998) to, 433
Lavoisier, Antoine Laurent, 27, 45
Lawrence, Ernest Orlando, 154, 175, 206, 214, 270, 271, 365, 470
   biographical notes on, 204
   cyclotron, 174, 205–206
   Nobel Prize (1939) to, 204
Lawrence Berkeley National Laboratory (LBNL), 5, 204, 233, 269–270, 307–308, 337, 369, 413
   Bevatron of, *see* Bevatron (LBNL)
Lebedev Institute of Physics, 5, 268, 291–294, 466
Lederman, Leon Max, 264, 273, 286, 363, 381–383, 397, 455, 473
   biographical notes on, 398
   bottom quark, 423
   Nobel Prize (1988) to, 396–397
   two-neutrino experiment, 285–286, 396–397
Lee, David, 284, 427, 464, 474
   biographical notes on, 427
   Nobel Prize (1996) to, 425
   superfluid helium-3, 425–426
Lee, Tsung Dao, 203, 261, 265, 267, 312, 362–363, 397, 471
   biographical notes on, 264–265
   Nobel Prize (1957) to, 260–261
   parity violation, 264

Leggett, Anthony J., 284, 474
   biographical notes on, 467
   Nobel Prize (2003) to, 463
   superfluidity in helium-3, 465–466
Lemaître, Georges, 166
Lenard, Philipp, 24, 29, 34, 35, 50–51, 59–61, 97, 116, 469
   biographical notes on, 49
   cathode rays, 48
   Nobel Prize (1905) to, 48–49
   photoelectric effect, 48–49
LEP (CERN), 407, 439
lepton families, 421, 440
leptons, 310, 356–357, 421–423, 440, 485
Levi-Cività, Tullio, 104–105
Lewis, Gilbert, 153
LHC (CERN), 356, 405, 409
Lifshitz, Evgenii, 283–311
light, 17, 485
   coherent, 293, 478
   diffraction of, 22, 479
   incoherent, 478
   interference of, 21–22, 485
   velocity (speed) of, 12, 20, 22, 54–55
   wave theory of, 13–14
light quanta, 49–50, 125–128
Lippmann, Gabriel, 34, 71, 469
   biographical notes on, 70
   Nobel Prize (1908) to, 70
Livermore National Laboratory, 433–435
Livingston, M. Stanley, 204
Lodge, Oliver, 71
Lofgren, E., 272
London, Fritz, 171, 177, 285
Lord Kelvin, 18, 28, 48, 303
Lord Rayleigh, 15, 37, 48, 59, 62, 78, 91, 102, 113, 118, 469
   argon, 47
   biographical notes on, 47–48
   Nobel Prize (1904) to, 45
   on the Cavendish Laboratory, 76
   Rayleigh scattering, 47, 167
Lorentz, Hendrik Antoon, 20, 23, 39–40, 41, 55, 57, 59, 84, 88, 119–120, 161, 469
   and the Zeeman effect, 40–41
   biographical notes on, 39
   electron theory, 20, 39–40
   Nobel Prize (1902) to, 38–39, 41
Lorentz–Fitzgerald contraction, 23, 39

Lorentz transformations, 39, 55, 485
Lovell, Bernard, 330
low-temperature physics, 369, 463–465
Lucent Technologies, 5
Lummer, Otto, 113

Mach, Ernst, 221
macromolecule, 410
magnetic moment, 476
   atomic nuclei, 219, 401
   atoms, 129, 219
   electron, 239, 403
   neutron, 308, 419
   proton, 216
magnetic resonance imaging (MRI), 218, 242
magnetism, 17, 19–20, 316, 342–346
magneto-hydrodynamics, 315
magneto-optical trap (MOT), 430, 450
magnetosphere, 269, 486
Maiani, Luciano, 339, 357
Maiman, Theodore, 277, 292
Majorana, Ettore, 188
Manhattan Project, 140, 154, 192, 204, 214, 240, 271, 298, 305, 308, 337, 365, 400, 420
Marconi, Guglielmo, 28, 29, 34, 63, 72, 73, 163, 224, 469
   biographical notes on, 71
   Nobel Prize (1909) to, 71–72
Marsden, Ernest, 80–81, 117, 122, 280
Marshak, Robert, 266
maser, 276, 290, 486
   hydrogen, 400–401
   solid-state, 369
Maskawa, Toshihide, 364
mass number, 192, 476
mass spectrograph, 121
Massachusetts Institute of Technology (MIT), 5, 173, 214, 218, 258, 293, 298–299, 305, 308, 310, 317, 324, 361, 404–405, 418–420, 431, 445, 448–453, 460
matrix mechanics, 143, 145, 151, 246, 486
matter wave, *see* de Broglie wave
Max Planck Society, 115, 387

Maxwell, James Clerk, 17, 19–20, 21, 22, 26, 29, 47, 50, 53, 54–56, 66, 71, 102, 113, 118, 143, 296, 417
  A Treatise on Electricity and Magnetism, 20
  electromagnetic waves, 20
  speed of light, 20
Maxwell equations, 19–20, 486
McIntyre, Peter, 381
McMillan, Edwin, 270
mechanics, 17
  classical (or Newtonian), 11, 22, 53–54
  relativistic, 56
  statistical, 18
Meer, Simon van der, 385, 411, 473
  biographical notes on, 385
  Nobel Prize (1984) to, 380–381
  W and *Z* bosons, 380–383
Meissner, Walter, 463
Meissner effect, 322, 324, 463, 486
Meitner, Lise, 170, 206–208
Mendeleev, Dmitri Ivanovich, 27, 121, 192
Mercury (planet), 105
  perihelion advance of, 107, 417
mesons, 193, 229, 231, 310, 314, 486
metal–insulator transition, 344–347, 487
Meyer, Julius Lothar, 27
Michelson, Albert Abraham, 22, 39, 66, 110, 134, 250, 469
  biographical notes on, 64–65
  Nobel Prize (1907) to, 64–65
Michelson interferometer, 23, 64
Michelson–Morley experiment, 23, 54, 57, 64, 66
Milky Way, 295, 329, 396, 415
Millikan, Greta, 136
Millikan, Robert Andrews, 103, 114, 126, 136, 163, 183, 193, 195–197, 470
  biographical notes on, 134
  cosmic rays, 194
  Nobel Prize (1923) to, 134
  oil-drop experiment, 134–135
  photoelectric effect, 102–103
Mills, Robert, 265, 355–356, 437
Minkowski, Hermann, 56, 70, 171, 247, 297
  space-time, 70–71, 441, 491
Misener, Donald, 284, 350, 426

MIT, *see* Massachusetts Institute of Technology (MIT)
MIT Francis Bitter National Magnet Laboratory, 433
MIT Radiation Laboratory, *see* Radiation Laboratory (MIT)
Mittag-Leffler, Gösta, 44, 70, 73
molecular-(atomic-) beam method, 216
molecular-beam magnetic resonance, 217, 218–219
molecule, 486
Moore, Gordon, 443
Moore, Ruth, 201
Moore, Walter, 185
Morel, Pierre, 347
Morley, Edward, 23, 39, 64–65
  see also Michelson–Morley experiment
Morse, Samuel, 71
Moseley, Henry, 109–110, 131, 138
  nuclear charge, 97–98
Mössbauer, Rudolf Ludwig, 471
  biographical notes on, 282
  Nobel Prize (1961) to, 280
Mössbauer effect, 281–282, 486
motor, electric, 29
Mott, Nevill Francis, 132, 345, 472
  biographical notes on, 343–345
  condensed-matter physics, 343–345
  Nobel Prize (1977) to, 340–341
Mott transition, *see* metal–insulator transition
Mottelson, Ben Roy, 472
  biographical notes on, 337
  Nobel Prize (1975) to, 333–334
  nuclear collective model, 334–336
Müller, Karl Alexander, 389, 391–396, 463, 473
  biographical notes on, 392
  high-temperature superconductors, 393–395
  Nobel Prize (1987) to, 391–392
Müller, Walther, 275
multiwire proportional chamber, 412, 487
muon, 58, 231–234, 262–264, 422, 487
  discovery of, 195
Mussolini, Benito, 201

Nambu, Yoichiro, 406
nanostructure, 448, 487

nanotechnology, 447–448, 452
nanotube, 448
Nafe, John E., 251
National Aeronautics and Space Administration (NASA), 413
National Bureau of Standards, *see* National Institute of Standards and Technology (NIST)
National Institute of Standards and Technology (NIST), 5, 263, 282, 428–431, 450–452
Nature, 36, 40, 89, 181, 187, 208, 231, 245, 277, 331, 350
Naturwissenschaften, 141, 208
Neddermeyer, Seth, 195, 196, 234, 422
Néel, Louis Eugène, 346, 419, 472
  biographical notes on, 316
  magnetism, 316
  Nobel prize (1970) to, 315
Néel temperature, 316
Ne'eman, Yuval, 296, 313
Nelson, Edward B., 251
Nernst, Walther, 83, 84, 114, 122
Neumann, John von, 288
neutral-current interaction, 356–358, 438, 487
neutrino, 169, 231–234, 357–358, 396–400, 420–422, 440, 455–459, 487, 502
  cosmic, 454
  cosmic-ray, 455–458
  detection of, 258–260
  electron, 286, 310, 400, 422, 457
  from SN 1987A, 396, 457
  muon, 286, 287, 310, 400, 422, 455–457
  oscillation of, 457–458, 487
  tau, 398–400, 409, 422–424, 457, 493
  see also solar neutrinos
neutrino astronomy, 455–457
neutron, 122, 173, 232, 404–406, 418–419, 487
  diffraction, 419
  discovery of, 181–182, 190–192
  scattering, 419
  spectroscopy, 419
neutron star, 333, 378, 414–418, 461, 487
New York Times, 36, 227, 278

Newton, Isaac, 7, 9–12, 13, 14, 15, 17, 20, 21–22, 24, 26, 51–52, 54–56, 66–67, 82, 105, 106–107, 143, 154, 186
　absolute space, 10, 54–55
　absolute time, 10, 54–55
　corpuscular hypothesis, 12–13
　laws of motion, 10–11
　Opticks, 10, 12–13, 14, 26, 154
　Principia, 10–12, 14, 54, 67, 186
　principle of inertia, 11
　principle of relativity, 11, 54, 489
　universal gravitation, 12
Niels Bohr Institute, 132, 325, 336–337, 355–360
Nishijima, Kazuhiko, 312
Nishina, Yoshio, 132, 336
NIST, *see* National Institute of Standards and Technology (NIST)
Nobel, Alfred Bernhard, 3–4, 302, 370
Nobel Foundation, 3
Nobel Institute of Physics, 138, 372
Nobel Physics Committee, 4, 34, 44, 45, 57, 68–70, 73–74, 78, 85, 112, 120–121, 126, 138, 140, 144–146, 153, 164, 167, 174–178, 187, 204, 220, 270, 279, 296, 302, 340, 362, 377, 400, 437
Nobel Prize, 3–4
　for physics, 4–6
　see also individual winners
non linear optics, 369, 370, 487
Noyce, Robert, 269, 443
nuclear magnetic resonance (NMR), 213, 237, 240, 241–242, 488
Nuclear Physics, 356, 437–438
nucleon, 402–403, 488
nucleus, atomic, 403, 488
　collective model, 335, 487
　liquid-drop model, 208, 485
　nuclear charge, 98
　proton–electron model, 122, 168
　shell model, 289–335, 488
numbers, magic, 289, 334, 485

Oak Ridge National Laboratory, 419–420
Occhialini, Giuseppe, 183, 194–196, 200, 225, 231, 460
Odhner, C. T., 34
Oersted, Hans Christian, 19, 28
omega-minus particle, 313
Oppenheimer, J. Robert, 132, 152, 171, 176–177, 203, 215, 252, 265, 296–299, 310–311
Oschsenfeld, Robert, 463
Oseen, Carl, 126, 164, 167, 174, 178, 220
Osheroff, Douglas Dean, 284, 464, 474
　biographical notes on, 427
　Nobel Prize (1996) to, 425
　superfluid helium-3, 425–426
Ostwald, Wilhelm, 46

pair production, electron–positron, 195, 226
Pais, Abraham, 73, 78, 112, 122, 228, 255, 266, 287, 299, 437
Palmaer, W., 188
Pancini, Ettore, 231
Panofsky, Wolfgang, 338
Panov, Vladimir, 67
parity, 261, 288, 362, 488
　conservation, 261–262
　violation, 264
　see also symmetry, mirror
particle, virtual, 494
particle accelerators, 271–273, 409, 412, 440, 475
　cyclic, 475
　linear, 174, 475
　synchrotron, 272
particle collider, 367, 475
particle detector, 274–276
　scintillation counter, 275–276, 491
　solid-state counter, 276, 491
　spark chamber, 286, 412, 491
particle physics, 237, 412, 466
particle trap, 400–403, 488
particles, strange, 195, 231–234
Paul, Wolfgang, 473
　biographical notes on, 403
　Nobel Prize (1989) to, 400–401
　particle traps, 402–403
Paul trap, 403
Pauli, Wolfgang, 58, 110, 125, 132, 141, 143, 148, 160–161, 170–171, 176–177, 178, 187, 190, 216, 218, 222, 232, 258–260, 266, 283, 297, 311, 336, 390, 392, 420, 471
　biographical notes on, 221
　exclusion principle, 142–143, 220
　neutrino hypothesis, 169
　Nobel Prize (1945) to, 220
Pauling, Linus, 132
Peebles, James, 351
Pegram, George, 201
Peierls, Rudolf, 180
Penning trap, 401
Penrose, Roger, 154
Penzias, Arno Allan, 352, 353, 371, 413, 472
　biographical notes on, 352–353
　cosmic microwave background radiation, 295, 351–352
　Nobel Prize (1978) to, 349
Pepper, Michael, 387
perihelion, 488
Perl, Martin Lewis, 340, 474
　biographical notes on, 423
　Nobel Prize (1995) to, 420
　tau lepton, 422
Perrier, Carlo, 270
Perrin, Jean Baptiste, 59–60, 84, 470
　biographical notes on, 146
　Nobel Prize (1926) to, 144–147
Perutz, Max, 102
PETRA (DESY), 405
phase-contrast microscope, 242–244
phase transition, 375–376, 488
Phillips, William Daniel, 431, 450, 474
　atom chilling, trapping, 429–430
　biographical notes on, 431
　Nobel Prize (1997) to, 428
Philosophical Magazine, 61, 76, 81, 91, 98, 118
　phosphorescence, 27, 42
photoelectric effect, 48–49, 50, 51, 61, 102–103, 126, 488
photon, 49, 153, 176, 232, 355–359, 436, 441, 488
　virtual, 297
Physical Review, 241, 249, 256, 260–266, 292, 312, 322, 335
Physical Review Letters, 277, 338, 364, 394, 397, 422, 426, 436, 460
Physical Society, German, 37, 113
Physics Committee, *see* Nobel Physics Committee
Physics Letters, 328, 383
Physics Section, *see* Swedish Academy of Sciences

Physics Today, 267, 347, 392, 411
Physikalische Zeitschrift, 87
Physikalisch-Technische Reichsanstalt (PTR), 37, 79, 113, 116, 153, 169, 181, 192, 245–248, 463–465
Picasso, Pablo, 49
Piccioni, Oreste, 231
Pierce, John, 259
pile, nuclear, *see* reactor, nuclear
pile, Voltaic, 19
pi-meson, *see* pion
pion, 231–234, 262–264, 488
  discovery of, 231
Pippard, Brian, 327
planar technology, 443
Planck, Max Karl, 33, 38, 50–51, 56, 69, 77, 78–79, 83, 84, 85, 91, 92–93, 97, 111, 114, 122, 151, 153, 161, 164–165, 185, 207, 470
  and the 1908 Nobel Prize, 70
  biographical notes on, 115
  Nobel Prize (1918) to, 112
  quantum of energy, 37–38
  radiation law, 37, 112–113
Planck constant, 38, 92–93, 103, 134, 489
Planck time, 354
plasma, 306–307, 315, 489
plasma physics, 315, 415
Plato, 178, 441
Plücker, Julius, 23
plutonium-239, 214, 215, 271
Poincaré, Henri, 44, 73, 84
Politzer, David, 406
polonium, 28, 41, 44
polymer, 410, 489
Pomeranchuk, Isaac, 425
Pontecorvo, Bruno, 188, 201, 395, 455–457
Pope, Alexander, 9
Poppins, Mary, 431
positron, 173, 176, 232, 422
  discovery of, 183–184
Pound, Robert, 282
Powell, Cecil Franck, 82, 230, 232, 234, 460, 471
  biographical notes on, 230
  Nobel Prize (1950) to, 230
  pions, 231
Priestley, Joseph, 14

Princeton University, 5, 154, 163, 174, 198, 229, 257, 282, 288, 295, 346, 351, 360, 363–366, 415–416, 434, 460–462
principle of equivalence, *see* relativity, general theory of
Pringsheim, Ernst, 113
Pritchard, David, 452–453
Proceedings of the Royal Society, 151, 184
Prokhorov, Alexander Mikhailovich, 292, 372, 472
  biographical notes on, 294
  maser, 291
  Nobel Prize (1964) to, 290
proton, 118, 168, 232, 401–404, 489
proton–antiproton collider (CERN), 381–385
proton–proton chain, 303–304, 489
PS (CERN), 272, 340
Ptolemy, Claudius, 7
PTR, *see* Physikalisch-Technische Reichsanstalt (PTR)
pulsar, 279, 319, 329–333, 334, 414–415, 461, 489
Pupin Physics Laboratory, 233, 335, 397
Purcell, Edward Mills, 214, 241, 369, 471
  biographical notes on, 240
  Nobel Prize (1952) to, 240
  nuclear magnetic resonance, 240, 241–242

QCD, *see* quantum chromodynamics (QCD)
QED, *see* quantum electrodynamics (QED)
quantum, 37, 489
  dots, 448
  jump, 93
  liquid, 284, 434
  numbers, 110
  of energy, 33, 37
  state, 92–93, 490
  theory, 37
  wells, 448
  wires, 448
quantum of action, elementary, *see* Planck constant
quantum chromodynamics (QCD), 361, 404–406, 441, 489

quantum electrodynamics (QED), 176, 213, 253–254, 296–300, 340, 355, 376, 401–402, 408, 440, 441, 489
quantum field theory, 440–441, 489
quantum Hall effect, 319, 344, 385–387, 433, 484
  fractional, 347, 433, 446, 484
quantum mechanics, 11, 33, 125, 143, 490
quantum weak dynamics, 441
quark families, 440
quarks, 279, 311, 314, 319, 401–406, 440–441, 490
  bottom, 319, 364, 423–424
  charm, 319, 337–339, 341, 423
  down, 314, 402, 423
  strange, 314, 423
  top, 409, 424–425, 439
  up, 314, 402, 423
quasars, 279, 330, 461, 490
quasiparticle, 285, 434

Rabelais, François, 358
Rabi, Isidor Isaac, 132, 214, 215, 218, 239, 242, 251–253, 291, 296–298, 337, 365, 372, 401–402, 423, 471
  biographical notes on, 217–218
  molecular-beam magnetic resonance, 218–219
  Nobel Prize (1944) to, 217
  on Albert Einstein, 128
radar, 213, 214, 490
Radiation Laboratory (MIT), 214, 218, 240, 299, 398, 400
radio, 63, 72, 163
radio astronomy, 329–332, 490
radio telescope, 415
radioactivity, 17, 28, 41–42, 187, 490
  artificial, 173, 187–189
  discovery of, 42–43
radium, 28, 41, 44
Rainwater, Leo James, 366, 472
  biographical notes on, 337
  Nobel Prize (1975) to, 333–334
  nuclear collective model, 334–336
Raman, Chandrasekhara Venkata, 168, 470
  biographical notes on, 167
  Nobel Prize (1930) to, 166–168
Raman effect, 167, 490

Ramsay, William, 47, 206
Ramsey, Elinor, 402
Ramsey, Norman Foster, 109, 214, 400, 473
    atomic clocks, 401
    biographical notes on, 401–402
    Nobel Prize (1989) to, 400–401
Rasetti, Franco, 188, 189, 201
reaction, nuclear, 377–379, 488
reactor, nuclear, 213, 289, 418, 419
Reale Scuola Normale Superiore, see Scuola Normale Superiore
Rebka, Glen, 282
red shift, gravitational, 108–109, 282, 401, 490
reductionism, 347–349
Rees, John, 340
Reid, Alexander, 157, 197
Reines, Frederick, 267, 285, 397, 421, 455, 474
    biographical notes on, 420–421
    neutrino detection, 260
    Nobel Prize (1995) to, 420
relativity, 33
relativity, general theory of, 12, 67, 77, 104, 106, 120, 128, 282, 319, 401, 414–418, 490
    principle of equivalence, 65–67; historic tests, 107–109, 480
    curved space-time, 106
    gravitational waves, 417–418, 483
relativity, special theory of, 10, 33, 51–56, 57–59, 106, 128, 490
    length contraction, 55
    Lorentz transformations, 55
    mass-energy equivalence, 56
    principle of relativity, 54
    simultaneity, 55, 57
    time dilatation, 55
    velocity (speed) of light, 54–55
renormalisation, 298, 376, 491
renormalisation group, 376, 491
resonance absorption, 281
resonance particles, 307–310, 491
Retherford, Robert, 250–251
Reviews of Modern Physics, 386–389
Ricci-Curbastro, Gregorio, 104–105
Richardson, John, 343
Richardson, Owen Willans, 40, 161, 170, 198, 470

biographical notes on, 163
Nobel Prize (1928) to, 162–163
thermionic effect, 162
Richardson, Robert Coleman, 284, 464, 474
    biographical notes on, 427
    Nobel Prize (1996) to, 425
    superfluid helium-3, 425–426
Richter, Burton, 339, 357, 402, 472
    biographical notes on, 339–340
    J/ψ particle, 337–339
    Nobel Prize (1976) to, 337
Riemann, Bernhard, 104–105, 171
Righi, Augusto, 71
Rochester, George, 231
Rohrer, Heinrich, 390–392, 407, 473
    biographical notes on, 390
    Nobel Prize (1986) to, 387–388
    scanning tunnelling microscope, 388–389
Roll, Peter, 351
Röntgen, Bertha, 36
Röntgen, Wilhelm Conrad, 5, 27, 35, 42, 86, 95, 244, 462, 469
    biographical notes on, 34
    Nobel Prize (1901) to, 34
    X-rays, 34–36
Röntgen rays, *see* X-rays
Roosevelt, Franklin Delano, 209, 214
Rosanes, Jakob, 179
Rossi, Bruno, 58, 194, 225, 248, 460
Rosenfeld, L., 133, 170
Royal Institution, 19, 47, 61, 101
Royal Society of London, 15, 40, 47, 62, 82, 101, 120, 122, 138, 156, 167, 186, 190–192, 199, 225, 239, 284, 305, 324, 327, 331–332, 343–345, 348, 350, 361, 372, 411, 420, 467
Royal Swedish Academy of Sciences, *see* Swedish Academy of Sciences
Royds, Thomas, 69
Rozsel, Alberta, 255
Rubbia, Carlo, 385, 411, 473
    biographical notes on, 384
    Nobel Prize (1984) to, 380–381
    W and Z bosons, 381–383
Rubens, Heinrich, 84, 113, 122
Ruska, Ernst August, 199, 473
    biographical notes on, 388
    electron microscope, 388

Nobel Prize (1986) to, 387
Russel, Henry Norris, 302, 379
Rutherford, Ernest, 15, 28, 33, 40, 68, 70, 76, 77, 84, 90–91, 97–98, 102, 121–122, 133, 155, 167, 168–169, 174, 181, 190–192, 197, 200, 206, 225, 230, 238–239, 247, 274–275, 280, 287, 304, 349, 405
    and the Cavendish, 118–119, 199–200
    alpha particles, 69
    biographical notes on, 82
    neutron, 122
    Nobel Prize for chemistry (1908) to, 68
    nuclear atom, 80–81
    October, 1937, 199
    on Max Planck, 113–114
    radioactivity, 41–42
    transmutation of atomic nuclei, 117–118
Rydberg, Johannes, 138
Rydberg constant, 93, 372
Ryle, Gilbert, 330
Ryle, Martin, 330–332, 472
    biographical notes on, 330–331
    Nobel Prize (1974) to, 329
    radio astronomy, 330–331

Sadler, Charles, 98
Sakharov, Andrei, 366
Salam, Abdus, 361, 362, 379–381, 436–438, 473
    biographical notes on, 361
    electroweak theory, 355–358
    Nobel Prize (1979) to, 355
Sandage, Allan, 330
scanning tunnelling microscope (STM), 369, 388–390, 405
Schawlow, Arthur Leonard, 276, 291–292, 373, 428, 473
    biographical notes on, 372–373
    laser spectroscopy, 372
    Nobel Prize (1981) to, 370
Schawlow Aurelia, Townes, 373
Schmidt, Gerhard, 41
Scholem, Gershon, 221
Schrieffer, John Robert, 257, 325, 426, 464, 472
    BCS theory, 321–323
    biographical notes on, 324–325
    Nobel Prize (1972) to, 321

Schrödinger, Anny, 185–186
Schrödinger, Erwin, 115, 125, 148, 151–152, 153–161, 164, 171, 173, 178, 186, 240, 246, 287, 297, 322, 433, 470
  biographical notes on, 184–185
  Nobel Prize (1933) to, 184
  wave function, 149, 246
  wave mechanics, 147–150
Schrödinger equation, 149, 177, 491
Schwartz, Laurent, 432
Schwartz, Melvin, 286, 363, 399, 455, 473
  biographical notes on, 398–399
  Nobel Prize (1988) to, 396–397
  two-neutrino experiment, 285–286, 397–398
Schwarzschild, Karl, 111, 247
Schweber, Silvan S., 300, 347
Schwinger, Julian Seymour, 214, 253, 297, 299, 337, 361, 437, 472
  biographical notes on, 299–300
  Nobel Prize (1965) to, 296
  QED, 297–298
Science, 204, 348, 451
scientific method, 8
Scientific Revolution, 7
Scorpio X-1, 460
Scuola Normale Superiore, 203, 384, 398
Seaborg, Glenn, 270
Segrè, Emilio Gino, 188–189, 199, 200, 202, 204, 208, 214, 216, 270, 271, 272, 471
  antiproton, 270, 507
  biographical notes on, 270–271
  Nobel Prize (1959) to, 269
  on Hans Bethe, 305
  on the Rome institute of physics, 211
semiconductor, 259, 491
Shakespeare, William, 185
Shapiro, Irwin, 108
Shapiro delay, 108
shell model, nuclear, *see* nucleus atomic
shell structure, atomic, 143, 289
Shimizu, Tadashi, 225
Shockley, William Bradford, 226, 257, 321, 443, 471
  biographical notes on, 258
  Nobel Prize (1956) to, 255
  transistor, 255–256

Shull, Clifford Glenwood, 473
  biographical notes on, 420
  neutron diffraction, 419
  Nobel Prize (1994) to, 418–419
Siegbahn, Kai Manne, 374, 473
  biographical notes on, 373–374
  electron spectroscopy, 373
  Nobel Prize (1981) to, 370
Siegbahn, Karl Manne, 138, 153, 204, 207, 373, 470
  biographical notes on, 138
  Nobel Prize (1924) to, 138–139
Silicon Valley, 258, 443–446
Silvera, Isaac, 450
Simon, Alfred, 153, 501
Sitter, Willem de, 119
SLAC, *see* Stanford Linear Accelerator Center (SLAC)
Slater, John, 248
Slipher, Vesto, 166
small physics, 369, 385, 409
Snow, Charles Percy, 211, 239
Soddy, Frederick, 42, 82
  isotopes, 121
soft-matter physics, 409–410
solar neutrinos, 279, 305–306, 455
solar neutrino problem, 456
Solvay, Ernest, 83–84
Solvay conference, 83–85, 161
Sommerfeld, Arnold, 83, 84, 95–96, 105, 117, 120, 125, 129, 130, 131, 138, 149, 177–180, 218, 220–221, 250, 304, 404
  extension of Bohr's model, 110–111
Sony Corporation, 325–326
Sorbonne, 43, 44, 46, 70, 75, 139, 164–165
space
  absolute, 10, 54–55
  relativity of, 55
space-time, 71, 106, 441, 491
  curved, 106
spatial quantization, 110, 129
SPEAR (SLAC), 337–340, 404, 422
specific heats, 83, 484
spectral lines, 24, 491
  fine structure of, 65, 250, 491
  hyperfine structure of, 401, 491
spectroscope, 24–25, 492

spectroscopy, 370, 400–402, 492
  electron, 370–373
  laser, 369, 370–372
  optical, 25
  radio-frequency, 219, 370
  single-ion, 401
  X-ray, 98, 138, 368–371
spectrum, electromagnetic, 20, 492
spectrum, light, 24–25
  absorption, 25
  continuous, 25
  emission, 25
  line, 25, 93, 492
spin, 492
  electron, 141–142
spin glasses, 347
SPS (CERN), 381–385
Sputnik I, 267, 279
SQUID, *see* superconducting quantum interference device (SQUID)
Stalin, Joseph, 283
standard model, 398, 440–441
standard solar model, 456–457
Stanford Linear Accelerator Center (SLAC), 5, 337–340, 366, 369, 404–406, 422–424
Stanford Positron–Electron Asymmetric Ring, *see* SPEAR (SLAC)
Stanford University, 5, 240, 253, 258, 280–282, 337–340, 353, 399, 404–406, 423, 431, 435, 452, 460
Stark, Johannes, 56, 97, 470
  biographical notes on, 116
  Doppler effect, 116–117
  Nobel Prize (1919) to, 116
Stark effect, 111, 117, 131, 148, 492
Stefan, Josef, 26
Stefan–Boltzmann law, 26
Steinberger, Jack, 203, 286, 363, 399, 455, 473
  biographical notes on, 399–400
  Nobel Prize (1988) to, 396–397
  on Enrico Fermi, 203
  two-neutrino experiment, 285–286, 397–398
Stern, Otto, 129, 170, 187, 207, 217–218, 239–240, 270, 471
  biographical notes on, 216
  Nobel Prize (1943) to, 215–216
  proton magnetic moment, 216–217

Stern–Gerlach experiment, 129–130, 131, 142
stochastic cooling, 381
Stokes, George Gabriel, 112
Stoney, George, 27
Störmer, Horst, 385, 435, 474
  biographical notes on, 434
  fractional quantum Hall effect, 433
  Nobel Prize (1998) to, 433
strangeness, 311, 313, 493
Strassmann, Fritz, 206–208
Strutt, John William, *see* Rayleigh Lord
Sudarshan, George, 266
superconducting quantum interference device (SQUID), 329
superconductivity, 77, 347, 426, 463–464, 493
  discovery of, 88–89
superconductors, 89
  high-temperature, 323, 392–395, 407, 463–466
  transition temperature of, 323, 464
  type-I, 463–465
  type-II, 463–465
superfluidity, 173, 347, 463–464, 493
Super-Kamiokande, 457–458
supernova, 333, 378, 417, 455, 493
  SN 1987A, 395, 396, 457
superstring theory, 441, 493
symmetry, 248, 288, 493
  charge-parity, 362–367
  charge reflection, 362, 477, 493
  gauge, 355, 359, 483
  mirror, 260–261, 493
  time-reversal, 364
symmetry breaking, 347, 356
  spontaneous, 356–359, 436–438
Swedish Academy of Sciences, 3–5, 34, 44–45, 64–65, 68–71, 73, 85, 104, 112, 121, 126–127, 138–140, 144–146, 153, 164, 167, 174–178, 187, 197–204, 215, 228, 242–245, 261, 302, 310, 333, 340–349, 358, 372–373, 374, 377, 381–385, 396, 410–414, 436–443, 454
  Physics Section of, 5, 68–70, 85, 145
Swiss Federal Institute of Technology (ETH), 34, 56, 104, 121, 128, 171, 216, 221, 240, 390, 392
Szilard, Leo, 171, 209, 233

Tamm, Igor Eugenevich, 282, 471
  biographical notes on, 268
  Nobel Prize (1958) to, 267
Tananbaum, Harvey, 461
τ lepton, 319, 420–423, 493
  discovery of, 422
Taylor, Joseph Hooton, 416, 473
  binary pulsar PS 1913+16, 414–415, 417–418
  biographical notes on, 415–416
  Nobel Prize (1993) to, 414, 415
Taylor, Richard Edward, 405, 406, 423, 473
  biographical notes on, 406
  Nobel Prize (1990) to, 403–404
  SLAC–MIT experiment, 402–403
technetium, 204, 270
Telegdi, Valentine, 201, 264
telegraph, 28
  wireless, 28, 29
telegraphy, wireless, 71, 163
telephone, 29, 163, 255
television, 72, 163, 255
Teller, Edward, 180, 265
Tevatron collider (Fermilab), 424
Texas Instruments, 269, 443–445
thermionic effect, 61, 162, 494
thermodynamics, 17, 18
  first law of, 18
  second law of, 18
The Times, 122
Thompson, Benjamin, *see* Count Rumford
Thomson, George Paget, 82, 156–157, 187, 470
  biographical notes on, 198–199
  electron diffraction, 157
  Nobel Prize (1937) to, 197
Thomson, Joseph John, 15, 24, 27–28, 29, 34, 48, 60, 62, 63, 68, 76, 81, 82, 86, 89–91, 102, 109, 117, 118–119, 120, 121, 133, 146, 155–156, 198, 200, 224, 247, 421, 469
  biographical notes on, 62
  electron, 59–61
  Nobel Prize (1906) to, 59
Thomson, William, *see* Kelvin Lord
't Hooft, Gerardus, 243, 357, 439, 474
  biographical notes on, 439–440
  Nobel Prize (1999) to, 435–436
  renormalizability of Yang–Mills theories, 436–439
thorium, 41
time
  absolute, 10, 54–55
  relativity of, 55
Ting, Samuel, 341, 357, 363, 365, 472
  biographical notes on, 340
  J/ψ particle, 338–339, 341
  Nobel Prize (1976) to, 337
Tomonaga, Sin-Itiro, 253, 297, 300, 437, 472
  biographical notes on, 300
  Nobel Prize (1965) to, 296
  QED, 297–298
Touschek, Bruno, 367
Townes, Charles Hard, 276, 292, 353, 371–373, 472
  biographical notes on, 292–293
  laser, 276–277, 292, 293
  maser, 276, 291–292
  Nobel Prize (1964) to, 290
Townes, Frances, 369
transistor, 213, 227, 237, 258, 259, 494
  high-speed, 443–446
  invention of, 226–227, 255–256
Trinity College, 47, 62, 100–101, 163, 198, 327, 349, 378
triode valve, 63
tritium, 306–307, 401
Tsui, Daniel Chee, 387, 435, 474
  biographical notes on, 434
  fractional quantum Hall effect, 433
  Nobel Prize (1998) to, 433
tunnelling, 176–177, 325–328, 494

Uhlenbeck, George, 132, 171
  electron spin, 141–142
uncertainty principle, 157, 159, 169, 229, 478
universality, 375
universe, 414, 461
  expanding, 165–166, 354
  static, 331
University of Berlin, Friedrich Wilhelms, 5, 37, 56, 64, 74, 79, 83, 97, 98, 104, 115, 131, 139–140, 180, 185, 247, 248, 288

University of California at Berkeley, 5, 174, 204, 214, 234, 252, 266, 271–272, 274, 293, 299, 308, 360, 413, 422, 430
University of Cambridge (England), 5, 10, 48, 62, 82, 147–150, 155–156, 163, 186, 223–224, 230, 239, 247, 325–327, 329–332, 349, 361, 376, 378, 414
University of Chicago, 5, 65, 102, 134, 140, 154, 203, 215, 237, 264, 271, 289, 308, 310, 364, 375–377, 398–400, 434
University of Copenhagen, 5, 89, 132, 133, 221, 336
University of Göttingen, 71, 79, 83, 115, 116, 132, 140–143, 152, 156, 178, 203, 221, 247, 288, 401, 447
University of Illinois at Urbana-Champaign, 5, 253, 257, 321–324, 467
University of Leiden, 5, 13, 20, 38–41, 88, 141, 369, 391, 463
University of Manchester, 68, 82, 102, 231, 239, 304
University of Munich, 34, 77, 79, 83, 95, 110, 115, 116, 140, 178–180, 220–221
University of Rome, 188, 203, 211, 231, 270–271
University of Zurich, 83, 97, 104, 128, 147, 184, 247, 390
Uppsala University, 41, 78, 121, 138, 315, 368
uranium
uranium-235, 209–210, 214, 215
uranium-238, 209–210
uranic rays, *see* Becquerel rays
Urey, Harold, 132, 179, 214
US National Academy of Sciences, 154, 284, 305, 311, 324–325, 343–345, 349, 360–361, 365–366, 377, 392, 411, 420, 452–453, 459, 466–467

vacuum, 254
Van Allen, James, 269
Van Allen radiation belts, 268–269
Van de Graaff, Robert, 174
Van Vleck, John Hasbrouck, 257, 345, 472

biographical notes on, 341–343
condensed-matter physics, 342–343
Nobel Prize (1977) to, 340–341
Veltman, Martinus J.G., 357, 439, 474
biographical notes on, 440
Nobel Prize (1999) to, 435–436
renormalisability of Yang–Mills theories, 436–439
Villard, Paul, 41
Volta, Alessandro, 19, 28
voltage multiplier, 174, 175, 182, 271
von Klitzing constant, 385

Waals, Johannes Diderik van der, 18, 34, 41, 88, 90, 469
biographical notes on, 75
Nobel Prize (1910) to, 74, 75–76
Wagner, Richard, 222
Waller, Ivar, 220
Walraven, Jook, 450
Walton, Ernest Thomas, 58, 82, 174, 175, 200, 204, 271, 471
artificial nuclear disintegration, 182–183, 238–239
biographical notes on, 239
Nobel Prize (1951) to, 238
voltage multiplier, 174
Warburg, Otto, 122
Ward, John, 356
Watson, James, 102, 244–245
Watt, James, 28
radio waves, 20
wave function, 149, 494
wave mechanics, 149, 151, 246–262, 494
wave–particle dualism, 160, 494
weak bosons, 356–359, 367, 441, 494
W bosons, 357–359, 382, 412, 436–439
discovery of, 380–383
Weber, Joseph, 291
Weber, Wilhelm, 31, 171
Weinberg, Steven, 311, 348, 359, 360, 381–383, 436–438, 473
biographical notes on, 360–361
Dreams of a Final Theory, 441–442
electroweak theory, 355–358
Nobel Prize (1979) to, 355
Weisskopf, Victor, 132, 143, 170, 171, 180, 217, 266, 290, 310, 440
on Wolfgang Pauli, 221–222

Weizsäcker, Carl, 180
Westgren, Arne, 68
Weyl, Hermann, 171, 176, 240
Wheeler, John, 266, 299
white dwarf, 378–379, 494
Wick, G., 170
Wiechert Emil, 59
Wiegand, Clyde, 269, 272
Wieman, Carl E., 451, 453, 474
biographical notes on, 452–453
Bose–Einstein condensate, 450–451
Nobel Prize (2001) to, 448
Wien, Max, 184
Wien, Wilhelm Carl, 26, 37, 57, 70, 79, 84, 113, 117, 151–152, 184, 469
biographical notes on, 79
displacement law, 26, 78
Nobel Prize (1911) to, 78–79
radiation law, 78
Wiener Press, 36
Wigner, Eugene Paul, 171, 209, 214, 215, 229, 257, 288, 289, 471
biographical notes on, 287–288
Nobel Prize (1963) to, 287–289
Wilczek, Frank, 408
Wilkins, Maurice, 245
Wilkinson, David, 351
Will, Clifford M., 109
Wilson, Charles Thomson Rees, 161, 167, 197, 200, 225, 230, 275, 470
biographical notes on, 155–156
cloud chamber, 85–86
Nobel Prize (1927) to, 153, 154–155
Wilson, H. A., 76
Wilson, Kenneth, G., 377, 473
biographical notes on, 376–377
critical phenomena, 375–376
Nobel Prize (1982) to, 374
Wilson, Robert Woodrow, 295–296, 352, 353, 371, 413, 472
biographical notes on, 353
cosmic microwave background radiation, 351
Nobel Prize (1978) to, 349
Wilson cloud chamber, 86, 118, 153, 154–155, 183, 230, 273–276, 478
wind, solar, 315
Wineland, David, 428
Wollan, Ernest, 419–420
Wollaston, William, 24

Wu, Chien Shiung, 263, 266

X-ray astronomy, 494
X-ray diffraction, 95–97, 160, 244–245, 419
X-rays, 17, 27, 34, 77, 371, 419, 460–461, 494
    characteristic, 98, 138–139
    discovery of, 34–36

Yang, Chen Ning, 203, 264, 265, 266–267, 312, 322, 355–356, 362–363, 365, 397, 437, 471
    biographical notes on, 265
    Nobel Prize (1957) to, 260–261
    parity violation, 264
Yang, Wei-T'e, 334
Yang-Mills theory, 436–438
Young, Thomas, 21–22, 451
Ypsilantis, Thomas, 269, 272
Yukawa, Hideki, 193, 195, 228, 231–234, 300, 471
    biographical notes on, 228
    meson theory, 229
    Nobel Prize (1949) to, 228

Z boson, 357–359, 412, 436–439
    discovery of, 379–381

Zeeman, Pieter, 34, 39, 88, 120, 270, 469
    biographical notes on, 41
    Nobel Prize (1902) to, 38–39, 41
    Zeeman effect, 40–41
Zeeman effect, 40–41, 111, 117, 131, 142, 148, 429, 494
    anomalous, 131, 220
Zeiger, Herbert, 291
Zeitschrift für Physics, 143, 157, 393
Zernike, Frits, 244, 439, 471
    biographical notes on, 243
    Nobel Prize (1953) to, 242–243
    phase-contrast microscope, 242–244
Zweig, George, 314